Communications
in Computer and Information Science 54

Sushil K. Prasad Harrick M. Vin
Sartaj Sahni Mahadeo P. Jaiswal
Bundit Thipakorn (Eds.)

Information Systems, Technology and Management

4th International Conference, ICISTM 2010
Bangkok, Thailand, March 11-13, 2010
Proceedings

Volume Editors

Sushil K. Prasad
Georgia State University
Atlanta, GA, USA
E-mail: sprasad@gsu.edu

Harrick M. Vin
Tata Research, Development
and Design Center (TRDDC)
Pune, India
E-mail: harrick.vin@tcs.com

Sartaj Sahni
University of Florida
Gainesville, FL, USA
E-mail: sahni@cise.ufl.edu

Mahadeo P. Jaiswal
Management Development Institute (MDI)
Sukhrali, Gurgaon, India
E-mail: mpjaiswal@mdi.ac.in

Bundit Thipakorn
King Mongkut's University of Technology
Thonburi, Bangkok, Thailand
E-mail: bundit@cpe.kmutt.ac.th

Library of Congress Control Number: Applied for

CR Subject Classification (1998): K.6, D.2, H.2.8, D.4.6, C.2

ISSN 1865-0929
ISBN-10 3-642-12034-2 Springer Berlin Heidelberg New York
ISBN-13 978-3-642-12034-3 Springer Berlin Heidelberg New York

springer.com

Printed in Germany

Typesetting: Camera-ready by author, data conversion by Scientific Publishing Services, Chennai, India
Printed on acid-free paper 06/3180 5 4 3 2 1 0

Message from the Technical Program Co-chairs

Welcome to the proceedings of the 4th International Conference on Information Systems, Technology and Management (ICISTM 2010), held at Bangkok, Thailand.

For the main program, a total of 32 regular papers were selected from 86 submissions. Additionally, 12 short papers were accepted. The submissions represented six continents and 25 different countries, with the top six being India, USA, Republic of Korea, Thailand, Taiwan, and Australia. These submissions were distributed among four track areas of: Information Systems (IS), Information Technology (IT), Information Management (IM), and Applications. The four Track Chairs, Han Hsiao (IS), Sham Navathe (IT), Praveen Pathak (IM), and Ramamurthy Badrinath (Applications), ably assisted us. The submissions went through a two-phase review. First, in coordination with the Track Chairs, the manuscript went through a vetting process for relevance and reassignment to proper tracks. Next, the Track Chairs assigned each paper to four Program Committee members for peer review. A total of 307 reviews were collected from 60 Program Committee members (drawn from 10 different countries) and 14 external reviewers. The top three countries which PC members belonged to were the USA, India, and Taiwan.

The conference proceedings are published in Springer's CCIS series for the second time. This required a review by Springer's editorial board, and represents a significant improvement in the publication outlet and overall dissemination —we thank Stephan Goeller and Alfred Hofmann, both from Springer, for their assistance. The regular papers were alloted 12 pages each. The short papers were allowed a maximum of six pages each.

For a newly created conference, it has been our goal to ensure improvement in quality as the primary basis to build upon. In this endeavor, we have been immensely helped by our Program Committee members and the external reviewers. They provided rigorous and timely reviews and, more often than not, elaborate feedback to all authors to help improve the quality and presentation of their work. Our sincere thanks go to the Program Committee members, the external reviewers, and the four Track Chairs. Another important goal has been to ensure participation by the international as well as India-based research communities in the organizational structure and in authorship, which we have also accomplished.

Throughout this process, we received the sage advice of Sartaj Sahni, our conference Co-chair, and help from Mahadeo Jaiswal and Bundit Thipakorn, two other conference Co-chairs; we are indeed thankful to them. We are also thankful to Akshaye Dhawan for taking up the important and tedious job of Publications Chair, interfacing with all the authors, ensuring that Springer's formatting requirements were adhered to, and that the requirements of copyrights, author registrations and excess page charges were fulfilled, all in a timely and professional manner. Finally, we thank all the authors for their interest in ICISTM 2010 and for their contributions in making this year's technical program particularly impressive.

January 2010

Sushil K. Prasad
Harrick M. Vin

Conference Organization

General Co-chairs

Mahadeo Jaiswal	Management Development Institute, India
Sartaj Sahni	University of Florida, USA
Bundit Thipakorn	King Mongut's Univeristy of Technology Thonburi (KMUTT), Thailand

Technical Program Co-chairs

Sushil K. Prasad	Georgia State University, USA
Harrick M. Vin	Tata Research, Development and Design Center (TRDDC), India

Track Chairs

Information Systems	Han C.W. Hsiao, Asia University, Taiwan
Information Technology	Shamkant Navathe, Georgia Tech., USA
Information Management	Praveen A. Pathak, University of Florida, USA
Applications	Badrinath Ramamurthy, HP Labs, USA

Workshop and Tutorial Chair

Sumeet Dua	Louisiana Tech University, USA

Publications Chair

Akshaye Dhawan	Ursinus College, USA

Publicity Co-chairs

Paolo Bellavista	Università degli Studi di Bologna, Italy
Chandan Chowdhury	IFS, India
Yookun Cho	Seoul National University, Korea
Mario Dantas	Federal University of Santa Catarina, Brazil
Koji Nakano	Hiroshima University, Japan
Atul Parvatiyar	Institute for Customer Relationship Management
Rajiv Ranjan	University of New South Wales, Australia
Akshaye Dhawan	Ursinus College, USA
Bhusan Saxena	Management Development Institute, India

Finance Chair

Vanita Yadav	Management Development Institute, India

Local Arrangements Chair

Suthep Madarasmi	King Mongkut's University of Technology Thonburi, Thailand

Web Chair

Sangeeta Bharadwaj	Management Development Institute, India

ICISTM 2010 Program Committee

Information Systems

Anurag Agarwal	University of South Florida, USA
Angela Bonifati	Icar CNR, Italy
Chun-Ming Chang	Asia University, Taiwan
Susumu Date	Osaka University, Japan
Amol Ghoting	IBM T.J. Watson Research Center, USA
Song Guo	University of Aizu, Japan
Han Hsiao	Asia University, Taiwan
Ruoming Jin	Kent State University, USA
Atreyi Kankanhalli	National University of Singapore, Singapore
Hsiu-Chia Ko	Chaoyang University of Technology, Taiwan
Sze-Yao Li	Asia University, Taiwan
Hsien-Chou Liao	Chaoyang University of Technology, Taiwan
Jim-Min Lin	Feng Chia University, Taiwan
Sumanth Yenduri	University of Southern Mississippi, USA

Information Technology

Rafi Ahmed	Oracle Corporation, USA
Janaka Balasooriya	Arizona State, USA
Rafae Bhatti	Purdue, USA
David Butler	Lawrence Livermore National Lab, USA
Amogh Dhamdhere	Cooperative Ass. for Internet Data Analysis, USA
Akshaye Dhawan	Ursinus College, USA
Xuerong Feng	Arizona State, USA
Kamal Karlapalem	IIIT, India
Zhen Hua Liu	Oracle, USA

Praveen Madiraju	Marquette University, USA
Srilaxmi Malladi	Georgia State University, USA
Shamkant Navathe	Georgia Tech, USA
Rupa Parameswaran	Oracle, USA
Animesh Pathak	INRIA Paris, France
Asterio Tanaka	Universidade Federal do Estado do Rio de Janeiro, Brazil
Wanxia Xie	Georgia Tech, USA
Wai Gen Yee	Illinois Institute of Technology, USA
Wenrui Zhao	Google, USA

Information Management

Haldun Aytug	University of Florida, USA
Subhajyoti Bandyopadhyay	University of Florida, USA
Fidan Boylu	University of Connecticut, USA
Michelle Chen	University of Connecticut, USA
Hsing Cheng	University of Florida, USA
Jason Deane	Virginia Tech, USA
Weiguo Fan	Virginia Tech, USA
Mabel Feng	NUS, Singapore
Hong Guo	University of Florida, USA
Jeevan Jaisingh	Hong Kong University of Science and Technology, China
Chetan Kumar	California State University San Marcos, USA
Subodha Kumar	Texas A&M University, USA
Kamna Malik	U21, India
Praveen Pathak	University of Florida, USA
Selwyn Piramuthu	University of Florida, USA
Ramesh Sankaranarayanan	University of Connecticut, USA
Chih-Ping Wei	National Tsing Hua University, Taiwan
Dongsong Zhang	University of Maryland Baltimore County, USA

Applications

Ramamurthy Badrinath	HP Labs, India
Geetha Manjunath	HP Labs, India
Mandar Mutalikdesai	International Institute of Information Technology, India
Ashish Pani	XLRI Jamshedpur, India
SR Prakash	National Institute of Technology, India
Santonu Sarkar	Accenture, India
Sudeshna Sarkar	IIT Kharagpur, India

Indranil Sengupta	IIT Kharagpur, India
Srinath Srinivasa	International Institute of Information Technology, India
Vasudeva Varma	IIIT Hyderabad, India

External Reviewers

Muad Abu-Ata
Carlos Alberto Campos
Lian Duan
Rafael Fernandes
Hui Hong
Kohei Ichikawa
Mohammad Khoshneshin
Shengli Li
Lin Liu
Sanket Patil
Aditya Rachakonda
Ning Ruan
Ravi Sen
Yang Xiang
Dongsong Zhang

Table of Contents

1. Keynote Talks

2. Information Systems

3. Information Technology

4. Information Management

5. Applications

6. Short Papers

7. Tutorial Papers

Service Discovery Protocols for VANET Based Emergency Preparedness Class of Applications: A Necessity Public Safety and Security*

Richard Werner Pazzi, Kaouther Abrougui, Cristiano De Rezende, and Azzedine Boukerche

PARADISE Research Laboratory,
SITE, University of Ottawa, Canada
{rwerner,abrougui,creze019,boukerch}@site.uottawa.ca
http://paradise.site.uottawa.ca/

Abstract. Emergency preparedness and response require instant collaboration and coordination of many emergency services in the purpose of saving lives and properties. Service discovery is used in this class of applications in order to increase the efficiency and effectiveness in dealing with a disaster. In this paper, we discuss on the importance of service discovery protocols for emergency preparedness and response class of application.

1 Introduction

In emergency preparedness and response class of application, service discovery plays an important role in the discovery of surrounding services that can help rescuers to achieve their tasks. Let us suppose that there is an incident at a site. Rescuers have to collaborate between each other in order to perform their tasks. A good collaboration between the different rescuers depends from the knowledge of services available and the capability to reach them even if the network is highly dynamic. Therefore, a good service discovery protocol that permits to rescuers to collaborate between each other is necessary. Besides, at an incident site, there is no fixed or static topology. Rescuers can be characterized by their high mobility, or low mobility, they can be moving in groups or individually. Many participants can exist at the same site (firefighters, policemen, nurses, victims...) and participants can use diverse devices (sensors, laptops, PDAs, palmtops...). In this paper, we focus on service discovery for emergency preparedness and response class of applications in vehicular networks. As the main goal of a vehicular network is to guarantee security on roads, several secondary objectives, or prerequisites, have to be achieved in order to reach the main objective. In fact, emergency preparedness and response class of application in vehicular networks

* This work is partially supported by Canada Research Chair Program, NSERC, Ontario Distinguished Research Award Program,PREA/ Early Research Award, and OIT/CFI fund.

S.K. Prasad et al. (Eds.): ICISTM 2010, CCIS 54, pp. 1–7, 2010.

heavily depends on service discovery for various reasons such as determining location and proximity of hospitals, service stations; determination of locations of accidents and disasters and user guidance to alternatives roads; automotive diagnosis and user warning.

As a consequence, the definition of a service discovery system [5] among the vehicular nodes is one of the prerequisites needed in order to make viable many of the emergency preparedness and response class of application in vehicular networks. The service discovery problem consists of finding services that satisfy a driver's request. Such a problem is very ample and extensive and it has been investigated in many areas such as wired networks, ad hoc networks, mesh networks, vehicular networks. The reminder of this paper is organized as follow. In section 2 we discuss the importance of service discovery for emergency preparedness and response class of application in vehicular networks. In section 3, we present some related works on service discovery protocols. Finally, in section 4 we present our conclusion.

2 Importance of Service Discovery for Emergency Preparedness and Response Class of Application in Vehicular Networks

In case of emergency, rescuers need the collaboration between many types of vehicular applications and services for the accomplishment of their tasks. Two major types of services can be found in a vehicular network: (i) safety services and ; (ii) convenience services. The former services deal with the enhancement of vehicle safety on roads, while the latter help provide services to passengers in their vehicles, such as entertainment or localization services. Both types of services are required for emergency preparedness and response class of application.

2.1 Discovery of Safety Services

Safety on the roads on Vehicular Networks heavily depends on service discovery and advertisement for various reasons such as discovery of collisions and obstacles on the road and avoiding eventual accident [3]. Car crashes is one of the most common traffic accidents. According to the National Highway Traffic Safety Administration there are about 43,000 people killed in fatal car accidents each year in the United States [1]. Different factors can cause car crashes, such as the mechanical malfunctioning of the vehicle. It has been noticed on the statistics that the most important cause of car crashes is related to the driver behavior. In fact, if a driver cannot react on time to an emergency situation, a potential collision chain can be created on the road involving vehicles following the initial vehicle. In order to avoid collisions, a driver relies on the break lights of the vehicle immediately ahead. However, if vehicles are following each other very closely, drivers may not have enough time to react and to avoid collisions. A better way to perform collision avoidance would be to have an efficient *Collision warning services* that would provide emergency discovery and advertisement protocols to reduce the delay of propagation of collision warnings compared to the traditional reaction to the break lights of the leading vehicle.

Another reason for car crashes is related to the bad weather. In fact, a nasty weather that hammers the streets with snow, freezing rain, rain, fog, etc, could leave the roads in a messy and dangerous situation: Roads are slippery, the visibility is reduced, the road lines and even the road signs are hidden mainly during the night. In these severe conditions, vehicle crashes and slippage become high if the drivers are not informed accurately about the quality of the road and the visibility in the coming area. Drivers can slow down and be more attentive or even change their way if the condition on the front road is very dangerous or there are vehicles crashes blocking the road. A better way to help drivers overcome these situations and to avoid collisions and accident would be to have *(i) Environmental monitoring services* with installed sensor on vehicles to monitor environmental conditions, such as temperature, humidity, voltage and other factors; and *(ii) Civil infrastructure monitoring services* by installing sensors on vehicles to monitor the state of roads and detect slippery roads, etc. Drivers can discover these services either passively or reactively and can take the appropriate decision or be more alert and preventive before it is too late.

Safety on the road is also improved by discovering the *Automotive diagnostics services*. In this kind of service, information is obtained from the vehicle on-board sensors that can detect and monitor bad driving tendencies. These latter are caused either by a mechanical malfunctioning on the vehicle or by a misbehavior of the driver because he is drunk or he falls asleep. If a problem is discovered with this kind of service, an alert is sent to the driver and to the surrounding drivers in order to be more careful and to avoid potential accidents.

Safety on the roads is also improved with the existence of (*(i) Disaster warning services* and *(ii) User guidance services in case of disasters*. Disaster warning services can help to detect disasters like bridge breakage, volcano eruption, flooding, avalanche, hurricane, etc. With the help of *Environmental monitoring services* and *Civil infrastructure monitoring services*, *Disaster warning services* can send warnings to drivers in the affected region such that, drivers can take the appropriate action before they are very close to the region affected. On the other hand, drivers that are already very close to the disaster can use the *User guidance services in case of disasters* in order to be rescued and evacuated from the affected region.

Road monitoring helps drivers in critic situations to be discovered by the closest service station and get the required help, mainly if the driver is in an isolated region (for example, a driver who is out of fuel in the middle of an isolated road).

From these examples, we conclude that it is clear that service discovery and advertisement is very important in order to improve the safety on the roads and prevent eventual accidents and deaths on the roads.

2.2 Discovery of Convenience Services

Convenience services could be entertainment services or informative services. Even if those types of services are not related to safety on roads, they are still important for emergency preparedness and response class of application in vehicular networks. Let us consider the *Free parking spots service discovery* [4], this type of service is considered as a convenience service that helps the rescuer to find a free parking spot in a congested city. Some studies have been performed in a congested city that highlight the annual damage caused by the parking space traffic search. This problem causes economical damages, gasoline and diesel wasted in search of free parking spots, wasted time for drivers, in addition to encumbering the entire traffic on the city [4]. Thus, an efficient *Free parking spots service discovery* protocol would overcome all the problems mentioned earlier.

Another type of convenience service discovery consists on *Location discovery* where a driver would like to discover the gas-stations on the road with their price list, in order to be able to chose the gas-station with the lowest price and take the appropriate exit. Another scenario could be the discovery a the nearby hospitals.

The *Traffic monitoring service* could also be discovered by drivers in order to help them to choose the less congested roads to their respective destinations.

Geo-imaging service discovery is another type of convenience service that consists on attaching cameras on cars that capture images embedding data GPS coordinates automatically. These images are then transferred to the appropriate applications like the landmark-based routing and driving direction applications that could be installed on other cars or centralized for a general use.

Service discovery can also be used by data muling applications. In data muling, mobile agents interact with static agents in order to upload, download and transfer data to other physical locations. Thus cars can be used to send data from remote sensornets to internet servers. *Service discovery of data muling agents* could help the communication and the interaction between the different agents.

The *Gateway Discovery Service* helps drivers to be able to connect to internet and retrieve information while they are in their cars.

3 Service Discovery Architectures for Vanets

In order to present the different service discovery protocols in vehicular networks, we classify them according to their architecture. In the literature, there are three different architectures of information dissemination for service discovery protocols: (i) *Directory-based architecture*, (ii) *Directory-less architecture*, and (iii) *hybrid architecture*. In the following we discuss each one of them with reference to existing service discovery protocols in the literature.

Vehicular networks comprise two main types of components: (a)*vehicles*, which are moving components and (b) *roadside components*, which are fixed components and consist of roadside routers, gateways, or fixed services on roads. In the following, we refer to both types of components as road components. We will use the terms vehicles and roadside components respectively, in order to differentiate between the mobile components and the fixed components.

3.1 Directory-Based Architecture

In a directory-based architecture, a road component can have one or more roles. It can be (1) a service provider, (2) a service requester and/or (3) a service directory. A service provider is a server that has a service to offer. A service requester is a client that needs to find a server for a specific service. A service directory is a facilitator that permits to service providers to register their services and permits to service requesters to find their desired services in the vehicular network through their direct queries to the service directory. A service directory can be centralized or distributed. Centralized service directories are implemented in a single road component, and distributed service directories are implemented in many road components. Centralized service directory approach has been used by service discovery protocols in wired networks or in wireless local area networks. It was not suitable for ad hoc networks due to all the constraints (limited energy, frequent disconnection, mobility, etc...) facing ad hoc networks. In this type of architecture, there is a central collection point for data and task management. The communication between the user and the network goes through the central point. In vehicular networks, the centralized approach has been adopted by some discovery protocols. In [6], the authors use a centralized approach for service discovery. They use "the portal" as the central point of data collection from service providers and query answering from client applications. Applications send their queries directly to the portal. The central portal answers the queries if the required information is provided in the portal. Otherwise, the portal starts a data collection process that permits to retrieve the required information from the network nodes.

Directory-based architecture requires additional traffic overhead for the selection or maintenance of the structure of directories, service information update in each directory and the replication of service information between directories when applied. Thus, the number of directories and the frequency of maintenance and update of directories are crucial parameters. For example, a high frequency of directory update and maintenance could incur high traffic overhead and congestion in the network. On the other hand, a low frequency of directory update and maintenance could result in the inconsistency of directory structures and the loss of accuracy of service information.

3.2 Directory-Less Architecture

In directory-less architecture [2, 7], there is no directory that coordinates between service providers and service requesters. Thus, additional costs related to service directory selection and maintenance are avoided. In the directory-less architecture, service providers advertise themselves in the network and service requesters broadcast their requests that should be answered by the appropriate service provider or by an intermediate node that has in its cache the requested service. The main concern with directory-less approaches consists of how to determine the advertisement frequency and range of service providers in order to reduce the bandwidth usage in the network and avoid redundant and useless service advertisements transmissions.

In [2], the authors proposed a gateway discovery protocol for vehicular networks. In order to deal with the advertisement range and frequency of gateways, they assumed that gateways do advertise themselves periodically, and vehicles learn passively about the available gateway such that they do not have to send their requests and flood the Vanet. According to the authors, the number of gateways in Vanets is much more lower than the number of vehicles. Thus, the incurred overhead would be less if only gateways advertise themselves. Thus the range of advertisement is the whole network and the advertisement periodicity was not investigated in the paper.

In [7], the authors use a *naive technique* for requests forwarding. They assume that for unorganized Vanets, vehicles do not have any information about the available resources in the Vanet. Thus, vehicles requesting information or services, flood the whole network with request messages.

The presented techniques are not promising. That is why, more investigated techniques have been proposed in the literature in order to deal with the advertisement frequency and range in directory-less approaches.

3.3 Hybrid Architecture

In hybrid architecture, a service provider registers its service with a service directory if there is any in its vicinity, otherwise it broadcasts service advertisement messages. The service requester sends its request to a service directory if there is one in its vicinity otherwise it sends its requests until it reaches the requested service provider or a service directory that would return the service reply.

In vehicular networks, there are no discovery protocols that combine hybrid architecture as defined earlier. Thus, this is an open issue, and it should be investigated further.

4 Conclusion

In this paper, we have discussed the importance of service discovery protocols for emergency preparedness and response class of application. Then, we have presented some service discovery protocols in vehicular networks classified based on their architecture.

References

1. http://www.car-accidents.com/pages/stats.html
2. Bechler, M., Wolf, L., Storz, O., Franz, W.J.: Efficient discovery of Internet gateways in future vehicular communication systems. In: The 57th IEEE Semiannual Vehicular Technology Conference, VTC 2003-Spring, vol. 2 (2003)

3. Biswas, S., Tatchikou, R., Dion, F.: Vehicle-to-vehicle wireless communication protocols for enhancing highway traffic safety. IEEE Communications Magazine 44(1), 74–82 (2006)
4. Caliskan, M., Graupner, D., Mauve, M.: Decentralized discovery of free parking places. In: Proceedings of the 3rd international workshop on Vehicular ad hoc networks, pp. 30–39. ACM, New York (2006)
5. Dikaiakos, M.D., Florides, A., Nadeem, T., Iftode, L.: Location-Aware Services over Vehicular Ad-Hoc Networks using Car-to-Car Communication. IEEE Journal on Selected Areas in Communications 25(8), 1590–1602 (2007)
6. Hull, B., Bychkovsky, V., Zhang, Y., Chen, K., Goraczko, M., Miu, A.K., Shih, E., Balakrishnan, H., Madden, S.: CarTel: A Distributed Mobile Sensor Computing System. In: 4th ACM SenSys, Boulder, CO (November 2006)
7. Wewetzer, C., Caliskan, M., Luebke, A., Mauve, M.: The Feasibility of a Search Engine for Metropolitan Vehicular Ad-Hoc Networks. In: IEEE Globecom Workshops, 2007, pp. 1–8 (2007)

High-End Analytics and Data Mining for Sustainable Competitive Advantage

Alok N. Choudhary, Ramanathan Narayanan, and Kunpeng Zhang

Dept. of Electrical Engineering and Computer Science, Northwestern University
2145 Sheridan road, Evanston, IL, 60208, USA

Abstract. In the rapidly changing business environment, with global competition and maturing markets, competitive advantage is extremely important. Furthermore, in most industries product differentiation is no longer a decisive edge over competition. Thus, there is fierce competition to find, grow and keep loyal and profitable customers, optimize processes, adapt quickly to rapidly changing environments, and discover actionable knowledge quickly as those are the only way to grow business and profitability. Transforming a product-centric organization into a customer-centric one to achieve the above objectives is another critical step. This talk will present vision, strategy, emerging technologies and analytics using which businesses can hope to obtain sustainable competitive advantage. The talk will also provide a synergistic perspective on strategy, organizational transformation, and technology enabled relationship management. Steps and strategies involved in a successful deployment of operational and analytical infrastructure along with a landscape of key business intelligence technologies will be presented. The talk will present many examples from targeted marketing, modern marketing and briefly discuss the future of marketing.

1 Introduction

The rapidly growing role of technology in our lives has prompted a paradigm change in the way businesses function. Business environments are continuously evolving and adapting to innovations in technology. In this highly competitive marketplace, it is not sufficient for a business to simply have an edge in product quality. Conventional marketing strategies, while easier to implement, are not effective in adapting to the rapidly changing demands of customers. In this highly dynamic environment, where businesses have to deal with local issues, as well as deal with global competitors with superior resources, it is imperative that a knowledge based strategy be used. The use of knowledge-based strategies can allow a organization to overcome multiple roadblocks, spread the customer base, and develop scalable optimized processes to increase profitability.

The most significant change brought about by the advent of technology is the massive amount of information that is easily accessible to a vast majority of customers. As a result, customers now have multiple marketplaces too choose from while purchasing a product. Many products that were only available through a

S.K. Prasad et al. (Eds.): ICISTM 2010, CCIS 54, pp. 8–15, 2010.

single source are now available via a wide variety of technology-enabled marketplaces. A striking example of this phenomena is the rapid proliferation of online retailers into our lives. While online shopping was considered a novelty just a few years ago, it has rapid become one of the most trusted and largest consumer marketplaces now. Online retailers like Amazon, Ebay, Zappos and Newegg sell virtually all kinds of consumer goods. Customers shopping online also exhibit strikingly different shopping patterns as compared to those purchasing goods from a brick-and-mortar store. Therefore, the kind of marketing strategies that need to be employed to attract and retain such customers are vastly different from those employed to provoke loyalty among traditional customers.

Communication is a key factor for many businesses to attract and retain customers, establish newer markets, and to improve profitability. Over the years, markets have adapted well to the newly available methods of mass communication. For example, the transition from radio to television, in terms of marketing strategies, is well known and researched. However, the advent of the internet age has unleashed a massive torrent of new communication platforms. In the last few years alone, Internet-based communication platform have revolutionized the way we communicate. The massive amount of changes taking place in this domain, have led to multiple technologies rising and falling within a few years. E-mail, which is a relatively new mode of mass communication, has risen to the peak of communication platforms in a very short while. However, its descent may be as quick as its rise was, since it is competing with highly innovative and personalized technologies. Social networks have emerged as the latest phenomena in the internet. They have a massive coverage and immense popularity among their users. A number of businesses have already started harnessing the power of social networks to spread their message, popularize their products, and get a strong foothold in this newly emerging domain. Websites like Facebook, Twitter, Myspace and Orkut are well known examples of successful social networks that have a massive following and ability to influence business, politics and governments, in addition to forming a critical part of the lifestyle of millions of users.

In this paper, we discuss a few approaches wherein technology can be used to gain useful information about customer behavior. To harness the power of online crowds, it is necessary to analyze what they are saying, and to transform this into actionable information. Sentiment Analysis is a newly emerging field in computer science which attempts the gauge the sentiment of an individual regarding a certain topic. It is a broad research area which encompasses techniques from Natural Language Processing, Text mining, Machine Learning and Computational Linguistics to automatically gauge sentiments of customers. The rise of social media (blogs, social networks, online reviews) has led to popularization of sentiment analysis. With the proliferation of reviews, ratings, recommendations and other forms of online expression, online opinion has turned into a kind of virtual currency for businesses looking to market their products, identify new opportunities and manage their reputations. As businesses look to automate the process of understanding what their customers are demanding, and why their competitors are doing better, they are using sentiment analysis to do this in a scalable manner. Sentiment analysis has various challenges: filtering out the noise, understanding the conversations, identifying the relevant content and actioning it appropriately, and finally understanding the sentiment expressed in the context of the product/service being discussed. The technical and computational challenges involving sentiment analysis are significant, and we will discuss a few contributions made by us in the following sections.

2 Methods

2.1 Improving Sentiment Analysis Using Linguistic and Machine Learning Approaches

Sentiment analysis (also called opinion mining) has been an active research area in recent years. There are many research directions, e.g., sentiment classification (classifying an opinion document as positive or negative) [1,2], subjectivity classification (determining whether a sentence is subjective or objective, and its associated opinion) [3,4,5,6,7], feature/topic-based sentiment analysis (assigning positive or negative sentiments to topics or product features) [8,9,10,11,12,13]. Formal definitions of different aspects of the sentiment analysis problem and discussions of major research directions and algorithms can be found in [14,15]. A comprehensive survey of the field can be found in [16].

Our work is in the area of topic/feature-based sentiment analysis or opinion mining [8]. The existing research focuses on solving the general problem. However, we argue that it is unlikely to have a one-technique-fit-all solution because different types of sentences express sentiments/opinions in different ways. A divide-and-conquer approach is needed, e.g., focused studies on different types of sentences. We have developed approaches that focus on one type of sentences, i.e., conditional sentences, which have some unique characteristics that make it hard to determine the orientation of sentiments on topics/features in such sentences. By *sentiment orientation*, we mean positive, negative or neutral opinions. By topic, we mean the target on which an opinion has been expressed. In the product domain, a topic is usually a product feature (i.e., a component or attribute). For example, in the sentence, *I do not like the sound quality, but love the design of this MP3 player*, the product features (topics) are *sound quality* and *design* of the MP3 player as opinions have been expressed on them. The sentiment is positive on *design* but negative on *sound quality*.

Conditional sentences are sentences that describe implications or hypothetical situations and their consequences. In the English language, a variety of conditional connectives can be used to form these sentences. A conditional sentence contains two clauses: the condition clause and the consequent clause, that are dependent on each other. Their relationship has significant implications on whether the sentence describes an opinion. One simple observation is that sentiment words (also known as opinion words) (e.g., *great, beautiful, bad*) alone cannot distinguish an opinion sentence from a non-opinion one. A conditional sentence may contain many sentiment words or phrases, but express no opinion.

Example 1: *If someone makes a beautiful and reliable car, I will buy it* expresses no sentiment towards any particular car, although beautiful and reliable are positive sentiment words. This, however, does not mean that a conditional sentence cannot express opinions/sentiments.

Example 2: *If your Nokia phone is not good, buy this great Samsung phone* is positive about the *Samsung phone* but does not express an opinion on the *Nokia phone* (although the owner of the *Nokia phone* may be negative about it). Clearly, if the sentence does not have *if*, the first clause is negative. Hence, a method for determining sentiments in normal sentences will not work for conditional sentences. The examples below further illustrate the point. In many cases, both the condition and consequent together determine the opinion.

Example 3: *If you are looking for a phone with good voice quality, dont buy this Nokia phone* is negative about the *voice quality* of the *Nokia phone*, although there is a positive sentiment word *good* in the conditional clause modifying *voice quality*. However, in the following example, the opinion is just the opposite.

Example 4: *If you want a phone with good voice quality, buy this Nokia phone* is positive about the *voice quality* of the *Nokia phone*.

As we can see, sentiment analysis of conditional sentences is a challenging problem. One may ask whether there is a large percentage of conditional sentences to warrant a focused study. Indeed, there is a fairly large proportion of such sentences in evaluative text. They can have a major impact on the sentiment analysis accuracy. Specifically, we determine whether a conditional sentence (which is also called a *conditional* in the linguistic literature) expresses positive, negative or neutral opinions on some topics/features. Since our focus is on studying how conditions and consequents affect sentiments, we assume that *topics* are given, which are *product attributes* since our data sets are user comments on different products.

Our study is conducted from two perspectives. We start with the linguistic angle to gain a good understanding of existing work on different types of conditionals. As conditionals can be expressed with other words or phrases than *if*, we study how they behave compared to *if*. We also show that the distribution of these conditionals based on our data. With the linguistic knowledge, we perform a computational study using machine learning. A set of features for learning is designed to capture the essential determining information. Note that the features here are data attributes used in learning rather than product attributes or features. Three classification strategies are designed to study how to best perform the classification task due to the complex situation of two clauses and their interactions in conditional sentences. These three classification strategies are *clause-based*, *consequent-based* and *whole-sentence-based*. Clause-based classification classifies each clause separately and then combines their results. Consequent-based classification only uses consequents for classification as it is observed that in conditional sentences, it is often the consequents that decide the opinion. Whole-sentence-based classification treats the entire sentence as a whole in classification. Experimental results on conditional sentences from diverse domains demonstrate the effectiveness of these classification models. The results indicate that the whole-sentence-based classifier performs the best.

Since our research only studies conditional sentences, a natural question is whether the proposed technique can be easily integrated into an overall sentiment analysis or opinion mining system. The answer is yes because a large proportion of conditional sentences can be detected using conditional connectives. Keyword search is thus sufficient to identify such sentences for special handling using the proposed approach. Further details of our approach and experimental results are available in [17].

2.2 Mining Online Customer Reviews for Ranking Products

The rapid proliferation of internet connectivity has led to increasingly large volumes of electronic commerce. A study by Forrester Research [18] predicted that e-commerce and retail sales in the US during 2008 were expected to reach $204 Billion, an increase of 17% over the previous year. As more consumers are turning towards online shopping over brick and mortar stores, a number of websites offering such services have prospered. Amazon.com, Zappos.com, ebay.com, newegg.com are a few examples of e-commerce retailers which offer consumers a vast variety of products. These platforms aim to provide the consumers a comprehensive shopping experience by allowing them to choose products based on parameters like price, manufacturer, product features etc. Since it is difficult for consumers to make their purchasing decisions based only on an image and (often biased) product description provided by the manufacturer, these e-commerce platforms allow users to add their own reviews. Consumer reviews of a product are considered more honest, unbiased and comprehensive than a description provided by the seller. They also relate to customers use of a product thereby linking different product features to its overall performance. A study by comScore and the Kelsey group [19] showed that online customer reviews have significant impact on prospective buyers. As more customers provide reviews of the products they purchase, it becomes almost impossible for a single user to read them all and comprehend them to make informed decisions. For example, there are several popular digital cameras at Amazon.com with several hundreds of reviews, often with very differing opinions. While most websites offer the customer the opportunity to specify their overall opinion using a quantitative measure, it leads to obfuscation of the multiple views expressed in a review. More importantly, reviews often contain information comparing competing products which cannot be reflected using a number measure. Also, different users have varying levels of quantitative measures(ex. easy grades vs. tough graders) thereby making the use of such numerical ratings even more difficult.

The widespread availability of customer reviews has led a number of scholars doing valuable and interesting research related to mining and summarizing customer reviews [21,22,23,24,25,26]. There has also been considerable work on sentiment analysis of sentences in reviews, as well as the sentiment orientation of the review as a whole [20]. In this work, we aim to perform a ranking of products based on customer reviews they have received. The eventual goal of our research

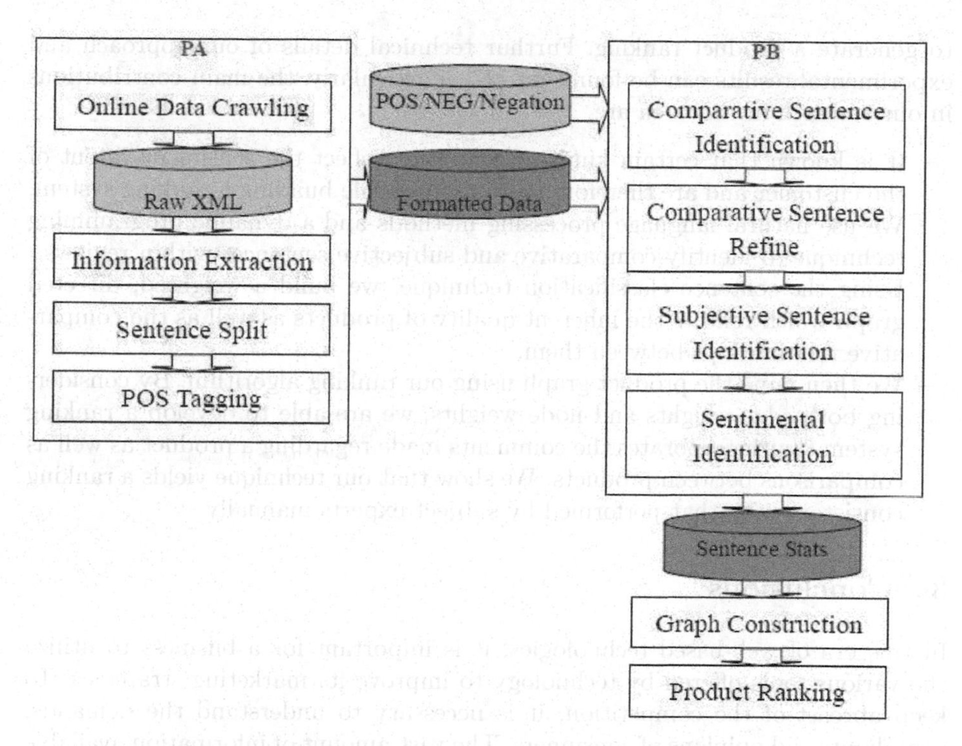

Fig. 1. Online Customer Review-Based Product Ranking

is to create a tool that will allow users to select products based on reviews they have received. To the best of our knowledge, there has been no comprehensive study using thousands of customer reviews to rank products. It is our hypothesis that this ranking will aid consumers in making better choices. Figure 1 shows the steps involved in gathering, processing and ranking products based on customer reviews. In a typical review, we identify two kinds of sentences that are useful in ranking products:

- *Subjective sentences:* Sentences containing positive/ negative opinions regarding a product. Examples:This camera has great picture quality and conveniently priced (positive subjective sentence), The picture quality of this camera is really bad. (negative subjective sentence).
- *Comparative sentences:* Reviewers often compare products in terms of the features, price, reliability etc. Comparisons of this kind are crucial in determining the relative worth of products. Examples: This camera has superior shutter speed when compared to the Nikon P40, This is the worst camera I have seen so far.

After developing techniques to identify such sentences, we build a product graph that captures the sentiments expressed by users in reviews. The advantage of using a directed graph whose edges indicate the preference of a user of pone product over another is that we can use several graph mining algorithms to generate a product ranking. Furthur technical details of our approach and experimental results can be found in [27]. Particularly, the main contributions in our research in this area are

- It is known that certain kinds of sentences reflect the sentiment/intent of the customer and are therefore more useful while building a ranking system. We use natural language processing methods and a dynamic programming technique to identify comparative and subjective sentences within reviews.
- Using the sentence classification technique, we build a weighted, directed graph which reflects the inherent quality of products as well as the comparative relationships between them.
- We then mine the product graph using our ranking algorithm. By considering both edge weights and node weights, we are able to develop a ranking system that incorporates the comments made regarding a product as well as comparisons between products. We show that our technique yields a ranking consistent with that performed by subject experts manually.

3 Conclusions

In this era of web-based technologies, it is important for a business to utilize the various tools offered by technology to improve its marketing strategies. To keep abreast of the competition, it is necessary to understand the demands, complaints and opinions of consumers. The vast amount of information available through social media outlets can be utilized to understand the sentiments of users towards products. The importance of the different features of a product and the effectiveness of marketing strategies can be known by systematically analyzing consumer generated text like blogs, reviews and forum posts. In this paper, we have demonstrated two techniques that advance the state-of-the-art in sentiment analysis and product ranking. We believe that the future holds exciting opportunities for businesses willing to use these new tools to understand customer sentiments and develop marketing strategies.

References

1. Pang, B., Lee, L., Vaithyanathan, S.: Thumbs up? Sentiment Classification Using Machine Learning Techniques. In: Proceedings of Conference on Empirical Methods in Natural Language Processing (2002)
2. Turney, P.: Thumbs Up or Thumbs Down? Semantic Orientation Applied to Unsupervised Classification of Reviews. In: Proceedings of ACL 2002 (2002)
3. Wiebe, J., Wilson, T.: Learning to Disambiguate Potentially Subjective Expressions. In: Proceedings of CoNLL-2002 (2002)
4. Wilson, T., Wiebe, J., Hwa, R.: Just how mad are you? Finding strong and weak opinion clauses. In: AAAI 2004 (2004)
5. Wiebe, J., Wilson, T.: Learning to Disambiguate Potentially Subjective Expressions. In: CoNLL 2002 (2002)
6. Kim, S., Hovy, E.: Determining the Sentiment of Opinions. In: COLING 2004 (2004)

7. Riloff, E., Wiebe, J.: Learning extraction patterns for subjective expressions. In: EMNLP 2003 (2003)
8. Hu, Liu, B.: Mining and summarizing customer reviews. In: KDD 2004 (2004)
9. Popescu, A.-M., Etzioni, O.: Extracting Product Features and Opinions from Reviews. In: EMNLP 2005 (2005)
10. Ku, L.-W., Liang, Y.-T., Chen, H.-H.: Opinion Extraction, Summarization and Tracking in News and Blog Corpora. In: AAAI-CAAW (2006)
11. Carenini, G., Ng, R., Pauls, A.: Interactive Multimedia Summaries of Evaluative Text. In: IUI 2006 (2006)
12. Kobayashi, N., Inui, K., Matsumoto, Y.: Extracting Aspect-Evaluation and Aspect-of Relations in Opinion Mining. In: EMNLP 2007 (2007)
13. Titov, I., McDonald, R.: A Joint Model of Text and Aspect Ratings for Sentiment Summarization. In: ACL 2008 (2008)
14. Liu, B.: Web Data Mining: Exploring Hyperlinks, Content and Usage Data. Springer, Heidelberg (2006)
15. Liu, B.: Sentiment Analysis and Subjectivity. In: Indurkhya, N., Damerau, F.J. (eds.) To appear in Handbook of Natural Language Processing, 2nd edn. (2009/2010)
16. Pang, B., Lee, L.: Opinion Mining and Sentiment Analysis. Foundations and Trends in Information Retrieval 2(1-2), 1135 (2008)
17. Narayanan, R., Liu, B., Choudhary, A.: Sentiment Analysis of Conditional Sentences. In: Proceedings of Conference on Empirical Methods in Natural Language Processing (EMNLP 2009), Singapore, August 6-7 (2009)
18. Forrester research, http://www.comscore.com/Press_Events/Press_Releases/2007/11/Online_Consumer_Reviews_Impact_Offline_Purchasing_Behavior
19. Comscore and Kelsey, http://www.shop.org/c/journal_articles/view_article_content?groupId=1&articleId=702&version=1.0
20. Liu, B.: Sentiment Analysis and Subjectivity. To appear in Handbook of Natural Language Processing, 2nd edn. (2010)
21. Hu, M., Liu, B.: Mining and Summarizing Customer Reviews. In: Proceedings of the 10th ACM International Conference on Knowledge Discovery and Data Mining (SIGKDD-2004), vol. 8, pp. 168–174 (2004)
22. Hu, M., Liu, B.: Mining Opinion Features in Customer Reviews. In: Proceedings of the 19th National Conference on Artificial Intelligence, vol. 7, pp. 755–760 (2004)
23. Popescu, A., Etzioni, O.: Extracting product features and opinions from reviews. In: Proceedings of the conference on Human Language Technology and Empirical Methods in Natural Language Processing, pp. 339–346 (2005)
24. Liu, B., Hu, M., Cheng, J.: Opinion Observer: Analyzing and Comparing Opinions. WWW 5, 342–351 (2005)
25. Kim, S., Pantel, P., Chklovski, T., Pennacchiotti, M.: Automatically Assessing Review Helpfulness. In: EMNLP, vol. 7, pp. 423–430 (2006)
26. He, B., Macdonald, C., He, J., Ounis, I.: An Effective Statistical Approach to Blog Post Opinion Retrieval. CIKM 10, 1063–1069 (2008)
27. Zhang, K., Narayanan, R., Choudhary, A.: Mining Online Customer Reviews for Ranking Products. Technical Report CUCIS-2009-11-001, Center for Ultra-scale Computing and Information Security (November 2009)

Resource Allocation for Energy Efficient Large-Scale Distributed Systems

Young Choon Lee and Albert Y. Zomaya

Centre for Distributed and High Performance Computing
School of Information Technologies
The University of Sydney
NSW 2006, Australia
{yclee,zomaya}@it.usyd.edu.au

Abstract. Until recently energy issues have been mostly dealt with by advancements in hardware technologies, such as low–power CPUs, solid state drives, and energy–efficient computer monitors. Energy–aware resource management has emerged as a promising approach for sustainable/green computing. We propose a *holistic* energy–aware resource management framework for distributed systems. This framework will seamlessly integrate a set of both site–level and system–level/service–level energy–aware scheduling schemes. One of our previous works as an example implementation of the framework is presented in this paper. The performance of the algorithm in this example implementation has been shown to be very appealing in terms particularly of energy efficiency.

Keywords: scheduling, energy efficiency, resource allocation, large-scale systems, distributed computing.

1 Introduction

Scheduling and resource allocation in distributed systems play a major role in finding the best task–resource matches in time and space based on a given objective function without violating a given set of constraints. This is an important and but computationally intractable problem (i.e. NP–hard) [1]. If one is to add energy or power as an added constraint to create a more sustainable computing system then the problem becomes much more complex. Energy–aware scheduling is a software based power saving approach which has great potential due to its inherent adaptability, cost–effectiveness and applicability. In this approach, given a set of tasks and a set of resources (e.g., processors and disks)—typically energy consumption/savings associated with each element in both sets is identified—these preprocessed tasks are then prioritized, matched with resources, and scheduled while complying with constraints if any. Each of these steps is devised to maximize the objective function. The quality of output schedule is determined using an appropriate objective function that takes into account energy consumption in addition to other conventional metrics, such as, task completion times and load balance. Much of the existing energy–aware

S.K. Prasad et al. (Eds.): ICISTM 2010, CCIS 54, pp. 16–19, 2010.

scheduling algorithms are aimed at reducing energy consumed by processors (e.g., [2–3]), because earlier studies primarily focused on battery–powered devices.

In this paper, we first describe our energy-aware scheduling and resource allocation (S&RA) framework—as a holistic approach—for large-scale distributed systems (LSDSs) followed by one of our previous works as an example implementation of the site-level component in the framework.

2 Holistic Energy-Aware S&RA Framework

2.1 Overview of the Framework

With the emergence of LSDSs (e.g., grids, clouds) as viable alternatives to supercomputers to meet the ever increasing need for performance, the reduction of energy consumption should be derived from resources in multiple levels not independently, but collaboratively and collectively. Site–level scheduling agents can deal with parallel computations and data accesses within individual autonomous administrative domains. The collaboration between these site–level scheduling agents shall be enabled by using a system–level coordinator. The main purpose for this collaboration is energy–load balancing to reduce indirect energy consumption. The system–level resource manager performs this coordination with information obtained from both site–level scheduling agents and several other system–wide services developed as part of our energy–aware resource management framework.

This holistic approach is a natural choice for LSDSs, since resource sites are autonomous and scheduling decisions in each resource site are made independently from others. The use of software approach to the minimization of energy consumption is of great practical importance.

2.2 Key Characteristics

Our energy-aware S&RA framework is novel for a number of reasons as follows:

- **Holistic approach.** A comprehensive set of resources (CPUs, disks, services, etc) are exploited at different levels to maximize the improvement of their energy efficiency using collaborative site–level and system–level energy–aware scheduling schemes.
- **Interweaving energy consumption with other performance objectives.** Scheduling decisions are determined by performance models devised with energy consumption as their integrated variable, not as a separate performance metric, which limits the reduction of energy consumption.
- **Wide scope.** The dynamic nature of availability and capacity in LSDSs is explicitly incorporated into scheduling.
- **Applicability and portability.** Distributed systems are the platform of choice for many applications these days. This implies that the energy–aware resource management framework and its components can easily be ported and applied to a wide range of computing systems.
- **Cost effectiveness.** The software framework is malleable and can rapidly adapt to changes in both application models and hardware upgrades.

The above holistic approach in itself is very sustainable, and will open up new research directions.

3 Preliminary Findings

3.1 Energy-Conscious Task Scheduling Using Dynamic Voltage Scaling

The task scheduling problem we address is how to allocate a set of precedence-constraint tasks to a set of processors aiming to minimize makespan with energy consumption as low as possible. The grounding of our algorithm using DVS can be found in the power consumption model in complementary metal–oxide semiconductor (CMOS) logic circuits, which is typically used as an energy model. The energy consumption of the execution of a precedence–constrained parallel application can be defined as $E = \sum_{i=1}^{n} ACV_i^2 f \cdot w_i^* = \sum_{i=1}^{n} \alpha V_i^2 w_i^*$ where V_i is the supply voltage of the processor on which task n_i executed, and w_i^* is the computation cost of task n_i (the amount of time taken for n_i's execution) on the scheduled processor.

Our energy-conscious scheduling (*ECS*) algorithm [4] attempts to exploit the DVS technique beyond the conventional slack reclamation method [5]; that is, the quality of schedules (makespans) and energy consumption are investigated in terms of the gain/loss relationship. Specifically, our algorithm is devised with relative superiority (RS) as a novel objective function, which takes into account these two performance considerations (makespan and energy consumption). More formally,

$$RS(n_i, p_j, v_{j,k}, p', v') = -\left(\left(\frac{E(n_i, p_j, v_{j,k}) - E(n_i, p', v')}{E(n_i, p_j, v_{j,k})}\right) + \left(\frac{EFT(n_i, p_j, v_{j,k}) - EFT(n_i, p', v')}{EFT(n_i, p_j, v_{j,k}) - \min(EST(n_i, p_j, v_{j,k}), EST(n_i, p', v'))}\right)\right)$$

where $E(n_i, p_j, v_{j,k})$ and $E(n_i, p', v')$ are the energy consumption of n_i on p_j with $v_{j,k}$ and that of n_i on p' with v', respectively, and similarly the earliest start/finish times of the two task-processor allocations are denoted as $EST(n_i, p_j, v_{j,k})$ and $EST(n_i, p', v')$, and $EFT(n_i, p_j, v_{j,k})$ and $EFT(n_i, p', v')$. For a given ready task, its *RS* value on each processor is computed using the current best combination of processor and voltage supply level (p' and v') for that task, and then the processor—from which the maximum RS value is obtained—is selected.

3.2 Experimental Results

Experimental results (Figure 1)—obtained from a large number of simulations with random task graphs—show the competent energy-saving capability of *ECS*. Results are presented based on average makespans (solid fill) and energy consumption (checked fill) and compared with two existing algorithms (*DBUS* and *HEFT*). Specifically, for a given task graph, we normalize both its makespan and energy consumption to lower bounds—the makespan and energy consumption of the tasks along the CP (i.e., CP tasks) without considering communication costs. Specifically, the 'schedule length ratio' (SLR) and 'energy consumption ratio' (ECR) were used as the primary performance metrics for our comparison.

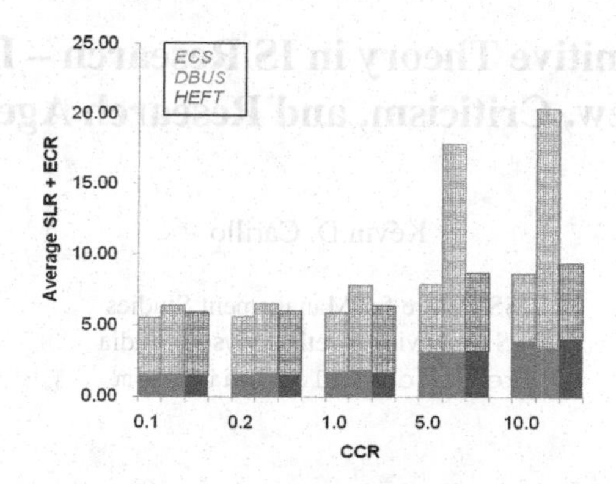

Fig. 1. Simulation results with respect to different algorithms

4 Conclusion

The ever increasing demand for performance from distributed systems brings about a number of serious operational and environmental issues. While several significant energy–saving technologies and techniques have been developed, they are mostly hardware–based approaches or low–level software approaches, such as thread scheduling at the operating system level. In this paper, we have presented a holistic energy-aware resource management framework as a software approach to energy issues in LSDSs. The framework covers various facets of energy consumption in LSDSs. We have demonstrated the feasibility of our framework with our energy-conscious task scheduling algorithm (*ECS*).

References

1. Garey, M.R., Johnson, D.S.: Computers and Intractability, pp. 238–239. W.H. Freeman and Co., New York (1979)
2. Goh, L.K., et al.: Design of fast and efficient energy–aware gradient–based scheduling algorithms heterogeneous embedded multiprocessor systems. IEEE Trans. Parallel Distrib. Syst. 20(1), 1–12 (2009)
3. Zhong, X., Xu, C.-Z.: Energy–aware modeling and scheduling for dynamic voltage scaling with statistical real-time guarantee. IEEE Trans. Computers 56(3), 358–372 (2007)
4. Lee, Y.C., Zomaya, A.Y.: Minimizing Energy Consumption for Precedence-constrained Applications Using Dynamic Voltage Scaling. In: Int'l Symp. Cluster Computing and the Grid (CCGRID 2009), pp. 92–99. IEEE Press, Los Alamitos (2009)
5. Zhu, D., Melhem, R., Childers, B.R.: Scheduling with dynamic voltage/speed adjustment using slack reclamation in multiprocessor real-time systems. IEEE Trans. Parallel Distrib. Syst. 14(6), 686–700 (2003)

Social Cognitive Theory in IS Research – Literature Review, Criticism, and Research Agenda

Kévin D. Carillo

JSS Centre for Management Studies
JSS Mahavidyapeetha, Mysore, India
kevin.carillo@gmail.com

Abstract. A multitude of research studies have been published investigating individual behavior from the viewpoint of Social Cognitive Theory. We have now reached a point where making sense of such a large number of studies has become a difficult task and where future research efforts must integrate past SCT findings but also express the full potential of SCT in IS research. The aim of the present paper is to organize the literature to provide a clear depiction of the use of SCT in IS research. A review the IS literature which used Social Cognitive Theory of the past 14 years yielded 62 papers that investigated individual behavior using the SCT perspective. This vast literature is mapped into the SCT framework, thus highlighting the main successes but also pitfalls of past research in using the theory. Future research directions are then identified and discussed.

Keywords: Social Cognitive Theory, individual behavior, literature review.

1 Introduction

Since the mid-seventies, the study of the factors leading to the adoption and use of information technology has been largely represented in MIS research, concentrating ever-increasing research efforts. This line of research emerged when both organizations and researchers started realizing that in spite of the immense promises of IT, the level of adoption of information technology did not match by far the level of expectations. Lucas [1,2] was among the first IS researchers to investigate the influence of individual and behavioral factors on IT adoption. The first theory that started providing evidence in the IT adoption cause was the Theory of Reasoned Action [3]. TRA posits that a person's behavioral intention depends on the person's attitude about the behavior and subjective norms. This theory gave birth to one of the most important IS-grounded theories: the Technology Acceptance Model [4,5]. The Diffusion of Innovations (DOI) theory [6] also provided successful insights by providing a complementary view to the IT adoption and use cause [7,8].

Drawn from social psychology, Social Cognitive Theory [9] has been another insightful and widely used theory in IS research. Focusing on individual learning, SCT relies mainly on the assumption that all individual behavior, cognition and other personal factors, and environmental influences operate as interacting determinants

S.K. Prasad et al. (Eds.): ICISTM 2010, CCIS 54, pp. 20–31, 2010.

and influence each other bi-directionally. IS academics started using SCT in the early nineties when realizing the relevance of the concept of self-efficacy (a central notion in SCT) in understanding the use and adoption of information technology. Ever since, SCT has been particularly insightful in IS research leading to numerous findings in the context of computer/software training and use [10,11] but also internet [12,13], electronic commerce-related issues [14,15], and e-learning [16]. However, the importance and success of the notion of self-efficacy in IS studies seem to have progressively reduced the consideration of the other essential SCT concepts leading to the loss of the full potential that SCT has promised in helping understanding individual behavior in IS research.

Past research had already warned researchers about the contradictory and equivocal nature of the results found in IS studies when using the concept of computer self-efficacy [17]. The authors justified their claim by highlighting a "general lack of attention to the dynamic, multileveled, and multifaceted nature of the computer self-efficacy construct" [17]. This research paper strives to strengthen this claim by extending it to the entire framework of SCT.

Based on a thorough literature review of the use of Social Cognitive Theory in IS research, this paper strives to highlight the main findings by mapping the reviewed studies into the SCT framework. By doing so, this theoretical paper reveals key aspects which emphasize direct contradictions between the approach used in certain studies and the core SCT assumptions. Furthermore, unexplored areas are then identified and future research avenues are defined. The study insists on the fact that, in IS research, Social Cognitive Theory has been often reduced to a sub-part of the theory itself: Self-efficacy theory. In addition, the paper posits that by having been used in a restrictive context, SCT has still an unrevealed but promising potential in IS research, justifying the need for clarification of the theory and guidance for future research. Drawing insights from a thorough literature review of the use of SCT in IS empirical studies focusing on individual behavior, this paper is organized as follows. The first section introduces SCT in social psychology and IS research. The research method is described next and is followed by our analyses. The paper concludes with the delineation of the state of our knowledge in a comprehensive framework which allows for the identification of future research directions in the area.

2 Social Cognitive Theory: An Overview

Social Cognitive Theory [9,18,19,20] introduces a model of individual behavior that has been widely accepted and empirically validated in various fields of research, and which focuses on learning experience. SCT was initially used in the context of therapeutic research [21,22], mass media [23,24], public health [25,26] and education [27,28]. Drawing insights from the findings in mass media, marketing was the first branch of business which started using the SCT approach to adopt a different towards the study of customer behavior [29]. Social Cognitive Theory posits that individual behavior is part of an inseparable triadic structure in which behavior, personal factors and environmental factors constantly influence each other, reciprocally determining each other [11] (See Figure 1). Environmental factors are seen as the factors that are physically external to the person and that provide opportunities and social support

[30] such as social pressure or situational characteristics [11]. Personal factors are any cognitive, personality, or demographic aspects characterizing an individual. In other words, individuals choose the environment in which they evolve, but they also shape their surrounding environment. Pajares [31] affirms that "How individuals interpret the results of their performance attainments informs and alters their environments and their self-beliefs, which in turn inform and alter their subsequent performances". Furthermore, individual behavior in a certain learning situation both influences and is influenced by environmental (or situational) and cognitive/personal factors. Social cognitive theory explains how people acquire and maintain certain behavioral patterns, while also providing the basis for intervention strategies [21].

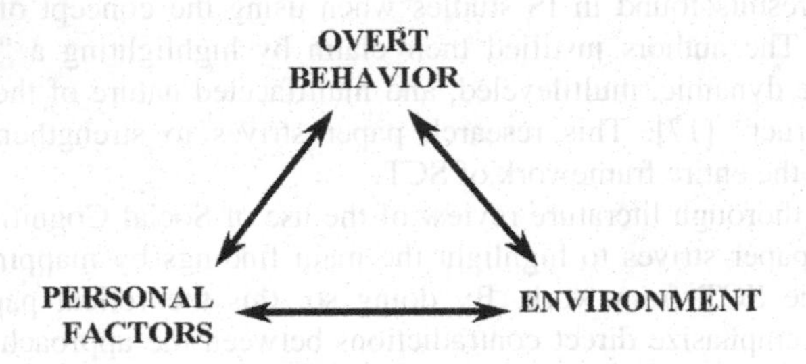

Fig. 1. The triadic nature of Social Cognitive Theory

Evaluating behavioral change thus depends on factors qualifying: environment, people and behavior. SCT provides a framework for designing, implementing and evaluating programs. Environment refers to the factors that can affect a person's behavior. There are social and physical environments. Social environment include family members, friends and colleagues. Physical environment is the size of a room, the ambient temperature or the availability of certain foods. Environment and situation provide the framework for understanding behavior [32]. The situation refers to the cognitive or mental representations of the environment that may affect a person's behavior. Social Cognitive Theory rotates around a central concept entitled self-efficacy which Bandura [9] defines as "*People's judgment of their capabilities to organize and execute courses of action required to attain designated types of performances. It is concerned not with the skills one has but with judgments of what one can do with whatever skills one possesses.* (p. 391)". In addition, Bandura posits the importance of a second individual factor that is closely inter-related to self-efficacy: outcome expectations defined as the extent an individual will undertake a certain behavior only if he/she perceives that it will lead to some valued outcomes or else favorable consequences [11].

2.1 Social Cognitive Theory in IS Research

The Technology Acceptance Model [4], Theory of Planned Behavior [3] and Diffusion of Innovations [6] have been particularly insightful in IS research. In such theories, behavior is viewed as a set of beliefs about technology, and a set of affective responses (typically measured in terms of attitude towards using) to the behavior [33].

Nonetheless, while TAM and DOI focus solely on beliefs about the technology, TPB and SCT integrate the notion of perceived outcomes when forecasting behavior. Indeed, both theories posit that the use of a certain technology is directly influenced by the perception that it will allow the individual to achieve positive outcomes. For instance, a person's motivation might be increased if he/she perceives that learning to use a certain system will lead to a higher level of performance for his/her job. The consideration of individuals' environment as playing an important role is another common concept in both theories. A key aspect differentiates SCT from TAM, DOI, and TPB. The last three theories adopt a unidirectional perspective towards causal relationship, in which environmental factors influence cognitive beliefs, which influence attitudes and behaviors. On the contrary, SCT relies on the bidirectional nature of causation in which behavior, cognitive and emotional factors, and environment constantly and mutually influence each other. For instance, an individual's self-efficacy towards using a certain software is influenced by prior successful experiences and performance with the software. In such a case, self-efficacy is seen as an effect whereas behavior acts as a cause. Such an implication creates a vast and original research avenue in which behavior plays the role of a predictor having an influence on either cognitive and emotional factors or else an individual's environment.

When computers started being more and more present in the workplace, SCT was found to be particularly insightful in understanding individual behavior in computer training [11,33,34,35,36]. Two key personal-level factors of SCT have been largely investigated: self-efficacy, and outcome expectations. The latter concept has been widely explored by IS researchers through the concept of perceived usefulness which was defined and measured by Davis [4]. Further comments about the matching of the two concepts will be found in the discussion section.

Compeau and Higgins [11] and Compeau et al. [33] were acknowledged to be among the pioneers in IS research to first integrate the Social Cognitive Theory framework by using a social psychology perspective in investigating computer training success. Focusing on knowledge workers, they first validated a measure of computer self-efficacy. Second, they found that computer efficacy plays a significant role in influencing individuals' expectations of computer use expectations, individuals' emotional reactions, as well as computer use itself. Both computer self-efficacy and individuals' expectations towards outcome where found to be positively influenced by encouragement of others and others' use confirming the need to integrate social and environmental factors in the understanding of individual behavior. Computer self-efficacy was overall found to significantly influence an individual's decision to use a computer. Later, the authors used a similar model on a sample of 394 end-users to predict computer use. The study confirmed that both self-efficacy and outcome expectations impact on an individual's affective and behavioral reactions to information technology [33].

With the advent of the Internet which lead to the adoption and use of web-based technologies as well as B2C electronic commerce practices, SCT started being used from a different perspective in which the use of the Internet in general or else the use of internet-based applications or services have been modeled as learning processes. Such processes are acquired by an individual in which his/her behavior, cognitive and emotional characteristics, and environment are inter-related, mutually influencing each other [37,38].

3 Research Method and Data Analysis

In order to broadly capture most of the different approaches and findings provided by Social Cognitive Theory in relation to IS research; it was decided to include all top business academic journals in the review process. In addition, during the initial investigation stage of this research project, a broader review revealed that SCT had been widely used in the top journals of online education research which interest in the theory was triggered by the spread of computers in educational institutions and the training it involved. Such research projects adopted an approach strictly identical to the one adopted in IS research, and were then included in the review process.

First, articles from the top IS journals (including electronic commerce journals) were selected when at least one construct of the SCT framework was integrated in the theoretical model, and when the use of SCT was clearly referred. Second, articles from business academic journals were also selected in which an IS approach was used, focusing on the adoption or use of a certain technology and drawing insights from SCT. Finally, articles from online education peer-reviewed journals were eventually chosen in which researchers use an "IS research approach" and investigate individual behavior towards using a certain technology in the context of education using SCT. It is important to highlight that only empirical papers were selected, and which adopted the level of analysis of individuals.

The procedure used to identify the articles was carried out in several steps. An initial search of keywords was performed. The table of contents and abstracts of journals and proceedings were reviewed. All studies involving any individual aspect related to SCT were retained. The process resulted in a total of 62 empirical papers (Contact author for a detailed list of the reviewed articles and review findings). The first step in data analysis consisted of identifying the variables taken into account in each study. The variables were then mapped into the SCT framework.

4 Results

Out of 62 articles, 17 addressed computer use and/or training, 20 are related to software training and/or use (either adopting a general software approach or else specific such as ERP, ESS …). Finally, 26 articles were found to have used SCT in the context of Internet based applications or services. Seven types of interaction were found to have been investigated (see Table 1). It is important to note that some studies were found to adopt a direct effect approach in which for instance personal factors directly influence individual behavior [35,39,40, for instance].

Such studies were classified into the P-B category (personal factors-behavioral factors). Other studies considered mediating effects in which, for instance, environmental factors influence personal factors, which in turn influence behavior. Such studies were classified in the E-P-B category (environment-personal-behavior). Such an approach was used to integrate all the perspectives adopted in articles.

Table 1. Various SCT interactions investigated in IS research

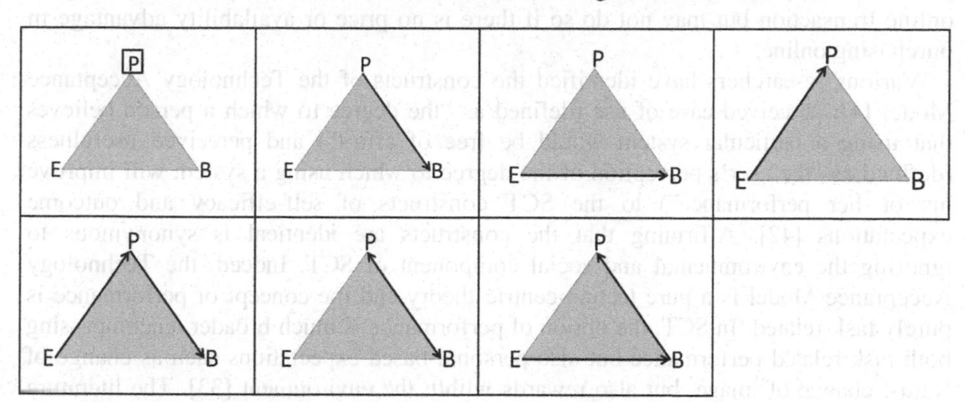

It was not rare to find an article combining several approaches such as P-B and E-P-B for instance, in which some personal factors have a direct effect on behavior whereas others are mediators between environmental factors and behavior.

In terms of dependent behavioral variable, it was found that behavior intention was the most represented by far (24 studies) followed by use (11 studies), task performance (10 studies), continuance intention (7 studies), task effectiveness (2 studies), and adoption (1 study). Out of 62 articles, 51 explored the Personal factors-Behavior relationship from which 30 did not include any environmental/situational factors which is in direct contradiction with the core assumptions of SCT. Such findings revealed that only 20 articles from the selected 62 (that is to say nearly one third) considered personal factors, behavior, and environment all together, respecting the triadic nature of learning processes stated by SCT. From the remaining 11 articles which did not focus on the personal factors-behavior interaction, 6 studies investigated the relationship among various emotional and cognitive individual factors (self-efficacy, outcome expectations and more), 4 studied solely the Environment-Personal factors relationship, (excluding the behavioral dimension of SCT). The remaining study encompassed both the Environment-Behavior and Environment-Personal factors links. It was also astounding to find that in spite of the importance of emotional characteristics during learning processes posited in SCT, only 11 studies integrated such individual considerations (nearly 18% of the selected papers) highlighting a clear lack in IS research.

A detailed review of the SCT constructs used in the selected papers revealed insightful considerations. From the 62 selected articles, all of them integrated the construct of self-efficacy in their research model emphasizing the relevance of the construct in IS research in the case of software, computer, internet, and electronic commerce learning processes. However, it was particularly interesting to have found that 20 studies did not integrate the concept of outcome expectations even though SCT strongly posits the synergic nature of both constructs in their interaction to an individual's environment and behavior [41]. SCT clearly affirms that self-efficacy and cognitive simulation (in other words the perception of potential outcomes) affect each other bi-directionally [18]. In other words, the feeling that one can perform a given task will only motivate oneself to perform it solely in case one perceives any kind of gain from the task outcome. An online customer may feel capable of conducting an online transaction but may not do so if there is no price or availability advantage in purchasing online.

Various researchers have identified the constructs of the Technology Acceptance Model [4]: perceived ease of use (defined as "the degree to which a person believes that using a particular system would be free of effort") and perceived usefulness (defined as "the user's perception of the degree to which using a system will improve his or her performance") to the SCT constructs of self-efficacy and outcome expectations [42]. Affirming that the constructs are identical is synonymous to ignoring the environmental and social component of SCT. Indeed, the Technology Acceptance Model is a pure techno-centric theory and the concept of performance is purely task-related. In SCT, the notion of performance is much broader, encompassing both task-related performance but also personal-based expectations such as change of status, change of image, but also rewards within the environment [33]. The literature review highlighted that some researchers noted the difference between the notions of perceived ease of use and self-efficacy. Indeed, from the 27 studies out of the 62 which integrated both TAM and SCT theories, 26 of them had used separate constructs: perceived ease of use and self-efficacy. Nevertheless, none of the studied used both perceived usefulness and outcome expectations together highlighting the overall techno-centric approach adopted by researchers, omitting the social dimension of performance.

Finally, among the other individual factors that were investigated, it was found that the notion of past experience was used in 19 research papers with successful results in most cases. Such findings insist on the temporal dimension of learning processes which should be seen as both evolving and dynamic [41]. The importance of past experience in the overall information system context but also e-commerce is thus highly emphasized.

5 Discussion and Future Research

The results issued from the literature review have allowed to highlight various research issues and avenues.

5.1 Self-efficacy Theory versus Social Cognitive Theory

It has been commonly acknowledged that the concept of self-efficacy plays a central role in Social Cognitive Theory in which the belief that a person is capable of using a certain system influences significantly his/her performance. However, it is very restrictive to summarize SCT as being solely the concept of individual self-efficacy leading to the definition of a "self-efficacy theory". The power of SCT does not reside in discovering the importance of self-efficacy but rather in making explicit the complexity of learning processes in which self-efficacy is intimately intertwined with other cognitive factors such as outcome expectations, emotional factors (such as affect or else anxiety) but also individual behavior and environmental factors such as (others' use or encouragements). At a higher consideration level, having realized that researchers often affirmed to use SCT in solely importing one construct from SCT, such findings raise an important issue in IS research: is it relevant to claim the use of a certain theory in a research project by simply integrating one or few concepts form a theory in an overall theoretical framework?

5.2 Inter-influence between Self-efficacy and Outcome Expectations

SCT posits that the belief that a person can perform a certain task influences the perception of potential positive or negative outcomes and reciprocally. In other words, the ability to envision the likely outcomes of prospective actions has a strong influence on human motivation and action [41]. As a consequence, self-efficacy and outcome expectations are inseparable notions when using SCT, because of their synergic nature. However, the literature review revealed that too few studies encompassed both concepts in their studies. Furthermore, it was also found that the integration of both SCT and TAM lead in most cases to merge the concepts of perceived usefulness (which is purely techno-centric) and outcome expectations which integrates all SCT dimensions: individual, behavioral, but also social and environmental. When using both TAM and SCT, it seems that the notion of perceived usefulness is a sub-construct of outcome expectations. Such findings raise the issue that integrating several theoretical backgrounds in a single framework may lead to relevance issues in which researchers are encouraged to be particularly cautious.

5.3 Lack of Emotional Considerations

SCT emphasizes the importance of the role played by emotional factors. Indeed, it is said that people's self-beliefs of efficacy affect how much stress and depression they experience in threatening situations, as well as their level of motivation [41]. Few studies considered emotional issues, suggesting future research to investigate the role played by emotional considerations on cognitive and behavioral factors. Prior results suggested that anxiety and stress reactions are low when people cope with tasks in their perceived self-efficacy range [41] whereas self-doubts in coping efficacy produce substantial increases in subjective distress and physiological arousal. Such insights deserve further investigation in IS.

5.4 Key Role of Past Experience

IS research has started acknowledging the importance of past experience in learning processes. Indeed, the bidirectional nature of causality stated in SCT indicates that performance has effects on both cognitive and emotional factors which in turn will affect future performance. Such a statement reminds that any learning process must be seen as a temporal and evolving phenomenon in which past outcomes become future inputs. Too few studies have considered such temporal aspects when investigating the ability of an individual to learn to perform a certain task. For instance, the belief that an individual is capable of conducting a certain online transaction (self-efficacy) constantly varies based on past actions and environmental influences (such as having a peer who encountered a certain fraud when conducting an online purchase). Such a claim insists on the relevance and importance of longitudinal studies in IS research when studying learning-based individual processes.

5.5 Triadic Reciprocality

Bandura postulates that "the person, the behavior, and the environment were all inseparably entwined to create learning in an individual" [9, p. 18]. In the social cognitive view people are neither driven by inner forces nor automatically shaped and controlled by external stimuli. Rather, human functioning is explained in terms of a model of triadic reciprocality in which behavior, cognitive and other personal factors, and environmental events all operate as interacting determinants of each other. Consequently, using SCT in an IS study consists of acknowledging the concept of triadic reciprocality which means integrating both individual and environment-based variables to predict an individual's behavior. In other words, SCT encourages researchers to encompass both levels of factors in order to effectively understand human behavior. Such considerations raise issues when considering studies that focus solely on either technological or individual factors when striving to understand and predict the use or adoption of a certain system.

5.6 Dependent Variables and Unexplored Effects

Social Cognitive Theory indicates that behavioral, cognitive, emotional, and environmental factors constantly influence each other. It was found that an overwhelming majority of studies used individual behavior as the dependent variable. A certain number of studies adopted an individual cognitive or emotional variable as the dependent variable. Based on the triadic reciprocality, several interactions in the SCT triangle have not been explored and deserve future research efforts. For instance, SCT indicates that an individual's behavior shapes his/her environment, but no IS study was found to have used an environmental variable as the dependent variable in order to capture such influence (B-E interaction). Similarly, the impact of personal cognitive and emotional factors on an individual's environment has also never been explored. The SCT framework thus highlights new research avenues which may help providing a more complete understanding of human behavior.

5.7 Towards an SCT Meta-framework for Future Guidance

The integration of all the findings stated in this research paper can be summarized in a meta-framework in which past and future studies may be mapped (see Figure 2) and which may provide guidance for future research efforts in line with SCT. In the proposed framework and for each dimension of the triadic model, some variables that have been commonly accepted and validated in IS research are presented in order to illustrate the mapping process. It is important to note that, according to SCT, environmental factors are twofold. First, social factors pertain to issues related to the influence of family members, peers, or co-workers, for instance such as others' encouragements, others' use [11,33] echoing the concepts of imitation and social persuasion of SCT. Other factors drawn from other theories such as social norms or else social support belong to this category. Second, situational factors are related to issues that characterize the learned task itself. All system and technological-based factors fall within this sub-category. In the behavioral dimension, IS research has mainly focused on issues such as use, adoption, or performance. Among personal factors, the literature review revealed the important role of self-efficacy, outcome

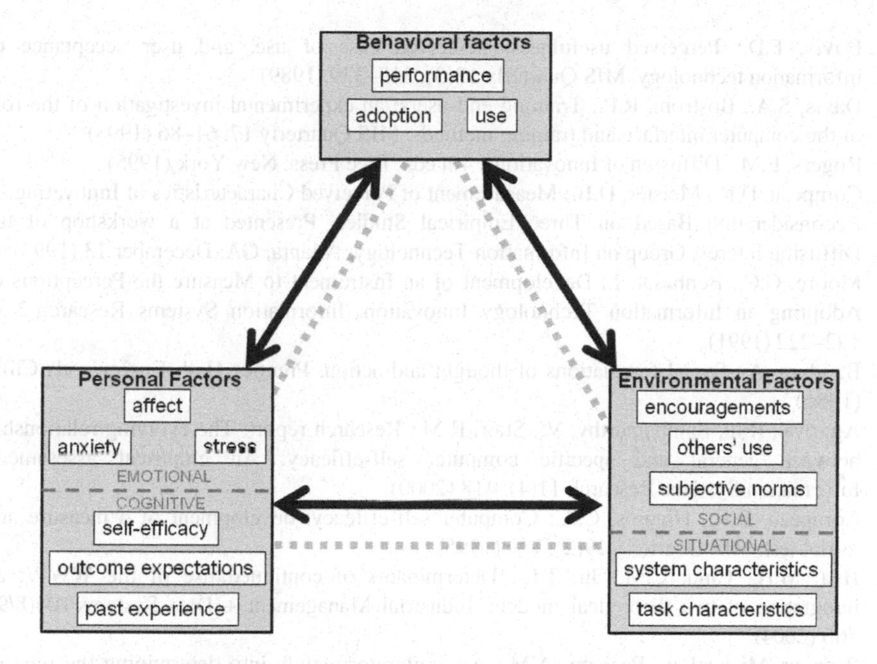

Fig. 2. SCT meta-framework for future IS research projects

expectations or even past experience. Among emotional factors, SCT and IS research have investigated the concepts of affect, anxiety or else stress.

6 Conclusion

A theory has always limitations and will not entirely explain and predict the complex issue of individual behavior. However, this paper defends the viewpoint that the use of the SCT framework in its integrality will shed some new light on the individual behavior issue in IS research. The full potential of SCT has still not been revealed due to a lack of consideration of the complete SCT model. This research paper has provided a thorough review of past IS studies which integrated SCT and has highlighted numerous unexplored areas that will help and guide researchers for future studies. We hope this paper will trigger such new research efforts.

References

1. Lucas Jr., H.C.: Behavioral Factors in System Implementation. In: Schultz, R.L., Slevin, D.P. (eds.) Implementing Operations Research/Management Science. American Elsevier, New York (1975)
2. Lucas Jr., H.C.: Empirical Evidence for a Descriptive Model of Implementation. MIS Quarterly 2(2), 27–41 (1978)
3. Fishbein, M., Ajzen, I.: Belief, Attitude, Intention and Behavior: An Introduction to Theory and Research. Addison-Wesley, Reading (1975)

4. Davis, F.D.: Perceived usefulness, perceived ease of use, and user acceptance of information technology. MIS Quarterly 13(3), 319–339 (1989)
5. Davis, S.A., Bostrom, R.P.: Training end-users: an experimental investigation of the role of the computer interface and training methods. MIS Quarterly 17, 61–86 (1993)
6. Rogers, E.M.: Diffusion of Innovations, 4th edn. Free Press, New York (1995)
7. Compeau, D.R., Meister, D.B.: Measurement of Perceived Characteristics of Innovating: A Reconsideration Based on Three Empirical Studies. Presented at a workshop of the Diffusion Interest Group on Information Technology, Atlanta, GA, December 13 (1997)
8. Moore, G.C., Benbasat, I.: Development of an Instrument to Measure the Perceptions of Adopting an Information Technology Innovation. Information Systems Research 2(3), 192–222 (1991)
9. Bandura, A.: Social foundations of thought and action. Prentice Hall, Englewoods Cliffs (1986)
10. Agarwal, Ritu, Sambamurthy, V., Stair, R.M.: Research report: The evolving relationship between general and specific computer self-efficacy: An empirical assessment. Information Systems Research 11(4), 418 (2000)
11. Compeau, D.R., Higgins, C.A.: Computer self-efficacy: development of a measure and initial test. MIS Quarterly 19(2), 189 (1995)
12. Hsu, M.H., Chiu, C.M., Ju, T.L.: Determinants of continued use of the WWW: an integration of two theoretical models. Industrial Management + Data Systems 104(8/9), 766 (2004)
13. Pearson, Michael, J., Pearson, A.M.: An exploratory study into determining the relative importance of key criteria approach. The Journal of Computer Information Systems 48(4), 155 (Summer 2008)
14. Hernandez, J.M.C., Mazzon, J.A.: Adoption of internet banking: proposition and implementation of an integrated methodology approach. International Journal of Bank Marketing 25(2), 72–88 (2007)
15. Klopping, I.M., McKinney Jr., E.: Practice makes a difference: experience and e-commerce. Information Technology, Learning, and Performance Journal 24(1), 25 (Spring 2006)
16. Hayashi, A., Chen, C., Ryan, T., Wu, J.: The Role of Social Presence and Moderating Role of Computer Self Efficacy in predicting the continuance usage of e-learning systems. Journal of Information Systems Education 15(2), 139 (Summer 2004)
17. Marakas, G.M., Yi, M.Y., Johnson, R.: The multilevel and multifaceted character of computer self-efficacy: Toward a clarification of the construct and an integrative framework for research. Information Systems Research 9(2), 126–163 (1998)
18. Bandura, A.: Self-efficacy: Toward a unifying theory of behavioral change. Psychological Review 84(2), 191–215 (1977)
19. Bandura, A.: Relections on self-efficacy. In: Rashman, S. (ed.) Advances in Behavioral research and Therapy, pp. 237–269. Pergamon Press, Oxford (1978)
20. Bandura, A.: Self-efficacy mechanism in human agency. American Psychologist 37(2), 122–147 (1982)
21. Bandura, A.: Self-efficacy: the exercise of control. W.H. freeman and Company, New York (1997)
22. Langlois, M.A., Petosa, R., Hallam, J.S.: Why do effective smoking prevention programs work? Student changes in social cognitive theory constructs. Journal of School Health 69(8), 326–331 (1999)

23. Bandura, A.: Social cognitive theory of mass communications. In: Bryant, J., Zillman, D. (eds.) Media effects: Advances in theory and research, 2nd edn., pp. 121–153. Lawrence Erlbaum, Hillsdale (2001)
24. Cantor, J.: Fright Reactions to Mass Media. In: Bryant, J., Zillman, D. (eds.) Media Effects. Lawrence Erlbaum, Hillsdale (1994)
25. Bandura, A.: Health promotion from the perspective of social cognitive theory. Psychology and Health (13), 623–649 (1998)
26. Holden, G.: The relationship of self-efficacy appraisals to subsequent health related outcomes: a meta-analysis. Social Work in Health Care 16, 53–93 (1991)
27. Dai, D.Y., Moon, S.M., Feldhusen, J.R.: Achievement motivation and gifted students: A social cognitive perspective. Educational Psychologist 33, 45–63 (1998)
28. Zimmerman, B.J.: A social cognitive view of self-regulated academic learning. Journal of Educational Psychology 81, 329–339 (1989)
29. Wang, G., Netemeyer, R.G.: The effects of job autonomy, customer demandingness, and trait competitiveness on salesperson learning, self-efficacy, and performance. Academy of Marketing Science Journal 30(3), 217 (2002)
30. Glanz, K., Rimer, B.K., Lewis, F.M.: Health Behavior and Health Education. Theory, Research and Practice. Wiley & Sons, San Fransisco (2002)
31. Pajares, F.: Self-efficacy beliefs in academic settings. Review of Educational Research 66, 543–578 (1996)
32. Parraga, I.M.: Determinants of Food Consumption. Journal of American Dietetic Association 90, 661–663 (1990)
33. Compeau, D.R., Higgins, C.A., Huff, S.: Social cognitive theory and individual reactions to computing technology: A longitudinal study. MIS Quarterly 23(2), 145–158 (1999)
34. Bolt, M.A., Killough, L.N., Koh, H.C.: Testing the interaction effects of task complexity in computer training using the social cognitive model. Decision Sciences 32(1), 1 (2001)
35. Hasan, B., Ali, J.H.: The impact of general and system-specific self-efficacy on computer training learning and reactions. Academy of Information and Management Sciences Journal 9(1), 17 (2006)
36. Yi, M.Y., Davis, F.D.: Developing and validating an observational learning model of computer software training and skill acquisition. Information Systems Research 14(2), 146 (2003)
37. Amin, H.: Internet Banking Adoption Among Young Intellectuals. Journal of Internet Banking and Commerce 12(3), 1 (2007)
38. Chang, S.-C., Lu, M.T.: Understanding Internet Banking Adoption and Use Behavior: A Hong Kong Perspective. Journal of Global Information Management 12(3), 21 (2004)
39. Gong, M., Xu, Y., Yu, Y.: An Enhanced Technology Acceptance Model for Web-Based Learning. Journal of Information Systems Education 15(4), 365 (Winter 2004)
40. Guriting, P., Ndubisi, N.O.: Borneo online banking: evaluating customer perceptions and behavioural intention. Management Research News 29(1/2), 6 (2006)
41. Wood, R.E., Bandura, A.: Social cognitive theory of organizational management. Academy of Management Review 14(3), 361–384 (1989)
42. Venkatesh, V., Davis, F.D.: A model of the antecedents of perceived ease of use: development and test. Decision Sciences 27(3), 451 (1996)

Knowledge Intensive Business Processes: A Process-Technology Fit Perspective

Uday Kulkarni and Minu Ipe

W. P. Carey School of Business, Arizona State University, Tempe, Arizona, 85287, USA
uday.kulkarni@asu.edu, minu.ipe@asu.edu

Abstract. We take a business process centric view of an organization and study knowledge intensive business processes (KIBPs). Using a case study research methodology, we studied a diverse set of business processes in multiple organizations to understand their basic ingredients. We find that, although materially dissimilar, KIBPs share a common framework at an abstract level, in that, the tasks that comprise them, the information/knowledge that they need, and the decision situations that are encountered within them have similar characteristics. We draw upon the task-technology fit (TTF) theory and expand it to consider the process-technology fit in the context of KIBPs. In doing so, we prescribe guidelines for organizations for planning for IT support in the form of knowledge management systems for decision making scenarios that are commonly encountered in KIBPs.

Keywords: Knowledge management, Knowledge management systems, Decision support, Knowledge intensive business processes, Task-technology fit.

1 Introduction

Prudent investments in Knowledge Management (KM) can play a significant role in improving organizational performance. However, there is persistent lack of understanding about where and when to make such investments so that they can truly benefit the organization. As a result, the KM literature contains numerous cases where serious efforts in implementing KM projects have not fully met the expectations of their sponsors and architects. While there are myriad factors that have a bearing on success, including organizational culture, knowledge worker commitment and motivation, technological sophistication and operationalization of initiatives [1], [14], [15], [16], a major factor that determines success is the decision context where KM investments are likely to be most successful and return "the biggest bang for the buck".

Substantial effort in prior literature has been devoted to the KM-centric view of workplaces [3], [4], [11], [20], [22]. This view emphasizes the importance of KM processes – from creation and capture of knowledge to its storage, organization, transfer and reuse. In this view, KM is viewed as the art of managing these KM processes and a KM System (KMS) is viewed as a system that facilitates these KM processes by

S.K. Prasad et al. (Eds.): ICISTM 2010, CCIS 54, pp. 32–43, 2010.

providing technology enabled knowledge discovery, knowledge elucidation and mining, search and retrieval, communication, coordination, and collaboration capabilities.

The current study is motivated by the need for a complementary view—a business process centric view of an organization. We focus on what has now come to be known as knowledge intensive business processes [7], [12], [17]. Although routinely executed, KIBPs tend to be complex and time consuming, requiring collaboration and the sharing of knowledge within specific work contexts. KIBPs exist in every organization. New product development, financial planning, partner relationship management are a few examples of KIBPs. While there appears to be an intuitive awareness of processes that are more knowledge intensive than others, the characteristics that constitute knowledge intensity have not been well documented in the research literature.

This research was rooted in our observation that KIBPs are often those that result in high value outcomes in organizations. Many times such processes are part of their core competencies giving them strategic and tactical advantages and making them more competitive. Seldom are these processes outsourced. Investments in KM initiatives designed specifically for KIBPs are thus likely to be worthwhile. Although the value of these processes is not easily measured, partly because their impacts tend to be long-term and indirect [2], [13], [18] and partly because their benefits tend to be perceptual, analyzing these processes and their characteristics will allow organizations to understand the link between knowledge sharing and process outcomes. By analyzing some of its knowledge intensive business processes, an organization can discover opportunities for improving the working and outcomes associated with them. Critical information on the characteristics and tasks within KIBPs will allow organizations to evaluate how information technology can be better deployed to facilitate the delivery of knowledge they need.

In this paper, we present the results of an in-depth investigation of some routine KIBPs in multiple organizations. We find that, although materially dissimilar, KIBPs share a common framework at an abstract level, in that, the tasks that comprise them, the information/knowledge that they need, and the decision situations that are encountered within them have common characteristics. It is these characteristics that define the need for a different type of decision support environment, termed as Knowledge Management Systems (KMS), for KIBPs.

Hence, beyond the characterization of KIBPs, the greater objective of this research is to inform organizations on how information technology can be used to enhance their effectiveness. For example, organizations can identify what knowledge can be codified and retained, what knowledge will be expert dependent, what collaborative technologies might increase the efficiency of these processes. For this purpose, we define a typology of KMSs and discuss a match between the specific characteristics of KIBPs and the types of KMSs that may be suitable for those. Just as the early research on decision support systems was informed by the task-technology fit (TTF) theory, we use the TTF theory as a basis for our prescriptions.

The next section describes the multi-case study methodology adopted for this research. After that, we present the empirical evidence as a summarized description of the KIBPS we studied in multiple organizations. Based on our detailed observations, in Section III, we draw the framework for defining KIBPs in terms of their common characteristics with examples of tasks and decision situations from the cases studied. In

Section IV, we adapt the TTF theory to consider the process-technology fit in the context of KIBPs. We also present a typology of KMSs describing how common types of knowledge intensive tasks and decisions encountered in KIBPs can be supported by these systems. Finally, the concluding section highlights our contribution and discusses the implications for researchers.

2 Research Method

Three dissimilar KIBPs across different organizations were studied for this research. We picked a case study approach for the following reasons: a) this is a very new area of study and knowledge had to be built from a close examination of KIBPs; b) the purpose of the study is to understand the phenomenon of KIBPs within their organizational contexts; c) an in-depth understanding of these processes is necessary to develop a robust model of KIBPs.

We followed a two-phase data collection strategy. The first phase involved choosing the appropriate KIBPs to study from multiple organizations. 144 mid-level managers enrolled in graduate professional degree programs at one of the largest urban universities in the U.S.A participated in this phase. As a part of the required course work, the subjects were asked to prepare a description of a KIBP in which they routinely participated in their own organizations. In the write-up, the subjects identified key stages of the process, decisions made, knowledge needs of the decision makers and a knowledge management system, if any, that facilitated the work. The authors chose half a dozen briefs for the second phase of data collection based on the perceived knowledge intensity of the process as well as the differences in processes and industries that they represented. Five out of the six organizations contacted by the researchers were willing to participate in this study. Our preliminary analysis led us to drop two of the five processes we studied from further consideration because they did not adequately meet the criteria for KIBPs that was developed inductively from our analysis. The second phase involved in-depth face-to-face structured interviews with the selected participants from the organizations identified in the earlier phase. Multiple interviews were conducted with individuals from each KIBP. Additional documentation supporting the process such as manuals, methodology documents, documents detailing operational requirements for the process, resource allocation details, organization charts, and workflow diagrams was collected from participants. After this initial session, the authors mapped the entire process using a process flow diagramming tool in great detail.

Case Studies. We studied the following three cases:

Information Technology Audit (IT Audit) process - at one of the "Big Four" firms, the four largest international accountancy and professional services firms. The IT audit process evaluates the controls within a company's IT systems, especially those that are connected to its financial transactions. It also evaluates a company's ability to protect its information assets and properly dispense information to authorized parties. IT audit is conducted annually for each client.

The Business Process Outsourcing (BPO) Transition process - at a medium sized U.S. based IT organization. The end-to-end Business Process Outsourcing process consists of three broad stages: Mutual Discovery, BPO Transition, and Relationship Management. The mutual discovery stage falls within the sales cycle, with the major activities being prospecting, qualifying potential clients and conducting a corporate assessment. The relationship management stage focuses on maintaining an on-going relationship between the client organization and the outsourcing vendor. Since this is a very large and complex process, we chose to focus on stage two (BPO transition) for more in-depth analysis of knowledge intensity.

Product Engineering (PE) process - at a medium sized European semiconductor manufacturing organization. The new product development process consists of the following stages: Conceptualization, Architecture, Design, Product Engineering, Manufacturing, and Shipping.

Analysis. We started data analysis with the process maps and supporting documentation from all the three cases. The first step involved a detailed examination of each process for direct evidence of knowledge intensity. After that, the second step involved the comparison of the evidence for knowledge intensity across the different processes. This allowed us to identify clear characteristics that distinguished knowledge intensive processes.

3 Results

The analysis resulted in the identification of three primary criteria to define the concept of knowledge intensity in business processes: Process complexity, Knowledge complexity and Decision Making complexity. Each primary criterion is manifested via multiple characteristics. Table 1 provides a description of each of these criteria along with examples from the processes. All the characteristics of knowledge intensity described in Table 1 apply to each of the three cases (albeit, to different degrees). However, to limit the paper length, we have restricted the description to one example per characteristic. All three criteria—process complexity, knowledge complexity and decision-making complexity—contributed to enhancing the complexity of the processes that we studied. While they may independently apply to any number of business processes, together they create a unique environment for KIBPs.

The characteristics under each of the three criteria describe the most common themes that emerged from the data relating to process, knowledge and decision-making complexity. While each of the three main criteria can be considered independent, they influence each other in numerous ways. For example, under some circumstances, uncertainty and ambiguity in the process may impact both knowledge and decision-making complexities. Likewise, the need for multiple sources of information may create not just knowledge complexity, but also has the potential to increase decision-making complexity. While the possibilities of interaction between the various characteristics exist, Table 1 represents the most significant characteristics that emerged from our data as they pertain to the three major criteria.

Table 1. Characteristics of KIBPs with data from case studies

I. Process Complexity
a. Higher number of stages: KIBPs have a greater number of distinct stages, each with multiple sub-processes and tasks. IT Audit: The process can be broken up into three sequentially occurring stages: Planning, Execution, and Closing, which together span almost the entire financial year of the client. BPO: The process end-to-end consists of three broad stages: Mutual Discovery, BPO Transition, and Relationship Management. Each of these stages is comprised of several tasks. PE: The new product development process consists of the following stages: Conceptualization, Architecture, Design, Product Engineering, Manufacturing, and Shipping. The product engineering process consists of the following stages: (a) introduction to the product, (b) introduction to design, (c) test program development, and (d) wafers to test.
b. Greater levels of uncertainty and ambiguity: In KIBPs, ambiguity and uncertainty apply to both the outcomes of various stages of the process and the process itself. Outcomes of stages while broadly predefined, are ultimately constructed through collaboration between agents during the process itself. Models and tools provide rules and specifications but changing dynamics within the organization as well as in the external environment create significant uncertainty. PE: Product engineers worked with design engineers, manufacturing experts and in some cases with marketing staff in order to complete their part of the process. Uncertainty was embedded in the design process, which was not totally completed by the design team before the product got to the PE team. So the PE team had to start their testing procedures with the knowledge that aspects of the design may change before the final design was approved. Uncertainty also lied in how best to test the product, and how the tests led to the final manufacturing process.
c. Greater interdependencies between and within stages: KIBPs are marked by an array of tasks within each stage of the processes. Often, tasks within stages are dependent on each other. This leads to the need for information, perhaps in multiple locations, to be current and readily accessible. IT Audit: The planning stage of the process involved budgeting, scoping, and scheduling of the IT audit, resulting in an audit plan. Developing a budget included estimating the time and resources needed to perform the audit, including training of audit staff, if needed. In order to do this, the audit team had to work with the team that executed the previous year's audit to get accurate information for their planning.
II. Knowledge Complexity
a. Multiple sources of information/knowledge needed: KIBPs rely on both tacit and codified sources of information. Each stage of the process requires a variety of experts who can make appropriate decisions and complete tasks .The tacit knowledge of multiple experts is combined with explicit information from project databases, documents, manuals and other external sources. BPO: Various teams from both the client and vendor organizations were involved in pooling their knowledge during this process. Both codified and non-codified information was used throughout the process.
b. Constant creation of new knowledge: Extensive interaction with people across earlier and later stages of the process results in *continuous sensemaking,* leading to the creation of new knowledge. Nonaka and Takeuchi's [19] four modes of knowledge conversion—socialization, externalization combination and internalization—were evident across the processes we studied. This continuous cycle of new knowledge creation also results in greater knowledge obsolescence as the process evolved.

Table 1. *(continued)*

PE: Based on the specification of the product, product engineers created test programs to measure performance and functionality of the chips. This process ass unique to each new design, involving not just the creation of test programs, but also the design, purchase and deployment of new test equipment for each design. Additionally, product engineers worked in close coordination with both the design and manufacturing teams to make adjustments to the product before the final designs were approved and manufacturing got underway. *c.* ***Context Dependency***: Knowledge used and created within KIBPs was relevant and useful only within the context of the process, much of which is also unstructured and not easily coded. Because of the context dependency of knowledge, the context needed to be recreated in order for the knowledge to be shared across stages. BPO: The specific process selected for outsourcing, the setting within the company in which the process resided, the relationship with the vendor organization, and the teams that were involved at various stages of the process, all formed a unique context in which knowledge was generated and used.
III. Decision Making Complexity
a. ***Higher agent impact on the decision making process and its outcomes:*** Decision makers have a significant role to play in determining both the trajectory and outcomes of the process. Decision making within these processes is characterized by higher order problem solving, and there is significant room for creativity and innovation. Experts have the ability to shape the tasks and outcomes of each stage as the process is not fully standardized. IT Audit: The audit team made decision about when exceptions found in the IT processes needed to be further examined to determine if they impacted the organization's financial statements. Auditors also made decisions about which samples were chosen for assessment by them from the many preventive and detective controls instituted by the client organization. *b.* ***Greater levels of collaboration in decision making:*** There is a need for high levels of collaboration between agents within and across the stages of the process. Although use of communication tools was available, face-to-face communication was deemed irreplaceable in the three processes, due to the uncertainly and ambiguity in the process and context dependency of the knowledge involved. BPO: This process involved extremely high levels of collaboration between teams within the vendor organization as well as with the client's transition team. Most of the key decisions at the early stages of the process were made collaboratively by the client and vendor assessment teams. Subsequently, collaboration was highest between the vendor's teams as they coordinated the tasks that needed to be completed before the process could be successfully transitioned. *c.* ***Greater numbers of criteria need to be considered during decision making:*** Since each stage of the process tends to be complex within itself, decision making involves considering a variety of criteria. Examples include quality versus quantity decisions, cost-savings vs. customer satisfaction. This characteristic of KIBPs also leads to the heterogeneity of agents involved in each stage of the process. PE: When testing the chip for performance, product engineers needed to collect and consider data related to: yields, deviations from specifications stipulated at the design stage, different temperatures and voltages needed for long term reliability, etc.

4 Theoretical Background – Task-Technology Fit and KIBPs

The common characteristics of KIBPs shed some light on the "requirements" of the technological support that can facilitate their efficient and effective conduct. We draw upon the task-technology fit (TTF) theory that was first introduced in the IS literature to help understand the link between information systems and individual performance [8]. The TTF model focuses on the task-system fit and specifies that fit is defined by the correspondence (matching) between the requirements of the task and the capabilities of the technology. The fit has a direct impact on performance and is also a predecessor of utilization of the technology (when the technology use is discretionary), which in turn also impacts performance.

Majority of the TTF literature has focused on specific atomic tasks performed by individuals, such as, tasks performed by students for course-related work (supported by word processors and spreadsheets), by librarians (supported by a central cataloguing system) [21], and by researchers including literature search and data analysis. Several TTF studies have been performed with specific technologies: tools to support software maintenance [5], [6], database query languages [10], and group support systems [23], [24].

A few examples of tasks that need access to knowledge and collaboration between individuals also exist in the literature, e.g., software maintenance tasks requiring common understanding and coordination [5], [6], managerial tasks requiring decision-making using quantitative information [8], [9], [10] and collective decision-making tasks [23], [24]. However, the literature shows no prior TTF studies that focus on knowledge intensive decision-making tasks and the types of KMSs that may be appropriate for them.

Typology of KMSs. We first describe the common types of KMSs that can be specialized to support decision-making tasks in KIBPs. We then discuss the ability of each type of KMS to resolve the complexity introduced by the specific characteristic of KIBPs discussed earlier. KMSs can be divided into six broad categories as follows:

Systems for locating expertise. These systems include Directories, Expertise Databases and Corporate Yellow Pages. They are designed to assist employees of an organization to locate the right expertise needed at the right time. The information in the system is populated by registering the experts' expertise profiles categorized according to a well-understood classification scheme. Multiple search criteria may be used to locate the right expertise. An expert's information may include recent projects worked on, problems tackled, etc., to assist the seeker in locating the right expert. Typically, these systems do not facilitate collaboration or consulting with the expert.

Collaboration Systems. Group Decision Support Systems and Electronic Meeting Systems are examples of such systems. They assist decision-making by supporting discussions, voting, ranking of alternatives, etc., especially in larger groups, typically in an asynchronous manner. Shared workspaces where artifacts can be jointly worked upon by multiple teams is another feature of such systems.

Tele-presence / video conferencing systems. Virtual Meeting Systems are an example of this type. More recent versions, called tele-presence systems are designed to simulate conference meeting rooms giving the touch and feel of a true face-to-face meeting experience. High bandwidth communication capabilities transfer multimedia information in real-time.

Lesson-Learned databases. A lesson-learned database is a repository that has knowledge about prior relevant experiences, solutions to common problems, lessons from successes and failures, answers to FAQs, etc. The repository is generally populated over time by knowledge workers with their real-life experiences which they think may be applicable to a wider set of circumstances. Expert moderators may maintain the quality and relevance of lessons learned. A ranking system may be used as a self-governing control mechanism. The system may provide a link to the original contributor of knowledge for further consultation.

Systems for real-time knowledge request. White Boards and Urgent Request Systems are examples of these type systems. These provide an opportunity to broadcast urgent requests for knowledge via specific questions or problems. Appropriate experts can then provide the answers or solutions to problems. The message requesting assistance is typically broadcast to only those experts who are capable of providing a solution. For this purpose, the system may work in conjunction with an Expertise Database.

Knowledge bases. Knowledge bases are repositories that include procedures to be followed for complex tasks, operating manuals, policies, etc., and also reports such as technical papers, project reports, white papers. The artifacts may be in rich media such as drawings, pictures, presentations, webcasts, video clips of intricate operations, audio instructions, etc. The classification schemes used to archive knowledge documents needs to be well-understood. Efficient search and retrieval mechanisms are an essential component of these systems.

Decision Support for KIBPs. We try to match the characteristics of the processes with the types of KMS that are most likely to provide the needed support using the TTF theory as a backdrop. Table 2 summarizes this prescribed matching. These prescriptions are meant as a guide for planning a systematic solution beginning with the most appropriate type of KMS. Specific tasks and decision scenarios will need to be considered while designing and deploying the KMS.

Process complexity. Higher number of stages in a process (Ia in Table 2) requires more information to be easily accessible to all the stages. To control the flow of the process effectively, tasks within each stage need to follow well-documented procedures for knowledge exchange. Also, the complexity of tasks increases the need for such information flow. Such explicit knowledge may be stored in a knowledge database. Moreover, inter-dependency between/within stages (Ic) intensifies the need for the handover of information between stages. Information such as exceptions or issues created in one task may be documented for use in the following tasks. Again, knowledge bases are an appropriate starting point for such information. Uncertainty and ambiguity in various stages and their outcomes (Ib) calls for an enhanced level of collaboration between agents via discussions regarding assumptions, ranking and

Table 2. Characteristics of KIBPs and suitability of KMSs

KIBP Characteristic		**Suitability of type of KMS**
I.Process Complexity	*a. Higher number of stages*	Knowledge Bases
	b. Greater levels of uncertainty and ambiguity	Collaboration Systems Systems for real-time knowledge request
	c. Greater interdependencies between stages and agents	Knowledge Bases
II.Knowledge Complexity	*a. Multiple sources of information/knowledge*	Knowledge Bases Systems for Locating Expertise
	b. Constant creation of new knowledge	Lesson-Learned Databases Systems for Collaboration
	c. Context dependency	Knowledge Bases Lesson-Learned Databases
III.Decision Making Complexity	*a. Higher agent impact on the decision making process and its outcomes*	Systems for Locating Expertise Knowledge Bases Tele-presence / Video Conferencing Systems
	b. Greater levels of collaboration in decision making	Lesson-Learned Databases Systems for Collaboration
	c. Greater number of criteria need to be considered during decision making	Collaboration Systems

voting among alternatives, etc. collaboration systems can provide the appropriate support as they can offer shared workspaces. In case of an urgent need for clarification/solution, near instant access to domain expert may be needed. In such cases, systems for real-time knowledge request can provide the needed technology support.

Knowledge complexity. Multiple sources of knowledge (IIa), both tacit and explicit, call for systems that make such knowledge easily available. Systems for locating the right expertise can help identify experts when needed, whereas for codified information, knowledge bases are an essential component. Some instances of processes tend to create new knowledge (IIb). Such knowledge is generally in the form of solutions to common problems, valuable observations made from successes and failures, etc. A lessons learned database may be the most appropriate format for capturing and disseminating such knowledge. Coding of context (IIc) is one of the most difficult problems for technology to tackle. Generally, contexts are embedded in the minds of people and are shared through specialized vocabulary and shared understanding. Nevertheless, for persons new to a domain, a good reference knowledge base, especially with multimedia content, as well as a lessons learned database is good places to start building the context.

Decision making complexity. In situations where decision makers have a high impact on various aspects of the process (IIIa), e.g., its course, outcome, and/or the options considered, they need to use all the resources at their disposal. They should have access to more and varied knowledge sources; hence, systems for locating expertise, as well as references to knowledge bases become key components of the technology infrastructure. Moreover, real-time discussions with tacit knowledge exchange provided by tele-presence/video conferencing systems can provide the face-to-face feel needed for rich exchanges. In some processes, periodic, intense level of face-to-face collaborations in decision making is needed (IIIb). Typically, these situations require teams of people from different stages to thrash out a complex decision together. Tele-presence/video conferencing systems can fulfill this need as they not only provide the touch and feel of a face-to-face encounter, but also allow high band-width communication capabilities in real-time. When decisions need heterogeneity of agents and hence greater numbers of criteria to be considered (IIIc), a robust infrastructure consisting of a variety of collaboration systems is a necessity. Such systems are designed to encourage discussion that involves idea generation, ranking/voting, critical evaluation, generally in asynchronous mode, and allow team members to contribute anonymously.

5 Contributions and Conclusions

This paper makes three important contributions to the literature on knowledge intensity in organizations. First, this paper is among the few that have examined knowledge intensity from a process perspective. Approaching knowledge from a process perspective allows us to see both the broader knowledge related issues that impact the process in its entirety as well as the more modest, but equally important issues that impact stages and tasks within the process. This is also arguably the first paper to explicitly characterize knowledge intensive business processes with case studies and providing a new framework for continuing research and dialogue in this area. Second, in an area where the existing literature is relatively inadequate, we have adapted the TTF theory—designed primarily for a task level analysis—and applied it to a much higher level of analysis with knowledge intensive business processes. Doing so allowed us to draw a link between the task-technology fit perspectives and expand it to consider the process-technology fit in the context of KIBPs. Third, this paper provides a framework for organizations in terms of planning for IT support systems for KIBPs. The prescribed KMSs are constructed as guidelines for decision making that have emerged from our understanding of the complexity of KIBPs.

While the strengths of this paper reside in opening up a relatively new area of inquiry, it also presents its limitations. The exploratory nature of this study and the nascent field we are studying has resulted in a weak theoretical foundation for this research. The choice of research methodology, while providing rich data for our framework, also limits our conclusions to three KIBPs. We acknowledge the difficulties in generalizing results from case study research. Yet the diverse nature of the processes and the different industries considered have allowed us to develop a framework that serves as the preliminary basis on which future work in this area can be developed. Future research can use the KIBP framework presented in this paper to develop a research model that can be tested across a larger sample of organizations, thereby lending strength to our understanding of knowledge intensity in organizations. Since KIBPs present a yet unexplored terrain, research can add value in the following areas as well: understanding how the effectiveness of KIBPs can be measured and evaluating the fit between KMSs and the KIBPs for which they were designed, ultimately leading to a process-technology fit theory in this area.

References

1. Alavi, M., Kayworth, T., Leidner, D.E.: An empirical examination of the influence of organizational culture on knowledge management practices. J. of Management Information Systems 22(3), 191–224 (2006)
2. Anantatmula, V., Kanungo, S.: Structuring the underlying relations among the knowledge management outcomes. J. of Knowledge Management 10(4), 25–42 (2006)
3. Bhatt, G.: Knowledge management in organizations: Examining the interaction between technologies, techniques and people. J. of Knowledge Management 5(1), 68–75 (2001)
4. Birkinshaw, J., Sheehan, T.: Managing the knowledge life cycle. Sloan Management Review 44(1), 75–83 (2002)
5. Dishaw, M.T., Strong, D.M.: Supporting software maintenance with software engineering tools: A Computed task-technology fit analysis. J. of Systems and Software 44(2), 107–120 (1998)
6. Dishaw, M.T., Strong, D.M.: Extending the technology acceptance model with task-technology fit constructs. Information & Management 36(1), 9–21 (1999)
7. Eppler, M.J., Seifried, P.M., Röpnack, A.: Improving knowledge intensive processes through an enterprise knowledge medium. In: ACM Conference on Managing Organizational Knowledge for Strategic Advantage, New Orleans, Louisiana, USA (1999)
8. Goodhue, D.L., Thompson, R.L.: Task-technology fit and individual-performance. MIS Quarterly 19(2), 213–236 (1995)
9. Goodhue, D.L.: Development and measurement validity of a task-technology fit instrument for user evaluations of information systems. Decision Sciences 29(1), 105–138 (1998)
10. Goodhue, D.L., Klein, B.D., March, S.T.: User evaluations of IS as surrogates for objective performance. Information & Management 38(2), 87–101 (2000)
11. Gold, A.H., Malhotra, A., Segars, A.H.: Knowledge management: An organizational capabilities perspective. J. of Management Information Systems 18(1), 185–214 (2001)
12. Gronau, N., Weber, E.: Defining an infrastructure for knowledge intensive business processes. In: Proceedings of I-KNOW 2004, Graz, Austria (2004)
13. Kiessling, T., Richey, G., Meng, J., Dabic, M.: Exploring knowledge management to organizational performance outcomes in a transitional economy. J. of World Business 44(4), 421–433 (2009)
14. Kulkarni, U.R., Ravindran, S., Freeze, R.: A knowledge management success model: Theoretical development and empirical validation. J. of Management Information Systems 23(3), 309–347 (2007)
15. Malhotra, A., Majchrzak, A.: Enabling knowledge creation in far-flung teams: Best practices for IT support and knowledge sharing. J. of Knowledge Management 8(4), 75–88 (2004)
16. Marks, P., Polak, P., McCoy, S., Galletta, D.F.: Sharing knowledge: How managerial prompting, group identification, and social value orientation affect knowledge-sharing behavior. Communications of the ACM 51(2), 60–65 (2008)

17. Massey, A.P., Montoya-Weiss, M.M., O'Driscoll, T.M.: Performance-centered design of knowledge- intensive processes. J. of Management Information Systems 18(4), 37–58 (2002)
18. Nesta, L.: Knowledge and productivity in the world's largest manufacturing corporations. J. of Economic Behavior & Organization 67(3-4), 886–902 (2008)
19. Nonaka, I., Takeuchi, H.: The knowledge-creating company. Oxford University Press, USA (1995)
20. Satyadas, A., Harigopal, U., Cassaigne, N.P.: Knowledge management tutorial: An editorial overview. IEEE Transactions on Systems Man and Cybernetics Part C-Applications and Reviews 31(4), 429–437 (2001)
21. Staples, D.J., Seddon, P.: Testing the technology-to-performance chain model. J. of Organizational and End User Computing 16(4), 17–36 (2004)
22. Zack, M.H.: Managing Codified Knowledge. Sloan Management Review 4(40), 45–58 (1999)
23. Zigurs, I., Buckland, B.K.: A theory of task/technology fit and group support systems effectiveness. MIS Quarterly 22(3), 313–334 (1998)
24. Zigurs, I., Buckland, B.K., Connolly, J.R., Wilson, E.V.: A test of task-technology fit theory for group support systems. The DATA BASE for Advances in Information Systems 30(3-4), 34–50 (1999)

A Web Service Similarity Refinement Framework Using Automata Comparison

Lu Chen and Randy Chow

E301 CSE Building, P.O. Box 116120, Gainesville, FL 32611, United States
{luchen,chow}@cise.ufl.edu

Abstract. A key issue in web service based technology is finding an effective web service similarity comparison method. Current solution approaches include similarities comparisons both on signature level and behavior level. The accuracy of comparison is generally satisfactory if the web services are atomic. However, for composite web services with complicated internal structures, the precision of similarity results depends on the comparison approaches. To enhance this precision for composite services, the paper proposes a two-stage framework to integrate the signature level and behavior level similarities. The first stage adopts the traditional WordNet-based interface measuring method to generate a set of possible similar services. Then in the second stage, an automaton is built for each service. By characterizing and comparing the languages generated from those automata, we can enhance the precision of similarity measures of similar services from the previous stage.

Keywords: web service similarity, nondeterministic finite automaton, OWL-S control constructs, regular languages.

1 Introduction

Web services are becoming widely used for online distributed computing. Rather than building a system from scratch, it is often more desirable for the developers to choose publicly available online services and compose them into a composite/complex service to satisfy a user's requirements. To support this composite service generation process, numerous literatures concentrate on web service modeling, service discovery, service composition and integration, service similarity comparison, service communication, and etc.

Among the above topics, we consider comparing the similarities of different services is highly important due to three reasons. Firstly, with the growing number of online web services, a critical step for web-based application is locating/discovering web services, which identifies the functionalities of existing services against users' requirements. It is valuable to notice that evaluating web service similarities facilitates the process of web service discovery [1, 3]. Secondly, though various solutions are proposed to automatically generate composite services, the one utilizing similarities of web services can achieve considerable optimization with respect to different service similarity, interval time and search times [8]. Thirdly, the traditional UDDI only

S.K. Prasad et al. (Eds.): ICISTM 2010, CCIS 54, pp. 44–55, 2010.

provides a simple browsing-by-business-category mechanism for developers to review and select published services, which hardly supports the concept of web service substitution. Thus, a more accurate infrastructure should be proposed to classify web services into different clusters based on service similarities.

Many solutions are proposed for deciding service similarities [3, 4, 5, 6, 9, 10, 11, 12], which are classified into signature level and behavior level similarity comparisons. The former solution compares the interface description of two web services to determine their semantic similarities, while the later one considers the internal structures of web services. The accuracy of comparison is generally satisfactory if the web services are atomic. However, for composite web services with complicated internal structures, the precision of similarity results depends on the comparison approaches. To enhance the precision of similarity measures between composite web services, this paper proposes a two-stage framework to integrate the interface/signature level and functional/behavior level similarities. The first stage adopts the traditional WordNet-based interface measuring method to generate a set of possible similar services. Then in the second stage, an automaton, which is suitable for modeling the internal execution flow of composite web services, is built for each OWL-S. By characterizing and comparing the languages generated from those automata, we can enhance the precision of similarity measures of similar services from the previous stage.

The paper is organized as following. Section 2 introduces a motivating example of our work. To translate OWL-S descriptions to automata representation, Section 3 first describes OWL-S and an OWL-S API [17] implemented at CMU. The paper uses this tool to parse and generate tokens which represent the basic units in the automata proposed for web service formalizations. In this section we also illustrate the automata representation for each OWL-S control construct and how to integrate the control constructs to generate an automaton for an entire OWL-S description. Section 4 points out that the equivalence issue of OWL-S Process Model can be addressed by proving the equivalence of automata languages. Besides, a web service similarity measuring framework is proposed by integrating the signature level and behavior level similarities. It also refines similarity analysis using language comparison. Our work is compared with related ones in Section 5, and Section 6 concludes the paper with a summary of significance of the research and its relevance with some ongoing work on web service composition framework.

2 Motivating Example

In the real world, it is very common that two complex web services are composed by similar or exactly the same atomic services while their composition sequences are different. In this case, if the two services are compared for similarity purpose using only textual descriptions, a wrong result could be concluded. For example, two composite web services descriptions are shown in Figure 1. Fig. 1(a) describes a service process of a composite service, which is implemented by four sequential sub-processes, Browsing, Add_to_Cart, Billing and Free_Shipping. Similarly, Fig. 1(b) describes a service process which has the same sub-processes as in Fig. 1(a). However, in Fig. 1(b), the execution order is Browsing, Add_to_Cart, Shipping_With_Fee and

Billing, which is logically different from Fig. 1(a). Using the existing approaches of comparing services via text information or interface descriptions, the two web services would be categorized as similar. But in fact, they are not.

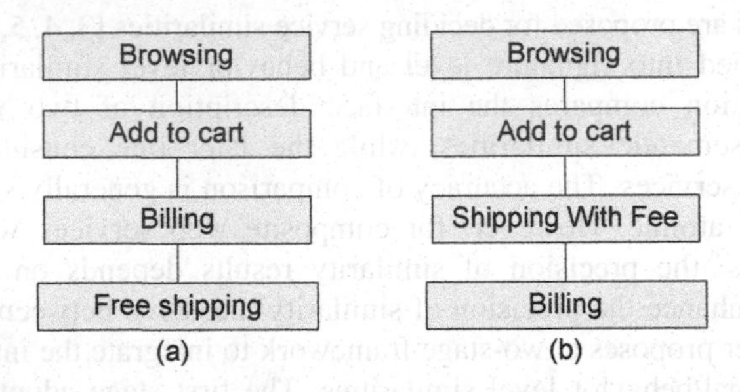

Fig. 1. Figure 1(a) describes a web service with free shipping while Figure 1(b) requires that different shipping methods have different costs

Enhancing the similarity analysis to include execution sequences in the composite web service is the primary motivation for incorporating an automaton for OWL-S description. An automaton allows us to describe more precisely the control flow in the composite web services. The execution flow can be represented by some regular expressions, and the similarity or dissimilarity can be decided by the comparison of regular expressions. Thus, the similarity decision process is refined into two steps. First, we can reuse the existing techniques on web service similarities to generate a set of *possibly* similar web services. This set of services forms the initial web service community. Second, we can eliminate dissimilar web services to improve the precision of comparison. Our work differs from the others in that instead of finding the similarity between web services, we try to find the dissimilarity of a set of possibly similar web services. Therefore, more accurate clustering of similar web services can be achieved.

3 Translation of OWL-S Representation into Automata

A composite web service is iteratively composed with atomic/composite web services. Therefore, an automaton for each composite web service can be generated by composing the automata for its component web services. Before introducing how the composition process works, it is necessary to recognize/abstract the control constructs from an OWL-S description, which is realized by using the OWL-S API tools.

3.1 OWL-S and OWL-S Parser

To describe a web service's functionality formally and precisely, many languages exist such as WSDL, SWRL, WSMO and OWL-S [1, 8]. This paper uses the OWL-S specification to describe semantic web service composition. OWL-S describes a semantic web service by providing Service Profile, Service Model and Service

Grounding. Current work compared the Input, Output, Precondition and Effect (IOPE) of two web services to decide the behavior similarities among web services. However, our work focuses more on the control constructs of a composite OWL-S Process Model. A composite process specifies how its inputs are accepted by particular sub-processes, and how its various outputs are produced by particular sub-processes. Besides, it indicates its interior control structure by using a ControlConstruct.

OWL-S API is a Java API for reading, writing, and executing OWL-S web service descriptions. It provides an execution engine that can invoke atomic processes and composite processes that uses control constructs such as Sequence, Unordered, and Split [3]. OWL-S API provides four different parsers, OWLSProfileParser, OWLSProcessParser, OWLSGroundingParser and OWLSServiceParser, which can parse OWL-S Profile files, OWL-S Process Model files, OWL-S Grounding files and OWL-S Service files, respectively.

Currently, the API does not support conditional process execution such as If-Then-Else and Repeat-Until in its default implementation. However, this does not present a problem to our implementation since we only utilize the parser, not the execution engine. The list of control constructs can still be generated from the OWLSProcessParser.

3.2 Automata for OWL-S Control Constructs

Given a composite web service w, which is composed of sub-web services w_1, w_2... w_n, to generate the automaton for w, we first define the automata for each control construct in OWL-S. Then the entire automata representation for web service can be generated by integrating sub-web services with control construct. Of all the OWL-S control constructs, the characteristics of Choice, and Split/Split + Join are intrinsically nondeterministic, thus, we choose the nondeterministic finite automaton (NFA) as the basis for modeling in this paper.

Definition 1. The NFA model for OWL-S construct is a 5-tuple (Q, $\sum\varepsilon$, δ, q_0, F), where

- Q is a finite set called the states. Each state is a pair (w_i, p_i) of a sub-atomic service w_i with preconditions p_i;
- $\sum\varepsilon$ is a finite set called the alphabet. Each element in the set is a 3-tuple (i_i, o_i, e_i), which means if p_i is true, the sub service w_i will accept i_i and then pass the output o_i to the next service w_{i+1} and generate an effect of e_i to the world;
- δ: $Q \times \sum\varepsilon \rightarrow Q$ is the transition function. This function describes that only if p_i is satisfied under input i_i, w_i is activated to generate information o_i and effect e_i;
- $q_0 \in Q$ is the start state, (w_0, p_0), w_0 is the first atomic process in an OWL-S description;
- $F \subseteq Q$ is the set of accept states, which are also the last/finishing processes in OWL-S description.

The transition diagram for each control construct is given in Figure 2.

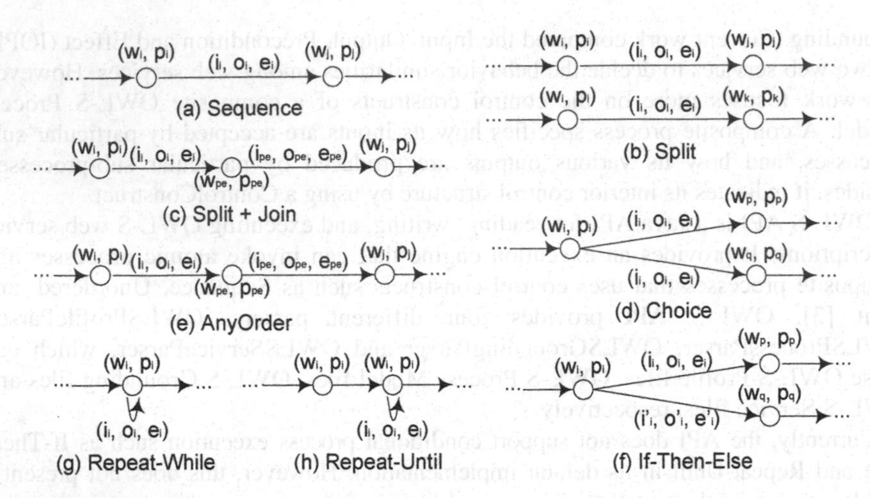

Fig. 2. The single circle describes the internal service. The previous and following services are omitted in this figure.

Sequence: If sub-services w_i and w_j are related via Sequence control construct, the automata is shown in Figure 2(a).

Split: In OWL-S, the components of a Split process are a set of sub-processes which are executed concurrently. The Split process completes as soon as all of its sub-processes have been scheduled for execution. If a sub-service w_i is split into two branches w_j and w_k, we create two automata, each represents one branch. It means both of the two branches are required for following execution. The automaton for Split construct is shown in Figure 2(b).

Split + Join: The Split + Join process consists of concurrent execution of a set of sub-processes with barrier synchronization. That is, Split+Join completes when all of its sub-processes have completed. Given a sub-service w_i which satisfies the following two conditions: 1) w_i splits into two branches, which start from web services w_p and w_q respectively; 2) the two execution branches will join at service w_j, then the automata for this branching execution is simulated as sequential execution, such as $w_i w_p w_q w_j$. This is reasonable because the execution order of w_p and w_q is not important as long as both of the two branches will finish. Therefore, we define a specific node PE standing for a set of nodes which can be interleaved without fixed execution sequences. For example, if PE includes w_p and w_q, then the language to describe PE can be either w_p w_q or w_q w_p. An example of such automata is shown in Figure 2(c).

Choice: *Choice* calls for the execution of a single branch from a set of following control constructs. Any of the alternative branches may be chosen for execution. After the sub-service w_i, if there are two possible branches w_p and w_q waiting for execution, then the choice between w_p and w_q is nondeterministic. The automaton is shown in Figure 2(d).

Any Order: *Any Order* allows a set of sub-processes to be executed without a predefined order. The execution and completion of all sub-processes is required. In addition, each sub-process execution cannot be overlapped or interleaved. Given a sub-web service w_i, sequentially followed by w_p and w_q with *Any Order* relationship, followed by w_j, the only constraint is that w_p and w_q are executed. Thus, similar to Split + Join, the automaton is shown in Figure 2(e) where PE includes w_p and w_q.

If-Then-Else: The automaton for *If-Then-Else* is very similar to Choice, but it is deterministic. As shown in Figure 2(f), which one of the continuous branches after w_i is executed is determined by the value of input.

Repeat-While: The web service in the *Repeat-While* construct is iteratively executed as long as the precondition p is true under input i. The automaton is like Figure 2(g).

Repeat-Until: The *Repeat-Until* construct does the operation, tests for the condition, exits if it is true, and otherwise loops. Since the process is executed at least once, the automata are shown in Figure 2(h).

Since Iterate can be implemented by Repeat-While and Repeat-Until, we do not need to include it. With the above automata definitions for all the control constructs in OWL-S description, we can recursively generate the automata for a composite web service.

3.3 Translation of Web Services Using Automata

The automata generation for an atomic web service is straightforward, and the automata generation for the composite services is based on it. Given an atomic web service w with input i, precondition p, output o and effect e. An automaton for an atomic web service w is described in Figure 3.

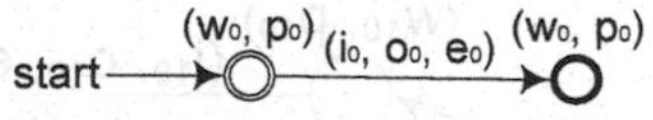

Fig. 3. The double circle describes the starting service while the black circle describes the ending service. Each state is a pair of service and preconditions, each transition is labeled with input, output and effect of the previous node.

In this automaton representation, since w has no sub-web services, it is straightforward that i = i0, o =o0, p = p0, e = e0 and w = w0. Thus we have W = {w0}. In addition, the start state and final state are the same since w0 is the starting service and the ending service in w, which means F= {w0} too.

A composite web service is recursively composed by atomic services with nested control constructs. Each control construct can be integrated with each other under the three regular operations defined in the traditional NFA theory: Union, Concatenation and Star. Given two automaton of control constructs C1= (Q1, $\sum\varepsilon 1$, $\delta 1$, q1, F1), and C2 = (Q2, $\sum\varepsilon 2$, $\delta 2$, q2, F2), we have the following definitions.

Definition 2. The *union* of control constructs C1 and C2 forms a new control construct C = (Q, $\sum\varepsilon$, δ, q_0, F), where

- $Q = \{q_0\} \cup Q_1 \cup Q_2$;
- $\sum\varepsilon = \{ (\varepsilon, \varepsilon, \varepsilon) \} \cup \sum\varepsilon_1 \cup \sum\varepsilon_2$;

- q_0 is the start state of C, $q_0 = (w_0, \text{true})$;
- $F = F_1 \cup F_2$;
- For any $q \in Q$ and $a \in \sum\varepsilon$, $\delta = \begin{cases} \delta_1(q,a),\ q \in Q_1 \\ \delta_2(q,a),\ q \in Q_2 \\ (q_1, q_2),\ q = q_0\ and\ a = (\varepsilon,\varepsilon,\varepsilon) \\ \emptyset,\ q = q_0\ and\ a \neq (\varepsilon,\varepsilon,\varepsilon) \end{cases}$

Definition 3. The *concatenation* of control constructs C1 and C2 forms a new control construct $C = (Q, \sum\varepsilon, \delta, q_0, F)$, where

- $Q = Q_1 \cup Q_2$;
- $\sum\varepsilon = \sum\varepsilon_1 \cup \sum\varepsilon_2$;
- The start state $q_0 = q_1$;
- The accept states F of C is the same as F_2;
- For any $q \in Q$ and $a \in \sum\varepsilon$, $\delta = \begin{cases} \delta_1(q,a), q \in Q_1\ and\ q \neq F_1 \\ \delta_2(q,a), q \in Q_2 \end{cases}$

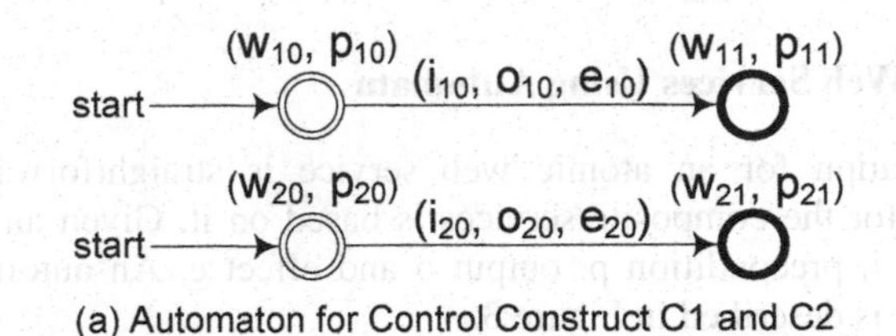

(a) Automaton for Control Construct C1 and C2

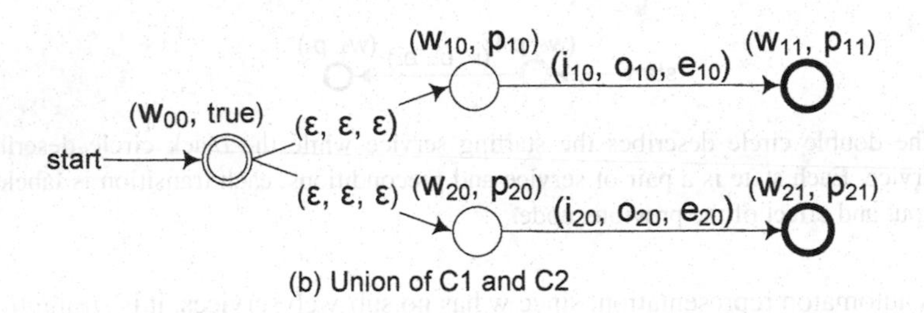

(b) Union of C1 and C2

(w10, p10) (i10, o10, e10) (w20, p20) (i20, o20, e20) (w21, p21) start

(c) Concatenation of C1 and C2

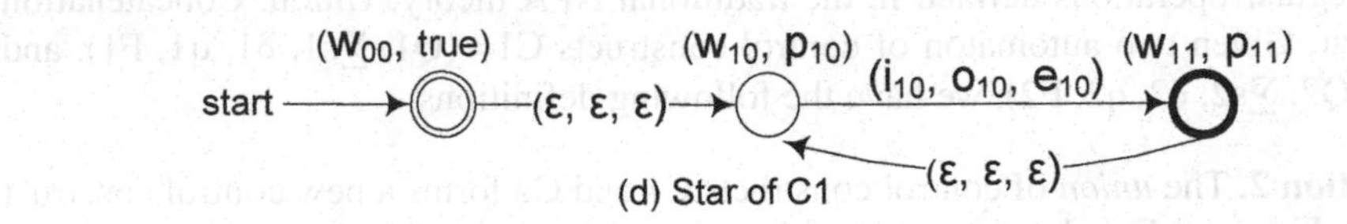

(d) Star of C1

Fig. 4. The figure describes the automaton composition of OWL-S constructs. Fig 4 (a) depicts the automaton for a single control construct C1 and C2, where Fig. 4(b) ~ (d) shows the resulting automaton after union, concatenation and star, respectively.

Definition 4. The *star* of control constructs C1 forms a new control construct C = (Q, ∑ε, δ, q0, F), where

- $Q = \{q_0\} \cup Q_1$, which are the old states of C1 plus a new start state;
- $\sum\varepsilon = \{(\varepsilon, \varepsilon, \varepsilon)\} \cup \sum\varepsilon_1$;
- The start state q_0 is a new state;
- The accept state $F = \{q_0\} \cup F_1$, which is the old accept states plus the new start state;
- For any $q \in Q$ and $a \in \sum\varepsilon$, $\delta = \begin{cases} \delta_1(q,a), q \in Q_1 \; and \; q \neq F_1 \\ \delta_1(q,a), q \in F_1 \; and \; a \neq (\varepsilon,\varepsilon,\varepsilon) \\ \delta_1(q,a) \cup \{q_1\}, q \in F_1 \; and \; a = (\varepsilon,\varepsilon,\varepsilon) \\ \{q_1\}, q = q_0 and \; a = (\varepsilon,\varepsilon,\varepsilon) \\ \emptyset, q = q_0 and \; a \neq (\varepsilon,\varepsilon,\varepsilon) \end{cases}$

4 Web Service Similarity Measuring Framework

Many researching work is proposed to measure similarity. Rather than proposing a new measuring method, our work introduces a framework which adopts the behavior similarity approach as the first stage to generate a set of possible similar web services, then structural matching is operated on this set to improve the precision of similarity measurement.

4.1 Equivalence of Web Services and Regular Languages

Theorem: The Process Models of two OWL-S descriptions are semantically equal if and only if their automata are represented in the same languages.

Proof: Each automaton, generated from an OWL-S description, is a NFA which can be translated to a DFA. Since EQDFA = {<A, B>| A and B are DFAs and L(A) = L(B)} is a decidable language, it is possible to determine the equality of the two DFAs. Thus, we can compare two OWL-S descriptions by comparing two corresponding DFAs. If the two DFAs are equal, we conclude that the two OWL-S descriptions are the same; otherwise, they are different.

Furthermore, each path in a DFA is a sequence of sub-processes which starts from the beginning web service and ends with the final web service. Since each path can be 1-1 mapped to a string, each DFA can uniquely determine a set of strings, which form the language of DFA. Therefore, it is reasonable to assert that two OWL-S descriptions are equivalent if and only if they share the same automata languages.

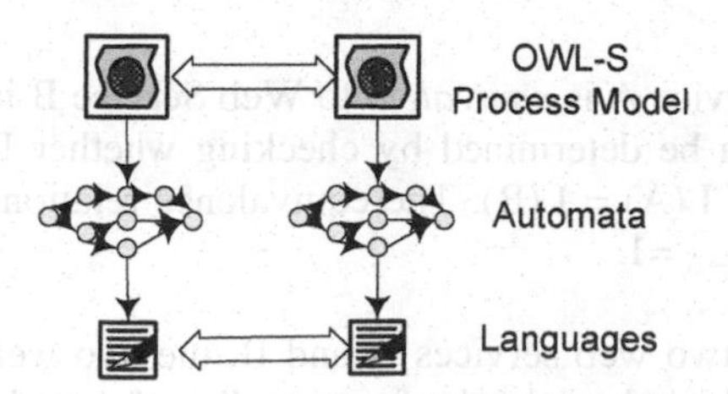

Fig. 5. Figure 5 shows the relationship between OWL-S Process Model and Language Equivalence, which make it reasonable to calculate dissimilarity to eliminate conflicted services in Section 4.2.

4.2 Similarity Measuring Framework

To calculate the similarity of web services with OWL-S specifications, our framework starts from the syntactic similarities for OWL-S Profile descriptions. As shown in Figure 6, this process begins with characterizing similar words/concepts in text and operation/process description of two web services. Similarities among text descriptions, input/output, preconditions, effects and operations can be obtained, which are accumulated to sum up the similarity value between two services. Using this process, a set of possible similar web services with mutual similarity values is produced.

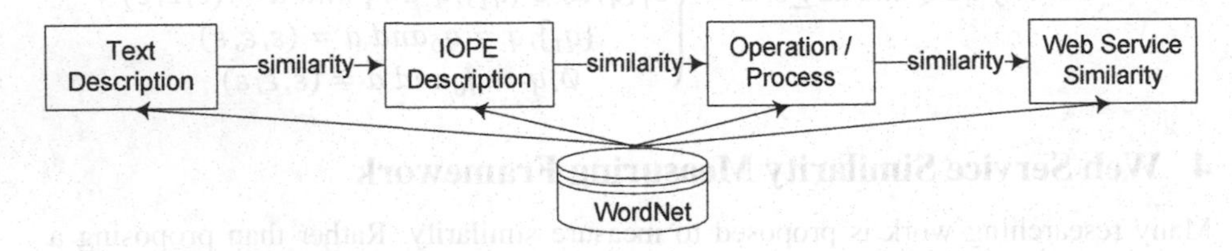

Fig. 6. Traditional WordNet-based Similarity Analysis for OWL-S Profile

However, this process is time-consuming and might not be precise due to two reasons: First, similarity values are not transitive. For example, supposing a, b, c are three web services, a is similar to b with value 0.8, b is similar to c with value 0.7, it is not precise to assert that a is similar to c with value 0.56. It is highly possible that a is similar to c with value 0.9. Second, a similarity analysis of two web services may be wrong as illustrated in Section 2. Thus, internal structural differences need to be considered to eliminate conflict services, which is the main goal for the second stage in our framework.

The input for the second stage is a set of possible similar web services. Then, an automaton is generated for each candidate in this set of services. By comparing the regular language for each automaton, it is able to find conflicted services with the following equivalence and dissimilarity definition.

For any two DFA A and B, a new DFA C can be constructed which satisfies:

L(A) is the language for DFA A, and L(B) is the language for DFA B.

$\overline{L(A)}$ is the complement of L(A).

$$L(C) = \left(L(A) \cap \overline{L(B)}\right) \cup \left(\overline{L(A)} \cap L(B)\right)$$

Equivalence: Web service A is *equivalent* to Web Service B if they provide the same functionality. This can be determined by checking whether L(C) is empty, because L(C) =φ if and only if L(A) = L(B). The equivalence relationship is symmetric, thus $\text{Similar}_{A\rightarrow B} = \text{Similar}_{B\rightarrow A} = 1$.

Dissimilarity: Given two web services A and B, the two web service are dissimilar when the languages which describe the functionality of A and B have no intersection. This is true when $L(A) \cap L(B) = \emptyset$..

In most cases, the equivalence and dissimilarity are too strong that some services cannot be classified into this two categories, thus, refinements are proposed as following. Similarly, we define another two DFAs, D and E, with following languages:

$$L(D) = (L(A) \cap \overline{L(B)}), \text{ and } L(E) = (\overline{L(A)} \cap L(B))$$

Strong Similar: The web service similarity *with respect to substitution* is described using containment relationship as defined in mathematics. The value for containment with respect to subset/superset is obtained from a Boolean function Contain(x, y) which has only two results, 0 and 1. This is because the containment relationship is not symmetric. Therefore,

1) $Similar_{A \to B} = Contain(A, B) = 1$ if $L(D) = \emptyset$ (or $L(A) \subseteq L(B)$), which means A is similar to B.
2) $Similar_{A \to B} = Contain(A, B) = 0$ if $L(E) = \emptyset$ (or $L(B) \subseteq L(A)$ if $L(E)$), which means A is not similar to B.

Weak Similar: Though A and B have some languages in common, they also contain conflict languages. Thus, we define their relationship as weak similar with respect to execution contexts. This consideration is practical when the Owls:Choice exist in the process definition.

For example, given the two following two automata:

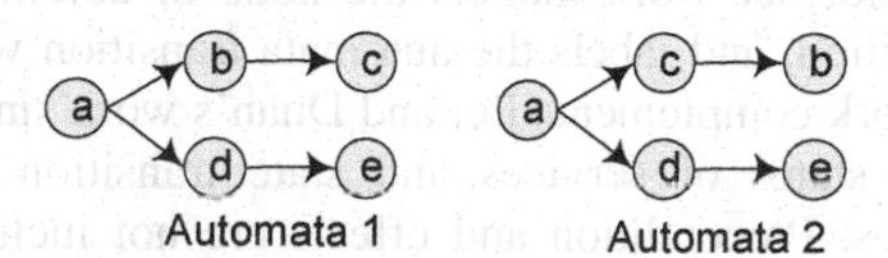

Due to the existence of the Owls: Choice construct, the language for Automata 1 is {abc, ade}, and the language for Automata 2 is {acb, ade}. It is clear that the two languages have conflict process execution orders, which are abc and acb. In this case, the two OWL-S descriptions are not similar to each other.

5 Related Work

There is an abundance of research work on deciding the similarity between two semantic web services. The Woogle search engine [3] computed the overall service similarities based on the comparison of the text descriptions of names, operations, and input/output of web services. Eleni Stroulia and Yiqiao Wang [4] used a comparison matrix to integrate similarities of data types, messages, operations and web services of WSDL descriptions to reflect similarities between two web services. Hau, Lee and Darlington [5] defined the intrinsic information value of a service description based on the "inferencibility" of each of OWL Lite constructs. Since OWL-S is based on OWL, the similarity between service profile, service model and service grounding could be computed to generate the overall similarities between two semantic web services. Wang, Vitar and Hauswirth et al [6] first built a new application ontology which was a superset of all the heterogeneous ontology. Then within this global application ontology, many existing methods, like semantic distances, can be reused to generate similarities between web services. All the above solutions only compared the functional semantics of two web services. The internal behavioral sequences of web services are not considered completely.

Recently, some researchers considered the structural sequences among web services in their web service discovery solutions. For example, to retrieve the exact/partial matched services for users' requirements, Grigori, Corrales and Bouzeghoub [12] proposed a behavior model for web services and reduced the behavior matching to a graph matching problem. Wombacher, Fankhauser and Neuhold [9, 11] translated the BPEL descriptions of web services to annotated deterministic finite state automata (aDFA) since BPEL itself cannot provide the state info of services. In this approach, the stateful matchmaking of web service is defined as an intersection of aDFA automata. Shen and Su [13] utilized automata and logic formalisms to build a behavior model, which integrates messages and IOPR model in OWL-S for service descriptions. They also designed a query language to integrate structural and semantic properties of services behaviors. Lei and Duan [10] proposed an extended deterministic finite state automaton (EDFA) to describe service, which also aimed to describe the temporal sequences of communication activities. Our work differed from the above ones in two aspects. First, since our goal focuses on building a web service community, it is more important to compare the descriptions between web services rather than to compare the specifications of user requirements against service description which is a key issue for service discovery. The advantage is that imprecise users' description will not affect the similarity results. Second, our work models the node of automata as a pair of sub-service and its precondition, and labels the automata transition with both input/output and effect. Thus, our work complements Lei and Duan's work since they only used the nodes to describe the states of services and state transition only considered the input/output of activities. Precondition and effects are not included in the automata model. Thus, the IOPE model of OWL-S is completely compatible with our model.

6 Conclusion

This paper proposes a formal approach based on automata to model the process behavior of web services. The primary objective is to enhance the similarity analysis of a set of substitutable services by taking the differences of the process model into consideration when comparing web services. In addition, we show that OWL-S similarity is equivalent to automata similarity. Compared with existing researches, the proposed modeling of web service functionalities is more comprehensive and the introduction of automata facilitates the refinement of similarity greatly. Currently, we are implementing the prototype to test correctness and efficiency of this proposed approach. By adopting this method, we plan to construct a set of mutually similar web services, which is critical for our ongoing work on building a web service community to support the web service substitution process in a web service composition framework [19].

References

1. Wang, Y., Stroulia, E.: Semantic Structure Matching for Assessing Web-Service Similarity. In: 1st International Conference on Service Oriented Computing, pp. 194–207. Springer, Heidelberg (2003)

2. Foster, H., Uchitel, S., Magee, J., Kramer, J.: Model-based verification of Web service compositions. In: 18th IEEE International Conference on Automated Software Engineering, pp. 152–161 (2002)
3. Dong, X., Halevy, A., Madhavan, J., Nemes, E., Zhang, J.: Similarity Search for Web Services. In: Very Large Data Bases Conference, pp. 372–383 (2004)
4. Stroulia, E., Wang, Y.: Structural and Semantic Matching for Assessing Web-Service Similarity. Intl. Journal of Cooperative Information Systems 14(4), 407–437 (2005)
5. Hau, J., Lee, W., Darlington, J.: A Semantic Similarity Measure for Semantic Web Services. In: Web Service Semantics Workshop at WWW (2005)
6. Wang, X., Vitar, T., Hauswirth, M., Foxvog, D.: Building Application Ontologies from Descriptions of Semantic Web Services. In: IEEE/WIC/ACM International Conference on Web Intelligence, pp. 337–343 (2007)
7. Wu, J., Wu, Z.: Similarity-based Web Service Matchmaking. In: IEEE International Conference on Services Computing, pp. 287–294 (2005)
8. Jiang, W., Hu, S., Li, W., Yan, W., Liu, Z.: ASCSS: An Automatic Service Composition Approach Based on Service Similarity. In: 4th International Conference on Semantics, Knowledge and Grid, pp. 197–204 (2008)
9. Wombacher, A., Frankhauser, P., Neuhold, E.: Transforming BPEL into annotated deterministic finite state automata for service discovery. In: 2nd IEEE International Conference on Web Services, pp. 316–323 (2004)
10. Lei, L., Duan, Z.: Transforming OWL-S Process Model into EDFA for Service Discovery. In: 4th IEEE International Conference on Web Services, pp. 137–144 (2006)
11. Wombacher, A., Fankhauser, P., Mahleko, B., Neuhold, E.: Matchmaking for Business Processes based on Choreographies. In: IEEE International Conference on e-Technology, e-Commerce and e-Service, pp. 359–368 (2004)
12. Grigori, D., Corrales, J.C., Bouzeghoub, M.: Behavioral matchmaking for service retrieval. In: 4th IEEE International Conference on Web Services, pp. 145–152 (2006)
13. Shen, Z., Su, J.: Web Service Discovery Based on Behavior Signatures. In: International Conference on Services Computing, pp. 279–286 (2005)
14. SWRL, `http://www.w3.org/Submission/SWRL/`
15. Web Service Modeling Ontology(WSMO), `http://www.w3.org/Submission/WSMO/`
16. OWL-S, `http://www.w3.org/Submission/2004/SUBM-OWL-S-20041122/`
17. OWL-S API, `http://www.mindswap.org/2004/owl-s/api/`
18. WSDL, `http://www.w3.org/TR/wsdl`
19. Chen, L., Chow, R.: Web Service Composition Based on Integrated Substitution and Adaptation. In: International Conferenece on Infomation Reuse and Integration, pp. 34–39 (2008)

A System for Analyzing Advance Bot Behavior*

JooHyung Oh, Chaetae Im, and HyunCheol Jeong

Korea Internet and Security Agency, Seoul, Korea
{jhoh,chtim,hcjung}@kisa.or.kr

Abstract. Bot behavior analysis is an essencial component in botnet detection and response. Recent reseach on bot behavior analysis is focus on idenyifing wheather analysis target file is bot or not by monitoring user-level API call information of bot process and discover their malicous behaviors. However, such research does not monitor the bot process which has kernel-rootkit, anti-VM and static-DLL/binary code injection capabilities. In this paper, we present an approach based on a combination of System Call Layer rebuilding and process executing that enables automatic thwarting static-DLL/binary code injection. Also, we have built a system for analyzing advance bot behavior that can monitor the behavior of bot process at kernel-level and thwart some anti-vm methods. For experiments and evaluation, we have conduct experiments on several recent bot samples which have kernl-rootkit, anti-VM and static-DLL/binary code injection capabilities and shown that our system can successfully extrat their API call information and malicious behaviors from them.

Keywords: Botnet, Bot, Malware, Behavior Analysis.

1 Introduction

Over the past several years, Botnet is recognized as one of the most serious threat in Internet. Recently, DDoS attacks or spamming using botnet has risen explosively and the damage caused by such attacks has been increasing exponentially [1]. Also, hackers lend botnet to the spammer or criminals in order to get financial benefit. Therefore, various analysis schemes have proposed to detect early and incapacitate botnet C&C server.

Anti-virus companies or researer typically have to deal with tens of thousands of new malware sample every day to know wheather malware is bot or not. Because of the limitation of static analysis, dynamic analysis systems are typically used to analyze these samples, with the aim of understanding how they connect to C&C sever and how they launch attacks. However, today's bot apply complex anti-behavior analysis techniques such as kernel rootkit, anti-VM and static DLL/binary code injection. Therefore, recent bot behavior analysis is bacame more difficult.

In this paper, we propose and implement behavior analysis system that is able to extract the malicious API call information from the bot file through newly made

* This work was supported by the IT R&D program of MKE/KEIT. [2008-S-026-02, The Development of Active Detection and Response Technology against Botnet].

S.K. Prasad et al. (Eds.): ICISTM 2010, CCIS 54, pp. 56–63, 2010.

System Call Layer on virtual memory address in kernel and process executing scheme. In our implemented system, a copy of System Call Layer is made by kernel-driver at simulation environment OS boot time, then we change the jump address of INT 2E or SYSENTER interrupts to point to new made KiSystemService function. So, bot process's API call is forwared to the new KiSystemService function instead of original KiSystemService function. In this manner, we can monitor behavior of bot process at kernel-level. Also, we change a pointer saved in newly made SSDT to hook and intercept native API call which are used by bot process to find specific VMware related process. Then we modify the return value to thwart virtual machine detection.

This paper is structured as follow. We present background information related to bot behavior analysis systems in Section 2. In section 3, we describe the principles of our behavior monitoring approach and the architecture and details of our system. Experimental evaluation of our system with various recent bot samples are discussed in Section 4. Finally, we conclude our paper in Section 5.

2 Related Work

Until now, researches on bot behavior analysis focus on monitoring of window API call. However, most of the researches can not monitor and analysis API call of bot which uses Kernel-level rootkit, virtual machine detection and static-DLL/Binary code injection technology for hiding its' existence.

Sunbelt Software's CWSandbox[2] creates a new process for the to be analyzed bot and then injects the cwmonitor.dll for monitoring user-level API calls into the process's address space. With the help of the DLL, API call is hooked and then API call information is sent to cwsandbox.exe. After collecting information is completed, cwsandbox.exe analysis API call list whether specific behavior event is occurred or not. If Createfile API call is contained in API call list, cwsandbox send a message that can notify file create event is occurred to reporting module. In this manner, behaviors of bot related to file, registry, network is monitored by cwsandbox. However, if bot injects some codes of himself for malicious behavior such as write file into the Windows's System folder into normal process, cwsandbox can not detect these behaviors because of API call list does not contain these API call information. Also, some recent bots which has kernel rootkit could not be analysised. Kernel rootkit can easily hide activities from bot process at kernel-level. For examples, if bot modifys SSDT for changing Deletefile API to Createfile API, bot causes file create event is reported file delete events.

Ikarus software's TTAnalyze[3] and ZeroWine[4] analyzing window API call using monitoring bot's instruction set executed QEMU. QEMU is CPU emulator which can execute instruction set of processor virtually. Therefore, TTAnalyze and ZeroWine can not analyze behavior of the bot which has detecting virtual environment.

Artem Dinaburge[5] propose a malware analysis system, named Ether, focus on thwarting virtual machine detection. Ether executes a bot file in hardware based virtual environment which consists of Intel-VT and XEN 3.0 virtual machine. So Ether does not induce any unconditionally detectable side-effects by completely residing outside of analysis environment.

Table 1. A comparison of behaviour monitoring systems

Function	CWSandbox	ZeroWine	TTAnalyze	Ether
Kernel rootkit	no	no	yes	yes
Virtual machine detection	no	no	no	yes
DLL/Binary code injection	no	no	no	no

3 Behavior Analysis System of Advanced Bot

The overall architecture of our Behavior analysis system is shown in Fig.1. Behavior analysis of advanced bot is performed in three main phases: bot staring, kernel-level behavior monitoring, and analysis.

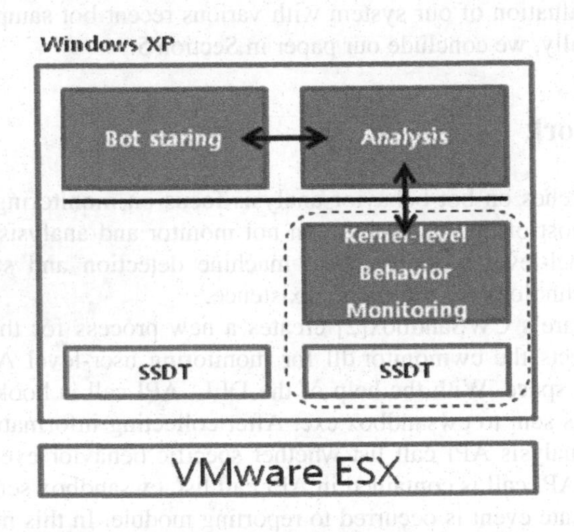

Fig. 1. Our bot behavior analysis system overview

The main goal of the *bot starting* is to execute a bot process, regardless of file type. If the bot file is dll, *bot starting* execute a dummy process such as svchost.exe and iexplorer.exe, then inject dll into the process. Finally, *bot starting* call loadlibrary API to load the bot file. In this manner, we execute a bot file in any case of type.

After, *bot starting* phase, *kernel-level behavior monitoring* collects API calls information caused by bot process at kernel level within a pre-configured timer period.

The final *analysis* phase is responsible for classifying malicious behaviors from API call lists. If any malicious file read or write API call to the system32 directory has occurred, *analysis* records that malilcious file is created by bot process. Also, *analysis* inspects whether bot process is injector or not. If malware did static injection during the monitoring period, target process is restarted to monitor an injected file or dll file.

3.1 Kernel-Level Behavior Monitoring and Thwarting Virtual Machine Detection

In order to monitor behaviors of bot at kernel-level, we make and install the kernel-driver at a guest OS as a windows service. The main goal of the driver is to make a copy System Call Layer on virtual memory address in Windows kernel. The system call layer consists of a common KiSystemService function, SSDT which contains pointer of window native API and window native API. It is used by the windows OS for handling system calls [6]. Therefore, during the window OS booting time, a new system call layer is made by kernel-driver.

In our implementation, Window XP is used as a guest OS for simulation environment and kernel-driver is pre-installed at Window XP for making a copy of System Call Layer.

When the guest OS is boot up, the System Call Layer is made on virtual memory address. After making System Call Layer is completed, *bot starting* changes the jump address of INT 2E or SYSENTER interrupts to point to new made KiSystemService function. So, API call of bot process is forwarded to the new KiSystemService function instead of original KiSystemService function. If our KiSystemService function (newly made) receives a forwared API call, it records and sends API call information to the *analysis* which is run at user space.

In this manner, we can monitor behavior of bot process at kernel-level as shown in Fig. 2.

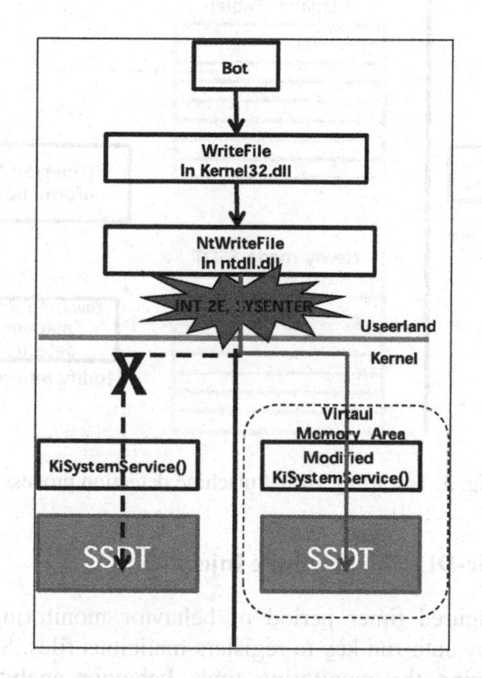

Fig. 2. Kernel-level behavior monitoring process

In figure 3, we depict the thwaring virtual machine detection method of our behavior monitoring system. Thwaring virtual machine detection is done by hooking the native API call by the bot process. We change a pointer (address of window native API) saved in newly made SSDT, to hook and intercept native API call called bot process. If native API call is hooked, we execute the API call and we modify the return value. This method is called SSDT Hooking and malware often use this to intercept private information such as banking account, and card number.

For example, bot usually use NtQuerySystemInformation system call looking for specific VMware related process such as VMTools service, VMwareService.exe, VMwareUser.exe. If bot detccts some vmware process, it shut off some of its powerful malicious function so that behavior analysis system cannot observe it and devises defenses. So, we hook some system calls which are user to detect virtual envirorment. Then, we modify return values. Also, there is another virtual machine detection method through checking IDT's base address. This method only use one native API to detect the existent of Virtual machine, because a lot of x86 software VM implementation choose to change IDT for guest OS. So, we hook the native API call which is used to know a IDT base address. Then, the modified IDT base address is returned to the bot process. These processes are described in Fig. 3.

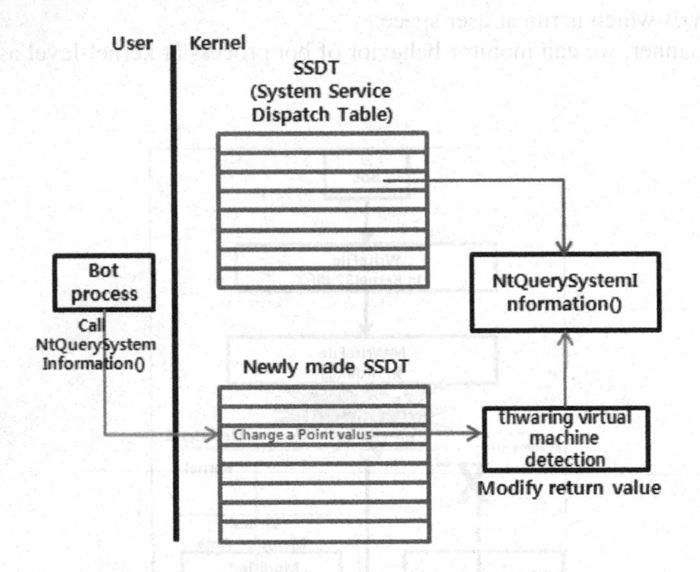

Fig. 3. Thwarting virtual machine detection process

3.2 Thwarting Static-DLL/Binary Code Injection

During the pre-configured timer period of behavior monitoring, bot process can modify system registry auto-run key to registers malicious files. Since registered files are not executed during the monitoring time, behavior analysis system can not monitoring these files.

In order to monitor static injected files, *analysis* inspects that registry related API calls are exist or not in API call lists after monitoring is completed. If auto-run registry event is occurred and registered file is not executed during the monitoring time, the file is executed in forced manner using injection target process executing.

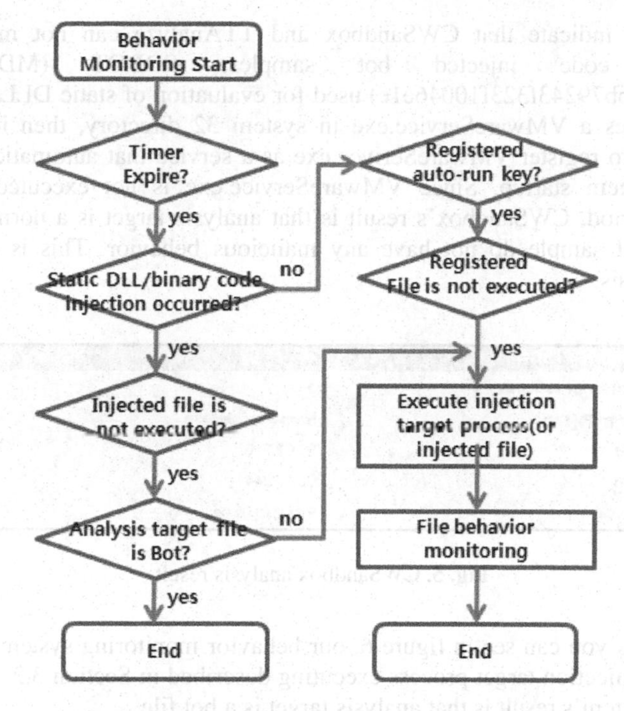

Fig. 4. Thwarting static-DLL/binary code injection

Fig. 4 depicts a flow of the thwarting static-DLL/binary code injection method.

4 Experiments and Evaluation

For performance comparison, we collect 290 recent bot samples via Honeynet implemented at Korea Information Security Agency in the period from October 1th 2008 to May 31st 2009. Then, we select unique 10 bot samples (based on their MD5 hashes) that contains a kernel-level rootkit or virtual machine detection or static-DLL/binary code injection capabilities. Then, we analyze it using CWSandbox, TTAnalyze, ZeroWine, and our behavior monitoring system.

The results are shown in Table 2.

Table 2. Results of bot behaviour analysis

Function	CWSandbox	TTAnalyze	Implemented System
Kernel-level rootkit	no	yes	yes
Virtual machine detection	no	no	yes
DLL/Binary code injection	no	no	yes

The results indicate that CWSandbox and TTAnalyze can not monitor static DLL/Binary code injected bot samples. SDBOT (MD5 hashes: 11ec6b462f1bbb79243f323f10046e1e) used for evaluation of static DLL/Binary code injection creates a VMwareService.exe in system 32 directory, then it creates the registry entry to register VMwareSerivce.exe as a service that automatically execute at a later system startup. Since VMwareService.exe is not executed during the monitoring period, CWSandbox's result is that analysis target is a normal file. This means that bot sample do not have any malicious behavior. This is clearly false negative analysis.

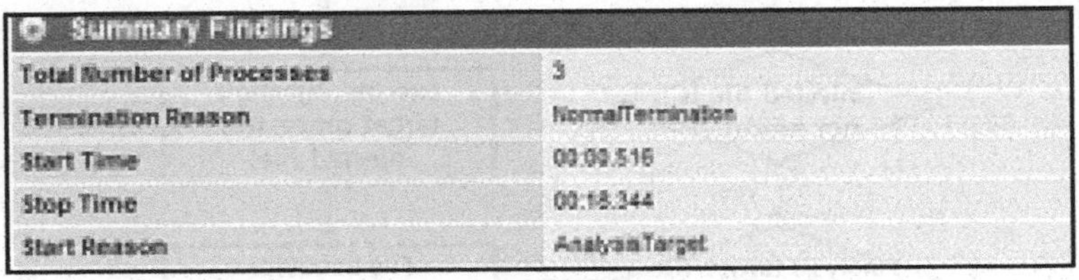

Summary Findings	
Total Number of Processes	3
Termination Reason	NormalTermination
Start Time	00:00.516
Stop Time	00:18.344
Start Reason	AnalysisTarget

Fig. 5. CWSandbox analysis result

However, as you can see in figure 6, our behavior monitoring system can monitor the sdbot via injection target process executing described in Section 3.2. our behavior monitoring system's result is that analysis target is a bot file.

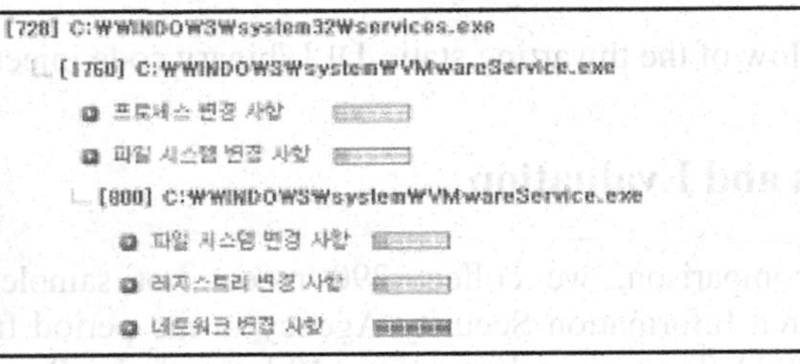

[728] C:WWINDOWSWsystem32Wservices.exe
[1760] C:WWINDOWSWsystemWVMwareService.exe
프로세스 변경 사항
파일 시스템 변경 사항
[800] C:WWINDOWSWsystemWVMwareService.exe
파일 시스템 변경 사항
레지스트리 변경 사항
네트워크 변경 사항

Fig. 6. Our behavior monitoring system result

5 Conclusion and Future Work

In this paper, we propose and implement behavior analysis system that is able to extract the malicious API call information from the bot file through newly made SSDT at kernel and process executing scheme. Also, we have applied our system to analyze several representative bot samples which have kernel-level rootkit or virtual machine detection or static-DLL/binary code injection capabilities and have shown that our system is able to automatically extract detailed information about their malicious capabilities. However, as bot use various virtual machine detection schemes in recently, researches related to thwarting virtual machine detection is needed. Therefore, our future work will focus on analyzing virtual machine detection technologies used recently and then we will especially research on thwarting virtual machine detection.

References

1. World Economic Forum, The internet is doomed, BBC News (January 2007)
2. CWSandbox, `http://www.cwsandbox.org`
3. TTAnalyze, `http://iseclab.org/projects/ttanalyze/`
4. ZeroWine, `http://sourceforge.net/projects/zerowine/`
5. Dinaburg, A., Royal, P., Sharif, M., Lee, W.: Ether: Malware Analysis via Hardware Virtualization Extensions. In: CCS 2008 (November 2008)
6. Yin, H., Liang, Z., Song, D.: HookFinder: Identifying and Understanding Malware Hooking Behaviors. In: NDSS 2008 (February 2008)

Two Approaches to Handling Proactivity in Pervasive Systems

Elizabeth Papadopoulou, Yussuf Abu Shaaban, Sarah Gallacher, Nick Taylor, and M. Howard Williams

School of Maths and Computer Sciences, Heriot-Watt University, Riccarton, Edinburgh, EH14 4AS, UK
{E.Papadopoulou,ya37,S.Gallacher,N.K.Taylor,M.H.Williams}@hw.ac.uk

Abstract. A key objective of a pervasive system is to reduce the user's administrative overheads and assist the user by acting proactively on his/her behalf. The aim of this paper is to present some aspects of how proactivity is handled in the approaches used in two different pervasive systems. The Daidalos system provides proactive behaviour based on the assumption that the user is responsible for requesting services and that proactivity is restricted to selecting and personalising these based on the user's preferences. The Persist system uses an extension of this approach combined with an analysis of user intent. The idea behind the latter is that, if the system knows what the user will do next, it can act on the user's behalf, initiating the actions that the user would normally perform. User intent predictions and those produced by the user preferences are used to determine the final action to be taken.

Keywords: Pervasive computing, Proactivity, User Preferences, User Intent.

1 Introduction

Pervasive computing [1, 2] is concerned with the situation where the environment around a user is filled with devices, networks and applications, all seamlessly integrated. As developments in communications and in sensor technologies are accompanied by a large expansion in services available, the result will soon become unmanageable and it is this problem that pervasive computing seeks to address by developing an intelligent environment to control and manage this situation [2].

To hide the complexity of the underlying system from the user, the system needs to take many decisions on behalf of the user. This can only be done if it knows what the user would prefer, i.e. it maintains a knowledge base that captures user preferences for each user and uses these to personalize the decision making processes within the pervasive system. This may be further enhanced with mechanisms for determining user intent and predicting user behaviour. Without this it is difficult for a pervasive system to identify accurately what actions will help rather than hinder the user. This is one of the major assumptions underpinning most pervasive system developments.

In tackling the problem of developing ubiquitous and pervasive systems over the past decade, different research projects have adopted different assumptions and

S.K. Prasad et al. (Eds.): ICISTM 2010, CCIS 54, pp. 64–75, 2010.

explored different approaches. As a result one class of system to emerge is that of the fixed smart space. This is generally focused on intelligent buildings – systems geared towards enhancing a fixed space to enable it to provide intelligent features that adapt to the needs of the user. On the other hand there are also systems that are focused on the mobile user, where the requirement is for access to devices and services in the user's environment wherever he/she may be. A number of prototype ubiquitous/pervasive systems have been emerging in recent years. Examples include [3 – 9]. Another example was the Daidalos project [10].

A novel approach that is currently being investigated in the research project, Persist, is that of the Personal Smart Space (PSS). The latter is defined by a set of services that are located within a dynamic space of connectable devices, and owned, controlled or administered by a single user or organisation. This concept has the advantage that it provides the benefits of both fixed smart spaces and mobile pervasive systems.

This paper is concerned with the problem of handling proactivity in a pervasive system and it describes the approach used in the Daidalos platform and compares it with that being developed for the Persist system. The former is based purely on user preferences whereas the latter uses a combination of user intent predictions and user preferences to provide input to decisions on proactivity. By incorporating user intent predictions, the system can identify situations that will require it to perform operations well in advance, thereby further reducing user involvement and enhancing user experience.

The remainder of this paper is structured as follows. The next section looks at related work on proactivity in pervasive systems. Section 3 introduces the Daidalos and Persist systems. Section 4 describes the approaches used in these two systems. Section 5 concludes and details future work.

2 Related Work

One of the major assumptions of pervasive systems is that they are adaptive to the needs of the individual user and able to personalise their behaviour to meet the needs of different users in different contexts. To do this such a system must retain knowledge about the user's preferences and behaviour patterns in an appropriate form and apply this in some way. This may be done proactively by identifying what actions the user might wish to take and performing these actions on the user's behalf.

Research on the development of fixed smart spaces identified the need for some form of proactivity from the outset. This was true both for smart homes (e.g. Intelligent Home [3], MavHome [4]) and smart office applications (e.g. Gaia[5], i-Room [6]). As interest extended to mobile users, nomadic applications such as Cooltown [7] used similar technologies.

Initial systems were based on the use of user preferences that were entered manually by the user. However, building up preferences manually is an arduous undertaking and experience has shown that the user soon loses interest and the resulting preference sets are incomplete and not very useful. In the case of the Aura project [8] manual input from the user was taken a step further when user intent was incorporated into the process of determining when to perform a proactive action and what action to carry out, as the user was required to provide not only user preferences but also information such as the user's current task.

As a result systems sought alternative approaches to creating and maintaining user preferences automatically. This inevitably involved some form of monitoring of the user's behaviour followed by some form of learning applied to the accumulated data. This approach was adopted in both fixed smart space developments (such as the MavHome [4] project) as well as mobile applications, e.g. Specter [9]. Another approach using Bayesian networks or Hidden Markov Models rather than rule based preferences to capture user behaviour and represent user needs was adopted by the Synapse project [10], which matches the context and current user task against past learnt patterns to select the most appropriate service for the user. Another system that uses context history for progressively learning patterns of user behaviour in an office environment is described by Byun et al [11].

3 Architectures of Daidalos and Persist Pervasive Systems

The Daidalos project [12] developed a pervasive system based on the mobile user. The Pervasive System Platform (illustrated in Fig. 1) includes a personalisation and preference management subsystem (including learning) which implicitly gathers and manages a set of preferences for the user by monitoring user behaviour and extracting preferences from the monitored user behaviour history. This pervasive system was successfully demonstrated in December 2008. The main focus of the personalisation subsystem was to help in the selection and personalisation of services. By so doing the system was able to personalise the user's environment in an unobtrusive and beneficial way (based on previous user behaviour).

The Persist project is another European research project which started in April 2008. It aims to create a rather different form of pervasive system based on the notion

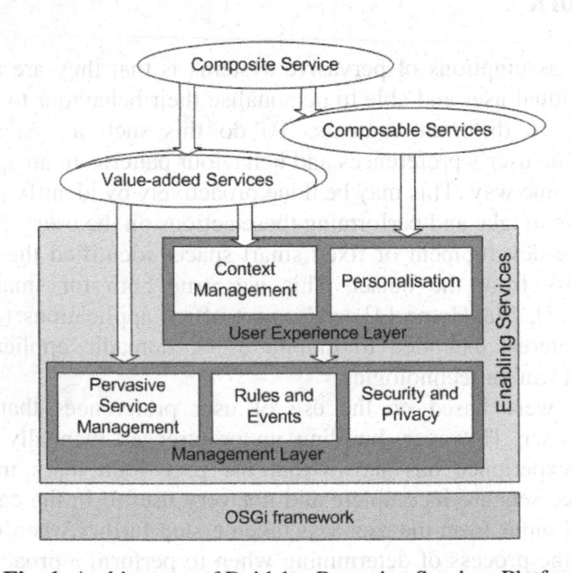

Fig. 1. Architecture of Daidalos Pervasive Services Platform

of a Personal Smart Space (PSS) [13]. A PSS is defined by a set of services that are running or available within a dynamic space of connectable devices where the set of services and devices are owned, controlled, or administered by a single user or organisation. In particular, a PSS has a number of important characteristics.

- The set of devices and services that make up the PSS have a single owner, whether this is a person or an organisation.
- A PSS may be mobile or fixed.
- A PSS must be able to interact with other PSSs (fixed or mobile). To do this a PSS must support an ad-hoc network environment.
- It must be able to adapt to different situations depending on the context and user preferences.
- It must also be self-improving and able to learn from monitoring the user to identify trends, and infer conditions when user behaviour or preferences change.
- Finally it must be capable of anticipating future needs and acting proactively to support the user.

The architecture of a PSS in Persist is shown in Fig. 2.

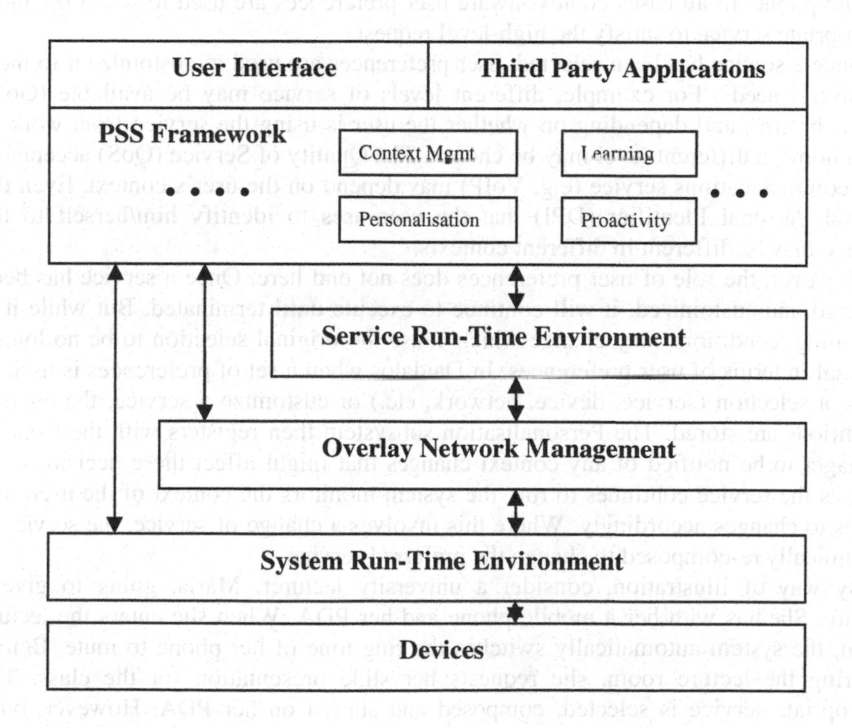

Fig. 2. The high level architecture of a Personal Smart Space

4 Approaches to Proactivity

This section describes two slightly different approaches used to handle proactivity in the Daidalos and Persist projects.

4.1 Proactivity in Daidalos

Proactivity in Daidalos is based on the assumption that the user will always initiate services and, where necessary, terminate them. Thus if a user wanted some service type, he/she would request this via a high-level GUI. However, because of the potentially wide range of services that may be selected to satisfy the user's request (which may, in turn, depend on a choice of devices and/or networks available), user preferences are used to help make this selection.

For example, if the user requests a news service, this request could be satisfied in a variety of ways. If the user is in his/her living room at home, the most obvious way to provide this might be by selecting an appropriate channel on the television set. If the user is at work, the best option might be to connect to a Web-based news service through the user's work PC. If the user is walking around town at the time of the request, the best option might be to connect to a phone-based service through his/her mobile phone. In all cases context-aware user preferences are used to select the most appropriate service to satisfy the high-level request.

Once a service has been selected, user preferences are used to customize it to meet the user's needs. For example, different levels of service may be available (Gold, silver, bronze) and depending on whether the user is using the service from work or from home, a different level may be chosen. The Quality of Service (QoS) acceptable in a communications service (e.g. VoIP) may depend on the user's context. Even the Digital Personal Identifier (DPI) that the user uses to identify him/herself to the service may be different in different contexts.

However, the role of user preferences does not end here. Once a service has been selected and customized, it will continue to execute until terminated. But while it is executing, conditions may change which cause the original selection to be no longer optimal in terms of user preferences. In Daidalos when a set of preferences is used to make a selection (service, device, network, etc.) or customize a service, the context conditions are stored. The Personalisation subsystem then registers with the Context Manager to be notified of any context changes that might affect these decisions. As long as the service continues to run, the system monitors the context of the user, and reacts to changes accordingly. Where this involves a change of service, the service is dynamically re-composed to change the preferred service.

By way of illustration, consider a university lecturer, Maria, going to give a lecture. She has with her a mobile phone and her PDA. When she enters the lecture room, the system automatically switches the ring tone of her phone to mute. Before entering the lecture room, she requests her slide presentation for the class. The appropriate service is selected, composed and started on her PDA. However, once inside the lecture room, as she approaches the front, the system discovers the large screen at the front of the lecture room and recomposes the service to use the large screen display. The first action is an example of changing the customisation in accordance with the preferences as the context of the user changes. The second is an example of reselection and re-composition of a service due to a change in context.

This form of proactivity is restricted in that it only applies to services initiated by the user. On the other hand it is sufficiently powerful to adapt the services either through parameters passed to the service affecting the way in which they are customised or through selection of an alternative service and re-composition to use this preferred service.

To assist the user in building up and maintaining their user preferences, several techniques are employed. These include:

(1) A user-friendly GUI is used to set up preferences manually.
(2) A set of stereotypes enables one to select an initial set of preferences.
(3) The user's actions are monitored and incremental learning employed to discover new preferences or refine existing ones.
(4) An offline data mining approach is used to analyse larger quantities of history data on user's actions and context.

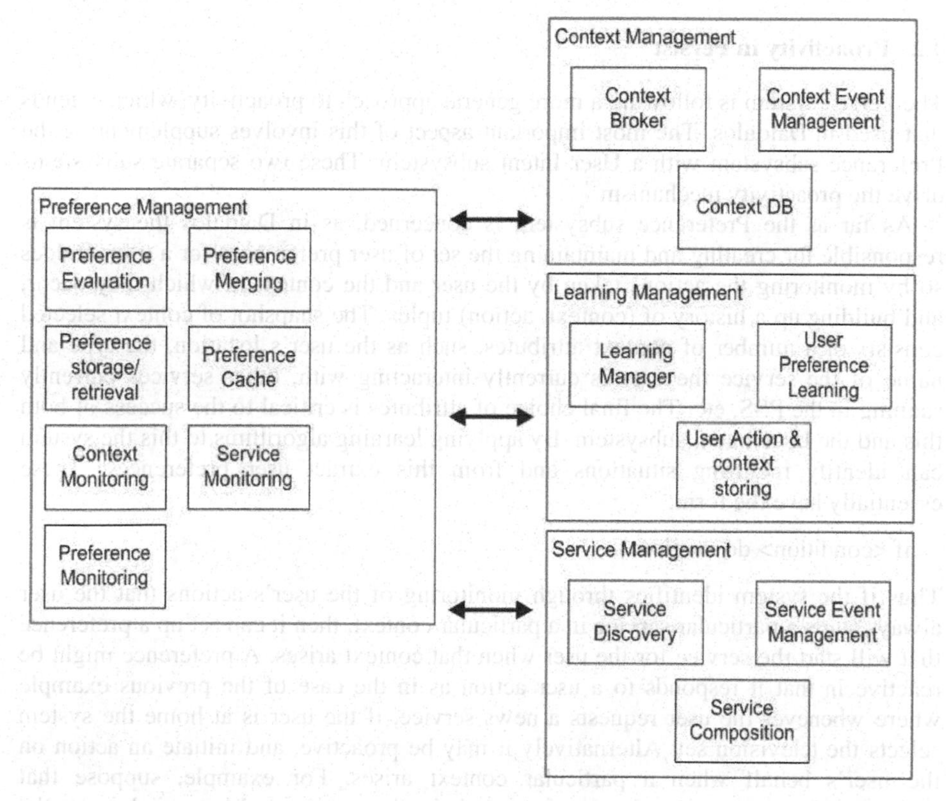

Fig. 3. Components of the Daidalos Pervasive Services Platform required for proactivity

As shown in Fig. 3, the Preference Management component includes the functionality that allows a preference to be created, stored, retrieved, merged with existing preferences, monitored and evaluated against current values of certain context attributes and current status of services. When a service is executed, any preference that affects the status of the service or a personalisable parameter of that service is an *active preference*. For each such active preference the Context Monitoring component registers for changes in specific context attributes affecting it. When a context event is fired from the Context Event Management, Context Monitoring is triggered to request the evaluation of these preferences by Preference Monitoring which cross-references the context condition with the preferences to locate all the active preferences affected by the change. The same process is performed by Service Monitoring which receives events from the Service Management component. If the status of a service changes, two tasks have to be performed. The first is to check whether this change affects any of the currently active preferences and order a re-evaluation of them. The second is to load or unload the conditions and preferences from the monitoring components if a service starts or stops accordingly and register or unregister for events that affect the service that started or stopped.

4.2 Proactivity in Persist

The Persist system is following a more general approach to proactivity, which extends that used in Daidalos. The most important aspect of this involves supplementing the Preference subsystem with a User Intent subsystem. These two separate subsystems drive the proactivity mechanism.

As far as the Preference subsystem is concerned, as in Daidalos the system is responsible for creating and maintaining the set of user preferences for a user. It does so by monitoring the actions taken by the user and the context in which they occur, and building up a history of (context, action) tuples. The snapshot of context selected consists of a number of context attributes, such as the user's location, the type and name of the service the user is currently interacting with, other services currently running in the PSS, etc. The final choice of attributes is critical to the success of both this and the User Intent subsystem. By applying learning algorithms to this the system can identify recurring situations and from this extract user preferences. These essentially have the form:

if <condition> do <action>

Thus if the system identifies through monitoring of the user's actions that the user always starts a particular service in a particular context, then it can set up a preference that will start the service for the user when that context arises. A preference might be reactive in that it responds to a user action as in the case of the previous example where whenever the user requests a news service, if the user is at home the system selects the television set. Alternatively it may be proactive, and initiate an action on the user's behalf when a particular context arises. For example, suppose that whenever the user arrives home after work in the winter, he/she switches up the temperature on the central heating. If the learning system identifies this, a preference rule could be set up to initiate this automatically for the user.

Unfortunately things are not quite as simple as this. When a recurring situation is identified, the action associated with the context condition may not always be performed. Equally, a preference may not have been used for a while and may have become out of date, eventually to be replaced by newer ones. To take account of these situations, the system maintains a set of attributes associated with each preference. These include a certainty factor representing the probability that this action is performed when the condition is satisfied, and a recency factor representing the date when this preference was last applied.

The second component responsible for driving the proactivity mechanism is the User Intent subsystem. This is similar to user preferences but in this case the aim is to look ahead and predict future actions based on past patterns of actions. The distinction between the two subsystems lies in the immediacy of the actions. User preferences are an immediate mechanism in that when a particular context condition arises, a corresponding action is taken. User intent is concerned with sequences of actions leading to a consequence at some later time in the future. Fig. 4 illustrates the main components of the PSS Framework for dealing with proactivity in a PSS.

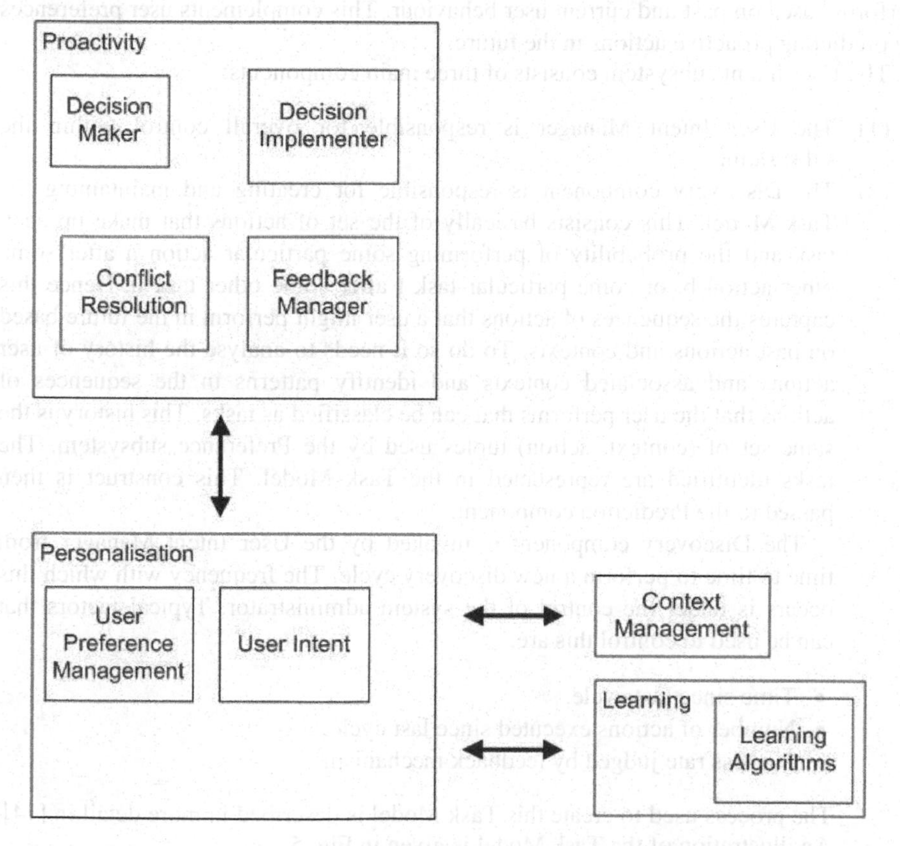

Fig. 4. Components of the PSS Framework for dealing with proactivity

To illustrate the idea behind user intent, consider the previous example of switching up the heating when the user arrives home after work in the winter. Suppose that the system recognises a pattern in the user's behaviour that starts with the user switching off his/her office computer at a time after 17.00. This may be followed by switching off the lights, exiting the building, starting the car, etc. If the system identifies that this pattern of actions always leads to the heating being switched on, it could store this as a user intent pattern. Thereafter whenever the system recognizes that the user is performing this sequence of actions, the resulting action can be carried out – in this case, it could send a message to trigger the user's heating system so that the house is at the required temperature for the user's arrival.

Hence the User Intent subsystem is responsible for the discovery and management of models of the user's behaviour. This is based on the actions performed by the user (in terms of interaction with services – starting a service, stopping a service, changing parameters that control the service, etc.), and tasks, where a task is a sequence of actions. In contrast to a user preference which specifies a single action to perform when a context situation is met, user intent may specify a sequence of actions to perform based on past and current user behaviour. This complements user preferences by predicting proactive actions in the future.

The User Intent subsystem consists of three main components:

(1) The User Intent Manager is responsible for overall control within the subsystem.
(2) The Discovery component is responsible for creating and maintaining the Task Model. This consists basically of the set of actions that make up each task and the probability of performing some particular action a after some other action b, or some particular task t after some other task u. Hence this captures the sequences of actions that a user might perform in the future based on past actions and contexts. To do so it needs to analyse the history of user actions and associated contexts and identify patterns in the sequences of actions that the user performs that can be classified as tasks. This history is the same set of (context, action) tuples used by the Preference subsystem. The tasks identified are represented in the Task Model. This construct is then passed to the Prediction component.

    The Discovery component is invoked by the User Intent Manager from time to time to perform a new discovery cycle. The frequency with which this occurs is under the control of the system administrator. Typical factors that can be used to control this are:

    - Time since last cycle.
    - Number of actions executed since last cycle.
    - Success rate judged by feedback mechanism.

    The process used to create this Task Model is described in more detail in [14]. An illustration of the Task Model is given in Fig. 5.

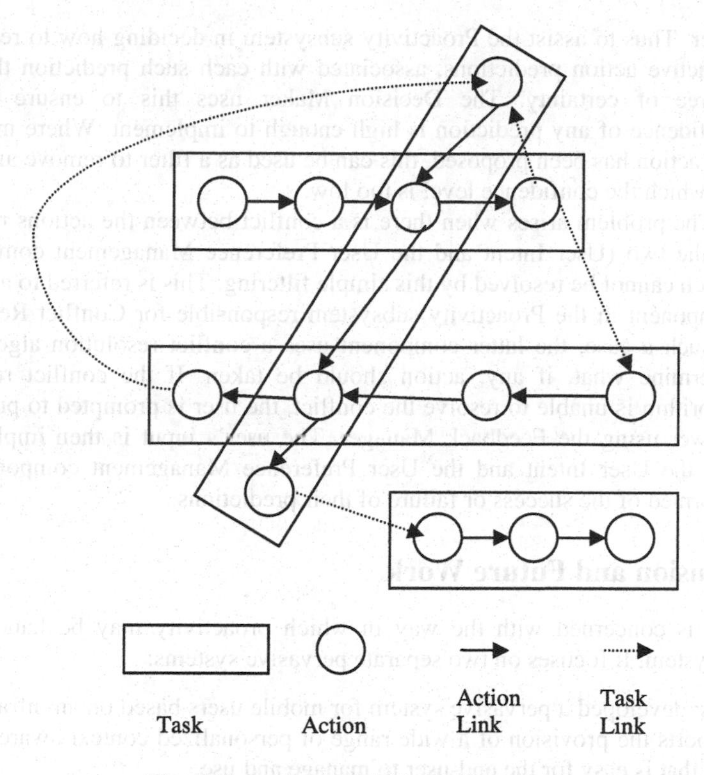

Fig. 5. Example of Task Model (from [14])

(3) The Prediction component is responsible for mapping the user's actions and associated context against the Task Model to identify potential tasks that the user might be engaged in, and once this is established with sufficient confidence, to infer any future action or actions that the user might be intending to perform. The result is passed to the Proactivity subsystem.

The Proactivity subsystem receives input from both the User Preference subsystem and the User Intent subsystem regarding the proactive actions recommended to be taken by the Proactivity subsystem. These inputs are passed to the Decision Maker component. As long as there is no conflict between these, the Decision Maker will trigger the Decision Implementer to initiate the recommended action. In doing so it also notifies the user and provides a means for the user to intervene if necessary and stop the action from proceeding. Whether the user does so or not, the Proactivity component informs the appropriate subsystem (Preference or User Intent) as to the user's acceptance or otherwise of the action concerned. This provides a useful feedback mechanism for updating the attributes associated with preferences and tasks. In the User Intent subsystem this is dealt with by the Feedback Manager.

However, the presence of two separate components (User Intent and User Preference Management) that can each provide predictions on future user behaviour, can lead to conflicting actions or actions that can cancel each other. Thus to assist the Proactivity subsystem in deciding how to respond to proactive action predictions, associated with each such prediction there is a degree of certainty. The Decision Maker uses this to ensure that the confidence of any prediction is high enough to implement. Where more than one action has been proposed, this can be used as a filter to remove any action for which the confidence level is too low.

The problem arises when there is a conflict between the actions requested by the two (User Intent and the User Preference Management components) which cannot be resolved by this simple filtering. This is referred to a separate component in the Proactivity subsystem responsible for Conflict Resolution. In such a case, the latter component uses a conflict resolution algorithm to determine what, if any, action should be taken. If the conflict resolution algorithm is unable to resolve the conflict, the user is prompted to provide an answer using the Feedback Manager. The user's input is then implemented and the User Intent and the User Preference Management components are informed of the success or failure of their predictions.

5 Conclusion and Future Work

This paper is concerned with the way in which proactivity may be handled in a pervasive system. It focuses on two separate pervasive systems:

- Daidalos: developed a pervasive system for mobile users based on an infrastructure that supports the provision of a wide range of personalized context aware services in a way that is easy for the end-user to manage and use.
- Persist: is developing a pervasive system based on the notion of a Personal Smart Space (PSS), which does not depend on access to complex infrastructure and can function independently wherever the user may be.

The Daidalos system uses a simple approach based on user preferences that can react to changes in the user's context to alter the customisation of a service or dynamically re-select and recompose a service if a change in user context needs this.

The Persist system extends this idea by including the ability to perform any user action. This includes the ability to start and stop services. This system also includes a user intent subsystem, based on a task model consisting of sequences of user actions and associated contexts. These two sources of proactive information are fed into the Proactivity subsystem, which determines when to perform an action on behalf of the user and what action is required.

A number of open issues still need to be tackled in this subsystem. These include the pattern discovery algorithm(s) to be used in task discovery. The issue of associating context with the actions and task discovered also needs further research. Work is also required on the format in which the discovered task model is stored as storing it as a graph with the associated context information rapidly becomes a problem for a mobile device with limited memory capabilities. Another possibility which could be of benefit is to use a priori knowledge of tasks, defined explicitly by the user or based on tasks performed by other users, to create or build up the Task Model. Such tasks could improve the accuracy of User Intent predictions at an early stage of system usage when there is insufficient history to create an accurate model.

Acknowledgments. This work was supported in part by the European Commission under the FP6 programme (Daidalos project) as well as under the FP7 programme (PERSIST project) which the authors gratefully acknowledge. The authors also wish to thank colleagues in the two projects without whom this paper would not have been possible. Apart from funding these two projects, the European Commission has no responsibility for the content of this paper.

References

1. Weiser, M.: The computer for the 21st century. Scientific American 265(3), 94–104 (1991)
2. Satyanarayanan, M.: Pervasive computing: vision and challenges. IEEE PCM 8(4), 10–17 (2001)
3. Lesser, V., Atighetchi, M., Benyo, B., Horling, B., Raja, A., Vincent, R., Wagner, T., Xuan, P., Zhang, S.X.Q.: The Intelligent Home Testbed. In: Anatomy Control Software Workshop, pp. 291–298 (1999)
4. Gopalratnam, K., Cook, D.J.: Online Sequential Prediction via Incremental Parsing: The Active LeZi Algorithm. IEEE Intelligent Systems 22(1), 52–58 (2007)
5. Román, M., Hess, C.K., Cerqueira, R., Ranganathan, A., Campbell, R.H., Nahrstedt, K.: Gaia: A middleware infrastructure to enable active spaces. IEEE Pervasive Computing 1, 74–83 (2002)
6. Johanson, B., Fox, A., Winograd, T.: The interactive workspaces project: Experiences with ubiquitous computing rooms. IEEE Pervasive Computing 1, 67–74 (2002)
7. Kindberg, T., Barton, J.: A web-based nomadic computing system. Computer Networks 35, 443–456 (2001)
8. Sousa, J.P., Poladian, V., Garlan, D., Schmerl, B., Shaw, M.: Task-based Adaptation for Ubiquitous Computing. IEEE Transactions on Systems, Man and Cybernetics, Part C: Applications and Reviews, Special Issue on Engineering Autonomic Systems 36(3), 328–340 (2006)
9. Kroner, A., Heckmann, D., Wahlster, W.: SPECTER: Building, Exploiting, and Sharing Augmented Memories. In: Workshop on Knowledge Sharing for Everyday Life, KSEL 2006 (2006)
10. Si, H., Kawahara, Y., Morikawa, H., Aoyama, T.: A stochastic approach for creating context aware services based on context histories in smart Home. In: 1st International Workshop on Exploiting Context Histories in Smart Environments, 3rd Int. Conf. on Pervasive Computing (Pervasive 2005), pp. 37–41 (2005)
11. Byun, H.E., Cheverst, K.: Utilising Context History to Provide Dynamic Adaptations. Journal of Applied AI 18(6), 533–548 (2004)
12. Williams, M.H., Taylor, N.K., Roussaki, I., Robertson, P., Farshchian, B., Doolin, K.: Developing a Pervasive System for a Mobile Environment. In: eChallenges 2006 – Exploiting the Knowledge Economy, pp. 1695–1702. IOS Press, Amsterdam (2006)
13. Crotty, M., Taylor, N., Williams, H., Frank, K., Roussaki, I., Roddy, M.: A Pervasive Environment Based on Personal Self-Improving Smart Spaces. In: Ambient Intelligence 2008. Springer, Heidelberg (2009)
14. Abu-Shabaan, Y., McBurney, S.M., Taylor, N.K., Williams, M.H., Kalatzis, N., Roussaki, I.: User Intent to Support Proactivity in a Pervasive System. In: PERSIST Workshop on Intelligent Pervasive Environments, AISB 2009, Edinburgh, UK, pp. 3–8 (2009)

A Spatio-Temporal Role-Based Access Control Model for Wireless LAN Security Policy Management

P. Bera[1], S.K. Ghosh[1], and Pallab Dasgupta[2]

[1] School of Information Technology
pbera@sit.iitkgp.ernet.in, skg@iitkgp.ac.in
[2] Department of Computer science and Engineering
Indian Institute of Technology, Kharagpur-721302, India
pallab@cse.iitkgp.ernet.in

Abstract. The widespread proliferation of wireless networks (WLAN) has opened up new paradigms of security policy management in enterprise networks. To enforce the organizational security policies in wireless local area networks (WLANs), it is required to protect the network resources from unauthorized access. In WLAN security policy management, the standard IP based access control mechanisms are not sufficient to meet the organizational requirements due to its dynamic topology characteristics. In such dynamic network environments, the role-based access control (RBAC) mechanisms can be deployed to strengthen the security perimeter over the network resources. Further, there is a need to incorporate time and location dependent constraints in the access control models. In this paper, we propose a WLAN security management system which supports a spatio-temporal RBAC (STRBAC) model. The system stems from logical partitioning of the WLAN topology into various security policy zones. It includes a *Global Policy Server* (GPS) that formalizes the organizational access policies and determines the high level policy configurations for different policy zones; a *Central Authentication & Role Server* (CARS) which authenticates the users (or nodes) and the access points (AP) in various zones and also assigns appropriate roles to the users. Each policy zone consists of an *Wireless Policy Zone Controller* (WPZCon) that co-ordinates with a dedicated *Local Role Server* (LRS) to extract the low level access configurations corresponding to the zone access points. We also propose a formal spatio-temporal RBAC (STRBAC) model to represent the security policies formally.

1 Introduction

The security management in wireless networks (WLAN) is becoming increasingly difficult due to its widespread deployment and dynamic topology characteristics. A few enterprise WLANs today are closed entities with well-defined security perimeters. Mobile users (with laptops and handheld devices) remotely access the internal network from a public network zone; hence may violate the organizational security policies. Typically, organizational security policy provides a set of rules to access network objects by various users in the network.

S.K. Prasad et al. (Eds.): ICISTM 2010, CCIS 54, pp. 76–88, 2010.

It requires a strong security policy management system with appropriate access control models to meet the organizational security need.

The policy based security management is useful for enforcing the security policies in organizational LAN. The basic idea of policy based security management lies in partitioning the network topology into different logical policy zones and enforcing the security policies in the policy zones through a set of functional elements. It requires distribution of the system functionalities (or functional rules) into various functional/architectural elements. However, the deployment of policy based security management in wireless network (WLAN) must require appropriate access control models (such as role-based access control (RBAC), temporal RBAC, spatio-temporal RBAC) for representing the security policies formally. This is due to the dynamic topology characteristics of wireless networks (wireless nodes may not bind to a specific IP address). The background and standards for policy based security management can be found in *RFC 3198* [5].

Role based access control (RBAC) mechanisms are already being used for controlled access management in commercial organizations. In RBAC, permissions are attached to roles and users must be assigned to these roles to get the permissions that determine the operations to be carried out on resources. Recently, temporal RBAC (TRBAC) and spatio-temporal RBAC (STRBAC) models are also evolved for location and time dependent access control. In wireless LAN security management, the STRBAC model can be used to add more granularity in the security policies. Here, the users associated to a role can access network objects, iff they satisfy certain location and time constraints. For example, in an academic network, Students are not allowed to access internet from their residential halls during class time (say, 08:00-18:00 in weekdays). However, they are always allowed to access internet from the academic departments.

In this paper, we propose a WLAN security policy management system supported by a spatio-temporal RBAC model. The present work evolves from the policy based security management architecture proposed by Lapiotis et al. [2] and has been extended by incorporating new functional elements and a formal spatio-temporal RBAC (STRBAC) model. Lapiotis et al. propose a Policy-based security management system through distributed wireless zone architecture with a central policy engine. But, the work does not state the type of policies enforced by the policy engine and also about the validation of the policies.

The major components of our WLAN security policy management system (refer Fig.1) are described as follows:

- We introduce a *Central Authentication & Role Server* (CARS) which authenticates the users (or nodes) and network access points (AP) and also assigns appropriate roles to the users based on user credentials.
- *Local Role Servers* (LRS) corresponding to the respective policy zones are populated with the user-role information from the CARS.
- The *Global Policy Server* formally models the organizational global security policy, GP; determines the high level policy configurations (represented as, $< GP_{Z_1}, ..., GP_{Z_N} >$) for various policy zones; and validates these configurations conform to the global policy.

- The distributed *Wireless Policy Zone Controllers* (WPZCons) determine the low level access configurations (represented as, $< LP_{Z_1}, ..., LP_{Z_N} >$) coordinating with the local role servers and validates these configurations.
- We propose a formal STRBAC model to represent the security policies in the system.

The rest of the paper is organized as follows. The related work in the areas of Wireless LAN policy based security management and spatio-temporal RBAC models has been described in section 2. In section 3, we describe the architecture and operational flow of the proposed WLAN policy management system. Section 4 describes the proposed spatio-temporal RBAC model to support our policy management system. The analysis of the framework with a case study has been presented in section 5.

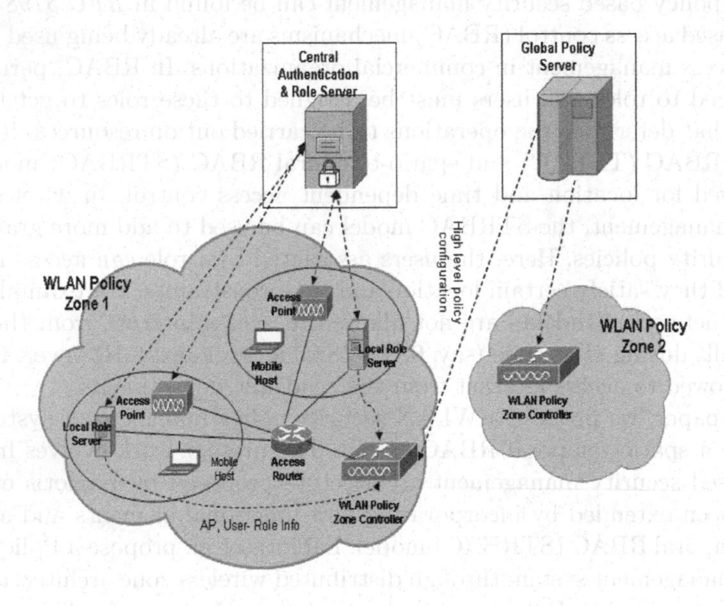

Fig. 1. Wireless LAN Security Policy Management System

2 Related Work

Several research has been performed in the area of network policy based security management but its deployment in wireless network is in a premature stage. Westrinen et al. [5] standardizes the terminologies and functional elements for policy based management. The IETF Policy working group developed a framework for network policy based admission control [4]. It consists of a central policy server that interprets the policies, makes policy decisions and communicates them to various policy enforcement points. Chandha et al. [6] propose a policy based management paradigm to support distributed enforcement of network policies from a logically centralized system. The work states that the policy-based approach is efficient in providing secured access to applications and network services. The research outcome of IST-POSITIF project [1] is policy-based security framework in local area networks. J Burns et al. propose a framework [3] for automatic management of network security policies based on central policy engine. The policy engine gets populated by the models of network elements and services, validates policies and computes new configurations for network elements when policies are violated. But, the framework considers very simple set of policy constraints. A recent work [2] has been proposed by Lapiotis et al. on policy based security management in wireless LAN. They propose a distributed policy based architecture which includes a central policy engine and distributed wireless domain managers with consistent local policy autonomy. But, they do not describe the type of security policies enforced and also do not describe the formal validation of the policies.

Role based access control (RBAC) model [7] is used for addressing the access requirements of commercial organizations. Several work has been done to improve RBAC functionalities incorporating time and location information. Joshi et al. [8] propose a Generalized Temporal Role Based Access Control Model (GTRBAC) incorporating time to the RBAC model. Temporal constraints determine when the role can be enabled or disabled. In this work, the authors introduce the concept of time-based role hierarchy. GEO-RBAC [9] is an extension RBAC incorporating spatial information. Here, the roles are activated based on location. Ray and Toahchoodee [10] propose a Spatio-Temporal Role-Based Access Control Model incorporating both time and location information. We introduce the notion of wireless policy zone to represent location in our model. The role permissions to access network objects are modeled through policy rules containing both policy zone(location) and temporal constraints.

The application of spatio-temporal RBAC model in wireless network security is in its infancy. Laborde et al. [12] presents a colored Petri Net based tool which allows to describe graphically given network topology, the security mechanism and the goals required. In this work, the authors model the security policies through generalized RBAC without considering time and location dependent service access. Moreover, the proposed tool is not applicable in wireless networks. To the best of our knowledge, the only work which uses spatio-temporal RBAC in wireless network is by Tomur and Erten [11]. They present a layered security architecture to control access in organizational wireless networks based STRBAC model using tested wired network components such as VPNs and Firewalls. However, this work does not describe the modeling of STRBAC policies using existing ACL standards. In our proposed WLAN policy management system, the global access policies are implemented through distributed wireless policy zone controllers which outsource the high level policy configurations from the global policy server. This makes the task of policy enforcement and validation easier and efficient. Moreover, the security policies are modeled using a formal STRBAC model.

3 WLAN Security Policy Management System

The proposed WLAN policy management system shown in Fig.1 stems from the notion of Wireless policy zones. A policy zone comprises of one or more wireless Access Points (AP), a dedicated *Wireless Policy Zone Controller* (WPZCon) and a *Local Role Server* (LRS). The authentication of the users and the access points are managed by a *Central Authentication & Role Server* (CARS). It also assigns appropriate roles to the authenticated users based on user credentials and policy zone (location) information. The LRS is responsible for maintaining the AP and user-role information in a policy zone. The *Global Policy Server* (GPS) formalizes the global security policy (GP) through a spatio-temporal RBAC model. The detail of the STRBAC model is described in section 4. It also determines and validates high level policy configurations for various policy zones. Each WPZCon coordinate with the local role server to derive low level access configuration for the policy zone and validates it with corresponding high level configuration. Finally, the implementation access rules corresponding to the low level access configurations are distributed to various zone access points. The operational flow of the system is shown in Fig.2. In our framework, the distributed policy zone architecture makes the task of policy enforcement and validation easier and efficient. We also propose a formal spatio-temporal RBAC model for representing the security policies described in the next section.

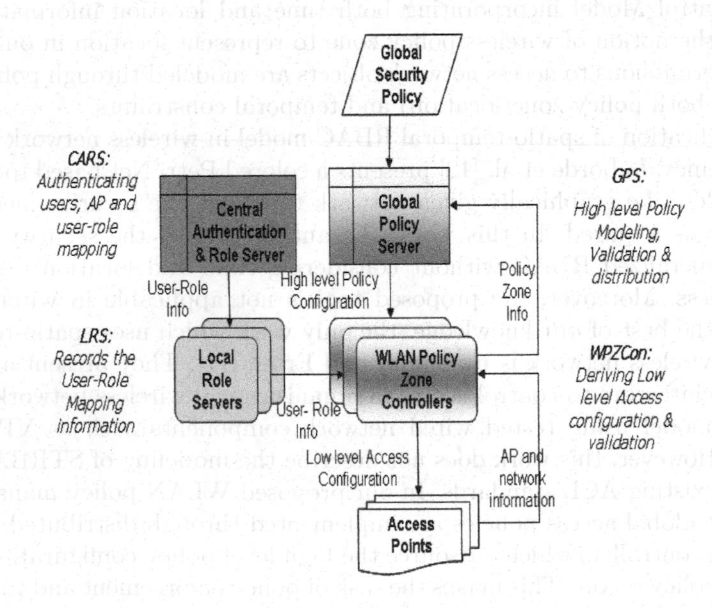

Fig. 2. Operational Flow of the WLAN Security Policy Management System

4 Proposed Spatio-Temporal RBAC Model for WLAN Policy Management

Typically, the spatio-temporal RBAC model incorporates the location and time information to the basic RBAC entities through various relations. The basic RBAC entities are users, roles, objects, permissions and operations. The modeling of location and time information to support the proposed WLAN policy management system has been described further below.

4.1 Modeling Location

In our model, the network location is represented in terms of policy zones. The policy zones physically represent different sections or units in an organizational WLAN. For example, in a typical Academic network, the policy zones can be Academic sections, Hostels or Administration. A policy zone is formally defined as follows:

Definition 1. [Policy Zone] *A Policy Zone $PZon_i$ is defined as a set of IP addresses or IP address block $\{IP_i, IP_j, ..., IP_n\}$. The IP addresses can be contiguous or discrete.*

Example of a contiguous IP address block is [10.14.0.0 – 10.14.255.255]. Example of a discrete IP address block is [10.14.0.0 – 10.14.255.255, 10.16.0.0 – 10.16.255.255]. A policy zone can contain another policy zone which is formalized as follows:

Definition 2. [Policy Zone Containment] *A policy zone Z_i is contained in another policy zone Z_k, denoted as $Z_i \subseteq Z_k$, if the following condition holds: $\forall IP_j \in Z_i, IP_j \in Z_k$. The Z_i and Z_k are referred as contained and containing policy zones respectively.*

4.2 Modeling Time

The time must be modeled with appropriate granularity to provide temporal object access. The granularity of time may depend on the organizational access control requirements. To represent time in our model, we use the notion of time instant and time interval.

A time instant is a discrete point on the time line. A time interval is a set of time instances. The interval can be continuous and non-continuous. Example of a continuous interval is 09:00-15:00 on 24th July. Example of a non-continuous time interval is 09:00-19:00 on Monday to Friday in the month of July. A time instant t_i in the interval T is indicated as $t_i \in T$.

Two time intervals can be related by any of the following relations: *disjoint*, *equality*, and *overlapping* as mentioned in [10]. The time intervals T_i and T_j are disjoint if the intersection of the time instances in T_i with those of T_j results in a null set. Two time intervals T_i and T_j are equal if the set of time instances in

T_i is equal to those of T_j. The time intervals T_i and T_j are overlapping if the intersection of the time instances in T_i with those of T_j results in a non-empty set. A special case of overlapping relation is referred to as *containment.* A time interval T_i is contained in another interval T_j if the set of time instances in T_i is a subset of those in T_j. The containment relation is denoted as $T_i \preceq T_j$.

4.3 Modeling Basic RBAC Entities in the Proposed System

This section describes the basic RBAC entities in the proposed model. The common RBAC entities are Users, Roles, Objects, Permissions and Operations. In our model, Permissions and Operations associated to various roles are modeled as Policy Rules; whereas Users, Objects and Roles are modeled according to the context of the network.

Users: The users (or nodes) interested in accessing some network object tries to communicate to an wireless access point (AP) from a policy zone. The central authentication & role server (CARS) authenticates the users and the AP(s) based on user credentials (login, password and device MAC address), locations (policy zones) and AP credentials (device ID and network-ID). The location of an user is the policy zone from which it communicates and that can change with time. The policy zone of an user u during time interval T can be identified by the function $UserPZone(u, T)$. Multiple users can be associated to a single policy zone at any given point of time.

Network Objects: In the proposed model, the network objects are logical. A network object is identified by a network service and a service policy zone.

Definition 3. [**Network Object**] *A network object* $Obj_i < Serv_j, Z_k >$ *represents a network service* $Serv_j$ *associated to a service policy zone* Z_k.

Network services refer to any network applications conforming to TCP/IP protocol. For example, some of the known network services are *ssh, telnet, http* etc. The service policy zone is the destination location associated to the service. For example, *ssh* service access to a policy zone Z_d can be represented by a network object $Obj_i < ssh, Z_d >$.

Roles: Roles represent group of users. For example, typical roles for an academic institution may be *faculty, student, administrator, guest* etc. In our model, the assignment of roles to the users is location and time dependent. For example, an user can be assigned the role of *faculty* in *academic* policy zone at *any* time. Thus, valid users must satisfy the spatial and temporal constraints before role assignment. $RoleAssignZone(r_i)$ represents the policy zone(s) where the role r_i can be assigned. $RoleAssignTime(r_j)$ represents the time interval when the role r_j can be assigned. Some role r_k can be allocated anywhere. In that case $RoleAssignZone(r_k) = Any$. Similarly, we specify $RoleAssignTime(r_k) = Always$, if some role r_k can be assigned any time.

The predicate $UserRoleAssign(u_i, r_j, T, Z_k)$ states that the user u_i is assigned to role r_j during the time interval T and policy zone Z_k. This predicate must satisfy the property: $UserRoleAssign(u_i, r_j, T, Z_k) \Rightarrow (UserPZone(u_i, T) = Z_k) \wedge (Z_k \subseteq RoleAssignZone(r_j)) \wedge (T \subseteq RoleAssignTime(r_j))$.

4.4 Modeling of Global Policy

The global policy of an organization can be modeled through a set of policy rules that *"permit"/"deny"* user accesses to various network objects from different policy zones during specified time intervals. A policy rule represents the network object accessibility permissions (*"permit"* or *"deny"*) of a role from a policy zone to the network objects during certain time interval.

Definition 4. [Policy Rule] *A Policy Rule* $PR_i < r_j, Z_l, Obj_k, T, p >$ *defines that the role* r_j *is assigned the* permission p *("permit"/"deny") to access the object* obj_k from the policy zone $PZon_l$ *during the time interval* T.

Each policy rule must satisfy the following predicates:

(1) $T \subseteq RoleAssignTime(r_j)$, i.e., time interval T must be contained in $RoleAssignTime(r_j)$;
(2) $Z_l \subseteq RoleAssignZone(r_j)$, i.e., source zone Z_l contained in $RoleAssign Zone(r_j)$. The global policy is represented as ordered set of policy rules $\{PR_1, ..., PR_N\}$.

Inter-rule Subsuming Conflicts and Resolution: The global policy model may contain inter-rule subsuming conflicts due to rule component dependencies. The rule components are source zone, object, role, time and permission. We define the inter-rule subsuming conflicts as follows.

Definition 5. [Inter-rule Subsuming Conflict] *A pair of policy rules* PR_x *and* PR_y *are subsume-conflicting, iff* $(PR_x[obj] = PR_y[obj]) \wedge (PR_x[role] = PR_y[role]) \wedge (PR_x[PZon] \subseteq PR_y[PZon]) \wedge (PR_x[T)] \preceq PR_y[T])$*, where* $\subseteq$ *and* $\preceq$ *indicate policy zone containment and time containment respectively.*

Here, two cases may occur based on *permission* component of the rules.

Case 1: $PR_x[permission(p)] \neq PR_y[permission(p)]$ and
Case 2: $PR_x[permission(p)] = PR_y[permission(p)]$.

Under each case, following subcases may appear.

subcase(a): $(PR_x[Z] \subset PR_y[Z]) \wedge (PR_x[T] = PR_y[T])$
subcase(b): $(PR_x[Z] = PR_y[Z]) \wedge (PR_x[T] \preceq PR_y[T])$
subcase(c): $(PR_x[Z] \subset PR_y[Z]) \wedge (PR_x[T] \preceq PR_y[T])$
subcase(d): $(PR_x[Z] = PR_y[Z]) \wedge (PR_x[T] = PR_y[T])$

To resolve the conflicts from a global policy model, we introduce the concept of rule-order majority. For example, considering a pair of conflicting rules PR_x and PR_y, PR_x has higher relative order than PR_y iff $x < y$; where the suffix indicate the relative rule index (positions) in the rule base. Thus for each such pair of rules, we state PR_x as order-major and PR_y as order-minor rule.

Now, in all the subcases under **Case 2** and **subcase 1(d)**, resolving the inter-rule conflicts require deletion of the order-minor rules, PR_y. Whereas, in **subcase 1(a)**, rule PR_x must be replaced by the rule $PR_x^{'}$ keeping PR_y unchanged where,

$PR_x^{'} :< PR_x[role], (PR_x[Z] - PR_y[Z]), PR_x[obj], PR_x[T], PR_x[p] >$.

Similarly, in **subcase 1(b)**, the rule PR_x must be replaced by the rule $PR_x^{''}$, where, $PR_x^{''} :< PR_x[role], PR_x[Z], PR_x[obj], (PR_x[T] - PR_y[T]), PR_x[p] >$.

In **subcase 1(c)**, PR_x must be replaced by the rule $PR_x^{'''}$, where,

$PR_x^{'''} :< PR_x[role], (PR_x[Z] - PR_y[Z]), R_x[obj], (PR_x[T] - PR_y[T]), PR_x[p] >$.

The global policy rule base generated after the removal of the inter-rule conflicts is represented as GP.

High Level Policy Configuration: To enforce the organizational security policy in the wireless LAN, the rules in the conflict-free global policy model GP must be properly distributed to various policy zone controllers (WPZCon). Thus, GP is represented as a distribution of zonal rule sets $< GP_{Z_1}, GP_{Z_2}, ..., GP_{Z_N} >$, where GP_{Z_i} represents the zonal rule set for the policy zone Z_i. To make this distribution correct, the following property must be satisfied: $(GP_{Z_1} \wedge GP_{Z_2} \wedge ... \wedge GP_{Z_N}) \Rightarrow GP$. A policy rule PR_i is included in the zonal rule set GP_{Z_k} corresponding to the policy zone Z_k, iff the policy zone of PR_i is contained by the policy zone Z_k. This is formally represented as follows:

$\forall PR_i \in GP, \exists Z_k \subseteq Any, (Z_k \subseteq PR_i[Z] \Rightarrow (PR_i[Z] \Uparrow GP_{Z_k}))$. Here, $(PR_i \Uparrow GP_{Z_k})$ indicates the inclusion of PR_i in GP_{Z_k}. Thus, $\forall k, GP_{Z_k} \subseteq GP$. In our model, $< GP_{Z_1}, GP_{Z_2}, ..., GP_{Z_N} >$ represents the high level policy configuration corresponding to the global policy GP.

Low Level Access Configuration: The global policy server (GPS) distributes the zonal rule sets of the high level policy configuration to various policy zone controllers (WPZCon). Each WPZCon translates the zonal rule set to low level configuration based on the local policy zone information. A WPZCon coordinates with the local role server (LRS) and access points (AP) for getting populated with the local policy states. The low level access configuration LP_{Z_k} represents a collection of implementation rules $\{IR_1, IR_2, ..., IR_N\}$ corresponding to the zonal rule set GP_{Z_k} of policy zone Z_k.

Definition 6. [Implementation Rule] *An Implementation rule* $IR_x < u_i, r_j, Serv_k, Z_s, Z_d, T, p, net_l >$ *defines that an user* u_i *associated to the role* r_j *is assigned the permission* p *to access the network service* $Serv_k$ *from the source zone* Z_s *to destination zone* Z_d *during time interval* T; where, net_l *represents the access point or the network to which the rule is physically mapped.*

Here, the access points are identified by the network-ID and the device credentials. These information are conveyed by the access points to the associated policy zone controllers prior to extract low level access configuration. For each implementation rule, IR_i, the service $Serv_k$ and destination policy zone Z_d can be determined from the associated network object ($PR_i[Obj]$) corresponding to the policy rule PR_i. More importantly, the relation $UserRoleAssign(u_i, r_j, T, Z_k)$ [described in section 4.3] ensures the correct user-role mapping which satisfies the following property:

$$UserRoleAssign(u_i, r_j, T, Z_k) \Rightarrow (UserPZone(u_i, T) = Z_k) \wedge$$
$$(Z_k \subseteq RoleAssignZon(r_j)) \wedge (T \subseteq RoleAssignTime(r_j)).$$

The validation of the low level access configuration is ensured by the property: $\forall(LP_{Z_i}, GP_{Z_i}), LP_{Z_i} \Rightarrow GP_{Z_i}$. It states that each low level implementation rule set or access configuration, LP_{Z_i} must conform to the corresponding high level policy rule set GP_{Z_i}.

5 Analysis with Case Study

In this work, we propose a WLAN policy management system supported by a spatio-temporal RBAC model. In this section, we analyze the framework with a case study. We consider a typical *Academic WLAN* conforming to Fig 1 with a global policy ***Example_policy***. The network consists of four wireless policy zones, namely *Hall*[refers to student hall of residences], *Academic*[refers to academic departments], *Administration*[refers to administrative block] and *Web_Proxy*[refers to the policy zone consisting of web-proxy servers]. The *internet(http)* access to the external world is processed through *Web_Proxy* zone.

Example_policy:

- The Academic network consists of five roles: *student, faculty, administrative staff, network administrator* and *guest.*
- The network services considered: *ssh, telnet* and *http* conforms to TCP/IP protocol.
- The *network administrator* can *always* access *internet* from *any* zone and can *always* access *ssh* and *telnet* from *any* zone to *any* zone.
- *faculty* can always access *internet* from *any* zone and can always access *ssh* and *telnet* from *any* zone to only *Academic* zone.
- *administrative staffs* can always access *internet* from *any* zone and can always access *ssh* and *telnet* from *any* zone to only *Administration* zone.
- *students* can not access *internet* from the *Hall* zone during *working hours*, i.e., *08:00-18:00, Monday to Friday*
- *students* can *always* access *internet* from *Academic* zone and can *always* access *ssh* and *telnet* only from *Academic* zone to the same zone.
- *guests* can access *internet* only from *Academic* zone during *working hours.*

Example_STRBAC Model [STRBAC model for ***Example_policy:***]

1. `Policy Zones = {Hall, Academic, Admin, Web_Proxy}`
2. `Network Objects = {O1<ssh,Academic>, O2<ssh,Admin>, O3<ssh,Any>, O4<telnet,Academic>, O5<telnet,Admin>, O6<telnet,Any>, O7<http,Web_Proxy>}`

3. Working hours(WH) = (2,3,4,5,6).Day@(<8.hr+00.min>:<17.hr+59.min>)
4. Always = (1,2,3,4,5,6,7).Day@(<01.hr+00.min>:<23.hr+59.min>)
5. Non-working hours(NWH) = {(2,3,4,5,6).Day@(<1.hr+00.min>:<7.hr+59.min>), (2,3,4,5,6).Day@(<18.hr+00.min>:<23.hr+59.min>), (1,7).Day@(<1.hr+00.min>:<23.hr+59.min>)}
6. Roles = {r1<student>, r2<faculty>, r3<administrative staff>, r4<network administrator>, r5<guest>}
7. RoleAssign ={(r1,<Hall,NWH>,<Academic,WH>),(r2,Any,Always), (r3,Any,Always), (r4,Any,Always),(r5,Academic,WH)}
8. Users = {user1,user2,user3,user4,user5}
9. Global Policy(GP) ={PR1<r4,Any,07,Always,permit>, PR2<r4,Any,03,Always,permit>, PR3<r4,Any,06,Always,permit>, PR4<r2,Any,01,Always,permit>, PR5<r2,Any,04,Always,permit>, PR6<r2,Any,07,Always,permit>, PR7<r3,Any,02,Always,permit>, PR8<r3,Any,05,Always,permit>, PR9<r3,Any,07,Always,permit>, PR10<r1,Academic,01,Always,permit>, PR11<r1,Academic,04,Always,permit>, PR12<r1,Hall,07,NWH ,permit>, PR13<r1,Hall,07,WH,deny>, PR14<r1,Academic,07,Always,permit>, PR15<r5,Academic,07,WH,permit>}
10. User-Role Assignment = {(user1,r1), (user2,r2),(user3,r3), (user4,r4)}
11. High level policy config = (GP_{Hall}{PR1,PR2,..,PR9,PR12,PR13}, $GP_{Academic}${PR1,..,PR9,PR10,PR11,PR14,PR15}, GP_{Admin}{PR1,..,PR9})
12. Low level access config corresponding to policy zone Hall(LP_{Hall}) = {IR1<user4,r4,http,Any,Web_Proxy,Always,permit,net_1>, IR2<user4,r4,ssh,Any,Any,Always,permit,net_1>,..., IR9<user3,r3,http,Any,Web_Proxy,Always,permit,net_1}>, IR10<user1,r1,http,Hall,Web_Proxy,NWH ,permit,net_1>, IR11<user1,r1,http,Hall,Web_Proxy,WH ,deny,net_1>}.

Property Validation

Property1: $UserRoleAssign(u_i, r_j, T, Z_k) \Rightarrow (UserPZone(u_i, T) = Z_k) \wedge (Z_k \subseteq RoleAssignZon(r_j)) \wedge (T \subseteq RoleAssignTime(r_j))$.

Property2: $GP_{Hall} \wedge GP_{Academic} \wedge GP_{Admin} \Rightarrow GP$.

Property3: $((LP_{Hall} \Rightarrow GP_{Hall}) \wedge (LP_{Academic} \wedge \Rightarrow GP_{Academic}) \wedge (LP_{Admin} \Rightarrow GP_{Admin}))$.

The ***Example_STRBAC*** represents the model corresponding to ***Example_policy*** in the academic network. Here, the model shows the low level access configuration corresponding to Hall zone. Similarly, the configurations for other policy zones are derived. Although the example considers one user corresponds to one role, multiple users can be assigned to one role also. The security of the proposed STRBAC model is ensured by *Property1*, *Property2* and *Property3*. It may be noted that the ***Example_STRBAC*** satisfies *Property1* and *Property2*.

We have shown, how the global policy of an organizational WLAN can be formally modeled using STRBAC and then described the hierarchical formulation of the high level and low level configurations. The model supports the proposed policy management system architecture. The proposed system is scalable as the task of policy configurations and validations are managed in a distributed manner. The system also capable of providing secure policy and role management through centralized authentication & role server which coordinates the local role servers to record the user-role information. This further helps in extracting the low level access configurations.

6 Conclusion

In this paper we present a security policy management system for Wireless Network (WLAN) supported by a formal spatio-temporal RBAC model. In the proposed system, the global security policy is enforced through distributed wireless policy zone controllers (WPZCons) which are populated by extracting the high level policy configurations from the global policy server (GPS). This makes policy enforcement and validation simple and efficient. The system also uses a centralized authentication & role server (CARS) which authenticates the users and the wireless access points. It also maps the users to appropriate roles and distributes the information to the local role servers. We present a spatio-temporal RBAC model to support the policy management system which ensures the time and location dependent access to the network objects and hence provides strong security perimeter over an organizational WLAN. The present work can be extended by implementing the STRBAC model using boolean logic and checking the enforcement of the access policies using boolean satisfiability analysis.

References

1. Basile, C., Lioy, A., Prez, G.M., Clemente, F.J.G., Skarmeta, A.F.G.: POSITIF: a policy-based security management system. In: 8th IEEE International Workshop on Policies for Distributed Systems and Networks (POLICY 2007), Bologna, Italy, June 2007, p. 280 (2007)
2. Lapiotis, G., Kim, B., Das, S., Anjum, F.: A Policy-based Approach to Wireless LAN Security Management. In: International Workshop on Security and Privacy for Emerging Areas in Communication Networks, Athens, Greece, September 2005, pp. 181–189 (2005)
3. Burns, J., Cheng, A., Gurung, P., Rajagopalan, S., Rao, P., Rosenbluth, D., Martin, D.: Automatic Mnagement of Network Security Policy. In: Proceedings of the 2nd DARPA Information Survivability Conference and Exposition (DISCEX II), Anaheim, California, June 2001, pp. 12–26 (2001)
4. Yavatkar, R., Pendarakis, D., Guerin, R.: RFC 2753: A Framework for Policy-based Admission Control. Internet Society, 1–20 (January 2000)
5. Westrinen, A., Schnizlein, J., Strassner, J., Scherling, M., Quinn, B., Herzog, S., Carlson, M., Perry, J., Wldbusser, S.: RFC 3198: Terminology for Policy-Based Management. Internet Society, 1–21 (November 2001)
6. Chandha, R., Lapiotis, G., Wright, S.: Special Issue on Policy-based Networking. IEEE Network Magazine 16(2), 8–56 (2002)

7. Ferraiolo, D.F., Sandhu, R., Gavrila, S., Kuhn, D.R., Chandramouli, R.: Proposed NIST standard for Role-Based Access Control. ACM Trnsactions on Information and Systems Security 4(3) (August 2001)
8. Joshi, J.B.D., Bertino, E., Latif, U., Ghafoor, A.: A Generalized Temporal Role-Based Access Control Model. IEEE Transactions on Knowledge and Data Engineering 17(1), 4–23 (2005)
9. Bertino, E., Catania, B., Damiani, M.L., Perlasca, P.: GEO-RBAC: a spatially aware RBAC. In: SACMAT 2005: Proceedings of the tenth ACM symposium on Access control models and technologies, NY, USA, pp. 29–37 (2005)
10. Ray, I., Toahchoodee, M.: A Spatio-Temporal Role-Based Access Control Model. In: Barker, S., Ahn, G.-J. (eds.) Data and Applications Security 2007. LNCS, vol. 4602, pp. 211–226. Springer, Heidelberg (2007)
11. Tomur, E., Erten, Y.M.: Application of Temporal and Spatial role based access control in 802.11 wireless networks. Journal of Computers & Security 25(6), 452–458 (2006)
12. Laborde, R., Nasser, B., Grasset, F., Barrere, F., Benzekri, A.: A Formal Approach for the Evaluation of Network Security Mechanisms Based on RBAC policies. Electronic Notes in Theoritical Computer Science 121, 117–142 (2005)

Proposing Candidate Views for Materialization

T.V. Vijay Kumar[1], Mohammad Haider[1,2], and Santosh Kumar[1,3]

[1] School of Computer and Systems Sciences, Jawaharlal Nehru University, New Delhi-110067, India
[2] Mahatma Gandhi Mission's College of Engineering and Technology, Noida, UP-201301, India
[3] Krishna Institute of Engineering and Technology, Ghaziabad, UP-201206, India

Abstract. View selection concerns selection of appropriate set of views for materialization subject to constraints like size, space, time etc. However, selecting optimal set of views for a higher dimensional data set is an NP-Hard problem. Alternatively, views can be selected by exploring the search space in a greedy manner. Several greedy algorithms for view selection exist in literature among which HRUA is considered the most fundamental. HRUA exhibits high run time complexity primarily because the number of possible views that it needs to evaluate is exponential in the number of dimensions. As a result, it would become infeasible to select views for higher dimensional data sets. The Proposed Views Greedy Algorithm (PVGA), presented in this paper, addresses this problem by selecting views from a smaller set of proposed views, instead of all the views in the lattice as in case of HRUA. This would make view selection more efficient and feasible for higher dimensional data. Further, it was experimentally found that PVGA trades significant improvement in time to evaluate all views with a slight drop in the quality of views selected for materialization.

Keywords: Materialized View Selection, Greedy Algorithm.

1 Introduction

The advent of Internet has resulted in enormous amount of data being generated rapidly and continuously by operational data sources spread across the globe. This data, which is of utmost importance for decision making, is usually heterogeneous and distributed and therefore needs to be integrated in order to be accessed. There are two approaches to access this data namely the lazy or on-demand approach and the eager or in-advance approach [20]. Since decision making requires historical data in an integrated form, the eager or in-advance approach is used. This approach is followed in the case of data warehouse [20].

A data warehouse is a repository that contains subject-oriented, integrated, time-variant and non-volatile data to support decision making [9]. The data in a data warehouse grows continuously with time. Most of the queries for decision making are analytical and complex, and it may take data warehouse hours or even days to provide answers to such queries. That is, query response time of analytical queries are high. This response time can be reduced by creating materialized views [12].

S.K. Prasad et al. (Eds.): ICISTM 2010, CCIS 54, pp. 89–98, 2010.

A materialized view is a view that contains data along with its definition. It stores pre-computed and summarized information with the aim of reducing the response time for analytical queries. This necessitates the materialized view to contain relevant and required information. The selection of such information is referred to as views selection problem [5], which is the focus of this paper.

View selection deals with selecting appropriate set of views for efficient query processing. View selection problem has been formally defined as "Given a database schema D, storage space B, Resource R and a workload of queries Q, choose a set of views V over D to materialize, whose combined size is at most B and resource requirement is at most R" [5]. The space consideration translates the view selection problem as an optimization problem that is NP-Complete [8]. Alternatively, the views can be selected empirically or heuristically.

An empirical way to select the views is based on monitoring the queries posed by users [4, 10, 18]. It considers parameters like query frequency or data volume to evaluate and materialize a set of views [16]. Pattern of user queries posed on the data warehouse serve as a valuable statistics for selecting materialized views. On the other hand, heuristic based approaches prune the search space to select the best set of views to materialize that can improve the average query response time. The commonly used heuristics like greedy, evolutionary etc. provide a reasonably good solution for selecting views to materialize for many real world problems. This paper focuses on the greedy based selection of materialized views.

The greedy based approach, at each step, selects the most beneficial view that fits within the available space for materialization [7, 8]. The algorithm in [8], referred to as HRUA hereafter in this paper, is considered as the most fundamental greedy based view selection algorithm. This algorithm is based on linear cost model where the cost is in terms of size of the view. HRUA is shown to achieve a solution lower bound of 63% of the optimal solution [8]. The algorithm in [6] extended HRUA by considering index along with the size of the view to select views for materialization under space constraint. The algorithm in [4] improves the process of view selection by reducing the size of data cube based on user specified relevant queries. The views are then greedily selected from this reduced space. An algorithm Pick By Size is proposed in [15], which selects views in increasing order of their size. This algorithm assumes aggregate queries on all views to be of equal likelihood. An algorithm that uses update and access frequency of views to select views to be materialized is proposed in [18]. This algorithm does not take into consideration space constraints. An approach presented in [1] selects materialized views and indexes from the table-sets for large workloads of queries. The algorithm first identify interesting set of tables for the given workload of queries and then use them to select traditional indexes and materialized views that answer multiple queries in the workload. A general framework using an AND/OR DAG representation for the views selection under user oriented and system oriented constraints is proposed in [17]. This approach considers both static and dynamic issues of data warehouse design. A scalable solution to view selection problem was presented as Polynomial Greedy Algorithm in [11]. The algorithm selects views for materialization in polynomial time. In [7], a polynomial time greedy heuristic is proposed that selects views for OR view graphs, AND view graphs and AND-OR view graphs under maintenance and space constraints. A hybrid approach is presented in [14], which categorizes all the views into static views and dynamic views. This approach selects views from the static set that persist over multiple queries whereas views from the dynamic set are computed on the fly. The approach proposed in [13] uses the dependent relationships among views, and the probability of change to these relationships, to select views to materialize. The algorithm does not take into consideration query evaluation cost and space constraints. In [2], a clustering based approach for view selection is proposed that groups closely related workload queries into set of clusters and views are greedily selected from each of these clusters. In [3], a dynamic approach has been proposed that considers view and index selection. The approach has been applied for isolated materialized view selection, isolated index selection and joint materialized view and index selection wherein index-view interaction is incorporated in its cost model. The approach uses data mining to select views for materialization.

Most of the greedy based view selection algorithms are focused around HRUA, which selects top-T views from a multidimensional lattice. This algorithm considers cost in terms of size of the view. This cost is then used to compute the benefit of the view. The benefit value of a view is computed as the product of its size difference, with its smallest nearest materialized ancestor view and the number of its dependent views. HRUA initially assumes the root view of the lattice to be materialized. In each of the iterations, HRUA evaluates the benefit value of each non-selected view and selects among them the most beneficial view for materialization. This continues till a pre-specified number of views i.e. T views are selected. Since the number of views is exponential with respect to the number of dimensions, the number of views requiring evaluation of benefit values will increase with increase in dimension. As a result, it would become infeasible for HRUA to select views for higher dimensional data sets. This problem has been addressed by the Proposed Views Greedy Algorithm (PVGA) presented in this paper. PVGA selects views in two phases namely Proposed Views Set (PVS) generation and View Selection. PVGA selects views from a smaller set of PVS instead of all possible views, as in HRUA. This would make view selection more efficient and feasible for higher dimensional data.

The paper is organized as follows: the proposed algorithm PVGA is given in section 2. Section 3 gives an example-based comparison of HRUA and PVGA followed by their experiment-based comparison in section 4. Section 5 is the conclusion.

2 Algorithm PVGA

HRUA, discussed above, has an exponential runtime complexity primarily because in each of its iterations, benefit values of nearly all views of the lattice have to be evaluated. This exponentially high runtime complexity makes selection of views using HRUA infeasible for higher dimensional data sets. The proposed algorithm PVGA addresses this scalability problem by selecting views from a proposed set of views i.e. PVS, instead of all the views in the lattice, as in HRUA. The proposing of views into PVS is based on a heuristic defined below:

"For a view, its smallest sized parent view is proposed for view selection"

The basis of this heuristic is that the smallest sized parent view, among all parent views, will have maximum benefit with respect to the views above it. Thus, proposing the smallest sized parent view would maximize the benefit and will have a high likelihood of being selected for materialization. To understand the basis of this heuristic used for proposing views, consider the lattice shown in Fig. 1. The size of each view (in million rows) is given alongside the node in the lattice.

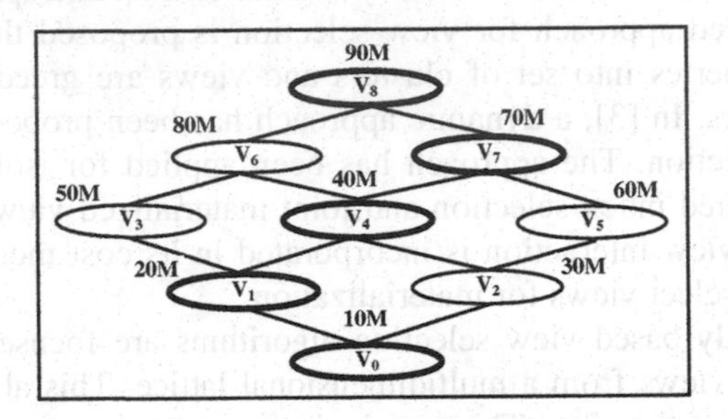

Fig. 1. Lattice

Now as per the heuristic, for view V_0, the parent view V_1 having the smallest size is proposed and added to PVS. The view V_1, being smaller in size, than the other view V_2 at the same level of the lattice, will have a higher likelihood of being materialized, as it will have maximum benefit in respect of views above it. Further, for view V_1, smallest sized parent view V_4 is added to PVS. Again it can be seen that view V_4 is the smallest in size among all the parent views of V_1 and therefore has a high likelihood of being materialized. In the similar manner, views are proposed for selection. The views in bold, shown in the lattice of Fig. 1, are the proposed views in PVS.

The algorithm PVGA then selects views greedily from PVS, generated using the above heuristic. The algorithm PVGA is given in Fig. 2. PVGA takes lattice of views, along with the size of each view, as input and produces top-T views as output.

The algorithm PVGA consists of two phases namely PVS generation and view selection. In the PVS generation phase, views are proposed for selection in each of the iterations, and are added to PVS. The proposing of views starts at the bottom of the lattice. In the first iteration, the bottommost view of the lattice is added to PVS. Then as per the heuristic, the smallest sized parent view of the bottommost view of the lattice, which has not yet been proposed, is proposed and added to PVS. Next, its smallest sized parent view, which has not yet been proposed, is added to PVS. This process of adding views to the PVS continues till the root view is reached. In the selection phase, the benefit of each view, in PVS, is computed as computed in HRUA and the most beneficial view among these is selected. The selected view is then removed from PVS. The next iteration again starts from the bottom of the lattice by proposing the smallest sized parent view, which has not yet been proposed, and is added to PVS. This process is continued, as in the first iteration, till the root view is reached. The most beneficial view is then selected for materialization. The iterations continue in a similar manner till top-T views are selected for materialization.

It can be observed that using the heuristic, the views are proposed in a greedy manner, where at each step, starting from the bottom of the lattice, the view having the smallest sized parent view is proposed for selection. This heuristic is similar to the one in [19], but proposes candidate views for selection.

```
Input: Lattice of views L
Output: top-T views V_T
Method:
L = {L_max, L_max -1,.....,0} where L is the set of Levels in the lattice L_max is the root level and 0 is the
bottommost level of the lattice
View Proposal Set (PVS) is a set of views proposed for materialization; PVS = ϕ
V_T is the set of top-T views; V_T = ϕ
VS is the number of views selected; VS=0
  While (VS ≤ T)
  Begin
  1. // Compute the proposed view set PVS
     Level L=0
     View set V=ϕ
     While (level(V) < L_max - 1)
     Begin
        V'={nodes at level L}
        If |V'| =1 then
             V = V'
        else
             Find set of immediate parent views Vp in V' of V such that Vp ∉ PVS ∧ Vp ∉ V_T i.e.
                   V_p = V' - PVS - V_T
             Find minimum sized view V in V_p
                   V = { v | v ∈ V_p ∧ size(v) = Min_{v'∈V_p} (Size(v')) }
             If |V| > 1 then
                   Select any one element from V say V"
                        V = {V"}
             end if
        end if
        PVS = PVS ∪ V
        L=L+1
     End While
  2. // Greedy selection of a view from PVS
     Compute the benefit of each view V in PVS as computed in HRUA i.e.
     Benefit(V) = (SizeNMA(V) – Size(V)) × Dependents(V)
     Select the most beneficial view say V_b i.e.
        V_b = { v | v ∈ PVS ∧ Benefit(v) = Max_{v'∈PVS} (Benefit(v')) }
     If |V_b| > 1 then
        Select any one element from V_b say V_b'
        V_b = {V_b'}
     end if
     V_T = V_T ∪ V_b
     PVS=PVS - V_b
     VS = VS + 1
  End While
  3. Return top-T views V_T
```

Fig. 2. Algorithm PVGA

The algorithm PVGA selects views from a small PVS instead of all possible views, as in HRUA. As a result, lesser number of views would require evaluation for benefit values thereby making view selection efficient and feasible for higher dimensional data sets.

3 Examples

Consider the following three examples where top-3 views are to be selected, using HRUA and PVGA, from a 3-dimensional lattice. The size of each view (in million rows) is given alongside the node in the lattice. Considering the lattice in each of the three examples, HRUA and PVGA assume view ABC as materialized, as queries on it cannot be answered by any other views in the lattice. If ABC is the only view materialized, then queries on other views in the lattice will be evaluated using ABC, thereby incurring a cost of ABC i.e. 50 million rows for each view resulting in total cost of evaluating all the views, referred to as Total View Evaluation Cost (TVEC), as 400 million rows.

The selection of top-3 views using HRUA and PVGA, from the lattice of Fig. 3(a), is given in Fig. 3(b) and Fig. 3(c) respectively.

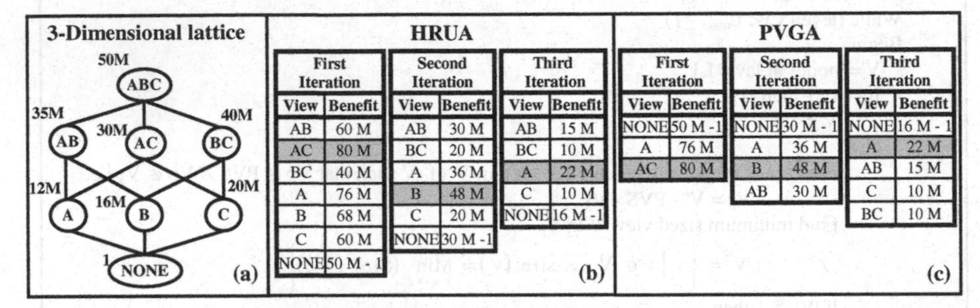

Fig. 3. Selection of top-3 views using HRUA and PVGA

The HRUA and PVGA select the same top-3 views i.e. AC, B and A. While selecting the top-3 views, the Total Benefit Computations (TBC) required by HRUA are 18 whereas PVGA required only 12. This difference in TBC would increase with the increase in dimension and would become significantly high for higher dimensional data sets. This is due to the fact that PVGA selects top T views from a relatively smaller set of proposed views, instead of all the views of the lattice, as in the case of HRUA. As a result, the total time to evaluate all the views, referred to as Total View Evaluation Time (TVET), would be less in case of PVGA, as compared to HRUA i.e. PVGA would be more efficient in selecting views. Further, materializing AC, B and A along with ABC reduces TVEC from 400 million rows to 250 million rows.

PVGA need not always select the same views as HRUA. As an example consider the selection of top-3 views, from the lattice of Fig. 4(a), using HRUA and PVGA. These selections are shown in Fig. 4(b) and Fig. 4(c) respectively.

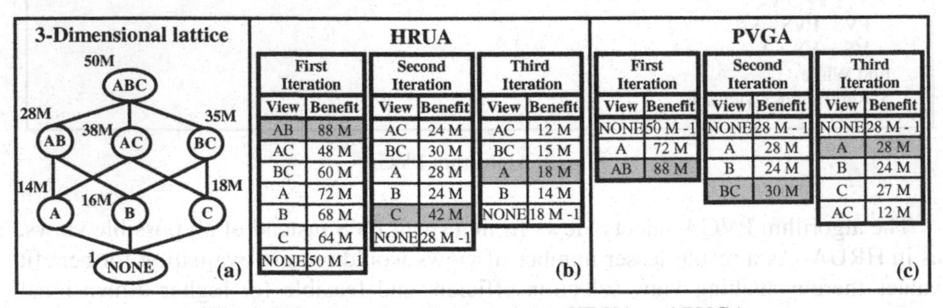

Fig. 4. Selection of top-3 views using HRUA and PVGA

HRUA selects AB, C and A as the top-3 views whereas PVGA selects AB, BC and A as the top-3 views. Although the views selected by the two algorithms are not the same, the heuristic followed by the PVGA is fairly good, as two out of the top-3 views selected by PVGA, are identical to those selected by HRUA. The reason behind this deviation is that in each of the iteration HRUA scans the entire search space in order to identify the beneficial view whereas PVGA performs scan on a very limited search space, over only the proposed views. Further, views selected using HRUA reduce the TVEC from 400 million rows to 252 million rows whereas PVGA is able to reduce this TVEC from 400 million rows to 254 million rows. That is, TVEC of HRUA is less than that of PVGA showing that HRUA is able to select slightly better quality views than PVGA.

PVGA can also select views that are of better quality than those selected using HRUA. As an example consider the selection of top-3 views, from the lattice of Fig. 5(a), using HRUA and PVGA. These selections are shown in Fig. 5(b) and Fig. 5(c) respectively.

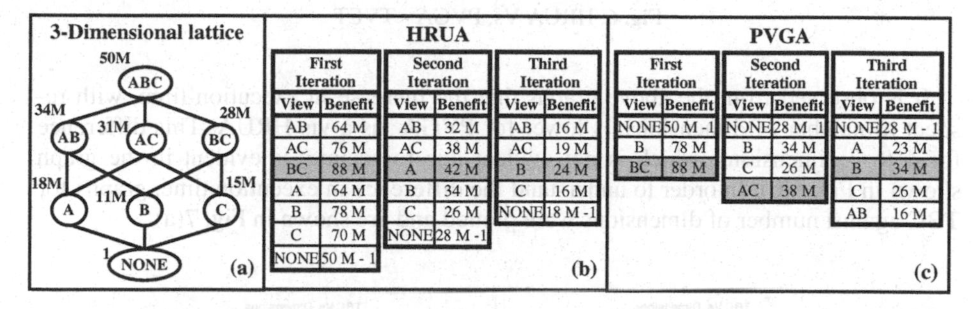

Fig. 5. Selection of top-3 views using HRUA and PVGA

HRUA selects BC, A and B as the top-3 views as against BC, AC and B selected by PVGA. The TVEC due to views selected using HRUA is 246 and PVGA is 240. PVGA selects views which reduce TVEC cost more than those selected using HRUA. Thus, it can be said that PVGA can select better quality views in comparison to HRUA.

From the above, it can be inferred that the heuristic used by PVGA is a fairly good as the views selected using PVGA are as good as the views selected using HRUA.

In order to compare the performance of PVGA with respect to HRUA, both the algorithms were implemented and run on data sets with varying dimensions. The experiment based comparisons of HRUA and PVGA are given next.

4 Experimental Results

The HRUA and PVGA algorithms were implemented using JDK 1.6 in Windows-XP environment. The two algorithms were compared by conducting experiments on an Intel based 2 GHz PC having 1 GB RAM. The comparisons were carried out on parameters like TVET, TBC and TVEC. The tests were conducted by varying the number of dimensions of the data set from 4 to 10. The experiments were performed for selecting top-10 views for materialization.

First, graphs were plotted to compare HRUA and PVGA algorithms on TVET (in milliseconds) against the number of dimensions. The graphs are shown in Fig. 6(a).

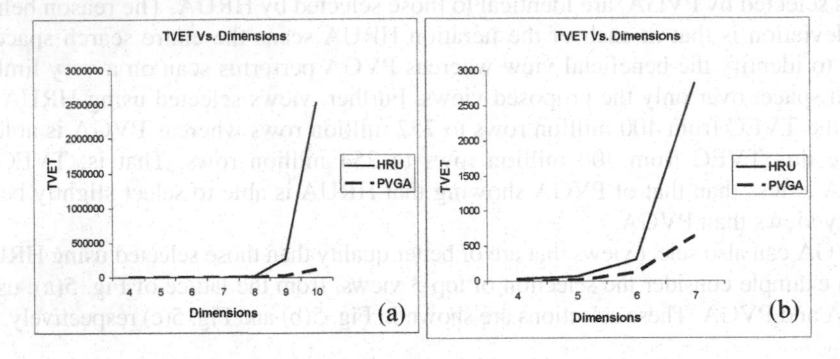

Fig. 6. HRUA Vs. PVGA - TVET

It is observed from the above graphs that the increase in execution time, with respect to number of dimensions, is lower for PVGA vis-à-vis HRUA. This difference for 4 to 7 dimensions, which is not evident due to scaling, is evident in the graph shown in Fig. 6(b). In order to understand the difference in execution time, graphs for TBC against number of dimensions were plotted and are shown in Fig. 7(a).

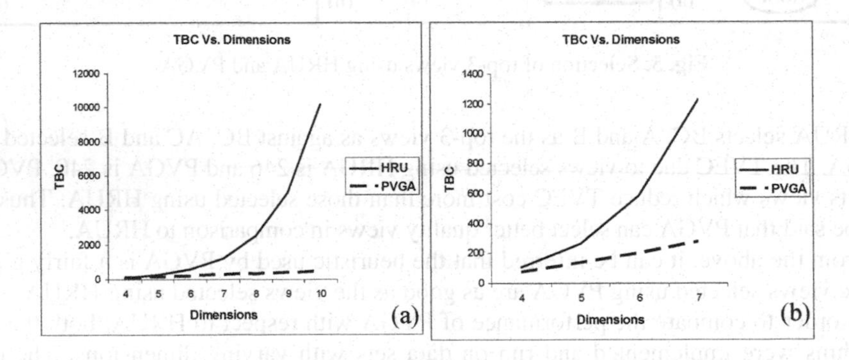

Fig. 7. HRUA Vs. PVGA - TBC

The graphs show that PVGA requires fewer TBC, in comparison to HRUA, and this is also visible for dimensions 4 to 7 as shown in Fig. 7(b). This may be due to the fact that in PVGA, views are selected from a smaller set of proposed views instead of all the views in the lattice. This difference may be the cause of difference in TVET between PVGA and HRUA. The difference in execution time would become more significant for higher dimensional data sets as HRUA would require significantly more TBC than PVGA. As a result, for higher dimensional data sets it would become infeasible to select views using HRUA whereas PVGA would continue to select views for materialization.

Next, in order to ascertain the quality of the views being selected by PVGA, in comparison with those being selected by HRUA, graphs were plotted capturing the TVEC against the number of dimensions. The graphs are shown in Fig. 8(a).

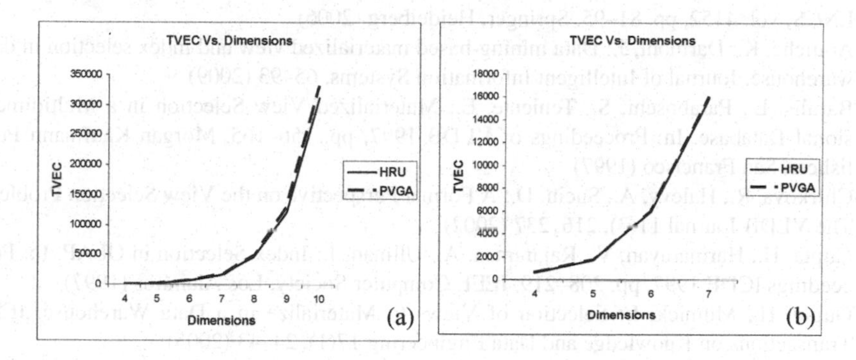

Fig. 8. HRUA Vs. PVGA - TVEC

It can be observed from the graphs that TVEC of PVGA is slightly more than that of HRUA. This difference is almost negligible for dimensions 4 to 7 as shown in Fig. 8(b). This small difference shows that PVGA selects views which are almost same in quality as selected by HRUA.

It can be reasonably inferred from the above graphs that PVGA trades significant improvement in TVET with a slight drop in TVEC of views selected for materialization.

5 Conclusion

This paper focuses on greedy based view selection algorithm with emphasis on HRUA. HRUA exhibits high run time complexity. One reason for it is that the number of possible views that can be materialized, and that it needs to evaluate, is exponential in the number of dimensions. Algorithm PVGA, proposed in this paper, addresses this problem by first proposing a set of views, among all the views in the lattice, and then greedily selecting views from these proposed views. PVGA selects beneficial views from a smaller set of proposed views instead of all the views in the lattice. As a result the search space from which views are selected is considerably smaller in the case of PVGA, in comparison to HRUA. This enables PVGA to select views efficiently for higher dimensional data sets.

Further, to validate the claim of PVGA, experiment based comparison of PVGA and HRUA was carried out on parameters like TVET, TBC and TVEC. The results showed that PVGA, in comparison to HRUA, was found to achieve significant improvement in TVET at the cost of a slight drop in the TVEC of the views selected for materialization. This shows that PVGA trades significant improvement in total time to evaluate all the views with a slight drop in quality of views selected for materialization.

References

1. Agarwal, S., Chaudhuri, S., Narasayya, V.: Automated Selection of materialized views and indexes for SQL Databases. In: Proceedings Of VLDB, pp. 496–505 (2000)
2. Aouiche, K., Jouve, P.-E., Darmont, J.: Clustering-Based Materialized View Selection in Data Warehouses. In: Manolopoulos, Y., Pokorný, J., Sellis, T.K. (eds.) ADBIS 2006. LNCS, vol. 4152, pp. 81–95. Springer, Heidelberg (2006)
3. Aouiche, K., Darmont, J.: Data mining-based materialized view and index selection in data warehouse. Journal of Intelligent Information Systems, 65–93 (2009)
4. Baralis, E., Paraboschi, S., Teniente, E.: Materialized View Selection in a Multidimensional Database. In: Proceedings of VLDB 1997, pp. 156–165. Morgan Kaufmann Publishers, San Francisco (1997)
5. Chirkova, R., Halevy, A., Suciu, D.: A Formal Perspective on the View Selection Problem. The VLDB Journal 11(3), 216–237 (2002)
6. Gupta, H., Harinarayan, V., Rajaraman, A., Ullman, J.: Index Selection in OLAP. In: Proceedings ICDE 1997, pp. 208–219. IEEE Computer Society, Los Alamitos (1997)
7. Gupta, H., Mumick, I.: Selection of Views to Materialize in a Data Warehouse. IEEE Transactions on Knowledge and Data Engineering 17(1), 24–43 (2005)
8. Harinarayan, V., Rajaraman, A., Ullman, J.: Implementing Data Cubes Efficiently. In: Proceedings of SIGMOD 1996, pp. 205–216. ACM Press, New York (1996)
9. Inmon, W.H.: Building the Data Warehouse, 3rd edn. Wiley Dreamtech, Chichester (2003)
10. Lehner, R., Ruf, T., Teschke, M.: Improving Query Response Time in Scientific Databases Using Data Aggregation. In: Proceedings of 7th International Conference and Workshop on Databases and Expert System Applications, September 1996, pp. 9–13 (1996)
11. Nadeau, T.P., Teorey, T.J.: Achieving scalability in OLAP materialized view selection. In: Proceedings of DOLAP 2002, pp. 28–34. ACM Press, New York (2002)
12. Roussopoulos, N.: Materialized Views and Data Warehouse. In: 4th Workshop KRDB 1997, Athens, Greece (August 1997)
13. Serna-Encinas, M.T., Hoya-Montano, J.A.: Algorithm for selection of materialized views: based on a costs model. In: Proceeding of eighth International conference on Current Trends in Computer Science, pp. 18–24 (2007)
14. Shah, B., Ramachandran, K., Raghavan, V.: A Hybrid Approach for Data Warehouse View Selection. International Journal of Data Warehousing and Mining 2(2), 1–37 (2006)
15. Shukla, A., Deshpande, P., Naughton, J.: Materialized View Selection for Multidimensional Datasets. In: Proceedings of VLDB 1998, pp. 488–499. Morgan Kaufmann Publishers, San Francisco (1998)
16. Teschke, M., Ulbrich, A.: Using Materialized Views to Speed Up Data Warehousing. Technical Report, IMMD 6, Universität Erlangen-Nümberg (1997)
17. Theodoratos, D., Bouzeghoub, M.: A general framework for the view selection problem for data warehouse design and evolution. In: Proceedings of DOLAP, pp. 1–8 (2000)
18. Uchiyama, H., Ranapongsa, K., Teorey, T.J.: A Progressive View Materialization Algorithm. In: Proceeding of 2nd ACM International Workshop on Data Warehousing and OLAP, Kansas City Missouri, USA, pp. 36–41 (1999)
19. Vijay Kumar, T.V., Ghoshal, A.: A Reduced Lattice Greedy Algorithm for Selecting Materialized Views. In: ICISTM 2009, March 12-13. CCIS, vol. 31, pp. 6–18. Springer, Heidelberg (2009)
20. Widom, J.: Research Problems in Data Warehousing. In: Proceedings of ICIKM, pp. 25–30 (1995)

A Fuzzy Reasoning Framework for Resolving Spatial Queries through Service Driven Integration of Heterogeneous Geospatial Information

Indira Mukherjee and S.K. Ghosh

School of Information Technology
Indian Institute of Technology, Kharagpur-721302, India
indira.mukherjee.23@gmail.com, skg@iitkgp.ac.in

Abstract. In today's decision support system, geospatial information is becoming an important component due to its capability of spatial referencing and visual analysis in form of map. Geospatial data are often collected and maintained by different organizations and has become a major bottleneck for spatial integration. Further, the properties and the relationships among the data are not always crisply defined. Thus in order to achieve more accuracy, the uncertainty associated with the spatial information should be captured. This paper proposes a methodology that integrates these diverse complex data using fuzzy geospatial web service. It starts with a fuzzy geospatial data modeling technique that captures the properties of the geospatial data, relationship among them and uncertainty associated with these properties and relationships. The issue of interoperability and integration of diverse data sets are realized through an Enterprise GIS framework which uses geospatial web services. The queries related to decision support processes are analyzed using fuzzy reasoning to enhance the accuracy.

1 Introduction

With increasing availability of hi-end computing systems and internet bandwidth at low cost, the geospatial information are being extensively used for various decision making processes. A geographic information system (GIS) or geographical information system captures, stores, analyzes, manages, and presents data that is linked to geographically referenced information. Geospatial data have some reference to location on earth and represents some geographic phenomenon (known as themes).For example, like *road network*, *land use map*, *population map*, *drainage network* etc.

Geospatial repository is an organized collection of thematic layers on several perspectives e.g. transport, land use, drainage etc. Each of these layers provides the corresponding thematic information about a region. Information of these layers are collected and maintained by various organizations and stored in their proprietary formats. Thus, collected data can be in different formats [1]. For sharing these diverse data sets across various organizations, it is necessary to have a

S.K. Prasad et al. (Eds.): ICISTM 2010, CCIS 54, pp. 99–110, 2010.

data models defined for the individual repository [2] [3] [4]. The Open Geospatial Consortium (OGC) [5] and ISO/TC 211 [6] are two geospatial modeling standards. Further, the data models have to be harmonized for sharing geospatial data enterprise-wide. This leads to requirement of *metadata* information, which describes the structure and the content of data to be exchanged [7]. Once the metadata of a repository is standardized, it can be a basis of data sharing.

The *geospatial data objects* are described by some set of attributes that captures the description of the *geospatial object*. Each *geospatial data object* is associated with one or more metadata that describes the format and the content. The metadata also represented as application schema. In the proposed methodology, the application schema is designed based on *Object Oriented Design* concept [8]. The attribute or properties of geospatial objects include spatial as well as non-spatial information about the geospatial object, which is basically an instance of a class in object oriented design. The geospatial information can be modeled using UML. An application schema can be derived from the UML class diagram, where the properties of objects are represented by class and association between classes are used to represent the relationship between the objects. The structure and content of the data that is to be exchanged is described in the application schema.

Further, the real world is always complex and full of uncertainties. The properties of geospatial object and relationship among them may be ambiguous. The standard geospatial data models do not consider the fuzziness of the geographic objects. Thus to improve the accuracy of the data sharing system fuzzy logic concepts can be used in developing the application schema. Attempts have been made in [9] [10] where the uncertainty in the objects has been mapped in the UML model using fuzzy class and fuzzy association. The concept of fuzzy class, fuzzy association, and fuzzy attributes are incorporated in geospatial data model. The fuzzy classes are used to represent uncertain instance of an object, the fuzzy attributes can be used to represent the uncertain properties of the objects and fuzzy association is used to represent the uncertain relationship between the objects.

Finally, service-based approach is used to integrate and operate on geospatial data of heterogeneous repositories from different data sources over the network. A service oriented architecture (SOA) tries to construct a distributed, dynamic, flexible and reconfigurable service system over the internet that could meet most of the user information requirements. The SOA for geospatial dataset approach uses GML (Geographic Markup Language), a variant of XML, and adheres to OGC service standards. The interoperability between the heterogeneous repositories and processing of queries related to decision making process can be realized through proper deployment of geospatial web services. The fuzzy reasoning can be integrated with this service framework for more accurate analysis and decision.

This paper proposes a fuzzy reasoning framework for integration of heterogeneous spatial data repositories which may facilitate decision making processes. The major contribution lies in the fuzzy modeling of geospatial data and its usage in the service framework for better query processing. The rest of the paper in organized as follows. Section 2 presents the overview of the proposed fuzzy geospatial system. The fuzzy geospatial modeling concepts has been given in section 3. Section 4 presents an overview of Enterprise GIS ($E-GIS$) framework for integrating on geospatial services. Section 5 describes the incorporation of fuzzy logic in decision making process. A detailed case study, demonstrating the proposed fuzzy reasoning approach, has been presented in section 6. Finally conclusion is drawn in section 7.

2 Overview of the Fuzzy Geospatial Reasoning Framework

This section gives an overview of the proposed fuzzy logic based geospatial information retrieval. Fig 1 shows the integration of spatial repositories for answering the queries. The datasets represent the storage of different heterogeneous repositories. The core of the system is the Enterprise-GIS (E-GIS), which provide the spatial web services to integrate the datasets and resolve users' or clients' queries. The clients may vary from human users to other systems and services. The client's query is presented to the E-GIS system in service mode. The E-GIS framework searches for the required datasets pertaining to the given query and access the data through geospatial web services. The base data model of the E-GIS captures the classes and the relationships between various classes. The data models of the individual datasets reflects the classes (or the data structures) and their relationships. In order to resolve the spatial queries, the individual data models need to be appropriately mapped with the base model. The corresponding data model of the query is matched with the base data model of the E-GIS to check whether the query is resolvable or not. Finally, the query is resolved by accessing the corresponding data from the datasets. In practice, the E-GIS

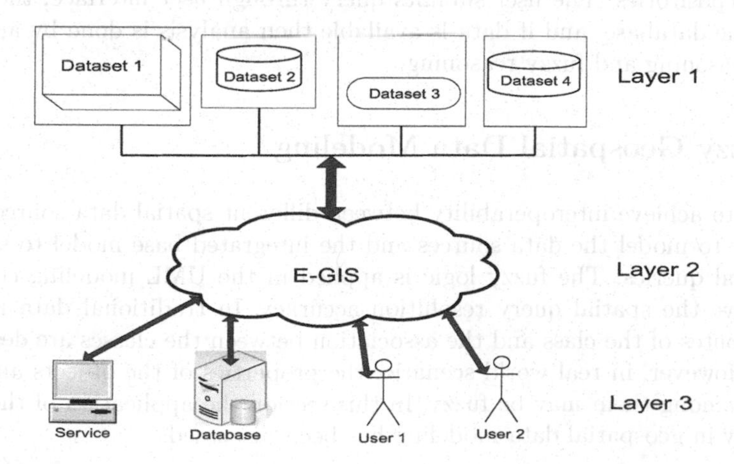

Fig. 1. Fuzzy Geospatial Information Integrating Framework

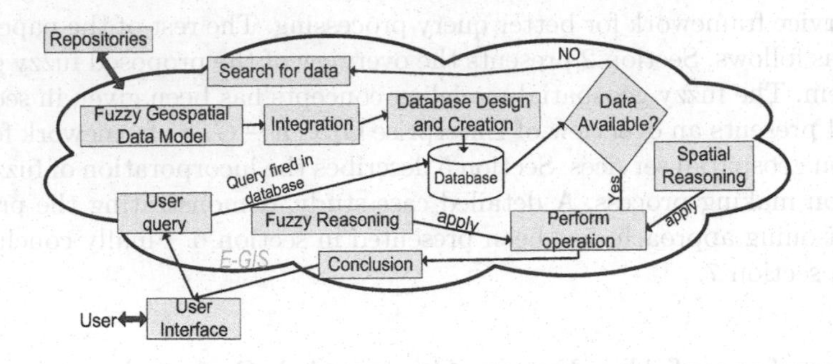

Fig. 2. Operation of the Enterprise-GIS system

framework usually contains a spatial database engine to perform various spatial operations.

The E-GIS framework helps in integration and interoperability between various heterogeneous datasets through OGC [5] complaint spatial web services. But in real world systems, the properties of the spatial and thematic data are often not well defined. Real world systems deal with many vital and complex decision making queries where these uncertainty related to geospatial and thematic data needs to be addressed. For example, queries like - "*finding the suitable location for establishing new schools which are near road but away from drainage network*". This sort of queries has lots of "fuzziness" or "uncertainties" and cannot be resolved by standard query resolving framework.

The proposed E-GIS system attempts to address these fuzziness or uncertainties in data, relationships and the queries. The operations preformed by E-GIS system are described in Fig 2. The data are integrated from diverse data repositories based on *fuzzy geospatial data model*; a spatial database is being designed and established based on the *fuzzy geospatial data model* and data retrieved from various repositories. The user submits query through user interface, the query fetches the database, and if data is available then analysis is done by applying spatial reasoning and fuzzy reasoning.

3 Fuzzy Geospatial Data Modeling

In order to achieve interoperability between different spatial data sources, it is necessary to model the data sources and the integrated base model to support the spatial queries. The fuzzy logic is applied in the UML modeling concepts to improve the spatial query resolution accuracy. In traditional data models, the attributes of the class and the association between the classes are described crisply. However, in real world scenario, the properties of the objects and relationship among them may be fuzzy. In this section the application of the fuzzy set theory in geospatial data modeling has been discussed.

3.1 Fuzzy Classes

The instance of a fuzzy class refers to an object type that contains fuzzy information in the description of attributes or the relations of object type. The fuzziness can be in the attributes of the class, attribute value of the class, relationship between the classes. In order to describe fuzziness in a class the keyword "Fuzzy" is used. The prefix "*fuzzy*" is added before attributes that have fuzzy values. To describe the possibility of the attribute in a class the attribute is followed by pair of words *WITH mem DEGREE* and $0 \leq mem \leq 1$. The *mem* refers to degree of membership that is the extent to which the attribute is the member of a class. For example (refer figure 3), there is a class named *Class 1* and an attribute *Attr b*. *Class 1* may or may not include the *Attr b*. Assume that we have *Attr b WITH* μ *DEGREE* means that *Attr b* belongs to *Class 1* with μ degree. Generally, the attribute or relationship will not be declared if the when its degree is 0. Attribute *Attr c* takes a fuzzy value therefore it is denoted by prefix FUZZY.

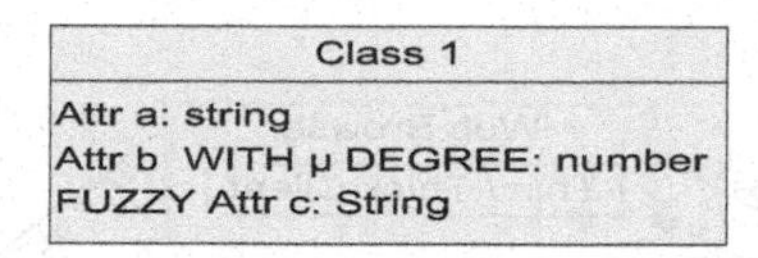

Fig. 3. Fuzzy Class

3.2 Fuzzy Association

The relationship among the objects in the data model can be fuzzy. In generic UML data model, the relationship among the objects is represented by association relationship. When the relationships are not well defined, the fuzzy association can be used. In such cases, the role name of the object is fuzzily defined and the association between two instances is connected with degree membership. For example (refer Fig 4), an instance of *Class 1* may or may not have spatial relationship with another geospatial object which is an instance of *Class 2*. The association relationship is written as "*Association Name WITH 0.6 DEGREE*", which means that *Class 1* is associated with *Class 2* with degree μ. The linguistic terms like, *very, mostly, slightly* can also be used to define relationship among the objects. Each linguistic term should be pre-defined, based on the degree of membership.

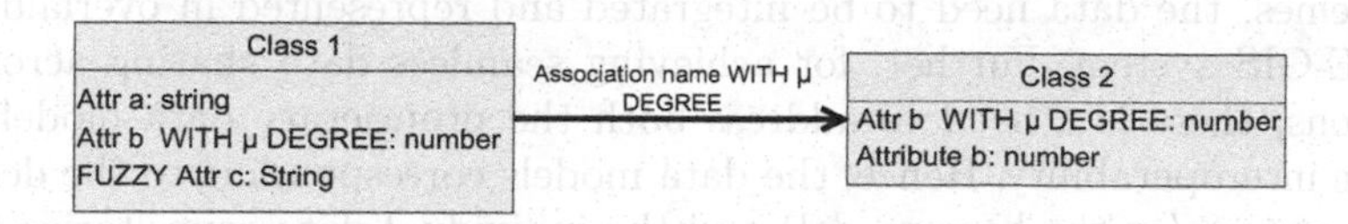

Fig. 4. Fuzzy Association

4 Enterprise GIS Framework for Geospatial Web Services

The geospatial information is usually collected and maintained by diverse organization in a distributed fashion. This necessitates the need for a flexible approach that will facilitate sharing of information which can ensure interoperability. The service-oriented architecture (SOA) tries to construct a distributed, dynamic, flexible, and re-configurable service system over Internet that can meet many different users' information requirements. Services are self-contained, stateless, and do not depend on the state of other services. The proposed Enterprise GIS (E-GIS) framework aims at integrating the independent geospatial datasets and service providers distributed over the web.The service-oriented GIS model uses OGC complaint service standards, namely, Web Feature Services (WFS), Web Map Services (WMS), Web Registries Services (WRS) etc. The specifications allow seamless access to geospatial data in a distributed environment [11] using spatial web service technology. The overall service base integration framework of the E-GIS system is shown in Fig 5.

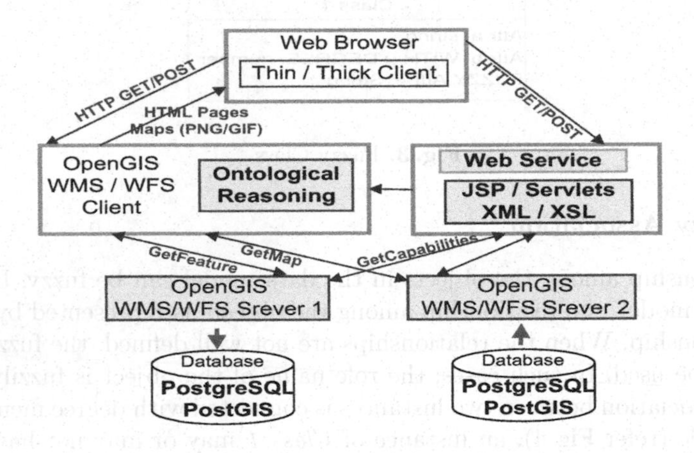

Fig. 5. Service based Geospatial Computing

The geospatial information of a region can be categorized into different themes (or thematic layers) like land-use, settlements, transport etc. For any decision making process, these themes act as information sources, which are stored and collected independently by various organizations. For queries involving one or more themes, the data need to be integrated and represented in overlaid form by the E-GIS system. Further, for achieving seamless data sharing across organizations, there is a need to address both the proprietary data models and syntactic interoperability. Hence, the data models corresponding to the decision support system (or the base model) and the individual data repositories are to be generated and advertized using web services. The appropriate data models or schemas make the backbone of the data sharing and interoperability framework.

5 Fuzzy Reasoning

The E-GIS system accepts client's query, process by applying some predefined logic and generates output in form of map and textual output. The predefined logics are framed assuming that the information related to the spatial and attribute data are crisp or well defined in nature. However, in real world scenario the properties of the spatial and attribute data may be ill defined. In this work a fuzzy logic concepts has been applied in process of spatial data integration and spatial query processing.

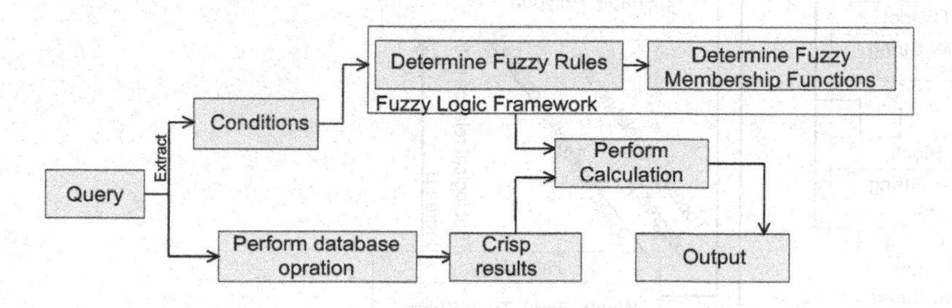

Fig. 6. Fuzzy Reasoning system

In Fig 6 the proposed fuzzy reasoning system has been described. It takes the user query as input, parses the query and then forms the fuzzy rules and fuzzy membership functions. The database operations are performed based on the user query and the spatial thematic relationship among the objects mention the user query are determined. The fuzzy logic concepts are applied to each spatial and thematic relationship and the fuzzy reasoning is carried out. The case study described in the following section explains the concept and its applicability in decision making process.

6 Case Study

In this section, a case study has been presented to demonstrate the efficacy of the proposed fuzzy reasoning system. It aims at finding the prospective location for a new educational unit (say, school) based on some selection criteria or requirements. For analyzing the requirements which involve spatial information, it is required to collect the data from various data repositories. In this case study, four repositories (or themes) have been considered, namely, administrative boundary, transport, hydrographic and educational (school locations) data. The data from different repositories are integrated through fuzzy geospatial data model. The base fuzzy data model has been developed (refer to Fig 7) conforming to the decision making system. The base model is advertized through registry service of E-GIS. The data models of the different repositories are mapped with the base model. Further, the data model maps the spatial objects in it and the relationship among the objects is defined using fuzzy association. Integrated map

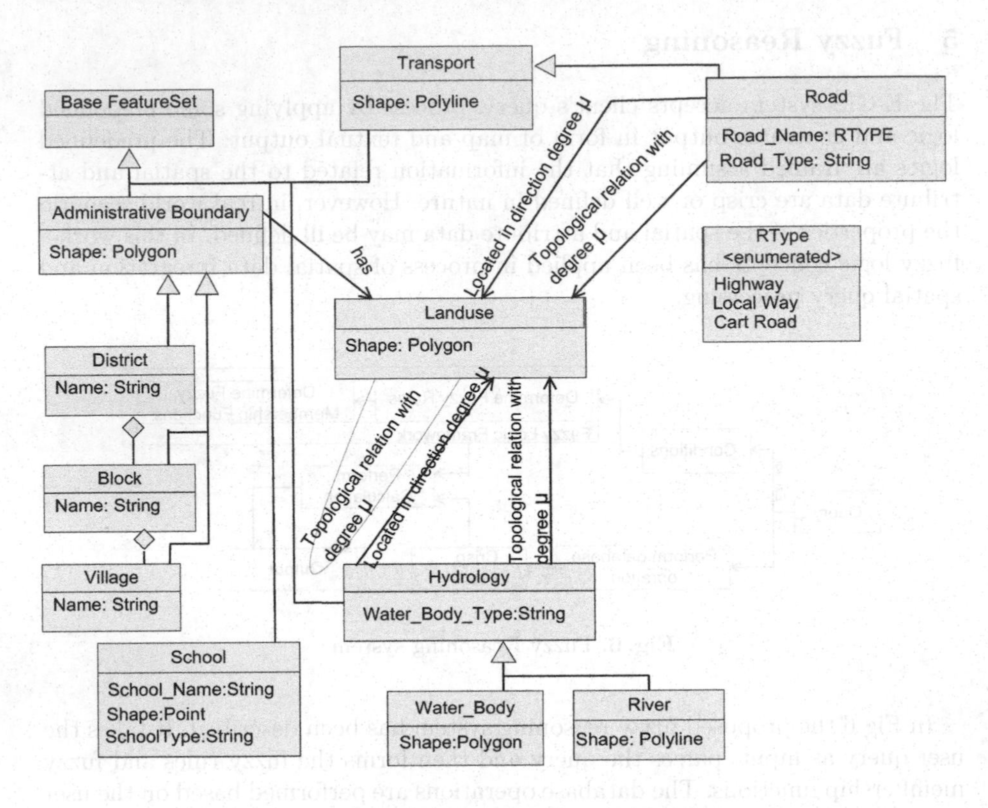

Fig. 7. Fuzzy Geospatial Data Model

is generated using the proposed E-GIS framework (refer Fig 1) and is shown in Fig 8.

Say, the process of finding the prospective location the new education unit (or school) is governed by following conditions:

- School should be at least 2000 meter away from an existing school.
- School should be at least 500 meter away from road but not more than 1000 meter.
- School should be at least 500 meter away from a drainage but not more than 1000 meter.

For the above conditions, the layers (or themes) representing educational information, transport, hydrology (or drainage) and the administrative boundary can be overlaid to generate the layer containing the prospective locations for the new school. Thus any location or point in the resultant layer satisfies the given constraints for establishing a new school. In order to find the suitability of location the fuzzy reasoning framework can be applied. The determination of a suitability value of a particular location using fuzzy reasoning has been presented in this section. This may help in indexing the locations with respect to their suitability values.Let us consider two prospective locations, namely, *Location A* and

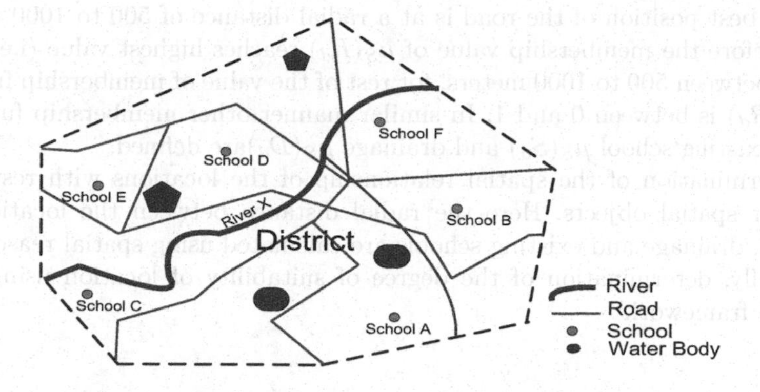

Fig. 8. Integrated Geospatial Data

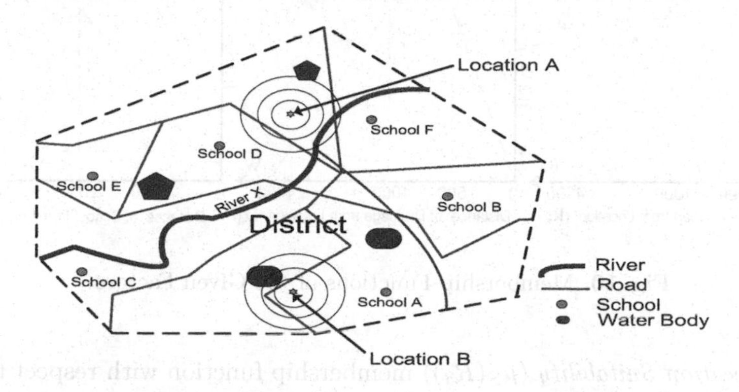

Fig. 9. Analyzing About a Location for New School

Location B, (refer Fig) and compare their suitability value for establishing a new school. A fuzzy logic concept has been applied to find out - "*to what extent (or degree) thc given locations are suitable for establishing a school*". The following steps are to be followed to determine the degree of suitability of location A.

1. Determine the fuzzy membership functions according to given conditions (refer figure 10). The universe of discourse of *Road Distance* in *Location Suitability* membership function $\mu_S(R_d)$ w.r.t *Road Distance*(R_d) is 0 to 4000 meters; i.e. the road is should not be considered beyond 4000 meters away from a school. However, the constraint states that the road should between 500 meters and 1000 meters from the school, i.e optimal distance of road from a given location is within 500 meter to 1000 meter. The membership function $\mu_S(R_d)$ w.r.t is R_d can be given as follows.
   - $\mu_S(R_d) = 0$ for $R_d = 0$
   - $\mu_S(R_d) = [0,1]$ for $0 \leq R_d \leq 500$
   - $\mu_S(R_d) = 1$ for $500 \leq R_d \leq 1000$
   - $\mu_S(R_d) = [0,1]$ for $1000 \leq R_d \leq 4000$

The best position of the road is at a radial distance of 500 to 1000 meters, therefore the membership value of $\mu_S(R_d)$ reaches highest value (i.e. 1) for Rd between 500 to 1000 meters, for rest of the value of membership function $\mu_S(R_d)$ is between 0 and 1. In similar manner other membership functions for existing school $\mu_S(S_d)$ and drainage $\mu_S(D_d)$are defined.

2. Determination of the spatial relationship of the locations with respect to other spatial objects. Here the radial distance between the location and road, drainage and existing schools are calculated using spatial reasoning.
3. Finally, determination of the degree of suitability of location using fuzzy logic framework

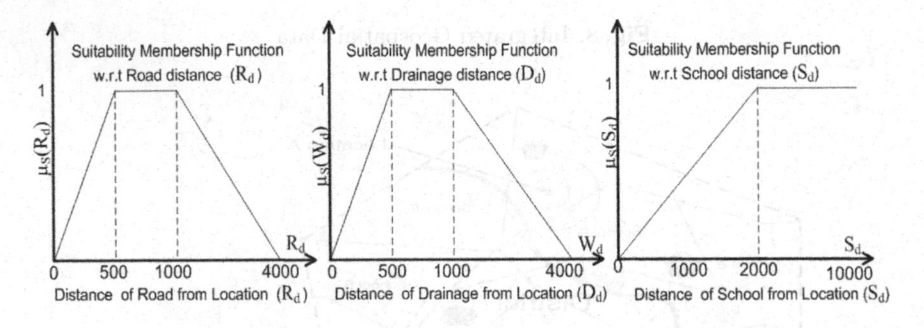

Fig. 10. Membership Functions of the Given Factors

The *Location Suitability* ($\mu_S(R_d)$) membership function with respect to *Road Distance*(R_d) is given by

$$\mu_S(\mathcal{R}_d) = \begin{cases} 1\text{-}(\mathcal{R}_d/500) & \text{if } 0 \le \mathcal{R}_d \le 500, \\ 1 & \text{if } 500 \le \mathcal{R}_d \le 1000, \\ (4000-\mathcal{R}_d)/3000) & \text{if } 1000 \le \mathcal{R}_d \le 4000, \end{cases} \tag{1}$$

The *Location Suitability* ($\mu_S(D_d)$) membership function with respect to *Drainage Distance*(D_d) is given by

$$\mu_S(\mathcal{D}_d) = \begin{cases} 1\text{-}(\mathcal{D}_d/500) & \text{if } 0 \le \mathcal{D}_d \le 500, \\ 1 & \text{if } 500 \le \mathcal{D}_d \le 1000, \\ (4000-\mathcal{D}_d)/3000) & \text{if } 1000 \le \mathcal{D}_d \le 4000, \end{cases} \tag{2}$$

The *Location Suitability* ($\mu_S(S_d)$) membership function with respect to *Drainage Distance*(S_d) is given by

$$\mu_S(\mathcal{S}_d) = \begin{cases} 1\text{-}(\mathcal{S}_d/20000) & \text{if } \mathcal{S}_d \le 2000, \\ 1 & \text{if } 2000 \le \mathcal{S}_d \le 10000, \end{cases} \tag{3}$$

The spatial relationship of the location A with respect to other spatial objects, namely road, drainage and existing schools can be calculated as follows

- Nearest road is at a distance of 800 meters from *Location A*.
- Nearest drainage is at distance of 1800 meters from *Location A*.
- Nearest school is at a distance of 4000 meters from *Location A*.

Now, considering the membership functions of each factor $\mu_S(R_d)$, $\mu_S(D_d)$ and $\mu_S(S_d)$ and the spatial relationships, the membership value can be determined as

- Rd = 800 according to Road Distance membership function $\mu_S(R_d)$ =1, 500≤Rd≤1000 therefore, $\mu_S(R_d)$ =1
- Wd= 800 according to Road Distance membership function $\mu_S(D_d) = (4000 - D_d)/3000$, $100 \leq D_d \leq 4000$, therefore, $\mu_S(D_d) = (4000 - 1800)/3000 = 0.73$.
- Similarly, $\mu_S(S_d)$=1

The product of the suitability membership value for *Location A* is determined by using product for T-norm operator in Mamdani fuzzy inference system [12].

- $\mu_S(R_d)\ \mu_S(D_d)\ \mu_S(S_d) = 10.731 = \mathbf{0.73}$

Thus, the degree of suitability for establishing the new school at *Location A* = **0.73**. Similarly, the degree of suitability for Location B can be determined as **0.9**. Hence, *Location B* is more suitable than A, though both of them satisfy the given constraints.

7 Conclusion

This paper presents a fuzzy logic framework for geospatial data modeling and spatial reasoning which may facilitate in designing an efficient decision support system. The proposed framework uses service oriented approach for integration and sharing of heterogeneous geospatial information among distributed data sources. Since uncertainties exist in spatial queries, the proposed fuzzy query resolution framework will help in resolving suck queries more accurately. A detailed case study has also been presented to demonstrate the efficacy of the system.

References

1. Worboys, M., Deen, S.: Heterogeneity in distributed geographic database. In: Proceedings of ACM SIGMOD International Conference on Management of Data, vol. 20, pp. 30–34 (1991)
2. Varies, M.: Recycling geospatial information in emergency situations: Ogc standards play an important role, but more work is needed. Directions Magazine (2005)
3. Gong, J., Shi, L., Du, D., Rolf, A.: Technologies and standards on spatial data sharing. In: Proceedings of 20th ISPRS:Geo-imagery bridging continents, vol. 34, pp. 118–128 (2004)

4. Abel, D., Gaede, V., Taylor, K., Zhou, X.: Smart: Towards spatial internet marketplaces. International Journal Geoinformatica 3, 141–164 (1999)
5. OGC: Open geospatial consortium, http://www.opengeospatial.org
6. ISO: Geographic information - Rules for application schema, Final text of CD 19109 (2001)
7. Nogueras, J., Zarazaga, F., Muro, P.: Geographic Information Metadata for Spatial Data Infrastructures Resources, Interoperability and Information retrieval, 1st edn. Springer, Heidelberg (2005)
8. Booch, G.: Object oriented design with applications. Benjamin-Cummings Publishing Company Inc. (1990)
9. Ma, Z.M., Yan, L.: Fuzzy xml data modeling with the uml and relational data models. Journal of Data & Knowledge Engineering 63, 972–996 (2007)
10. Sicilia, M., Mastorakis, N.: Extending uml 1.5 for fuzzy conceptual modeling: A strictly additive approach 3, 2234–2239 (2004)
11. Paul, M., Ghosh, S.: A service-oriented approach for integrating heterogeneous spatial data sources realization of a virtual geo-data repository. Journal of Cooperative Information System 17, 111–153 (2008)
12. Jang, J., Sun, C., Mizutani, E.: Neuro-fuzzy and Soft Computing. Pretice Hall, Upper Saddle River (1997)

Dynamic Web Service Composition Based on Operation Flow Semantics

Demian Antony D'Mello and V.S. Ananthanarayana

Department of Information Technology
National Institute of Technology Karnataka
Surathkal, Mangalore - 575 028, India
{demian,anvs}@nitk.ac.in

Abstract. Dynamic Web service composition is a process of building a new value added service using available services to satisfy the requester's complex functional need. In this paper we propose the broker based architecture for dynamic Web service composition. The broker plays a major role in effective discovery of Web services for the individual tasks of the complex need. The broker maintains flow knowledge for the composition, which stores the dependency among the Web service operations and their input, output parameters. For the given complex requirements, the broker first generates the abstract composition plan and discovers the possible candidate Web services to each task of the abstract composition plan. The abstract composition plan is further refined based on the Message Exchange Patterns (MEP), Input/Output parameters, QoS of the candidate Web services to produce refined composition plan involving Web service operations with execution flow. The refined composition plan is then transferred to generic service provider to generate executable composition plan based on the requester's input or output requirements and preferences. The proposed effective Web service discovery and composition mechanism is defined based on the concept of functional semantics and flow semantics of Web service operations.

Keywords: Abstract Composition Plan, Functional Semantics, Flow Semantics, Complex Request, Broker.

1 Introduction

The success of Web service technology lies in the effective & efficient dynamic discovery and compositions of advertised Web services. A Web service is an interface, which describes a collection of operations that are network accessible through standardized XML messaging [1]. At present, the Web service architecture is based on the interactions between *three* roles i.e. service provider, service registry and service requester. The interactions among them involve publish, find and bind operations [1]. Web service discovery is the mechanism, which facilitates the requester, to gain an access to Web service descriptions that satisfy his functional needs. The dynamic composition process assembles the available services to build new service to satisfy the requester's complex demand. UDDI

S.K. Prasad et al. (Eds.): ICISTM 2010, CCIS 54, pp. 111–122, 2010.

[2] is the early initiative towards discovery, which facilitates both keyword and category based matching. The main drawback of such mechanism is that, it is too syntactic and provides no support for dynamic composition of Web services. There is a need to describe the Web services in a natural way to improve the effectiveness of the discovery and composition mechanism. The conceptual Web service architecture [1] involving service registry (UDDI) does not provide infrastructure or the mechanism for effective dynamic Web service discovery and composition. To enable effective dynamic Web service discovery and composition, the existing architecture has to be augmented by introducing new roles and new operations.

The rest of the paper is organized as follows. The next subsection provide the brief survey of Web service composition techniques. In section 3, we present a model to describe the Web service operation functionality and operation execution flow. Section 4 presents the broker based architecture for the discovery and composition. Section 5 presents the Web service composition mechanism which generated composition plan for execution. In section 6, we discuss prototype implementation and experiment results. Section 7 draws the conclusions.

2 Literature Survey and Brief Review

In literature, different architectures are proposed for dynamic Web service discovery and composition. We classify the architectures based on the storage of Web service information and processing component of discovery and composition as follows: agent based architectures, broker based architectures, peer-to-peer architectures and hybrid architectures. In agent based Web service architectures [4], the service agents are used to initiate the request, terminate the request and to process the messages. In the broker based architectures [5] [6], the broker is used for the optimal selection of Web services for the composition plan towards dynamic integration of QoS-aware Web services with end-to-end QoS constraints. The peer-to-peer composition architecture is an orchestration model which is defined based on the peer-to-peer interactions between software components hosted by the providers participating in the composition [7]. Such architectures are also capable of composing Web services across wide area networks with the service composition based on the interface idea integrated with Peer to Peer technologies [8]. In hybrid architectures, along with service registry, other roles (for example third party provider in [9] and composition engine in [10]) are defined for the abstract composition plan generation and execution.

A variety of techniques have been proposed in literature which integrates existing services based on several pieces of information. Most of the composition strategies are defined based on the output and input matching of available Web services [11]. Such composition mechanisms use chain [12], graph (tree) [13], vector [14] data structures for the dynamic composition of concrete services. The main problem with this approach is that, the repository search time is quite more for the matching of output parameters with the inputs. Also domain ontology has to be used for effective matchmaking. The requester's constraints are useful to build composition involving concrete Web services [15]. The atomic or composite Web services are composed to satisfy the complex demand based on business rules or policy information [16]. The context (process/user) or process view also plays a major role in effective Web service composition [17]. The goal [18], service behavior [19], user satisfaction and interaction patterns [20] guide the effective dynamic Web service composition.

The Web service composition can be modeled in different ways. The Petri nets [21], Labeled behavior diagrams (LBD) [22], Mathematical model [23], UML Activity [24] and state chart diagrams [25], Workflow Model [26] and Finite automata [27] are the major modeling methods used to represent the composite Web service. In this paper, we propose a methodology to build an abstract composition plan which is defined based on the operation flow graph. The abstract composition plan (graph) is then refined by selecting suitable candidate Web services and their operations. The paper also proposes the broker based architecture for the dynamic Web service composition which facilitates the requester to discover the suitable Web service(s) for his simple or complex functional need.

3 A Model for Web Service Description, Discovery and Composition

The Web service can be viewed as a collection of interdependent or independent operations. For example the travel Web service may offer its services through the following *three* interdependent operations. They are - (i) check train ticket availability (ii) reserve train ticket (iii) cancel train ticket. The operation-ii is executed after operation-i and operation-iii is executed only after successful execution of operation-ii. The possible execution order (execution dependency) of Web service operations can be modeled as a graph structure called *Operation Dependency Graph (ODG)*

Operation Dependency Graph (ODG). Operation dependency graph is a directed acyclic graph with finite vertices which represent the number of operations of a Web service. A directed edge (u v) between any two vertices u and v indicate the possible order of execution such that, the activity v is executed after the successful execution of activity u i.e. the activity v is dependent on activity u.

As an illustration consider the Web service "travel agent" involving *four* core operations. The possible order of execution of activities can be modeled using ODG as shown in Fig. 1. The core operation are highlighted using thick circles whereas thin circles represent supplimentary operations. The operation dependency graphs of various Web services has to be represented in the form of directed graph called *Global Operation Dependecy Graph (GODG)*. The GODG represents the flow knowledge which provides operation execution flow information to generate composition plan.

Global Operation Dependency Graph (GODG). Global operation dependency graph is a directed graph with finite vertices which represent Web service

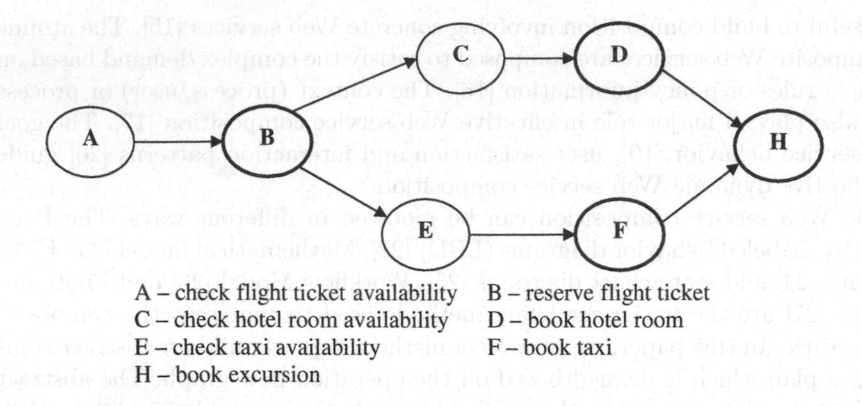

Fig. 1. Operation Dependency Graph of Travel Agent Web Service

operations. A directed edge (u v) between any two vertices u and v indicate the possible order of execution such that, the activity v is executed after the execution of activity u.

Thus global operation depedency graph represents the execution order of operations of all advertised Web services. Fig. 2 represents the ODG's of *four* Web services involving multiple operations. Fig. 3 represents the GODG of four advertised Web services. To represent the GODG effectively in computer memory for traversals, a well-known graph representation data structure called *Adjacency Matrix* is used. The GODG normally contains more number of directed edges than the nodes (dense graph) as the providers define different execution sequences for the set of operations. Thus adjacency matrix is selected for the GODG representation instead of adjacency list. Apart from the adjacency matrix of GODG (A_G), the flow knowledge also maintains another list called *Vertex List*. The vertex list is a sorted dynamic array with finite elements each having *three* fields. They are - Adjacency matrix index (A_G) Index, Operation Identifier (op-id), Operation Type. This list is used to distinguish the Web service operations based on the type information during composition process.

4 The Broker Based Architecture for Compositions

This section describes the broker based architecture for dynamic Web service discovery and composition by introducing *two* new roles to the conceptual architecture [1] with a few new operations (interactions) between different architectural roles.

4.1 Architectural Roles and Operations

The architecture involves a total of *five* roles. They are - Service Provider, Service Requester, Service Composer (Generic Service Provider), Service Registry

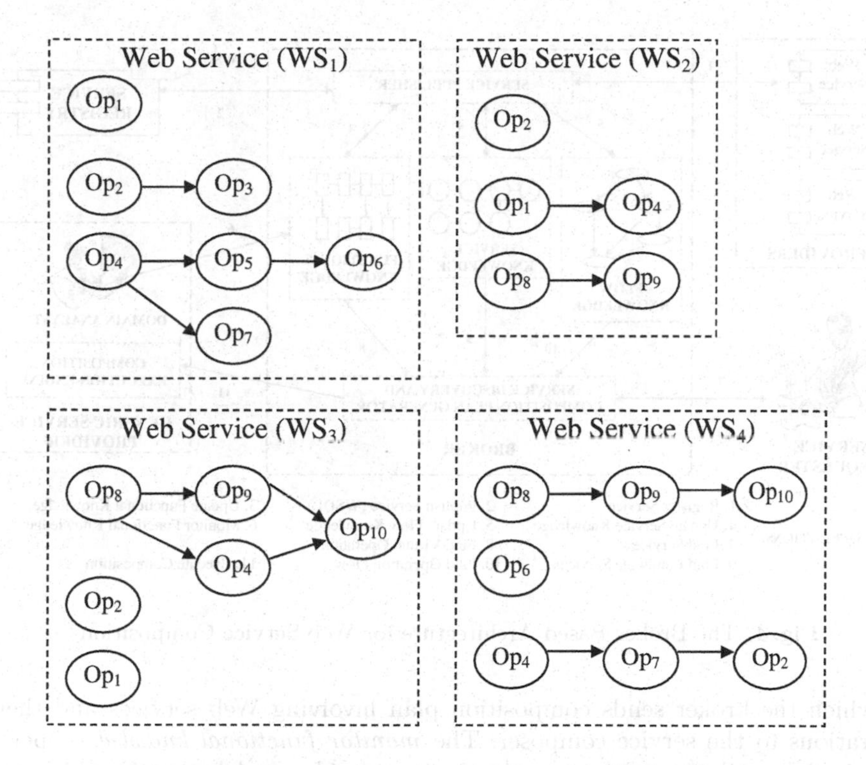

Fig. 2. Operation Dependency Graphs of four Web Services

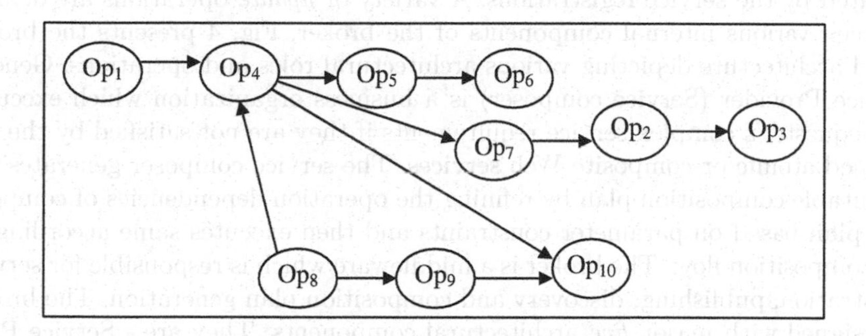

Fig. 3. Global Operation Dependency Graph of Web Services

and the Broker. The *register* operation is defined between the provider and broker. The provider registers service specific information including WSDL to the broker during Web service publishing. The *publish* operation is defined between the broker and service registry which saves the service binding and WSDL details into service registry. The *find service* operation is defined between the service requester and broker to obtain candidate Web services for a service request. The *execute composition* operation is defined between the broker and service composer

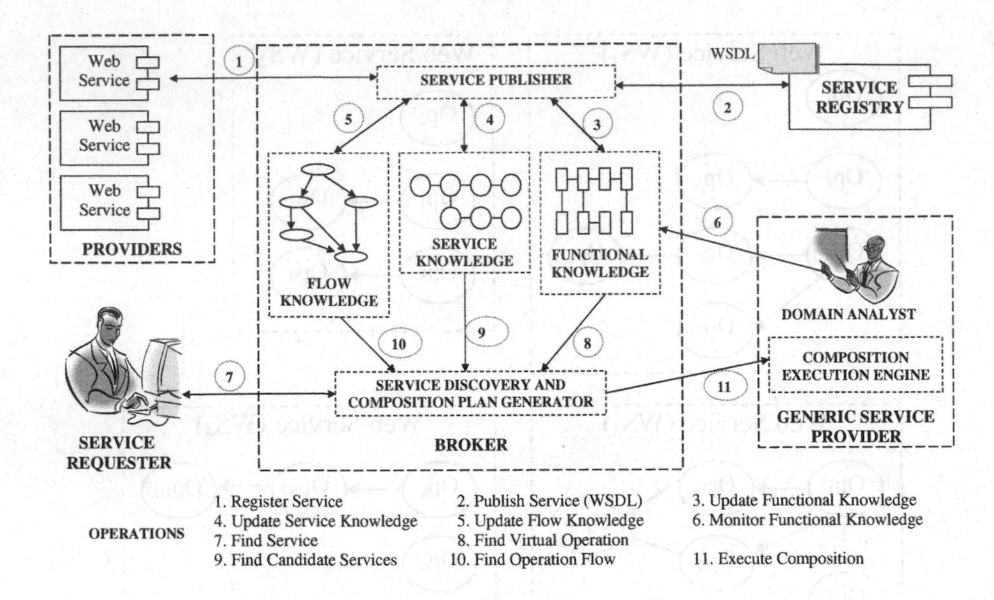

Fig. 4. The Broker Based Architecture for Web Service Composition

in which the broker sends composition plan involving Web services and their operations to the service composer. The *monitor functional knowledge* operation is defined between the generic service provider and broker in which the domain analyst updates the flow knowledge and monitors functional knowledge updated by the service registrations. A variety of *update* operations are defined between various internal components of the broker. Fig. 4 presents the broker based architecture depicting various architectural roles and operations. Generic Service Provider (Service composer) is a business organization which executes the requester's complex service requirements if they are not satisfied by the advertised atomic or composite Web services. The service composer generates the executable composition plan by refining the operation dependencies of composition plan based on parameter constraints and then executes same according to the composition flow. The broker is a middleware which is responsible for service registration, publishing, discovery and composition plan generation. The broker is designed with major *five* architectural components. They are - Service Publisher, Service Discovery & Composition Plan Generator, Service Knowledge, Functional Knowledge and Flow Knowledge.

4.2 Extending Service Knowledge Structure

The service knowledge consists of *two* interlinked structures called Web Service List (WSL) and Service Operation Tree (SOT) [28][29]. Towards the effective Web service composition, *three* more interlinked data structures are defined. They are - Input List (IL), Output List (OL) and Parameter List (PL).

Input List (IL). Input list is a dynamic sorted array with finite elements. Each element has *two* fields namely ws-id and op-id where, ws-id is Web service identifier and op-id is operation identifier. This list is sorted based on the Web service identifier.

Output List (OL). Output list is a dynamic sorted array with finite elements. Each element has *two* fields namely ws-id and op-id where, ws-id is Web service identifier and op-id is operation identifier.

Parameter List (PL). Parameter list is a dynamic sorted array with finite elements. Each element has *four* fields namely param-id, param-name, input-link and output-link. param-id is unique identifier generated by the broker, param-name refers to operation parameter name, input-link is a pointer to IL (Web service operations for which the parameter is input parameter) and output-link is a pointer to OL (Web service operations for which the parameter is output parameter).

Dependent Parameter List (DL). Dependent parameter list is a dynamic array with finite parameter names which are time dependent. For example the patermeters "date " and "time "are time dependent parameters.

5 Dynamic Composition of Web Services Based on Operation Flow

The service discovery and composition plan generator component of the broker is responsible for the discovery and composition of advertised Web services. The Web service discovery and composition mechanism for the simple or complex service request is described below.

1. The service request is preprocessed according to functional semantic rules (Chapter 4) to retrieve the functional requirement(s) of service request.
2. The action list, qualifier list (if required), object list and noun list (if required) of the functional knowledge are searched to get the corresponding identifiers. The unavailability of such identifiers result in discovery failure.
3. After obtaining required identifier(s) from the functional knowledge, the operation patterns for each operation of the service request are formed. After building the operation patterns, the patterns are searched in abstract operation list (AOL). If the pattern is found then the corresponding operation identifier is retrieved from the AOL otherwise, discovery failure is reported.
4. The abstract Web service information of all published Web services is searched by traversing SOT for the requested operation identifier(s). The discovered Web services are stored against the requested operations.
5. The Web services found common in all requested operation (s) are selected as candidate Web services for the service discovery and are returned to the requester.
6. The Web service selection mechanism (Chapter 5) may be executed to select the best Web services for the requester's based on his nonfunctional requirements.
7. The absence of a common Web service for all requested operations triggers the decrease and conquer based composition algorithm.

The composition alogorithm works on the principle of decrease and conquer methodology as it performs multiple depth first search (DFS) traversals on the GODG. The composition algorithm takes the transpose of GODG and set of requested operations (complex service request) as an input. The composition algorithm generates ODG of composition plan involving all requested Web service operations along with other supplimentary operations. The composition plan for the given complex service request is generated as follows. First the transpose of GODG is obtained and the DFS on the transpose of GODG is performed taking one requested operation as a start node. On traversal all encountered supplimentray operations are added to request set and ODG of composition plan along with respective edges. All encountered requested operations are added to the ODG of composition plan along with respective edges. Again the traversal is started from remaining requested or supplimentary operations and this process is terminated on visiting all requested and added supplimentary operations.

After generation of composition plan, the service knowledge (SOT) [29] is searched to discover the Web services for all supplimentary operations of composition plan. Now discovered Web services of all operations (requested core operations and supplimentary operations) are attached to the respective operation nodes of composition plan. The composition plan is now refined in based on Message Exchange Pattern (MEP) of Web service operations, relationship between core and supplementary operations, Input and Output parameters dependency and quality of service (QoS) like reliability.

(a) Refinement based on the Message Exchange Pattern

Let M+K be the nodes of abstract composition plan (M= number of requested core operations and K= number of supplementary operations explored), Let G be the ODG of abstract composition plan.

1. For each node S in G perform Step-2.
2. For each directed edge from node S to node D (S $\rightarrow$ D) perform Step-3.
3. Let C_S be the Web services selected at S and C_D be the Web services selected at D.
   – Eliminate all the Web services at S having MEP other than Out-Only, Out-In, In-Out and Robust-Out-Only.
   – Eliminate all the Web services at D having MEP other than In-Only, In-Out, In-Optional-Out, Robust-In-Only and Out-In.

(b) Refinement based on Supplementary and Core Operation Relationship

For every pair of nodes in ODG say, S and D with a dependency relation (S $\rightarrow$ D) or (D $\rightarrow$ S) where S represents supplementary operation and D is core operation, eliminate the Web services which are not found both in S and D.

(c) Refinement Based on the Input and Output Parameter Dependency

The dependent parameter list, parameter list, input list and output list of service knowledge are used for the plan refinement as follows. Let M_O be the set of nodes of ODG with zero out-degree and M_I be the set of nodes with zero in-degree.

1. For every node pair (X, Y) such that, X $\in M_O$ and Y $\in M_I$ perform step-2.
2. Let C_X be the Web services selected at X and C_Y be the Web services selected at Y. If there exists one pair of Web services (C_{Xi}, C_{Yj}) such that, any one time dependent output parameter of C_{Xi} is an input parameter of C_{Yj} then, insert an edge between node X and node Y.

(d) Refinement based on Quality of Service (QoS)
Let W be the number of Web services attached to any node of composition plan. The Web services with highest QoS score (based on reliablity) are retained for the individual operations of composition plan. Now the composition plan consists of nodes (operations) attached with reliable Web service.

After refinement of the composition plan, it is transferred to the generic service provider with plan identifier and requester identifier for further processing (flow refinement). The generic service provider transforms the recieved composition plan into executable composition plan involving concrete service operations. This transformation is performed based on the functional requirements (input parameter constraints) defined on the input and output parameters of the requested operations.

6 Implementation and Experiments

The prototype of the proposed broker based Web service discovery and composition mechanism is implemented on the Windows XP platform using Microsoft Visual Studio .NET development environment and Microsoft visual C# as a programming language. The broker is designed and implemented as a standalone visual program which interacts with the provider and requester through different interface forms. The service repository is implemented as a Web service which in turn communicates with the SQL server 2000 database. The database table is created to store the information (including WSDL link) of all published Web services.

The requester of a Web service can submit the simple request as well as complex request through different interface forms. The requester can also browse the functional knowledge for the successful discovery of requested functionality. The requester can browse the service knowledge i.e. operations supported by the specific Web service. The prototype also displays the refined composition plan to the requester along with the generic provider key and composition plan key. The authorized provider of Web services is allowed to browse and augment the functional knowledge in order to improve the effectiveness of Web service discovery. The interface form is created for the provider to publish the Web service along with the WSDL. The service publisher component of the broker processes the WSDL to extract functional and flow information of operations after publishing the Web service into service registry.

We have conducted several experiments involving simple and complex service requests. We use a collection of 35 Web services having total of 60 distinct operations from XMethods service portal (http://www.xmethods.com) [30] and divide them into SIX categories. The Web service operations are published to the broker using functional semantics. The simple Web service requests are created according to the functional semantic rules. The Recall and Precision of discovery process are recorded. The recall of discovery is less than 100% as some Web service descriptions take multiple functional semantic representations. The precision is 100% provided both the provider and requester precisely follow the functional semantic rules. The service operation tree of published Web services also take less memory which results in 30% compactness (compression ratio). The complex service requests are also created to test the effectiveness of Web service composition in travel domain. We define few Web services in travel domain and register them into broker with functional semantics, flow semantics, input parameter concepts and output parameter concepts of operations. The empirical results proved the correctness of proposed composition mechanism defined on functional and flow semantics of Web service operations.

7 Conclusion

The dynamic Web service discovery enables the requester to consume the desired Web services. The service composition satisfies the requester's complex need by integrating available Web services in a transparent way. The functional semantics and flow semantics of Web service operations enable the effective Web service discovery and composition mechanism. The service knowledge component of broker represents the abstract Web service information which facilitates the quick Web service discovery. The proposed broker based architecture for Web services facilitates effective Web service discovery and composition through functional and flow semantics of Web service operations. The broker based architecture is implemented on .NET platform and the empirical results prove the correctness of concepts proposed for the Web service discovery and compositions.

References

1. Kreger, H.: Web Services Conceptual Architecture (WSCA 1.0). Published (May 2001), `www.ibm.com/software/solutions/webservices/pdf/wsca.pdf` (visit: April 2007)
2. Riegen, C.V. (ed.): UDDI Version 2.03 Data Structure Reference. OASIS Open 2002-2003 (2002), `http://uddi.org/pubs/DataStructure-V2.03-Published-20020719.htm` (Accessed 8 November 2007)
3. Booth, D., Liu, C.K.: Web Services Description Language (WSDL) Version 2.0 Part 0: Primer, W3C Recommendation, June 26 (2007), `http://www.w3.org/TR/2007/REC-wsdl20-primer-20070626` (visit: December 2008)

4. Buhler, P.A.: Dominic Greenwood and George Weichhart, A Multi-agent Web Service Composition Engine, Revisited. In: Proceedings of the 9th IEEE International Conference on E-Commerce Technology and The 4th IEEE International Conference on Enterprise Computing, E-Commerce and E-Services (CEC-EEE 2007). IEEE, Los Alamitos (2007)
5. Yu, T., Lin, K.-J.: A Broker-Based Framework for QoS-Aware Web Service Composition. In: Proceedings of the 2005 IEEE International Conference on e-Technology, e-Commerce and e-Service, EEE 2005, pp. 22–29. IEEE, Los Alamitos (2005)
6. Zeng, L., Benatallah, B., Ngu, A.H.H., Dumas, M., Kalagnanam, J., Chang, H.: QoS-Aware Middleware for Web Services Composition. IEEE Transactions on Software Engineering 30(5), 311–327 (2004)
7. Benatallah, B., Sheng, Q.Z., Dumas, M.: The Self-Serv Environment for Web Services Composition. IEEE Internet Computing, 40–48 (January/ February 2003)
8. Liu, A., Chen, Z., He, H., Gui, W.: Treenet:A Web Services Composition Model Based on Spanning tree. In: Proceedings of the 2nd International Conference on Pervasive Computing and Applications, 2007 (ICPCA 2007). IEEE, Los Alamitos (2007)
9. Karunamurthy, R., Khendek, F., Glitho, R.H.: A Novel Business Model for Web Service Composition. In: Proceedings of the IEEE International Conference on Services Computing (SCC 2006). IEEE, Los Alamitos (2006)
10. Vuković, M., Kotsovinos, E., Robinson, P.: An architecture for rapid, on-demand service composition. Journal of Service Oriented Computing and Applications - SOCA 1, 197–212 (2007)
11. Lukasz, et al.: Large Scale Web Service Discovery and Composition using High Performance In-Memory Indexing. In: Proceedings of the 9th IEEE International Conference on E-Commerce Technology and The 4th IEEE International Conferenceon Enterprise Computing, E-Commerce and E-Services(CEC-EEE 2007). IEEE, Los Alamitos (2007)
12. Li, L., Jun, M., ZhuMin, C., Ling, S.: An Efficient Algorithm for Web Services Composition with a Chain Data Structure. In: Proceedings of the 2006 IEEE Asia-Pacific Conference on Services Computing (APSCC 2006). IEEE, Los Alamitos (2006)
13. Shin, D.-H., Lee, K.-H.: An Automated Composition of Information Web Services based on Functional Semantics. In: Proceedings of the 2007 IEEE Congress on Services (SERVICES 2007). IEEE, Los Alamitos (2007)
14. Jeong, B., Cho, H., Kulvatunyou, B., Jones, A.: A Multi-Criteria Web Services Composition Problem. In: Proceedings of the IEEE International Conference on Information Reuse and Integration (IRI 2007), pp. 379–384. IEEE, Los Alamitos (2007)
15. Zhao, H., Tong, H.: A Dynamic Service Composition Model Based on Constraints. In: Proceedings of the Sixth International Conference on Grid and Cooperative Computing (GCC 2007). IEEE, Los Alamitos (2007)
16. Orriëns, B., Yang, J., Papazoglou, M.P.: A Framework for Business Rule Driven Service Composition. In: Jeusfeld, M.A., Pastor, Ó. (eds.) ER Workshops 2003. LNCS, vol. 2814, pp. 52–64. Springer, Heidelberg (2003)
17. Maamar, Z., Moste faoui, S.K., Yahyaoui, H.: Toward an Agent-Based and Context-Oriented Approach for Web Services Composition. IEEE Transactions on Knowledge and Data Engineering 17, 686–697 (2005)
18. Vuković, M., Kotsovinos, E., Robinson, P.: An architecture for rapid, on-demand service composition. Journal of Service Oriented Computing and Applications - SOCA 1, 197–212 (2007)

19. Berardi, D., Calvanese, D., De Giacomo, G.: Automatic Composition of e-Services, Technical Report (2009), `http://www.dis.uniroma1.it/ mecella/publications/eService/BCDLM_techRport_22_2003.pdf` (visit: April 2009)
20. Wan, S., Wei, J., Song, J., Zhong, H.: A Satisfaction Driven Approach for the Composition of Interactive Web Services. In: Proceedings of the 31st Annual International Computer Software and Applications Conference (COMPSAC 2007). IEEE, Los Alamitos (2007)
21. Hamadi, R., Benatallah, B.: A Petri Net-based Model for Web Service Composition. In: Proceedings of the 14th Australasian database conference, pp. 191–200. ACM, New York (2003)
22. Preuner, G., Schrefl, M.: Requester-centered composition of business processes from internal and external services. Data & Knowledge Engineering 52, 121–155 (2005)
23. Li, B., Zhou, Y., Zhou, Y., Gong, X.: A Formal Model for Web Service Composition and Its Application Analysis. In: Proceedings of the 2007 IEEE Asia-Pacific Services Computing Conference. IEEE, Los Alamitos (2007)
24. Xu, Y., Xu, Y.: Towards Aspect Oriented Web Service Composition with UML. In: Proceedings of the 6th IEEE/ACIS International Conference on Computer and Information Science (ICIS 2007). IEEE, Los Alamitos (2007)
25. Gamha, Y., Bennacer, N., Romdhane, L.B., Vidal-Naquet, G., Ayeb, B.: A Statechart-Based Model for the Semantic Composition of Web Services. In: Proceedings of the 2007 IEEE Congress on Services (SERVICES 2007). IEEE, Los Alamitos (2007)
26. Bouillet, E., Feblowitz, M., Feng, H., Liu, Z., Ranganathan, A., Riabov, A.: A Folksonomy-Based Model ofWeb Services for Discovery and Automatic Composition. In: Proceedings of the 2008 IEEE International Conference on Services Computing. IEEE, Los Alamitos (2008)
27. Balbiani, P., Cheikh, F., Feuillade, G.: Composition of interactive Web services based on controller synthesis. In: Proceedings of the 2008 IEEE Congress on Services 2008 - Part I. IEEE, Los Alamitos (2008)
28. D'Mello, D.A., Ananthanarayana, V.S.: A Tree Structure for Efficient Web Service Discovery. In: Proceedings of the Second International Conference on Emerging Trends in Engineering and Technology (ICETET 2009). IEEE Computer Society, Los Alamitos (2009)
29. D'Mello, D.A., Ananthanarayana, V.S.: Effective Web Service Discovery Based on Functional Semantics. In: Proceedings of the International Conference on Advanced Computing (ACT 2009). IEEE Computer Society, Los Alamitos (in press, 2009)
30. XMethods, `www.xmethods.com` (visit: February 2009)

Distributed Scheduling of a Network of Adjustable Range Sensors for Coverage Problems

Akshaye Dhawan[1], Aung Aung[2], and Sushil K. Prasad[2]

[1] Department of Mathematics and Computer Science
Ursinus College
Collegeville, PA 19426

[2] Department of Computer Science
Georgia State University
Atlanta, GA 30030

Abstract. In this paper, we present two distributed algorithms to maximize the lifetime of Wireless Sensor Networks for target coverage when the sensors have the ability to adjust their sensing and communication ranges. These algorithms are based on the enhancement of distributed algorithms for fixed range sensors proposed in the literature. We outline the algorithms for the adjustable range model, prove their correctness and analyze the time and message complexities. We also conduct simulations demonstrating 20% improvement in network lifetime when compared with the previous approaches. Thus, in addition to sleep-sense scheduling techniques, further improvements in network lifetime can be derived by designing algorithms that make use of the adjustable range model.

1 Introduction

Wireless Sensor Networks (WSNs) are networks of low-cost sensing devices equipped with a radio. These sensors are deployed over an area of interest usually in large numbers, resulting in dense networks with significant overlaps. Once deployed, the sensor nodes monitor their environment and send this information to a base station. A critical constraint of WSNs is their limited energy supply. This is because sensors are powered by a battery that is non-replenishable. This limitation has given rise to various power saving techniques for WSNs. The most common of these is that of scheduling sensors into sleep-sense cycle [1, 2, 3, 4, 5, 6, 7]. The objective of these algorithms is to select a minimal subset of sensors that completely cover the region/targets of interest to switch on while the remaing sensors can enter a low-power sleep state.

In addition to switching off redundant sensors, another means of prolonging the network lifetime is proposed via the adjustable sensing and communication range model. To the best of our knowledge, this model was introduced in [8] and [9]. Algorithms based on this model were presented in [10, 11]. In this model, a

S.K. Prasad et al. (Eds.): ICISTM 2010, CCIS 54, pp. 123–132, 2010.

sensor has the capability to adjust its sensing as well as communication range. Each sensor has the option of different ranges $r_1, r_2, ..., r_p$ and corresponding energy consumptions of $e_1, e_2, ..., e_p$. Instead of operating at a fixed range, a sensor can now reduce its range to the needed amount and thus reduce its energy consumption.

A slightly different model was proposed by us in [12]. In this model, instead of selecting a range from one of p fixed values, a sensor has the ability to smoothly vary its range between 0 and r_{max} where, r_{max} is its maximum range. This has the advantage of allowing a sensor to precisely adjust its range to that needed to cover a target t, thus resulting in additional energy gains. In this paper, we utilize the smoothly varying range model.

Our Contributions: In this paper, we present two distributed algorithms for the adjustable range model - ALBP which is an extension of the fixed range protocol LBP [13] and ADEEPS which is an extension of DEEPS [14] to the adjustable range model. We compare the performance of these algorithms to their fixed range counterparts through extensive simulations that show lifetime improvements of 20% or higher. This shows that in general, the adjustable range model is an effective energy-efficient approach that adds to the lifetime improvements of existing sleep-sense scheduling algorithms. Part of this work was completed for meeting thesis requirements [15].

The remainder of this paper is as follows. In Section 2, we briefly survey existing work on the adjustable range model and then discuss the fixed range algorithms LBP [13] and DEEPS [14]. In Section 3, we present the adjustable range variations, ALBP and ADEEPS. We outline the operational details of these algorithms along with their time and message complexities. In Section 4, we present our simulation results, where we compare the adjustable range algorithms with their fixed range versions. Finally, we conclude in Section 5.

2 Background

The problem of scheduling sensors into minimal sensor covers that are activated successively so as to maximize the network lifetime has been shown to be NP-complete [3, 5].

Existing work on this problem has looked at both centralized and distributed algorithms to come up with such a schedule. A common approach taken with centralized algorithms is that of formulating the problem as an optimization problem and using linear programming (LP) to solve it [16, 5, 17, 12]. The distributed algorithms typically operate in rounds. At the beginning of each round, a sensor exchanges information with its neighbors, and makes a decision to either switch on or go to sleep. In most greedy algorithms [1, 2, 3, 4, 13, 14], the sensor with some simple greedy criteria like the largest uncovered area [2], maximum uncovered targets [13], etc. is selected to be on.

We now focus on the various algorithms that make use of the adjustable range model.

2.1 The Adjustable Range Model

The adjustable range model was proposed independently by [8] and [9]. In this subsection, we briefly survey the literature on extending the lifetime of WSNs using the adjustable range model.

In [8], the authors present two different coverage algorithms based on an adjustable model where the sensor can chose between one of two different ranges and one of three (maximum, medium and small) different ranges. Their objective was to minimize the overlapped area between sensor nodes, thereby resulting in energy savings.

[9] also address the problem of selecting a minimum energy connected sensor cover when nodes have the ability to vary their sensing and transmission radius. The authors present a number of centralized and distributed heuristics for this problem including greedy formulations, Steiner Tree and Vornoi based approaches. The Vornoi based approach was shown to result in the best gains.

In [10], the authors utilize a sensing model that allows a sensor to adjust its range from one of several different fixed values. The authors address the problem of finding the maximum number of sensor covers and they present a linear programming based formulation, a linear programming based heuristic and also greedy formulations for this problem. A more constrained case of the same problem is studied in [11]. The authors add the requirement of connectivity to the sensor covers and present distributed heuristics to maximize the total number of rounds.

We present a different model for adjusting the range in [12]. Instead of allowing a sensor to adjust its range between a number of fixed options, we allow the nodes to vary their range smoothly between 0 and r_{max} where, r_{max} is the maximum value for the range. We give a mathematical model of this problem using a linear program with exponential number of variables and solve this linear program using the approximation algorithm of [18], to provide a $(1+\epsilon)(1+2lnn)$ approximation of the problem.

More recently, [19] uses the adjustable range model to give two localized range optimization schemes based on a one-hop approximation of the Delaunay Triangulation. They also show that their approximation scheme achieves the same results as the original Delaunay Triangulation.

Next, we look at the two fixed range protocols for which we propose adjustable range variations. The main idea of the load balancing protocol (LBP) is to keep the maximum number of sensors alive for as long as possible by means of load balancing [13]. LBP is greedy on battery life. The Deterministic Energy-Efficient Protocol for Sensing (DEEPS) was presented by the same authors in [14]. The main intuition behind DEEPS is that they try to minimize the energy consumption rate for low energy targets while allowing higher energy consumption for sensors with higher total supply.

3 Distributed Algorithms Using Adjustable Range

3.1 Adjustable Range Load Balancing Protocol (ALBP)

In this section, we present a distributed load balancing protocol for sensors with an adjustable range called ALBP. This protocol extends the ideas of LBP [13] to the adjustable range model. As with LBP, the objective of the protocol is to maximize the time for which all targets in the network are covered. The intuition behind the protocol is also similar to LBP, the aim is to keep as many sensors alive by balancing their load so as to let them exhaust their batteries simultaneously. ALBP, however, differs from LBP in that while making a decision on switching a sensor to an active state, it also needs to decide what range this sensor should have.

We begin by defining the different states a sensor can be in at any point of time:

- **Active:** The sensor is monitoring a target('s).
- **Idle:** The sensor is listening to other neighboring sensors, but does not monitor targets.
- **Deciding:** The sensor is presently monitoring a target, but will change its state to either active or idle soon.

Setup: The setup phase starts as before. At the beginning of a round, each sensor exchanges information on its battery level and the set of targets it covers with its neighbors. However, after broadcasting this information, a sensor enters the *deciding* state with its *maximum* range.

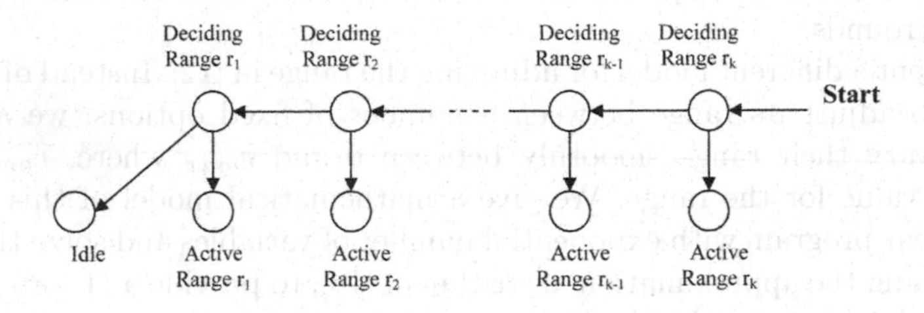

Fig. 1. State transitions for the ALBP protocol

Transitions: Figure 1 shows the state transitions for ALBP. At the end of the setup phase, each sensor is in the deciding state with its maximum range. A sensor then changes its state according to the following transition rules:

- **Active state with a range r:** A sensor transitions to the Active State with a range r, if there is a target at range r which is not covered by any other active or deciding sensors.
- **Deciding state with lower range:** A sensor in the deciding state with some range r can decrease its range to the next closest target if all its targets at range r are covered by another sensor in the active state or by a sensor that is in the deciding state and has a higher battery life.

- **Idle state:** A sensor is in the idle state if it has reduced its range to zero (i.e., all its targets are covered by active sensors or higher energy sensors in the deciding state).

  Every sensor uses these rules to make a decision on whether to enter the active or idle state. The sensors will stay in this state until the end of a round, upon which the process will repeat itself. When the network reaches a point where a target cannot be covered by any sensor, the network is considered dead.

Correctness: In ALBP, a sensor can enter the *idle* state only when its range reaches zero. To achieve this, all its targets had to have been covered by an active or deciding sensor that had a higher energy than it. Hence, all targets are always covered.

Time and Message Complexity: The time complexity of ALBP is $O(\Delta^2)$ and the message complexity is $O(\Delta)$ where, Δ is the maximum degree of the sensor graph.

At the start of a round, each sensor receives from every neighbor a message containing the targets that neighbor covers, and its battery life. If Δ is the maximum degree of the graph, a sensor can have no more than Δ neighbors. This means that a sensor can receive no more than Δ messages, which it can process in $O(\Delta)$ time.

In the worst case, a sensor may have to wait for all its neighbors to decide their state before it can make a decision. Thus, the waiting time accumulates as $O(\Delta^2)$, hence the time complexity.

Since each sensor has at most Δ neighbors and during a round a sensor sends at most two messages to its neighbors (its battery and targets covered information, and its status - on/off), at most $O(\Delta)$ messages are sent in the setup phase. Hence, the message complexity is $O(\Delta)$.

3.2 Adjustable Range Deterministic Energy Efficient Protocol (ADEEPS)

In this subsection, we present an adjustable range version of the DEEPS protocol [14]. Each sensor can once again be in one of three states - *active*, *deciding* or *idle*. The definition of these states remains as before.

We make use of similar concepts of *sink* and *hill* targets as DEEPS. However, instead of defining these using the total battery of the sensors covering a target (as DEEPS does), we define them with respect to the maximum lifetime of the target. Let the lifetime of a sensor with battery b, range r and using an energy model e be denoted as $Lt(b,r,e)$. Then, the maximum lifetime of a target would be $Lt(b_1,r_1,e)+Lt(b_2,r_2,e)+Lt(b_3,r_3,e_3)+...$ assuming that it can be covered by some sensor with battery b_i at distance r_i for $i=1,2,....$

Now, we can define the sink and hill targets as follows. A target t is a *sink* if it is the smallest maximum lifetime target for at least one sensor covering t. A *hill* is a target which is not a sink for any of the sensors covering it. We define the in-charge sensors for a target t as follows:

- If the target t is a sink, then the sensor s covering t with the highest lifetime $Lt(b, r, e)$ for which t is the poorest is placed in-charge of t.
- If target t is a hill then out of the sensors covering t, the sensor s whose poorest target has the highest lifetime is placed in-charge of t. If there are several such sensors, then the richest among them is placed in-charge of t.

Setup: Each sensor initially broadcast its lifetime and covered targets to all neighbors of neighbors. This is similar to [14]. After this, it stays in the deciding state with its maximum range.

Transitions: A sensor that is in the deciding state with range r changes its state according to the following rules:

- **Active state with a range r:** If there is a target at range r which is not covered by any other active or deciding sensors, the sensor enters the Active state with range r.
- **Deciding state with lower range:** A sensor in the deciding state with some range r can decrease its range to the next closest in-charge target if all its in-charge targets at range r are covered by another sensor in the active state or by a sensor that is in the deciding state and has a higher battery life.
- **Idle state:** When a sensor s is not in-charge of any target except those already covered by on-sensors, s switches itself to the idle state.

The algorithm again operates in rounds. When there exists a target that cannot be covered by any sensor, the network is considered to be dead.

Correctness: The correctness of ADEEPS can be proved from the fact that each target has a sensor which is in-charge of that target and the transition rule to active state assures that the resultant sensor cover is minimal in which each sensor s has a target covered only by s.

Time and Message Complexity: In each round, the time complexity of ADEEPS is $O(\Delta^2)$ and the message complexity is $O(\Delta^2)$ where, Δ is the maximum degree of the graph.

Each sensor has no more than Δ neighbors. At the start of each round, every sensor receives from its neighbors and their neighbors (2-hops) information about their lifetime and the targets covered. Thus a sensor can receive at most Δ^2 messages. Once this information has been received, all decisions on sink/hill targets, in-charge sensor and the active/idle state can be made locally. Thus the time complexity is $O(\Delta^2)$.

Also, since a sensor has at most Δ neighbors and it needs to communicate the setup information to two-hops, each sensor sends $O(\Delta^2)$ messages in the setup phase. This means that the message complexity is $O(\Delta^2)$.

4 Simulations

To evaluate the performance of the new algorithms and to make comparison with the algorithms in [13, 14], the new algorithms are implemented by using

C++. We built on the source code for [13, 14]. For the simulation environments, a static wireless network of sensors and targets which are scattered randomly, while ensuring that all targets can be covered, in $100m \times 100m$ area is considered. The location of the sensor nodes can be randomly generated and the targets can also be placed randomly. We assume that the communication range of each sensor is two times the sensing range. Simulations are carried out by varying the number of sensors and the lifetime is measured. We also vary the maximum range, energy models, and numbers of targets with various combinations. For these simulations, we use the linear energy model wherein the power required to sense a target at distance d is proportional to d. We also experiment with the quadratic energy model (power proportional to d^2). Note that to facilitate comparison, we follow the simulation setup of [13, 14].

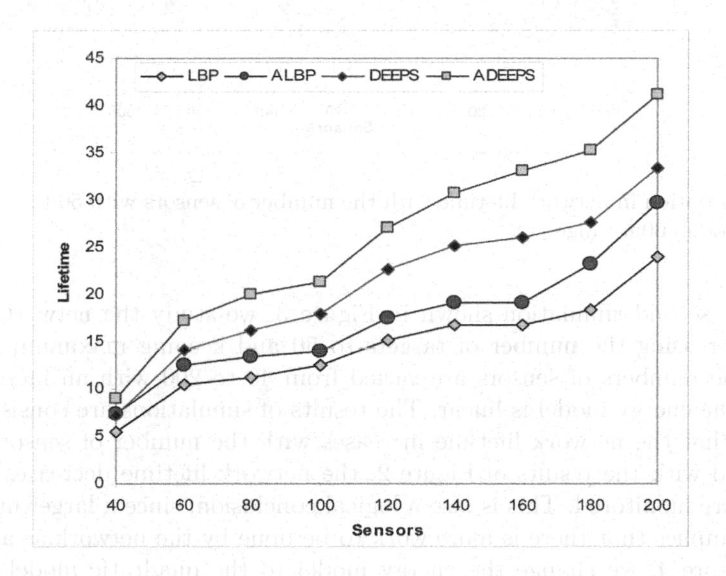

Fig. 2. Variation in network lifetime with the number of sensors with 25 targets, linear energy model, 30m range

In the first simulation shown in Figure 2, we limit the maximum range to $30m$. This means that a sensor can smoothly vary its range from 0 to $30m$. The simulation is conducted with 25 randomly deployed targets, 40 to 200 sensors with an increment of 20 and a linear energy model. As is expected, increasing the number of sensors while keeping the number of targets fixed causes the lifetime to increase for all the protocols. Also, using the adjustable range model shows performance improvements when compared to the fixed range model. As can be seen from the figure, ALBP outperforms LBP by at least 10% and ADEEPS outperforms DEEPS by around 20%.

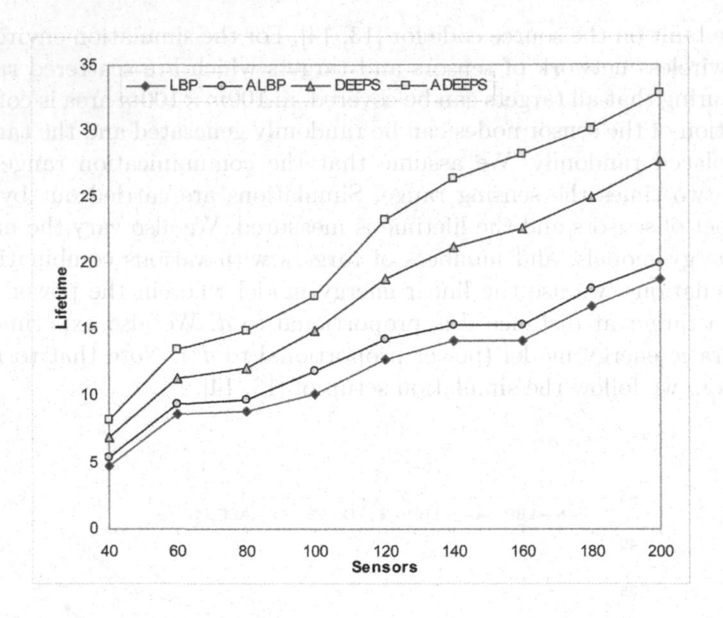

Fig. 3. Variation in network lifetime with the number of sensors with 50 targets, linear energy model, 60m range

In the second simulation shown in Figure 3, we study the network lifetime while increasing the number of targets to 50 and keeping maximum range at $30m$. The numbers of sensors are varied from 40 to 200 with an increment of 20 and the energy model is linear. The results of simulations are consistent and showed that the network lifetime increases with the number of sensors. When compared with the results of Figure 2, the network lifetime decreases as more targets are monitored. This is also a logical conclusion, since a larger number of targets implies that there is more work to be done by the network as a whole.

In Figure 4, we change the energy model to the quadratic model. We use the same number of sensors (40 to 200 with increment of 20), the maximum range is $30m$ and the energy model is quadratic. As in Figure 2, for both energy models, the result indicates that the network lifetime increases with the number of sensors. As is expected, the quadratic model causes all protocols to reduce in total lifetime when compared to the linear model of Figure 2. It can also be seen that the network lifetime is significantly improved with ALBP and ADEEPS in the quadratic model. This phenomenon is quite logical since in the fixed sensing model, each sensor consumes more energy than the adjustable range model. Improvements here are in the range of 35-40% when compared to their fixed range counterparts.

From the results, the overall improvement in network lifetime of ALBP over LBP is around 10% and ADEEPS over DEEPS is about 20% for linear energy model. For quadratic energy model, the improvements are even higher.

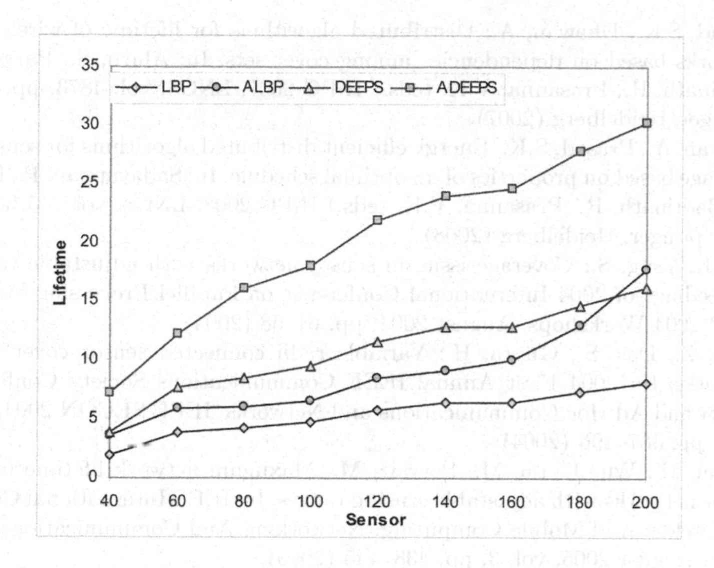

Fig. 4. Variation in network lifetime with the number of sensors, with 25 targets, quadratic energy model and 30m maximum range

5 Conclusion

In this paper, we present two distributed algorithms for maximizing the lifetime of wireless sensor networks when the sensors have adjustable ranges. These algorithms are built as an extension of earlier algorithms that were devised for the fixed range model. Simulations show that significant improvements in network lifetime can be obtained for comparable algorithms using the adjustable range model. Overall the new heuristics exhibit a 10-20% improvement over their fixed range versions for the linear energy model and 35-40% improvement for the quadratic energy model.

References

[1] Slijepcevic, S., Potkonjak, M.: Power efficient organization of wireless sensor networks. In: IEEE International Conference on Communications (ICC), vol. 2, pp. 472–476 (2001)

[2] Lu, J., Suda, T.: Coverage-aware self-scheduling in sensor networks. In: 18th Annual Workshop on Computer Communications (CCW), pp. 117–123 (2003)

[3] Abrams, Z., Goel, A., Plotkin, S.: Set k-cover algorithms for energy efficient monitoring in wireless sensor networks. In: Third International Symposium on Information Processing in Sensor Networks, pp. 424–432 (2004)

[4] Cardei, M., Thai, M., Li, Y., Wu, W.: Energy-efficient target coverage in wireless sensor networks. In: INFOCOM 2005, March 2005, vol. 3 (2005)

[5] Cardei, M., Du, D.Z.: Improving wireless sensor network lifetime through power aware organization. Wireless Networks 11, 333–340 (2005)

[6] Prasad, S.K., Dhawan, A.: Distributed algorithms for lifetime of wireless sensor networks based on dependencies among cover sets. In: Aluru, S., Parashar, M., Badrinath, R., Prasanna, V.K. (eds.) HiPC 2007. LNCS, vol. 4873, pp. 381–392. Springer, Heidelberg (2007)
[7] Dhawan, A., Prasad, S.K.: Energy efficient distributed algorithms for sensor target coverage based on properties of an optimal schedule. In: Sadayappan, P., Parashar, M., Badrinath, R., Prasanna, V.K. (eds.) HiPC 2008. LNCS, vol. 5374, pp. 269–281. Springer, Heidelberg (2008)
[8] Wu, J., Yang, S.: Coverage issue in sensor networks with adjustable ranges. In: Proceedings of 2004 International Conference on Parallel Processing Workshops, ICPP 2004 Workshops, August 2004, pp. 61–68 (2004)
[9] Zhou, Z., Das, S., Gupta, H.: Variable radii connected sensor cover in sensor networks. In: 2004 First Annual IEEE Communications Society Conference on Sensor and Ad Hoc Communications and Networks, IEEE SECON 2004, October 2004, pp. 387–396 (2004)
[10] Cardei, M., Wu, J., Lu, M., Pervaiz, M.: Maximum network lifetime in wireless sensor networks with adjustable sensing ranges. In: IEEE International Conference on Wireless And Mobile Computing, Networking And Communications (WiMob 2005), August 2005, vol. 3, pp. 438–445 (2005)
[11] Lu, M., Wu, J., Cardei, M., Li, M.: Energy-efficient connected coverage of discrete targets in wireless sensor networks. In: Lu, X., Zhao, W. (eds.) ICCNMC 2005. LNCS, vol. 3619, pp. 43–52. Springer, Heidelberg (2005)
[12] Dhawan, A., Vu, C.T., Zelikovsky, A., Li, Y., Prasad, S.K.: Maximum lifetime of sensor networks with adjustable sensing range. In: Proceedings of the International Workshop on Self-Assembling Wireless Networks (SAWN), pp. 285–289 (2006)
[13] Berman, P., Calinescu, G., Shah, C., Zelikovsky, A.: Efficient energy management in sensor networks. In: Ad Hoc and Sensor Networks, Wireless Networks and Mobile Computing (2005)
[14] Brinza, D., Zelikovsky, A.: Deeps: Deterministic energy-efficient protocol for sensor networks. In: Proceedings of the International Workshop on Self-Assembling Wireless Networks (SAWN), pp. 261–266 (2006)
[15] Aung, A.: Distributed algorithms for improving wireless sensor network lifetime with adjustable sensing range. M.S. Thesis, Georgia State University (2007)
[16] Meguerdichian, S., Potkonjak, M.: Low power 0/1 coverage and scheduling techniques in sensor networks. UCLA Technical Reports 030001 (2003)
[17] Berman, P., Calinescu, G., Shah, C., Zelikovsky, A.: Power efficient monitoring management in sensor networks. In: Wireless Communications and Networking Conference (WCNC), vol. 4, pp. 2329–2334 (2004)
[18] Garg, N., Koenemann, J.: Faster and simpler algorithms for multicommodity flow and other fractional packing problems. In: FOCS 1998: Proceedings of the 39th Annual Symposium on Foundations of Computer Science, Washington, DC, USA, p. 300. IEEE Computer Society, Los Alamitos (1998)
[19] Wang, J., Medidi, S.: Energy efficient coverage with variable sensing radii in wireless sensor networks. In: Third IEEE International Conference on Wireless and Mobile Computing, Networking and Communications, WiMOB 2007, October 2007, pp. 61–69 (2007)

Verifiable (t, n) Threshold Secret Sharing Scheme Using ECC Based Signcryption

Atanu Basu and Indranil Sengupta

Department of Computer Science and Engineering
Indian Institute of Technology, Kharagpur, India
{atanu,isg}@cse.iitkgp.ernet.in

Abstract. In this paper, a secured (t, n) threshold secret sharing scheme has been proposed which prevents cheating from participants as well as from trusted dealer (TD) (which distributes shared secrets and reconstructs shared secret key). Here, a signcryption scheme based on elliptic curve cryptography (ECC) which incorporates both the digital signature and encryption scheme in a single logical step is used which helps to protect authenticity of the participants of the scheme and confidentiality of the secret shares being transferred through the network. The participants transfer their secret shares to the dealer when they require reconstruction of the secret key after acquiring confidence that the dealer is not compromised or captured. Before reconstruction of secret key (using t out of n secret shares), the dealer detects and identifies the dishonest participants (cheaters) assuming that more than t participants will submit their secret shares to the dealer. The dealer uses the concept of *consistency* and *majority of secrets* as proposed by Lein Harn *et al.*[17]. The proposed scheme helps to prevent cheating of dishonest participants, captured or compromised trusted dealer. It also reduces computational cost and communication overhead as the proposed scheme uses ECC based signcryption scheme.

Keywords: Threshold secret sharing, elliptic curve cryptography, trusted dealer, participant, signcryption, *consistency* and *majority of secrets*.

1 Introduction

The basic idea of the secret sharing scheme is that a trusted party or *dealer* generates and distributes pieces of secret key among a group of participants and only specific subset of them defined by the *access structure* can recover or reconstruct the secret key at a later time. *Access structure* is the subset of all participants who can reconstruct the secret key. The secret sharing scheme, after its invention has been an important tool in the field of group cryptography and distributed computing. Shamir [1], Blakley [2], Chaum [3] independently formulated the concept of secret sharing.

Shamir's secret sharing scheme [1] is (t, n) threshold secret sharing scheme where a secret key D is divided into n pieces $D_1, D_2, \ldots, D_n$. The pieces or secret

S.K. Prasad et al. (Eds.): ICISTM 2010, CCIS 54, pp. 133–144, 2010.

shares are distributed to n-participants such that knowledge of any t or more D_i pieces leaves D easily computable and knowledge of any $(t-1)$ or fewer pieces leaves D completely undetermined. In the Shamir's secret sharing scheme, a dealer randomly selects $(t-1)$ number of coefficients $a_1, a_2, \ldots, a_{t-1}$ from a finite field, F_p where p is a prime number and constructs the polynomial $f(x)$ as

$$f(x_i) = S + \sum_{t=1}^{t-1} a_i \cdot x_i$$

where $f(0) = S$ is the secret key selected by the dealer.

The dealer computes the n shares $(x_1, y_1), \ldots, (x_n, y_n)$ as $y_i = f(x_i)$ for unique x_i and the values of (x_i, y_i) are distributed to n-participants.

For the recovery of the secret key, the *combiner* (who reconstructs shared secret key) receives at least t shares $(x_1, y_1), \ldots, (x_t, y_t)$ and reconstructs the polynomial $f(x)$ by the Lagrange's interpolation formula as

$$f(x) = \sum_{i=1}^{t} y_i \left(\prod_{j=1, j \neq i}^{t} \frac{x - x_j}{x_i - x_j} \right)$$

and recovers the secret key, S as S $= f(0)$. In another method, S can be derived by solving t linear equations.

Threshold secret sharing schemes are ideally applicable in a situation where a group of persons must cooperate for performing a certain task though they may be mutually suspicious. It is desirable that the cooperation should be based on mutual consent but the right to deny to provide secret share to the *combiner* given to each member can block the activities of the group. As an example, no employee alone in a bank can access the bank vault for security reason. Instead, some employees of the bank (manager, vice managers, cashiers) defined in the *access structure* deliver their secret shares (piece of the password) of the vault. At least, threshold (t) number of bank employees assemble together and reconstruct the secret key of the vault to open it. But, (t-1) corrupt employees cannot open the bank vault with their secret shares.

Some applications of threshold secret sharing scheme demand the reconstruction of the shared secret key in the trusted dealer *(TD)* at a later time after distribution of secret shares to the participants. The participants require the verification of the *TD* before delivery of their shares to the *TD*. But the participants may not be always in the trusted network environment to perform that task. Moreover, they may need to connect the *TD* by using their mobile devices being in the hostile and insure wireless environment. So, it requires that any secret sharing scheme should incorporate proper authentication and confidentiality features with much low computational cost and communication overhead. The scheme should also include the feature of verification of trustworthiness of the *TD* as well as detection of dishonest participants for the reconstruction of the shared secret key. The proposed scheme should also prevent various active and passive attacks.

Our proposed scheme uses elliptic curve cryptography (*ECC*) [4] based signcryption scheme which incorporates both the digital signature and encryption of the secret shares in a single logical step and this *ECC* based signcryption scheme incurs much less memory, computational cost and communication overhead compared to other conventional digital signature and encryption schemes. This helps to protect the authenticity and confidentiality of the secret shares of the participants transferred through the hostile and insecure wireless environment. At the time of reconstruction of the secret shares, before delivering the secret shares to the *TD*, the participants verify the trustworthiness of the *TD* by polynomial computation method [20]. After acquiring confidence about the trustworthiness of the *TD*, the participants deliver their shares to the *TD* through the *ECC* based signcryption scheme. The participants may also use mobile devices for delivering their secret shares connecting through the insecure wireless environment. The *TD* after successful recovery of the shares checks the validity of the secret shares, detects if there exists any malicious or cheating participant and reconstructs the shared secret key. The proposed scheme also prevents various active and passive attacks including attack from collusion among malicious participants.

2 Previous Work

Shamir's secret sharing scheme does not incorporate any security measures. The dishonest participants may deliver fake or invalid shares to the *combiner* for reconstruction of shared key. This will help the dishonest participants to get the original secret key while the honest participants will get the fake secret key. As an example, a secret key D can be constructed as a combination of three shares say D_1, D_2 and D_3 *e.g.*, $D = D_1 + D + D_3$. Now, during reconstruction phase if the first participant submits its fake share say, D_1^f to the *combiner*, then the reconstructed shared secret key, $D^f = D_1^f + D_2 + D_3$. Other two participants will get the invalid secret key D^f but not the original secret key D. The first participant will get the original shared secret key by calculating $D = D^f + D_1 - D_1^f$. This indicates if any one of the t participants is dishonest, he can easily retrieve the original secret key and other participants will get the invalid secret key. Tompa and Woll [5] first detected this type of attack may happen in secret sharing scheme and proposed a scheme to prevent such an attack. A lot of schemes have been evolved with different techniques for the prevention of cheating in secret sharing scheme and we mention here some of the schemes. In schemes [5],[6],[7], cheating is prevented without depending on any assumption of computational intractability. In schemes [8],[9], security has been achieved based on coding technique. The schemes [10],[11],[12] are based on computationally secure cryptographic techniques. Some secret sharing schemes [13],[14] also provide security where the threshold value (t) can be changed dynamically. A verifiable secret sharing is called *interactive verifiable secret sharing* if verification of the shares are performed through the exchange of messages among the participants. In a *non-interactive verifiable secret sharing* the dealer and the participants exchange messages themselves for the verification of the secret shares. The verifiable secret sharing schemes (*interactive* and *non-interactive*) [6],[14],[15] may be unconditionally secure or computationally secure. In these schemes, the participants verify their secret shares with some extra information whether the shares provided by the trusted dealer is valid or not and the trusted dealer also checks whether the shares submitted by the participants are valid or not.

The proposed scheme has been discussed in Section 3. Section 4 discusses the security analysis of the proposed scheme and finally Section 5 concludes the scheme.

3 Proposed Work

In the **attack model** of the proposed work, we consider that initially the trusted dealer (*TD*) is trusted by all the participants and the *TD* distributes the secret shares to all the participants securely. After that, at any time the *TD* may be compromised or captured by any adversary and also a participant or group of participants may be compromised or captured by any adversary. As the participants may deliver their shares through the insecure wireless environment to the *TD*, the adversary or eavesdropper may easily capture the t number of shares for the reconstruction of the secret key and may transfer the shares to the *TD* after modification (replay attack) to cheat the *TD*. A group of malicious participants may collude to reconstruct the shared secret key. So, there may be trusted dealer capture attack, malicious participant attack, passive or replay attacks by the eavesdropper, collusion attack by the malicious participants.

For the verification and detection of the dishonest participants, we propose to use Harn and Lin [17] work to be used by the *TD* which uses the concept of *consistency* and *majority of secrets*. This scheme [17] is simple and does not include any check vector to the participant's share or uses any cryptographic technique for the detection and identification of the cheaters. We briefly describe the scheme [17] below.

3.1 Harn and Lin [17] Scheme

In this scheme, dishonest participants can be detected and identified while reconstructing secure secret key using redundant (more than threshold number of participants) secret shares of the participants. Same type of concept has been incorporated in [16]. The scheme discusses the different types of attacks of dishonest participants and detection as well as identification of the dishonest participants under different conditions. The scheme assumes that the dealer is always trusted and will never cheat the participants. First of all, the trusted dealer (*TD*) selects a polynomial $f(x)$ of degree $(t-1)$ randomly :

$$f(x) = a_0 + a_1.x + \ldots + a_{t-1} \,.\, x^{t-1}$$

where the secret key $D = a_0 = f(0)$ and all the coefficients $a_0, a_1, \ldots, a_{t-1}$ are in a finite field F_p where p is a prime number.

The trusted dealer (*TD*) computes $D_1 = f(1)$, $D_2 = f(2), \ldots, D_n = f(n)$ and then distributes n number of D_i secretly to n number of participants P_i. During reconstruction of secret key, (*TD*) selects any t number of shares $(D_{i_1}, \ldots, D_{i_t})$

where $(i_1,\ldots,i_t) \subset \{1,2,\ldots,n\}$. After that, the secret key D is computed through Lagrange's Interpolation of polynomial. The scheme first explains the concept of *consistency* and *majority of secrets.*

Let j denotes the number of participants in the secret reconstruction where $(n \geqslant j > t)$ and $J = \{i_1,\ldots,i_j\} \subseteq \{1,2,\ldots,n\}$ denotes the participants which has participated in the secret key reconstruction phase. Let $T = \{T_1, T_2,\ldots,T_u\}$ denotes all subsets with t participants of J where $u = \binom{j}{t}$. A set of J shares are said to be consistent if any subset containing t shares of the set reconstructs the same secret. If D^i denotes the ith reconstructed share among u then $D^i = F(T_i) = F_{T_i}(D_{i_1},\ldots,D_{i_t})$ where $i = 1,2,\ldots,u$ and F is an interpolating polynomial function. If the secret shares $D_1, D_2,\ldots,D_j$ are *consistent* then according to the *consistency* the reconstructed shares $D^1 = \ldots = D^u$ *e.g.*, all the reconstructed shares will be identical.

If the secret shares $D_1, D_2,\ldots,D_j$ are *inconsistent* then the reconstructed shares D^i for $i = 1,\ldots,u$ reconstructed by combinations of t out of j shares are not identical. The set of all reconstructed secret keys $U = \{D^1,\ldots,D^u\}$ is divided into several mutually disjoint subsets U_i for $i = 1,\ldots,v$ where each U_i contains the same secret reconstructed keys *e.g.* $U = U_1 \cup U_2 \ldots \cup U_v$ where $U_i = \{D^{i_1},\ldots,D^{i_{w_i}}\}$. If $w_i = |U_i|$ and $w_z = \max_i\{w_i\}$, then the secret S^{w_z} is said to be the *majority of secrets.*

In the secret reconstruction phase, if the shares are found *inconsistent* then their exist dishonest participants or cheaters. It is assumed that majority of participants are honest. If c be the number of cheaters then the condition $(j-c) > t$ should be satisfied for the detection and identification of cheaters. By the *majority of secrets,* it is considered that the legitimate secret $D = D^{w_z}$ (as $w_z = \max_i\{w_i\}$) where $D = F_{T_K}(D_{i_{k_1}},\ldots,D_{i_{k_t}})$ and $T_k \in T$. The rest of the secret shares excluding the legitimate majority of shares may be defined as $R = J$ - $\{i_{k_1},\ldots,i_{k_t}\}$. The participant i_r is picked up from R orderly and the shared key is reconstructed as $D^r = F(D_{i_r}, D_{i_{k2}},\ldots,D_{i_{k_t}})$. If $D^r = D$, then the participant i_r is put into the honest set of participants H, otherwise it is put into the dishonest participant or cheater set C. The process is continued until the set R is null. All the participants in cheater set C is declared as dishonest participants or cheaters.

3.2 Signcryption Scheme

The standard approach to achieve both the authentication and confidentiality is digital signature followed by encryption before any message is transferred through the network. In this approach first the sender digitally sign the message. After that, the original message and the signed message is encrypted using the randomly chosen message encryption key. The message encryption key is then encrypted using the public key of the recipient. This is the two step approach *signature then encryption.* This current method of *signature then encryption* consumes much computational power of the system and the bandwidth of the network. Here, the cost for delivering a message in a secure and authenticated way is sum of the cost for digital signature and message encryption. As both the computational cost and communication overheads are becoming serious issues in today's various applications, a new paradigm called signcryption scheme [18] in public key cryptography has been evolved. A signcryption scheme is a cryptographic method which satisfies both the functions of digital signature and encryption but with lower cost compared to *signature then encryption* scheme. A signcryption scheme [18] has been implemented with elliptic curve cryptography where it has been shown that when compared with *signature then encryption* scheme, this scheme can save 58% in computational cost and 40% in communication overhead.

3.3 Description of the Proposed Scheme

In our scheme, trusted dealer (*TD*) chooses two secret random polynomials - one for generation of secret shares for the participants and another for verification of the dealer by the participants. The dealer chooses *ECC* domain parameters and generates *ECC* private and public keys. The n-participants of the scheme possess their own secret shares and the corresponding identifiers. In our proposed scheme, the *TD* will act as *combiner*. At the time of reconstruction of secret key, the *TD* supplies certain parameters to participants on request to verify the trustworthiness of the *TD* itself. The participants (more than t-participants) send all the secret shares and corresponding unique identifiers to the dealer through the signcryption scheme. After receiving the signcrypted messages, the dealer recovers the secret shares of the participants. The dealer reconstructs the secret key while running the algorithm as proposed in Section 3.1 for detection and identification of dishonest participants.

The schemes [20],[21] proposes *ECC* based threshold signcryption schemes. The scheme [20] adopts verifiable secret key split protocol which is based on the approach of verifiable secret sharing scheme [15],[22] and this scheme [20] allows each participant to verify his own secret share for the implementation of group signcryption scheme. Our proposed scheme uses this approach in modified form for the verification of trustworthiness of the *TD* and the signcryption scheme. The signcryption scheme which our proposed scheme uses does not use one way hash function [23]. It consists of three steps:

- Setup phase
- Verification phase of the *TD* by the participants.
- Detection as well as identification of cheaters and reconstruction of secret key.

a) Setup phase

First, the participants register themselves to the *TD*. The *TD* assigns a unique identification number (id_i) to each participant.

The *TD* performs the following tasks which includes choosing *ECC* parameters, selection of two polynomials, generation of *ECC* public and private keys, distribution of secret shares to the n-participants.

Step-1:

Let G be the set of n-participants, $G = \{P_1, P_2, \ldots, P_n\}$. The TD chooses a secure elliptic curve (C) over finite field F_p where p is a prime number, base point B of order q (where $q \geq 160$ bits).
The TD follows the steps below:

- It chooses a random integer $d_G \in [1, q-1]$ which is a private key of group G and the corresponding public key is $Q_G = d_G.B$. The TD distributes securely d_G to the members of group G.
- It also chooses its private key $d_{TD} \in [1, q-1]$ and corresponding public key $Q_{TD} = d_{TD}.B$.
- The TD publishes p, q, Q_G, Q_{TD}.

Step-2:

The TD randomly generates a secret polynomial:

$$f^v(x) = a_0^v + a_1^v \cdot x + \ldots + a_{t-1}^v \cdot x^{t-1} \ mod \ q$$

where $a_0^v = f^v(0) = d_G$.

The coefficients $a_1^v, \ldots, a_{t-1}^v$ are selected randomly from finite field F_p.
The TD follows the steps below:

- It computes $d_i^v = f^v(x_i)$ as private key of P_i where $(1 \leq i \leq n)$ and corresponding public key $Q_i^v = d_i^v.B$. The unique parameter x_i for each participant, P_i are selected randomly from the finite field F_p.
- It publishes Q_i^v and sends d_i^v securely to P_i where $(1 \leq i \leq n)$.
- The $(t-1)$ number of $(a_j^v.B)$s are kept in secured place by the TD. It sends $(t-1)$ number of $(a_j^v.B)$s to all active participants by encrypting with the corresponding P_i's public key Q_i^v as requested by the participants when they require to reconstruct the secret key for verifying the trustworthiness of the TD.

Step-3:

Now, again the TD generates a secret polynomial:

$$f(x) = a_0 + a_1.x + \ \ldots + \ a_{t-1}.x^{t-1} \ mod \ q$$

where $a_0 = f(0) = d_{secret}$ and d_{secret} is the shared secret key.

The coefficients $a_1, \ldots, a_{t-1}$ are selected randomly from finite field F_p.
The TD follows the steps below:

- It computes $d_i^{secret} = f(id_i)$ as secret share for P_i where $(1 \leq i \leq n)$ and generates a set of n unique random numbers PS_i corresponding to each id_i. It prepares a list of $[P_i, id_i, PS_i]$ for the n-participants and the list is kept securely by the TD.

- It securely distributes $[PS_i,\ d_i^{secret}]$ to all P_i where $(1 \leq i \leq n)$.
- The secret polynomials, private keys of the participants d_i^v, the secret shares d_i^{secret}, the shared secret key d_{secret} are erased from the *TD*.

It has been stated that the group key d_G, the secret share d_i^{secret} and the parameter $[PS_i,\ d_i^{secret}]$ are transferred to the participants securely. Any secure channel can be used for that purpose *e.g.*, postage system.

b) Trusted authority or dealer *(TD)* verification phase

Now, for the trustworthiness of the dealer as well as for verification of private keys of the participants, the participants request for the $(t-1)$ number of $(a_j^v.\ B)$s from the *TD* [20]. The active participants receive encrypted message of $(t-1)$ number of $(a_j^v.\ B)$s from the *TD* as requested by them. Each participant P_i decrypts using his private key d_i^v and computes the following equation as follows:

$$d_i^v.B = \sum_{j=0}^{t-1} i^j.(a_j^v.\ B) \quad \text{where } a_0^v = f^v(0) = d_G.$$

If this equation holds, the trustworthiness of the *TD* is accepted otherwise rejected.

c) Detection as well as identification of cheaters and reconstruction of secret key

The participants exchange messages among themselves for reconstruction of the shared secret key as it requires more than threshold number of participants should agree for the reconstruction of the secured shared secret key. The active participants now verify the trustworthiness of the *TD e.g.*, whether it has been compromised or captured by using the algorithm described in Section 3.3(b). After the participants are satisfied with the trustworthiness of the *TD*, the following steps are followed:

Signcryption:

Step-1:

Each participant, P_i chooses a random integer, $k_i \in [1,\ q-1]$ per each signcryption session and computes $Y_i^1 = k.B = (x_1,\ y_1)$ and $Y_i^2 = k_i.Q_{TD} = (x_2,\ y_2)$ where x_1, x_2 are the $x-$coodinates and y_1, y_2 are the $y-$coordinates of the corresponding points on C.

Step-2:

The participant, P_i computes $r_i = m_i.x_2 \ mod \ q$ where $m_i = [PS_i,\ d_i^{secret}]$ and $s_i = k_i$ - $d_A.r_i \ mod \ q$.

Step-3:

The signcrypted message $(r_i,\ s_i, Y_i^1)$ are sent to the *TD* through public channel by the participant, P_i. It is evident that the parameter, r_i includes the message m_i while the parameter, s_i includes the signature of the sender participant, P_i.

Unsigncryption:

Step-1:

After receiving the message (r_i, s_i, Y_i^1), the TD verifies validity of the received message and recovers the message m_i. The TD computes $Y_{TD}^1 = r_i.\ Q_i + s_i.B = (x_3, y_3)$ and $Y_{TD}^2 = d_{TD}.Y_{TD}^1 = (x_4, y_4)$ where x_3, x_4 are the $x-$coodinates and y_3, y_4 are the $y-$coordinates of the corresponding points on C. The TD verifies whether Y_i^1 matches with Y_{TD}^1 or not. If it matches, the signcrypted message is valid (r_i, s_i, Y_i^1), otherwise it is invalid.

Step-2:

If the signcryption message (r_i, s_i, Y_i^1) is valid, then TD recovers the message m_i by $m_i = r_i.\ (x_4)^{-1}\ mod\, q$.

It can be proved that

$$Y_{TD}^1 = r_i.\ Q_i + s_i.B = r_i.(d_i^{secret}.B) + (k_i - d_i^{secret}.r_i).B = k_i.B = Y_i^1.$$

Thus the validity of the signcrypted message (r_i, s_i, Y_i^1) can be verified by the condition $Y_i^1 = Y_{TD}^1$.

Again, it can be shown that

$$Y_{TD}^2 = d_{TD}.Y_{TD}^1 = d_{TD}.Y_i^1 = d_i^{secret}.(k_i.B) = k_i.\ (d_{TD}.B) = k_i.Q_{TD} = Y_i^2 = (x_2, y_2) = (x_4, y_4)$$

Now, we know that $r_i = m_i.x_2\ mod\ q = m_i.x_4\ mod\ q$.

So, the message $m_i = r_i.\ (x_4)^{-1}\ mod\ q$.

This signcryption scheme protects the security features like unforgeability, confidentiality, non-repudiation and forward secrecy.

Now, by using Lagrange's interpolation polynomial, the TD recovers secret key from the set of t out of n-participants. It also simultaneously detects and identifies the cheating participants by using the algorithm described in Section 3.1.

The pictorial diagram of the distribution of private keys, group key, shared secret, signcrypted messages of the proposed scheme is shown in the Figure-1.

4 Security Analysis

The $(t-1)$ number of $(a_j^v.B)$s and the list of $[P_i, id_i, PS_i]$ for the n-participants will be kept securely by the TD. The security analysis shows the prevention of various types of attacks below:

a) TD capture attack:

At the time of reconstruction of the secret key, the participants request $(t-1)$ number of $(a_j^v.B)$s from the TD. After receiving $(t-1)$ number of $(a_j^v.B)$s, the participants verify the trustworthiness of the TD as well as validity of their private keys as described in Section 3.3(b). Any compromised or fake TD cannot

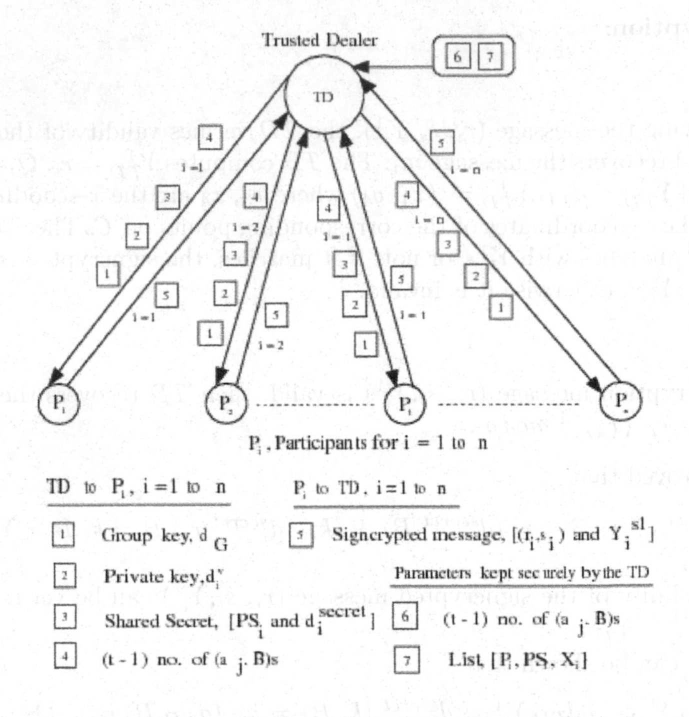

Fig. 1. Pictorial diagram of the proposed scheme

create original ($a_j^v.B$)s as uncompromised *TD* had selected the $(t-1)$ number of a_j^v randomly from the F_p in the initialization phase.

If the *TD* is captured or compromised at any stage after setup phase, it will not be able to reconstruct the secret key. The threshold t number of $[id_i,\, d_i^{secret}]$ will not be present in the *TD* for reconstruction of secret key as those information have been erased from the *TD* after the setup phase. The *TD* will obtain the secret shares of the participants only if the participants are convinced of the trustworthiness of the *TD*. Thus, any type of *TD* capture attack is prevented in this scheme.

b) Malicious participant attack and passive attack:

Each participant signcrypts his secret share and sends it to the *TD*. If a signer P_i (malicious participant) modifies his private key after the node is compromised or captured, it will be easily detected by the *TD* through this signcryption scheme. Even if any fake node eavesdrops (r_i, s_i) value of any participant, he will not be able to recover the message as the sender participant only knows the value k_i though it will be very difficult for the eavesdroppers to recover any message from the transmitted signcrypted messages. Thus, any type of malicious participant and passive attacks are prevented in this scheme. Any participant who sends any *inconsistent* share to deceive the *TD* can be easily detected by the *TD* by running the algorithm as discussed in Section 3.1.

c) Collusion attack:

If the t participants collude to construct the secret key, they will not be able to do that as they require t number of $[id_i, d_i^{secret}]$ as each participant only owns $[PS_i, d_i^{secret}]$. The trusted dealer TD can only reconstruct the secret key as it only owns the list of $[P_i, id_i, PS_i]$ which is also kept in a secured place by the TD. After collecting all the $[PS_i, d_i^{secret}]$ of the participants, it recovers from the list the corresponding $[id_i, d_i^{secret}]$ of the participants for the reconstruction of the secret key. Thus, any type of collusion attack by the malicious participants is prevented in this scheme.

After detection and identification of dishonest participants or cheater, the TD may blacklist (may not allow them in the future reconstruction process) the concerned cheaters as the cheaters would never able to deny the offence as in this scheme proper non-repudiation scheme (signcryption) has been adopted. Though in some cases, the sending of *inconsistent* or invalid shares may be accidental. In such cases, the TD has the liberty to ask again the erring participants to send the valid shares.

5 Conclusion

In our proposed scheme, the participants transfer their secret shares to the trusted dealer through the ECC based signcryption scheme for secure reconstruction of secret key. As the shares are transferred as signcrypted messages, eavesdroppers cannot be successful for decryption of secret shares. The threshold number of participants can never be able to reconstruct the secret key by collusion. When the active participants need to reconstruct the secret key after long time of generation of secret shares by the TD or when the participants communicate the TD from a hostile/non-trusted environment, they may need to verify the trustworthiness of the trusted dealer TD. The scheme makes the provision to verify the trustworthiness of the trusted dealer. This scheme also detects and identifies the dishonest participants. As the scheme uses the signcryption scheme, computational cost and communication overhead is low and the participants can use mobile systems connecting in wireless environment for reconstruction of secret key.

References

[1] Shamir, A.: How to share a secret. Communications of the ACM 22(11), 612–613 (1979)

[2] Blakley, G.: Safeguarding cryptographic keys. In: Proceedings of AFIPS 1979 National Computer Conference, vol. 48, pp. 313–317 (1979)

[3] Chaum, D.: Computer Systems Established, Maintained, and Trusted by Mutually Suspicious Groups. Tech. Rep., Memorandum No. UCB/ERL M/79/10, University of California, Berkeley, CA (February 1979)

[4] Koblitz, N., Menezes, A., Vanstone, S.: The state of elliptic curve cryptography. Designs, Codes and Cryptography 19(2-3), 173–193 (2000)

[5] Tompa, M., Woll, H.: How to Share a Secret with Cheaters. In: Odlyzko, A.M. (ed.) CRYPTO 1986. LNCS, vol. 263, pp. 261–265. Springer, Heidelberg (1987)
[6] Rabin, T., Ben-Or, M.: Verifiable Secret Sharing and Multiparty Protocols with Honest Majority. In: Proceedings of the twenty-first annual ACM symposium on Theory of computing (1989)
[7] Araki, T.: Efficient (k, n) Threshold Secret Sharing Schemes Secure Against Cheating from n-1 Cheaters. LNCS, vol. 4586. Springer, Heidelberg (2007)
[8] McEliece, R.J., Sarwate, D.V.: On sharing secrets and Reed-Solomon codes. Comm. ACM 24, 583–584 (1981)
[9] Blundo, C., De Santis, A., Gargano, L., Vaccaro, U.: Secret sharing schemes with veto capabilities. In: Cohen, G., Lobstein, A., Zémor, G., Litsyn, S.N. (eds.) Algebraic Coding 1993. LNCS, vol. 781, pp. 82–89. Springer, Heidelberg (1994)
[10] Lin, H.Y., Harn, L.: A generalized secret sharing scheme with cheater detection. In: Matsumoto, T., Imai, H., Rivest, R.L. (eds.) ASIACRYPT 1991. LNCS, vol. 739, pp. 149–158. Springer, Heidelberg (1993)
[11] Pieprazyk, J., Li, C.H.: Multiparty key agreement protocols. Proceedings of Computers and Digital Techniques 147(4) (July 2000)
[12] Tartary, C., Wang, H.: Dynamic threshold and cheater resistance for shamir secret sharing scheme. In: Lipmaa, H., Yung, M., Lin, D. (eds.) Inscrypt 2006. LNCS, vol. 4318, pp. 103–117. Springer, Heidelberg (2006)
[13] Wei, C., Xiang, L.: A new Dynamic Threshold Secret Sharing Scheme from Bilinear Maps. In: International Conference on parallel Workshops, ICPPW 2007 (2007)
[14] Feldman, P.: A practical scheme for non-interactive verifiable secret sharing. In: Proceedings of the 28th IEEE Symposium on Foundations of Computer Science, pp. 427–437. IEEE, Los Alamitos (1987)
[15] Pedersen, T.P.: Non-interactive and information-theoretic secure verifiable secret sharing. In: Feigenbaum, J. (ed.) CRYPTO 1991. LNCS, vol. 576, pp. 129–140. Springer, Heidelberg (1992)
[16] Simmons, G.: An introduction to shared secret schemes and their applications. Sandia Reports, SAND 88-2298 (1988)
[17] Harn, L., Lin, C.: Detecting and identification of cheaters in *(t, n)* secret sharing scheme. Designs, Codes and Cryptography 52, 15–24 (2009)
[18] Zheng, Y.: Digital signcryption or how to achieve cost (Signature & encryption) << cost(Signature) + cost(Encryption). In: Kaliski Jr., B.S. (ed.) CRYPTO 1997. LNCS, vol. 1294, pp. 165–179. Springer, Heidelberg (1997)
[19] Zheng, Y., Imai, H.: How to construct efficient signcryption schemes on elliptic curves. Information Processing Letters 68(5), 227–233 (1998)
[20] Chaggen, P., Xiang, L.: Threshold Signcryption Scheme Based on Elliptic Curve Cryptosystem and Verifiable Secret Sharing. In: Proceedings of International conference on Wireless Communications, Networking and Mobile Computing, September 23-26, vol. 2 (2005)
[21] Hu, C.-J., Zhang, Q.-F.: An Improved Authenticated (T,N) Threshold Signature Encryption Scheme Based on ECC. In: Proceedings of the Fifth International Conference on Machine Learning and Cybernetics, Dalian, 13-16, pp. 2674–2678 (2006)
[22] Pedersen, T.P.: Distributed provers with applications to undeniable signatures. In: Davies, D.W. (ed.) EUROCRYPT 1991. LNCS, vol. 547, pp. 221–242. Springer, Heidelberg (1991)
[23] Lee, W.B., Chang, C.C.: Authenticated encryption scheme without using a one-way hash function. Electronic letters 31(19), 1656–1657 (1995)

Model-Based Software Reliability Prediction

G. Sri Krishna and Rajib Mall

Department of Computer Science and Engineering
Indian Institute of Technology, Kharagpur, 721 302, India

Abstract. We investigate an approach for early estimation of the reliability of software products based on their design models. We start by computing several structural and behavioral metrics from Unified Modeling Language(UML) models. The choice of the selected metrics has been based on their ability to influence the final product reliability. A trained neural net is used to predict the reliability of individual modules. The final product reliability is obtained from these predicted values. Our approach can help to decide between design alternatives and also help a manager trade off between the cost of redesigning certain modules and increased testing effort to meet product reliability goals. *abstract* environment.

Keywords: Software reliability, object-oriented metrics, ANN, UML, early reliability prediction.

1 Introduction

A general trait that many present day software products share is that they are extremely expensive, large and object-oriented in structure. Reliability of such products is being accorded a high priority by the customers of these products. In this context, it is crucial for the development manager to get an estimation of the final product reliability early in the development cycle. The manager also needs to know the extent to which the final reliability might improve on account of increased testing of certain modules *vis a vis* redesign of the modules. Based on this information, the manager might decide to redesign certain modules and increase testing effort on certain(possibly other) modules to cost-effectively meet the final product reliability goals.

We have developed a neural network model that can predict system-level reliability by estimating the reliability of the individual modules of the system. Software metrics including both structural and behavioral metrics are computed from UML diagrams and a neural network model is used to estimate the number of defects in each module. Based on these values, the reliability of each use case is determined.

The rest of the paper is organized as follows. In Sec. 2, the related work have been very briefly surveyed. In Section 3, an overview of our approach has been presented. In Section 4, the metrics used by our approach have been defined. Section 5 discusses the neural network model used. Section 6, presents an empirical study. Section 7 presents a comparison with related work. Section 8 concludes this paper.

S.K. Prasad et al. (Eds.): ICISTM 2010, CCIS 54, pp. 145–155, 2010.

2 Related Work

Khoshgoftaar [14] introduced neural networks for predicting software development faults. They used 14 structured metrics as the input variables to the neural network and total number of faults found in the programs as the output variable. They found that neural network models are capable of better prediction than the statistical models. In a later work, Khoshgoftaar et al. [12] presented a neural network model for a large telecommunication system, classifying modules as either not fault-prone or fault-prone. They compared the neural-network model with a nonparametric discriminate model, and found the neural network model had better predictive accuracy. Cheung et al. [5] presented an architecture-level approach to predict the individual component reliabilities which can be "plugged into" existing system-level approaches. They used a discrete time Markov chain stochastic process with a state and transition matrix. They computed reliability of the individual components by solving for the steady state probability of not being in failure states by applying standard numerical techniques to solve the Markov chain model. Ravi Kumar et al. [21] presented linear and non-linear ensemble models. Their experiments revealed that non-linear ensemble models outperformed linear ensemble models.

Nagappan et al. [18] proposed a metric suite for assessment of software reliability and to provide feedback to developers on the quality of their testing effort. They collected information from the source code and test programs.

The applications of principal component analysis have been explored by Khoshgoftaar [13].They used neural network modeling as a way of improving the predictive quality of neural network models. Their results show that use of principal component analysis improves predictive quality and also reduces the training time of the neural network. Neural network model for predicting the testability of program modules was also found to be successful.

Two neural networks (Ward, GRNN) were introduced by Quah and Thwin [19] for quality prediction of classes developed in object-oriented environment using measures (mainly from CK metric suite). They applied this model for predicting number of faults and number of lines of change in the classes. Their results show that GRNN models work better than Ward neural network.

3 An Overview of Our Approach

Though many approaches for reliability prediction have been investigated by researchers, application of software design metrics to reliability prediction has largely remained unexplored. UML has become the standard technique for design and analysis model construction, we concentrate on metrics that can be computed from UML models. In the following we present an overview of our approach.

We first identify the metrics relevant for reliability estimation. Many metrics are available in the literature [6,7,9,15,16,17]. Based on this, only those metrics that have some bearing on the final product reliability are chosen.

We have designed a tool which computes the metric values from the UML diagrams of the system. UML diagrams are drawn using any of the available

CASE tools and are exported to XML format. The XML file is then parsed using XML parser and the relevant metric values are computed.

We seed each class of the systems developed with a known number of errors and test the modules. Testing reveals some seeded errors, some unseeded errors are also exposed. The number of seeded and unseeded errors detected are used in estimate the total errors latent in each class of the system.

Principal component analysis is performed on the obtained metric data to reduce the size of the data set. A stopping value is chosen to limit the data size which removes the redundancy present in the data. The components of reduced data set are called principal components and they are orthogonal to each other. This reduced data set along with number of defects present in each class is used to train the Neural network model. The neural net is trained on the computed metric data.

Reliability of the system is computed by determining the reliability of each use case of the system. This is done by the considering the estimate of operation profile given by design engineers and taking in to account the message passing among different modules for execution a particular use case.

4 Metrics Used in Our Approach

In the following we describe the metrics chosen for reliability estimation.

Depth of Inheritance Tree(DIT) [6,7] is the height of the inheritance tree. This determines the complexity of a class. The deeper a class is in the hierarchy, the greater the number of methods it is likely to inherit making it more complex to predict its behavior.

Number of Children(NOC) [6,7] measures amount of potential reuse of the class. The more reuse a class might have, the more complex it may be, and the more classes are directly affected by changes in it implementation.

Weighted Methods per Class(WMC) [6,7] is the summation of *McCabes cyclomatic* complexity of each local method. The more control flows a class methods have, the harder it is to understand them, thus, the harder it is to maintain them. A method with a low cyclomatic complexity is generally better.

Public Instance Methods(PIM) [16] of a class is a good measure of the amount of responsibility of the class.

Number of Instance Methods(NIM) and **Number of Instance Variables(NIV)** [16] in a class relates to the amount of collaboration being used.

Number of Class Methods(NCM) and **Number of Class Variables(NCV)** [16] are related to whole class. The number of methods available to the class and not its instances affects the size of a class.

Number of Methods Overridden(NMO) [16] is the total number of methods overridden by a subclass. This metric looks at the quality of the classes use of inheritance. It examines superclass-subclass inheritance relationships.

Number of Methods Inherited(NMI) [16] is the total number of method inherited by a subclass. This metric looks at the quality of the classes use of inheritance. It examines superclass-subclass inheritance relationships.

Number of Methods Added(NMA) [16] is the total number of methods defined in a subclass which are neither overridden nor inherited.

Number of Associations(NAssoc) [9] determines the associations of a class with other classes which affects the complexity of the class. This metric identifies which classes are important in a class model.

Height of Aggregation(HAgg), **Number of Direct Parts(NODP)** and **Number of Parts(NP)** [9] metric measures the class complexity due to aggregation relationship (whole-part relationship).

Number of Dependencies In(NDepIn) [9] tells the number of classes that depend on a given class.

Number of Dependencies Out(NDepOut) [9] tells the number of classes on which a given class depends.

4.1 Implementation of UML Metric Calculator

Fig. 1 shows a schematic representation of a tool that we have implemented for computation of the metric values from UML diagrams. We have used MagicDraw CASE tool[25] to draw UML diagrams. These UML diagrams are exported to XML format using the export functionality of the CASE tool. We have named our developed tool UML Metric Calculator(UMC). This has been implemented in java using JDOM parser.

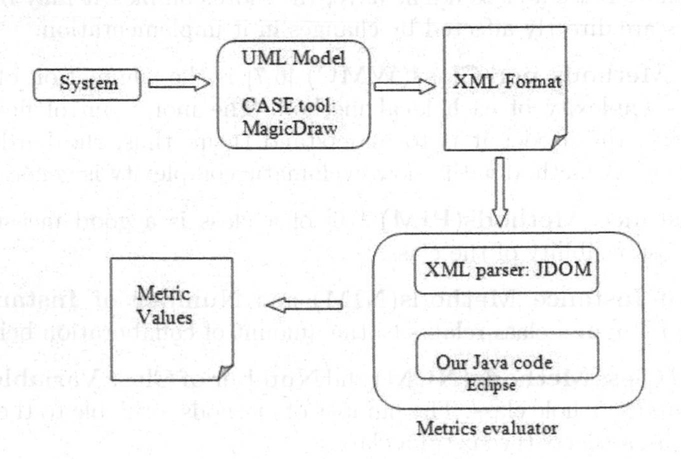

Fig. 1. Schematic Model of UML Metric Calculator

5 Neural Network

We use the metric values computed from UML models as our independent variables in predicting the number of defects of the system. We use GRNN which has been found to be effective for prediction purpose [19]. Fig. 2 shows an outline of our neural network modeling approach.

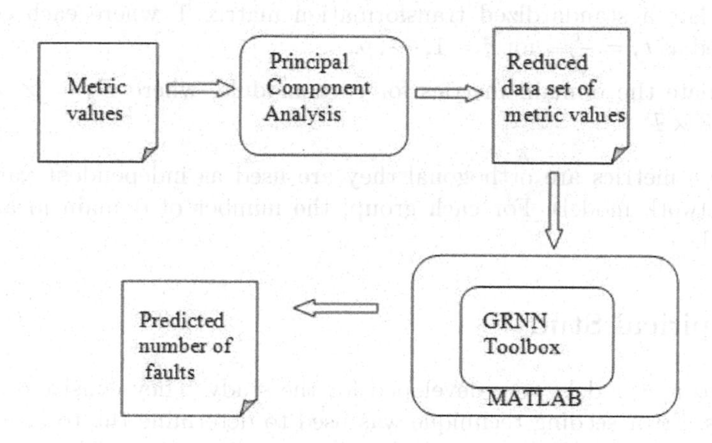

Fig. 2. Schematic model of Fault Prediction Process

5.1 General Regression Neural Network (GRNN)

GRNN is based on a one-pass learning algorithm with a highly parallel structure. GRNN is a powerful memory based tool that could estimate continuous variables and converges to the underlying regression surface. An important strength of GRNN is its ability to deal with sparse data effectively. For a GRNN, the number of neurons in the hidden layer is usually the number of patterns in the training set because each pattern in the training set is represented by ON neurons. The primary advantage to the GRNN is the speed at which the network can be trained. Training a GRNN is performed in one pass.

5.2 Principal Component Analysis

Many software design metrics have high correlation with each other. Principal components analysis can help to transform raw data into variables that are not correlated to each other. From the original object-oriented software metrics, principal component variables are computed and these are called as domain metrics. Principal components analysis is primarily a data reduction technique. The principal components are linear combinations of m standardized metrics Z1,· · ·,Zm. The principal components represent the same data in a new rotated coordinate system where the variability is maximized in each direction and the principal components are uncorrelated. Principal components analysis performs the following calculations, given an $Z(n \times m)$ matrix of standardized metric data

- Calculate the covariance matrix S of Z
- Calculate eigen values j and eigen vectors e_j of S, $j = 1, \cdots, m$.
- Reduce the dimensionality of the data. We choose the stopping rule as components with eigen value > 1.
- Calculate a standardized transformation matrix T where each column is defined as $t_j = \frac{e_j}{\sqrt{\lambda_j}}$ for $j = 1, \cdots, p$.
- Calculate the domain metrics for each module, where $D_j = Z \times T_j$ and $D = Z \times T$

Since these metrics are orthogonal they are used as independent variables in neural network models. For each group, the number of domain measures are calculated.

6 Empirical Study

Seven software models were developed for the study. They consisted a total of 37 classes. Error seeding technique was used to determine the total number of defects in each class. The following types of errors were introduced and tested.

1. Arithmetic errors
2. Logical errors
3. Absence of initialization
4. Improper type
5. Array index mismatch
6. Improper parameter types

UMC was used to compute metrics from the UML diagrams of the systems developed. Two-thirds of the data was used for training purpose and remaining for validation. After standardizing the metric data, we performed principal component analysis. Table 1 shows the relationship between original metrics and principal components. After performing PCA, it identified five principal components, which capture 83.446% of the data set variance as shown in Table 1. Metrics with values more than 0.3 have been shown in bold face in the table as they are the affect the corresponding principal component the most.

6.1 Goodness of Fit

To measure the goodness of fit of the neural network model, we use the coefficient of multiple determination (R-square), the coefficient of correlation (r), r-square, mean square error and p-values.

Coefficient of correlation (r) (Pearson's Linear correlation coefficient) is a statistical measure of the strength of the relationship between the actual versus predicted outputs. The r coefficient can range from $-1 to +1$. The closer r is to 1,

Table 1. Rotated principal components for the systems considered

Metrics	PC1	PC2	PC3	PC4	PC5
WMC	-0.3578	-0.1387	0.1277	-0.1669	0.0541
DIT	0.0325	**0.3575**	**0.3467**	-0.0357	0.0131
NOC	-0.1900	-0.2437	0.2905	0.3346	-0.0719
PIM	-0.3440	-0.1557	0.1820	-0.1640	0.0813
NIV	-0.3702	-0.0427	-0.1801	0.0319	0.0109
NOV	0.0042	**0.3242**	**0.3661**	-0.1832	0.0242
NCV	-0.1788	0.2349	-0.1157	-0.3263	-0.0869
NOA	-0.3731	0.0057	-0.1870	-0.0330	-0.0066
NIM	-0.3440	-0.1557	0.1820	-0.1640	0.0813
NMI	0.0132	**0.3437**	**0.3780**	-0.1145	0.0284
NAM	-0.3077	-0.1372	0.2549	-0.0607	-0.0716
NAbsm	-0.1350	-0.0753	0.1146	**0.6747**	0.1170
NP	**0.3049**	-0.3452	0.0692	-0.2098	0.0125
NODP	0.1953	-0.3687	0.0930	-0.1926	0.0004
NAssoc	-0.1751	-0.0010	-0.2357	-0.2193	0.1754
NDepIn	0.0785	-0.0747	-0.0345	-0.0778	**0.7784**
NDepOut	0.0431	-0.2089	0.1355	-0.1034	-0.5171
HAgg	**0.3249**	-0.3452	0.0692	-0.2098	0.0125
NCM	-0.1265	0.1299	-0.4286	-0.0306	-0.2151
Eigen Values	6.2889	4.6717	2.4310	1.4026	1.0796
% variance	33.0995	24.5880	12.7946	7.3820	5.6819
Cumulative % variance	33.0995	57.6875	70.4821	77.7641	**83.4460**

the stronger the positive linear relationship, and closer r is to -1, the stronger the negative linear relationship. When r is near 0, there is no linear relationship.

$$r = \frac{S_{xy}}{\sqrt{S_{xx} - S_{yy}}}$$

$$S_{xx} = \sum x^2 - \frac{1}{n} \star (\sum x)^2$$

$$S_{yy} = \sum y^2 - \frac{1}{n} \star (\sum y)^2$$

$$S_{xy} = \sum xy - \frac{1}{n} \star (\sum x) \sum y$$

where n equals the number of patterns, x refers to the set of actual outputs and $\hat{y}$ refers to the predicted output.

Coefficient of multiple determinations is a statistical indicator. It compares the accuracy of the model to the accuracy of a trivial benchmark model wherein the prediction is just the mean of all of the samples. It can range from 0 to 1, and a perfect fit would result in an R-square value of 1, a very good fit near 1, and a very poor fit less than 0, it is calculated as follows:

$$R^2 = 1 - \frac{\sum (y - \hat{y})^2}{\sum (y - \overline{y})^2}$$

where y is the actual value for the dependent variable,$\hat{y}$ is the predicted value of y and $\overline{y}$ is the mean of the y values.

Goodness of fit measures are shown in the Table 2.

Table 2. Experimental results for goodness of fit

Correlation coefficient (r)	0.8684
Mean square error	0.7574
R-square	0.7541
p-value	0.0024

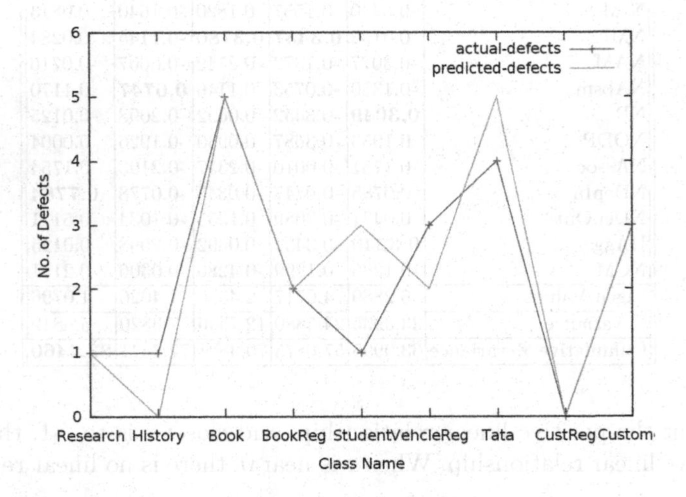

Fig. 3. Predicted and actual number of defects

6.2 Results

The plot in Fig. 3 shows the predicted and actual number of defects for each class. Table 3 shows the number of failures that originally occurred and the number of failures as predicted by the neural network model for each module. The predicted failures gives the reliability measure of modules and is used to determining the reliability of the system.

The reliability of each use case is determined by determining the execution path followed for the execution of the use case. If the execution path of a use case involves modules M_1 and M_2 with reliabilities r_1 and r_2 respectively and

Table 3. Reliability factors

System	failures occurred	predicted failures
Module 1	9	12
Module 2	12	13
Module 3	2	4
Module 4	5	5

has an operation profile value of p, then reliability of the use case is calculated as $r_1 \star r_2 \star p$. The reliability values for the system are shown in Table 4. It can be observed that there is a good correlation between predicted and actual number of defects of class as evident from r value of 0.8684 from Table 2.

Table 4. Predicted Reliability Values

Use case	Predicted $ROCOF \times 10^{-3}$
u1	9.4
u2	6.0
u3	10.2
u4	10.0

7 Comparison with Related Work

Several code based approaches have been reported in the literature[19,22]. However, design metrics based approaches have scarcely been reported. We have presented a model based approach which computes the metric values from UML diagrams of the system, which can be used to predict product reliability at early phases of a software life cycle model.

Ravi Kumar et al. [21] presented ensemble models for reliability prediction and found that non-linear ensemble models were good at prediction, but they did not consider design metrics and using this approach it would be difficult to determine how design changes can impact the product reliability.

Nagappan et al. [18] proposed metric suite for reliability estimation, but these metrics can only be computed at the testing phase of a software life cycle and cannot be used at an early stage.

Quah and Thwin [19] concentrated mainly on CK metrics[6,7] and Li-Henry metrics[15]. Sankarnarayanan et al. [22] computed metrics from C++ code of the system and predicted the number of faults. In comparison to these work, our approach considers a set of structural and behaviorial metrics computed from UML diagrams. Compared to these work, one important advantage of our work is that we first predict the module-level faults. Based on this, we predict the final product reliability. As a result, our approach not only predicts the final product reliability more accurately (verified through experiments), but also lets a project manager scope to choose between different design alternatives. Also, our approach can be used by a project manager to trade off between the cost of redesigning certain modules versus increased testing of modules to meet the final reliability goals.

8 Conclusion

A major goal of our work was to make a tool available to the project manger, which he can use early in the product life cycle to determine the modules that need to be redesigned and the ones that need to be tested with higher intensity to meet the final product reliability goals. Our approach can also be used to study the impact of a design change on the final product reliability. A trained neural network was used to predict the final product reliability using various structural and behavioral metrics. Results from the limited experiments that we have carried out show that our approach outperforms related approaches. We are now trying to use our approach on industry standard problems. We are also extending our approach to make it applicable to prediction of the reliability of embedded and safety critical systems.

References

1. Bansiya, J., Davis, C.: A Hierarchical Model for Object-Oriented Design Quality Assessment. IEEE Transactions on Software Engineering 28(1), 4–17 (2001)
2. Brito Abreu, F., Carapuça, R.: Object-Oriented Software Engineering: Measuring and Controlling the Development Process. In: 4th International Conference on Software Quality, McLean, VA, USA (1994)
3. Brito Abreu, F., Melo, W.: Evaluating the Impact of Object-Oriented Design on Software Quality. In: 3rd International Metric Symposium, pp. 90–99 (1996)
4. Specht, D.F.: A General Regression Neural Network. IEEE Transactions on Neural Networks 2(6), 568–576 (1991)
5. Cheung, L., Roshandel, R., Golubchik, L.: Early Prediction of Software Component Reliability. In: Proceedings of the 30th international conference on Software engineering, pp. 111–120 (2008)
6. Chidamber, S., Kemerer, C.: Towards a Metrics Suite for Object Oriented Design. In: Conference on Object-Oriented Programming: Systems, Languages and Applications (OOSPLA 1991). SIGPLAN Notices, vol. 26(11), pp. 197–211 (1991)
7. Chidamber, S., Kemerer, C.: A Metrics Suite for Object Oriented Design. IEEE Transactions on Software Engineering 20(6), 476–493 (1994)
8. Harold, E.R.: Processing XML with Java: A Guide to SAX, DOM, JDOM, JAXP and TRAX. Addison-Wesley, Reading (2004)
9. Genero, M., Piattini, M., Calero, C.: Early Measures for UML Class Diagrams. L'object 6(4), 489–515 (2001)
10. Harrison, R., Counsell, S., Nithi, R.: Coupling Metrics for Object-Oriented Design. In: 5th International Software Metrics Symposium, pp. 150–156 (1998)
11. Hu, Q.P., Dai, Y.S., Xie, M., Ng, S.H.: Early Software Reliability Prediction with Extended ANN Model. In: Proceedings of the 30th Annual International Computer Software and Applications Conference (COMPSAC 2006), vol. 2, pp. 234–239 (2006)
12. Li, W., Henry, S.: Object-Oriented Metrics that Predict Maintainability. Journal of Systems and Software 23(2), 111–122 (1993)
13. Khoshgoftaar, T.M., Allen, E.B., Hidepohl, J.P., Aud, S.J.: Application of Neural Network to Software Quality Modelling of a Very Large-Scale Telecommunication System. IEEE Transactions on Neural Networks 8(4), 902–909 (1997)
14. Khoshgoftaar, T.M., Szabo, R.M.: Improving Neural Network Predictions of Software Quality using Principal Component Analysis. In: Proceedings of IEEE International World Congress on Computational Intelligence, pp. 3295–3300 (1994)

15. Khoshgoftaar, T.M., Szabo, R.M., Guasti, P.J.: Exploring the Behaviour of Neural Network Software Quality Models. Software Engineering Journal, 89–96 (May 1995)
16. Cheung, L., Roshandel, R., Golubchik, L.: Early Prediction of Software Component Reliability. In: Proceedings of the 30th international conference on Software engineering, pp. 111–120 (2008)
17. Lorenz, M., Kidd, J.: Object-Oriented Software Metrics: A Practical Guide. Prentice Hall, Englewood Cliffs (1994)
18. Marchesi, M.: OOA Metrics for the Unified Modelling Language. In: 2nd Euromicro Conference on Software Maintenance and Reengineering, pp. 67–73 (1998)
19. Nagappan, N., Williams, L., Vouk, M.: Towards a Metric Suite for Early Software Reliability Assessment. In: 2nd Euromicro Conference on Software Maintenance and Reengineering, pp. 67–73 (2003)
20. Quah, T.S., Thewin, M.M.T.: Application of Neural Network for Software Quality Prediction. In: Proceedings of the International Conference on Software Maintenance (ICASM 2003), vol. 3 (2003)
21. Mall, R.: Fundamentals of Software Engineering. Prentice-Hall India, Englewood Cliffs (2003)
22. Raj Kiran, N., Ravi, K.: Software Reliability Prediction by Soft Computing Techniques. The Journal of Systems and Software 81, 576–583 (2008)
23. Sankarnarayanan, V., Kanmani, S., Thambidurai, P., Rhymend Uthariaraj, V.: Object Oriented Software Quality Prediction Using General Regression Neural Network. ACM SIGSOFT Software Engineering Notes 29(5) (2004)
24. Sarvana, K., Mishra, R.B.: An Enhanced Model for Early Software Reliability Prediction using Software Engineering Metrics. In: Proceedings of the 2008 Second International Conference on Secure System Integration and Reliability Improvement, pp. 177–178 (2008)

Real-Time Power-Aware Routing Protocol for Wireless Sensor Network

Omar Al-Jarrah[1], Ayad Salhieh[2], and Aziz Qaroush[1]

[1] Department of Computer Engineering, Jordan University of Science and Technology, Irbid 22110, Jordan
{aljarrah,qaroush}@just.edu.jo
[2] Department of Computer Science, Jordan University of Science and Technology, Irbid 22110, Jordan
salhieh@just.edu.jo

Abstract. Recent advances in wireless technology have enabled the design of tiny low-cost sensor modules. Many applications may need real-time data transmission such as imaging and video data transmission which requires provision of Quality of Service (QoS) of the communication network. In this paper, we have presented a Real-Time Power-Aware (RTPA) routing protocol that is energy efficient and support real-time application. The presented protocol uses a geographic approach and considers the remaining energy and the delay to selects the suitable or the best route from the set of available routes between a source node and a destination node. In addition, the protocol uses two route decision components: one for Real-Time (RT) traffic and the other for Non-Real-Time (NRT) traffic. For RT traffic, the protocol looks for a delay constrained path with the least possible cost based on packet urgency. In NRT traffic, the protocol looks for a path based on a cost function. Through simulation, we show that RTPA protocol improves the packet delivery ratio and the network lifetime compared to other protocols. Also, the performance of RTPA protocol remains very good even in large networks, and also scales with density.

Keywords: Wireless Sensor Networks, Routing Protocols, Real-time, Power.

1 Introduction

Wireless sensor networks (WSNs) represent a new trend in the past few years. Building sensors have been made due to the recent advances in micro-electro mechanical systems (MEMS) technology. Wireless ad hoc sensor network consists of large number of tiny, low-cost, low-power, and multifunctional sensor nodes. These nodes are densely deployed either inside the phenomena or very close to it [1]. Each sensor node consists of a sensing, data processing, and communicating component that leverage the idea of sensor network. The two most important operation of the sensor network are data dissemination, that is, the propagation of data/queries through the network, and data gathering, that is, the collection of the observed data from the individual sensor nodes to a sink or Base Station (BS) node using wireless radio signals [2].

S.K. Prasad et al. (Eds.): ICISTM 2010, CCIS 54, pp. 156–166, 2010.

Routing in WSNs is not an easy task due to the special characteristics such as the random deployment and the constrains in terms of transmission power, on-board energy, processing capability, and data storage that distinguish these networks from other wireless networks like mobile ad hoc networks or wireless LAN. It is extremely important to prolong the network lifetime, to be resilient to node failure, and to guarantee scalability, availability, and QoS [2][3].

Researchers have recently introduced suitable set of MAC layers and routing algorithm to handle data dissemination in WSNs [4-10]. Introducing applications that need real-time data transmission such as imaging and video data transmission requires careful handling in order to guarantee the QoS parameters such as bandwidth, minimum acceptable end-to-end delay, and jitter. These are called Quality of Service (QoS) of the communication network. QoS-aware routing protocols in sensor networks have several applications including real time target tracking in battle environments, emergent event triggering in monitoring applications etc [11].

Many approaches have been proposed to handle real-time data traffic in wired network. None of these approaches can be directly applied to wireless sensor network due to the resource constrained and the specific characteristics of this type of network. On the other hand, a number of protocols have been proposed for QoS-aware routing in wireless ad-hoc networks taking into account the dynamic and the mobility nature of the network. However, little of these protocols consider energy awareness along with the QoS parameters [11]. Therefore, supporting real-time data transmission in wireless sensor networks has recently attracted many research efforts.

There are some major differences in application requirements between WSNs and traditional networks [12]. First of all, all applications in WSNs are no longer end-to-end applications. Despite the fact that bandwidth may be an important factor to a group of sensors, it is not the main concern for a single sensor node. Finally, packet losses in traffic generated by one single sensor node can be tolerated to due to data redundancy. The primary real-time requirement is guaranteeing bounded end-to-end delays or at least statistical delay bounds.

One of the first routing protocols for WSNs is the Sequential Assignment Routing (SAR) which introduced the notion of QoS in the routing decisions [14]. Routing decision in SAR depends on three factors including: energy resources, QoS on each path, and the priority level of each packet. SPEED [15] is a QoS routing protocol that provides soft real-time end-to-end guarantee. It utilizes geographic routing and provides three different services for routing messages including: Regular Unicast, Area Multicast, and Area Anycast. Another protocol that was introduced for real-time communication is RAP [16]. It provides high level query and event services for distributed micro-sensing application. RAP uses Geographical Forwarding routing and a new packet scheduling policy called Velocity Monotonic Scheduling that inherently accounts for both time and distance constraints.

Real-time traffic can be generated by imaging or video sensors [17][18]. The proposed protocols proceed by first finding the least cost and energy efficient path using an extended version of Dijkstra's algorithm, and then finding the path from the least cost path founds that meets certain end-to-end delay requirement during the connection. In addition, a class-based queuing model and a Weighted Fair Queue (WFQ) model are employed to support both best effort and real-time traffic simultaneously.

These protocols are complex and centralized because the BS is responsible for calculating end-to-end delay and for assigning bandwidth.

DEAP [19] is a new delay-energy aware routing protocol that was introduced to handle tradeoffs between energy consumption and delay in heterogeneous WSAN. The protocol consists of two components: one for power management through a wakeup scheme and the other for geographical routing and load balancing through forwarding sets.

QoS-Based Geographic Routing for Event-Driven Image Sensor Networks supports two types of traffic: periodic low-bandwidth data and event driven high bandwidth data which is generated by image sensor node [13]. The protocol uses a geographic approach to forward the packets, and it introduces a cost function in which three parameters of neighboring nodes are considered when making next-hop routing decisions.

In this paper, we present a new routing protocol that is energy efficient and supports real-time applications. The presented protocol uses a geographic approach and considers the remaining energy and the delay to selects the suitable or the best route from the set of available routes between a source node and a destination node. In addition, the protocol uses two route decision components: one for Real-Time (RT) traffic and the other for Non-Real-Time (NRT) traffic. For RT traffic, the protocol looks for a delay constrained path with the least possible cost based on packet urgency. In NRT traffic, the protocol looks for a path based on a cost function. Also, we compare our protocol with the QoS-based geographic routing protocol [13] in terms of system lifetime, packet delivery ratio, and delay using several testing scenarios.

The remainder of this paper is organized as follows. The next section presents in details our RTPA protocol. The third section is devoted for simulations and performance evaluation and comparisons with other protocol. Finally, the fourth section concludes the paper.

2 Real-Time Power-Aware Routing Protocol for Wireless Sensor Network (RTPA)

In this section, we present our Real-Time Power-Aware Routing Protocol for Wireless Sensor Network (RTPA) to support real-time communication in large-scale sensor networks. Our protocol supports routing of two types of application: Real-Time (RT) application that is delay-constraint and elastic or Non-Real-Time (NRT) application in which the delay can be tolerated. To handle these applications, the protocol assigns each type of traffic to a separate queue, and uses a scheduling procedure to determine the order in which packets are transmitted from these queues based on the packet urgency. In addition, the routing protocol uses different routing decision for each type of traffic. In our protocol we assume the following:

1. The deployment of nodes is random and based on uniform distribution.
2. Each node knows its position or location through using a Global Position System (GPS) or using any of the localization protocols.
3. All sensors are stationary node.
4. The deployment is densely.

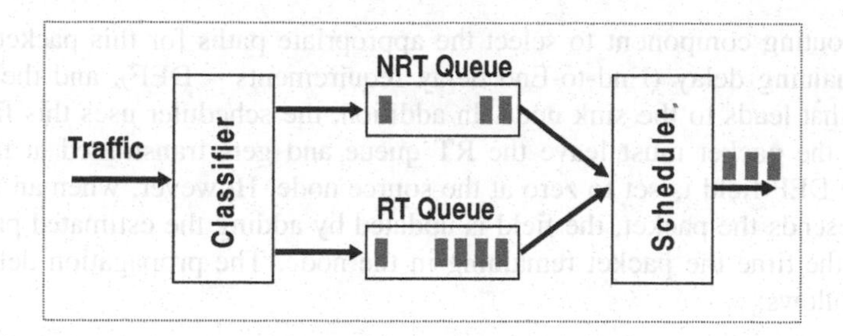

Fig. 1. Simplified queuing model

Given the unique characteristics of sensor networks, the objectives of our protocol include the following: maximize the number of messages meeting their end-to-end deadlines, be scalable with large number of nodes and hops, and increase network lifetime. The proposed protocol contains several components: Neighbor management, queuing model, path delay estimation, and routing decision.

2.1 Neighbor Management

Each node maintains a table of its neighbors. These neighbors are the next hop or the first node on a path towards the BS. Our protocol uses a greedy geographic forwarding approach, in which the neighbor set of a node i is the set of nodes that are inside the radio range R of node i and that are also closest to the BS in term of distance. The purpose of this approach is to eliminate loops and to reduce total energy consumption. It relies on neighborhood information to compute path delays.

In the presented protocol, each node periodically and asynchronously broadcasts a beacon packet to its neighbors. Each beacon packet has five fields including the node identification, the (x,y) coordinate or location of the neighbor node, RT_Delay, NRT_Delay, and the remaining energy of the node relative to the initial energy. The RT_Delay represents the estimated path delay for real-time data from the sending node to the BS, which is calculated as the estimated propagation delay plus the average queuing delay per node along the path. Likewise, the NRT_Delay is the estimated path delay for non-real-time data from the sending node to the BS. This delay is calculated as the estimated propagation delay plus the average queuing delay per node along the path.

2.2 Design Overview

We use two queues for the traffic; one for RT data and the other for NRT data. The order of packets in the RT queue is based on packet urgency (packet remaining delay) and that for NRT queue is based on FIFO approach. Each node has a classifier that checks the type of incoming packet and sends it to the appropriate queue. There is also a scheduler, which determines the order of packets to be transmitted from the queues according to a procedure as explained in Fig. 1.

For real-time traffic, each packet contains a cost field called Delay-Elapsed Field (DEF). This field contains the time elapsed while the packet is traveling from the source or target node to the current sensor node. This field has several advantages. It helps the routing component to select the appropriate paths for this packet based on packet remaining delay (End-to-End delay requirements – DEF), and the delay for each path that leads to the sink node. In addition, the scheduler uses this field to decide when the packet must leave the RT queue and gets transmitted at the current round. The DEF field is set to zero at the source node. However, when an intermediate node resends the packet, the field is updated by adding the estimated propagation delay and the time the packet remaining in the node. The propagation delay is estimated as follows:

$$\text{Propagation delay} = \frac{\text{distance to the next hop}}{\text{light speed}}. \tag{1}$$

2.3 Path Delay Estimation

We use an approach similar to that in MCFA [22] to estimate the path delay from the current sensor node to the BS. Due to the densely deployment, each node has several neighbors lead to the BS with different RT_Delay and NRT_Delay values. In our protocol, we use the average delay between neighbors to adjust RT_Delay and NRT_Delay fields.

When a node receives a beacon packet, it will search its neighbor-table for this neighbor. If the neighbor is not in the table, the node will create a new entry for this neighbor using the information received in beacon packet. Otherwise, if the neighbor is found, the node will update the RT_Delay and NRT_Delay for this neighbor using a low-pass filter approach. The general formula for the low-pass filter is:

$$Delay_{new} = (1\text{-}\alpha) \times Delay_{old} + \alpha \times Delay_{current}, \tag{2}$$

where $Delay_{new}$ is the new average delay for the path starting from this neighbor, $Delay_{old}$ is the old average delay obtained from the entry in the neighbor- table, and $Delay_{current}$ is the RT_Delay or NRT_Delay from the beacon packet. α is a constant that ranges from 0 to 1.

2.4 Routing Decisions

Our protocol uses different routing decisions for the two different types of traffic. For RT traffic, the decision or selecting next hop is based on the packet remaining delay and the estimated path delay.

For each packet, the routing component checks the DEF field. If it is greater than end-to-end delay, the routing component will drop the packet. Otherwise, the routing component will compute packet remaining delay and retrieve from the neighbor-table all paths that can handle this packet. If more than one path is found, the routing component will randomly select one of them. Otherwise, if no suitable path is found, the routing component will drop the packet.

Each node will have several neighbors because of the densely deployment. The protocol should select the next hop that is suitable for the packet in term of delay and that also reduces the number of hops. In order to do this, the node dynamically computes the *average distance* to its neighbors. The selection of a neighbor as a next hop for real-time traffic must fulfill the following two conditions:

1) The estimated delay on a path starting from this neighbor is less than the packet remaining delay.
2) The difference in distance between current node and neighbor is greater than or equal the average distance.

The routing component obtains two lists of the neighbor nodes: the first list contains the nodes that are satisfying the two conditions and the second list contains the nodes that are satisfying the first condition. To select the next hop, the protocol checks the first list. If more than one path is found, the routing component will randomly select one of them. Otherwise, the protocol checks the second list and if more than one path is found, it will randomly select one of them. Otherwise, the routing component will drop the packet.

A scheduler is introduced to determine the order in which packets are transmitted from the two queues. In each round (next time channel become free), we select one NRT packet from NRT queue and (1 + n) RT packets from the RT queue, where the value of n ranges from 0 to the size of the RT queue.

The scheduler decides whether the next RT packet in the RT queue should be transmitted at the current round based on packet urgency, i.e., Packet Remaining Delay (PRD). Therefore, the next RT packet should be transmitted if:

$$PRD \le Delay + (TCfree \times Hops), \tag{3}$$

where *Delay* is the average path delay needed to reach BS from this node, *TCfree* is the time between two successive transmissions, and hops is the average number of *Hops*.

For NRT traffic, we proposed a cost function in which three parameters of the neighboring nodes are considered when making next-hop routing decisions including: relative position to the base station, existing NRT_Delay, and remaining energy. A weighted cost function is calculated at the transmitting node to determine their relative importance. The cost function can be written as follows:

$$Cost = \frac{r_i^2}{X_{CN} - Y_i} + \alpha \cdot NRT\_Delay_i + \beta \cdot ECost_i, \tag{4}$$

where r_i is the distance between the current node (CN) and its neighbor N_i, X_{CN} is the distance from the current node (CN) to the base station, Y_i is the distance from the neighbor N_i to the base station, $ECost_i$ is a function that depends on the ratio between the remaining and the initial energy, and α and β are relative weights.

3 Simulation and Performance Evaluation

In the experiments described in this section, we conduct several simulations to adjust different parameters used in our RTPA protocol and to compare the performance of our protocol with the "QoS-Based Geographic Routing for Event-Driven Image Sensor Network" (QBGR) protocol [13].

Table 1. Other simulation parameters

Description	Value
Noise generation	Gaussian
Node deployment	Uniform distribution
Node transmission range (R)	75m
Traffic generation for RT traffic	Gaussian with mean = 0.1 and variance = 0.5.
MAC layer	IEEE 802.11
Base Station location	(0,0)m
Radio propagation speed	3×10^8 m/s

Table 2. Characteristics of the test network

Network Size	Number of Sensor Node	End-to-End Delay	Broadcast Rate	Load	
				Event	Periodic
(350×350)m2	225	1 s	Every 25 s	20	20

We use the J-Sim simulator to evaluate the performance of our protocol [20]. For the experiments described in this paper, we use the Seismic Propagation model to realize the sensor propagation model and 90% confidence interval. For the energy consumption, we use first order radio model that is used in [21].

For the data injection; in non-real-time traffic, the data is generated periodically every 10s by the periodic target node. In the real-time traffic, the data is generated by the event target node (image sensor) using Gaussian distribution function where the event can occur on an interval between 0-100 s. In our simulation, we assume that the packet size is 1kb, the periodic report fit in one packet, and the event report (image) is broken into 10 packets [13]. Table 1 shows other simulation parameters that are used in the simulation.

Before we make comparisons with other protocols, we need first to adjust and to test several parameters and models that are used in our RTPA protocol. We first test the use of the shortest and average delay paths and we find that the average delay path is preferred over the shortest delay path in the sense that it increases the number of sensors remaining alive. We also found that the best value for α that maximized the Packet Delivery Ratio (PDR) is 0.45.

We compare the RTPA protocol with the QBGR protocol [13] in terms of system or network lifetime, packet delivery ratio, latency or average delay, and network overhead using different testing scenarios. Table 2 describes the testing network parameters [13].

In the first experiment, we compare the two protocols in term of network lifetime. For network lifetime parameter, Fig. 2 shows that our protocol outperforms the QBGR protocol on average by 13%. We have compared the two protocols in term of the end-to-end delay. Figure 3 shows that as we increases the value of end-to-end delay, the total number of RT packets received at the base station on time is increased.

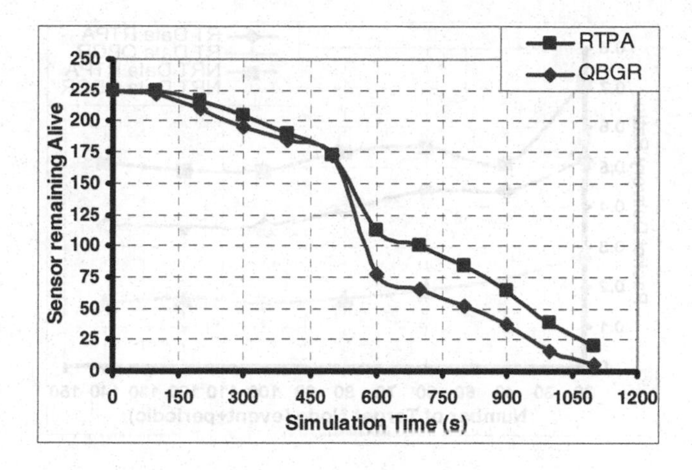

Fig. 2. Network lifetime

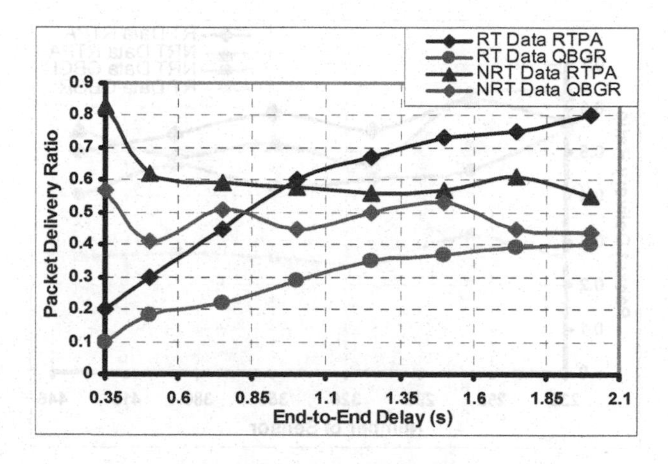

Fig. 3. End-to-End delay versus PDR

Also, the figure shows that our protocol outperforms the QBGR protocol on average by 95% in term of packet delivery ratio for RT traffic and 27% in term of packet delivery ratio for NRT traffic. Fig. 4 shows that the PDR of the two protocols decreases while increasing the network load and our protocol outperforms the QBGR protocol.

Density can be defined as the number of neighbors for a given sensor node or the number of sensor nodes per a given area. In this experiment we fix the network size and increase the number of sensor node. Fig. 5 shows that both protocols perform well while increasing the number of sensor nodes, and our protocol outperforms the QBGR protocol.

Network overhead can be defined as the influence of the control packets on the network. In both protocols, the control packet is the beacon packet that the node broadcasts every certain interval of time and each node broadcast a beacon packet

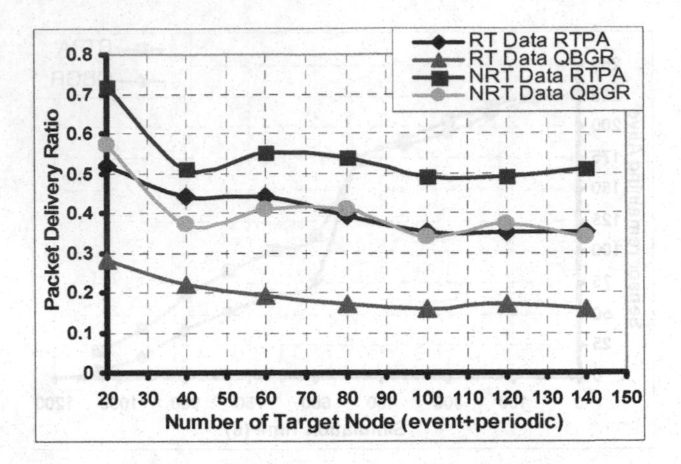

Fig. 4. Number of target node versus PDR

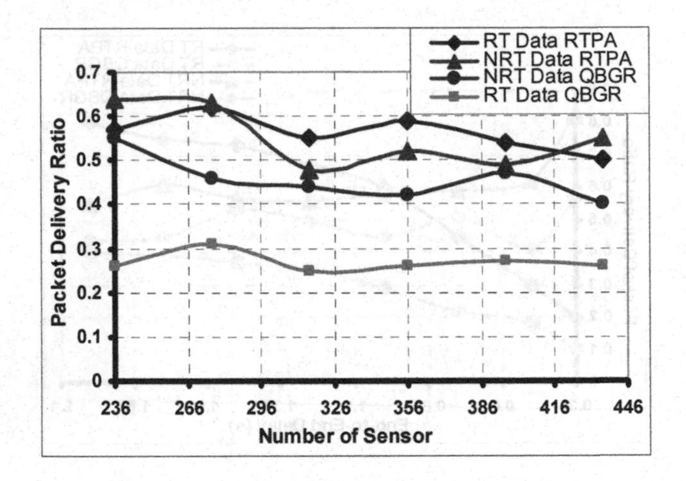

Fig. 5. Number of sensor node versus PDR

periodically every certain interval of time. The size of the beacon packet for the two protocols is the same (48 byte) with different contents. Therefore, the energy needed to transmit this packet is the same for the two protocols.

4 Conclusion

In this paper, we have presented a new routing protocol for wireless sensor network. It uses a geographic forwarding approach and takes into account the energy level in the neighboring nodes and the estimated delay on the path to select the next-hop node towards the sink node. The protocol supports routing of two types of application: Real-Time (RT) application that is delay-constraint, and elastic or Non-Real-Time

(NRT) application in which the delay can be tolerated. To handle these applications, the protocol assigns each type of traffic to a separate queue, and uses a scheduling procedure to determine the order in which packets are transmitted from these queues based on the packet urgency. In addition, the routing protocol uses different routing decision for each type of traffic. For RT traffic, the selection of the path depends on the packet urgency (packet remaining delay) and the estimated delay on the path. For NRT traffic, the protocol looks for a path based on a cost function that takes into account the remaining energy, delay, and position of the next-hop node. Simulation results show that our protocol outperforms recent protocols in terms of packet delivery ratio and network lifetime.

References

1. Akyildiz, F., et al.: Wireless sensor networks: a survey. Computer Networks 38, 393–422 (2002)
2. Al-Karaki, J.N., Kamal, A.E.: Routing Techniques in Wireless Sensor Networks: A Survey. IEEE Wireless Comm. 11, 6–28 (2004)
3. Akkaya, K., Younis, M.: A Survey on Routing Protocols for Wireless Sensor Networks. Elsevier Ad Hoc Network Journal 3(3), 325–349 (2005)
4. Ye, W., Heidemann, J., Estrin, D.: An Energy-Efficient MAC Protocol for Wireless Sensor Networks. In: Proc. of the IEEE INFOCOM, New York (June 2002)
5. Dam, T.V., Langendoen, K.: An Adaptive Energy-Efficient MAC Protocol for Wireless Sensor Networks. In: The First ACM Conference on Embedded Networked Sensor Systems, Los Angeles (November 2003)
6. Lu, G., Krishnamachari, B., Raghavendra, C.S.: An adaptive energy efficient and low-latency MAC for data gathering in wireless sensor networks. In: Proceedings of 18th International Parallel and Distributed Processing Symposium, April 26-30, p. 224 (2004)
7. Heinzelman, W., Kulik, J., Balakrishnan, H.: Adaptive Protocols for Information Dissemination in Wireless Sensor Networks. In: Proc. 5th ACM/IEEE Mobicom Conference (MobiCom 1999), Seattle, WA, August 1999, pp. 174–85 (1999)
8. Intanagonwiwat, C., Govindan, R., Estrin, D.: Directed diffusion: a scalable and robust communication paradigm for sensor networks. In: Proceedings of ACM MobiCom 2000, Boston, MA, pp. 56–67 (2000)
9. Heinzelman, W., Chandrakasan, A., Balakrishnan, H.: Energy-Efficient Communication Protocol for Wireless Microsensor Networks. In: Proceedings of the 33rd Hawaii International Conference on System Sciences (HICSS 2000) (January 2000)
10. Shah, R.C., Rabaey, J.: Energy Aware Routing for Low Energy Ad Hoc Sensor Networks. In: IEEE Wireless Communications and Networking Conference (WCNC), Florida, March 17-21 (2002)
11. Younis, M., Akayya, K., Eltowiessy, M., Wadaa, A.: On Handling QoS Traffic in Wireless Sensor Networks. In: Proceedings of the International Conference HAWAII International Conference on System Sciences (HICSS-37), Big Island, Hawaii (January 2004)
12. Chen, D., Varshney, P.K.: QoS Support in Wireless Sensor Networks: A Survey. In: International Conference on Wireless Networks, pp. 227–233 (2004)
13. Savidge, L., Lee, H., Aghajan, H., Goldsmith, A.: QoS-Based Geographic Routing for Event-Driven Image Sensor Networks. In: Proc. of Int. Conf. on Broadband Networks, Boston, MA (October 2005)

14. Sohrabi, K., Pottie, J.: Protocols for self-organization of a wireless sensor network. IEEE Personal Communications 7(5), 16–27 (2000)
15. He, T., et al.: SPEED: A stateless protocol for real-time communication in sensor networks. In: Proceedings of International Conference on Distributed Computing Systems, Providence, RI (May 2003)
16. Lu, C., et al.: RAP: A Real-Time Communication Architecture for Large-Scale Wireless Sensor Networks. In: Proc. of the IEEE Real-Time and Embedded Technology and Applications Symposium (RTAS 2002), San Jose, CA (September 2002)
17. Akkaya, K., Younis, M.: An Energy-Aware QoS Routing Protocol for Wireless Sensor Networks. In: Proc. of the IEEE Workshop on Mobile and Wireless Networks (MWN 2003), Providence, RI (May 2003)
18. Akkaya, K., Younis, M.: Energy-aware Delay-Constrained Routing in Wireless Sensor Networks. International Journal of Communication Systems 17(6) (2004)
19. Durresi, A., Paruchuri, V., Barolli, L.: Delay-Energy Aware Routing Protocol for Sensor and Actor Networks. In: Proceedings of the 11th International Conference on Parallel and Distributed Systems (2005)
20. Sobeih, A.: A simulation framework for sensor networks in J-Sim., University of Illinois at Urbana-Champaign (November 2003)
21. Salhieh, A., Schwiebert, L.: Power-Aware Metrics for Wireless Sensor Network. International Journal of Computers and Applications 26(4) (2004)

Document Clustering with Semantic Features and Fuzzy Association

Sun Park[1], Dong Un An[2,*], ByungRea Cha[3], and Chul-Won Kim[4]

[1] Advanced Graduate Education Center of Jeonbuk for Electronics and Information Technology-BK21, Chonbuk National University, Korea
sunbak@jbnu.ac.kr
[2] Division of Electronic & Information Engineering, Chonbuk National University, Korea
duan@jbnu.ac.kr
[3] Network Media Lab., GIST, Korea
brcha@nm.gist.ac.kr
[4] Department of Computer Engineering, Honam University, Korea
cwkim@honam.ac.kr

Abstract. This paper proposes a new document clustering method using the semantic features and fuzzy association. The proposed method can improve the quality of document clustering because the clustered documents by using fuzzy association values to distinguish well dissimilar documents, the selected cluster label term by semantic features, which is used in document clustering, can represent an inherent structure of document set better. The experimental results demonstrate that the proposed method achieves better performance than other document clustering methods.

Keywords: Document Clustering, LSA, Semantic Features, Fuzzy Association.

1 Introduction

Many documents in WWW with the same or similar topics were duplicated by the explosive increasing in internet access, and caused the problem that. This kind of data duplication problem increases the necessity for effective document clustering methods. Also, document clustering has been receiving more and more attentions as an important method for unsupervised document organization, automatic summarization, topic extraction, and information filtering or retrieval [1, 2, 4, 11, 13].

Generally, document clustering method can be classified into hierarchical methods and partitioning methods [1, 4]. Hierarchical clustering methods proceed successively by building a tree of clusters. On the other hand, partitioning clustering methods decompose the data set into k disjoint classes such that the data points in a class are nearer to one another than the data points in other classes [1, 2, 4, 9, 11].

* Corresponding author.

S.K. Prasad et al. (Eds.): ICISTM 2010, CCIS 54, pp. 167–175, 2010.

Recent studies for document clustering method use machine learning [6, 8, 18] techniques, graphs-based method [5, 13], and matrix factorization based method [10, 14, 15]. Machine learning based methods use semi-supervised clustering model with respect to prior knowledge and documents' membership [6, 8, 18]. Graphs based methods model the given document set using an undirected graph in which each node represents a document [5, 13]. Matrix factorization based method use semantic features of documents set for document clustering [10, 14, 15].

In this paper, we propose a document clustering method using semantic features by LSA and fuzzy association. LSA [1, 4, 17] is a theory and method for extracting and representing the contextual-usage meaning of words by statistical computations applied to a large corpus of text. Fuzzy association [3] uses a concept of fuzzy set theory [16] to model the vagueness in the information retrieval. The basic concept of fuzzy association involves the construction of a index terms form a set of documents [3].

The proposed method has the following advantages. First, it can extract important cluster label terms in document set using semantic features by LSA. So it can identify major topics and subtopics of clusters with respect to the semantic features. Second, it can remove the dissimilar documents in clusters using fuzzy association. So, it can improve the quality of document clustering since it helps us to remove dissimilarity information easily.

The rest of the paper is organized as follows: Section 2 describes the related works regarding document clustering methods. In Section 3, we review LSA and fuzzy association in detail. In Section 4, the proposed document clustering method is introduced. Section 5 shows the evaluation and experimental results. Finally, we conclude in Section 6.

2 Related Works

Traditional clustering method can be classified into partitioning, hierarchical, density-based, and grid-based. Most of these methods use distance functions as object criteria and are not effective in high dimensional spaces [1, 2, 4, 9, 11]. Li et al. proposed a document clustering algorithm ASI (adaptive subspace iteration) using explicitly modeling of the subspace structure associated with each cluster [9]. Wang et al. proposed a clustering approach for clustering multi-type interrelated data objects. It fully explores the relationship between data objects for clustering analysis [12]. Park et al. proposed a document clustering method using non-negative matrix factorization and cluster refinement [10].

Wang and Zhang proposed the document clustering with local and global regularization (CLRG). It uses a local label predictors and a global label smoothness regularizer [13]. Liu et al. proposed a document clustering method using cluster refinement and model selection. It uses GMM (Gaussian Mixture Model) and EM (Expectation Maximization) algorithm to conduct an initial document clustering. It also refines the initially obtained document clusters by voting on the cluster label of each document [8]. Ji et al. proposed a semi-supervised clustering model that incorporates prior knowledge about documents' membership for document clustering analysis [6]. Zhang et al. adopt a RL (Relaxation Labeling) based cluster algorithm to evaluate the effectiveness of the aforementioned types of links for document clustering. It use both content and linkage information in dataset [18]. Xu et al proposed a document partitioning method based on the non-negative factorization matrix (NMF) of the given document corpus [15]. Xu and Gong proposed a data clustering method models each cluster as a linear combination of the data points, and each data point as a linear combination of the cluster centers [14]. Li and Ding presented an overview and summary on various matrix factorization algorithms for clustering and theoretically analyze the relationships among them [7].

3 LSA and Fuzzy Association Theory

3.1 Latent Semantic Analysis

In this paper, we define the matrix notation as follows: Let X_{*j} be j'th column vector of matrix X, X_{i*} be i'th row vector and X_{ij} be the element of i'th row and j'th column.

Latent semantic analysis (LSA) is a decomposition method using singular value decomposition (SVD). This method decomposes matrix A into three matrices, U, D, and V^T [1, 4, 11, 17].

$$A = UDV^T \ , \ A \approx \tilde{U}\tilde{D}\tilde{V}^T \tag{1}$$

where U is an mxn orthonormal matrix of eigenvectors of AA^T (left singular vectors), and V is an nxn orthonormal matrix of eigenvectors of A^TA (right singular vectors). $D = diag(\sigma_1, \sigma_2, ..., \sigma_n)$ is an nxn diagonal matrix whose diagonal elements are non-negative eigen values sorted in descending order. $\tilde{U}$ is a mxr matrix, where $\tilde{U}_{ij} = U_{ij}$ if $1 \le i \le m$ and $1 \le j \le r$. $\tilde{D}$ is a rxr matrix, where $\tilde{D}_{ij} = D_{ij}$ if $1 \le i, j \le r$. $\tilde{V}$ is a nxr matrix, where $\tilde{V}_{ij} = V_{ij}$ if $1 \le i \le n$ and $1 \le j \le r$. Finally, $r << n$. In the method using LSA, the i'th column vector A_{*i} of matrix A is the weight vector of the i'th document, and is represented as the linear combination of the left eigenvectors U_{*j}, which are semantic feature vectors, as shown in Equation (2). That is, the weight of the j'th semantic vector U_{*j} corresponding to the document vector A_{*i} is $\sigma_j V_{ij}$.

$$A_{*i} \approx \sum_{j=1}^{r} \sigma_j \tilde{V}_{ij} \tilde{U}_{*j} \tag{2}$$

We assume that f clusters are constructed and then matrix A can be represented as follows:

$$\{A_{*i} \mid i = 1,...,n\} = \bigcup_{k=1}^{f} \{C_l^k \mid l = 1,...,s_k\} \ , C^p \cap C^q \neq \phi \ , \ p \neq q \tag{3}$$

where C^k is the matrix of k'th the cluster of documents, s_k is the number of column of C^k, or the number of document in cluster C^k, n is the number of documents, and f is the number of clusters.

3.2 Fuzzy Association Theory

In this section, we give a brief review to the fuzzy association theory [1, 3, 16] that is used in document clustering. The fuzzy set is defined as follows.

Definition 1. A fuzzy association between two finite sets $X=\{x_1, \ldots, x_u\}$ and $Y=\{y_1, \ldots, y_v\}$ is formally defined as a binary fuzzy relation $f: X \times Y \rightarrow [0, 1]$, where u and v are the numbers of elements in X and Y, respectively.

Definition 2. Given a set of index terms, $T=\{t_1, \ldots, t_2\}$, and a set of documents, $D=\{d_1, \ldots, d_v\}$ each t_i is represented by a fuzzy set $h(t_i)$ of documents; $h(t_i)=\{F(t_i,d_j) \mid \forall\ d_i \in D\}$, where $F(t_i, d_j)$ is the significance (or membership) degree of t_i, in d_j.

Definition 3. The fuzzy related terms (RT) relation is based on the evaluation of the co-occurrences of t_i and t_j in the set D and can be defined as follows.

$$RT(t_i,t_j) = \frac{\sum_k \min(F(t_i,d_k), F(t_j,d_k))}{\sum_k \max(F(t_i,d_k), F(t_j,d_k))} \quad (4)$$

A simplification of the fuzzy RT relation based on the co-occurrence of terms is given as follows.

$$r_{i,j} = \frac{n_{i,j}}{n_i + n_j - n_{i,j}} \quad (5)$$

where $r_{i,j}$ represents the fuzzy RT relation between terms i and j, $n_{i,j}$ is the number of documents containing both ith and jth terms, n_i, is the number of documents including the ith term, and n_j is the number of documents including the j'th term.

Definition 4. The membership degrees between each document to each of the cluster sets can be defined as follows.

$$\mu_{i,j} = \max_{\forall t_a \in d_i} [1 - \prod_{\forall t_b \in CT^j} (1 - r_{a,b})] \quad (6)$$

where $\mu_{i,j}$ is the membership degree of d_i belonging to C^j, $r_{a,b}$ is the fuzzy relation between term $t_j \in d_j$ and term $t_b \in CT^j$.

4 The Proposed Document Clustering Method

In this section, we propose a method that clusters documents by semantic features by LSA and fuzzy association. The proposed method consists of the preprocessing phase, cluster label extraction phase, and the document cluster phase. We will give a full explanation of three phases as Figure 1.

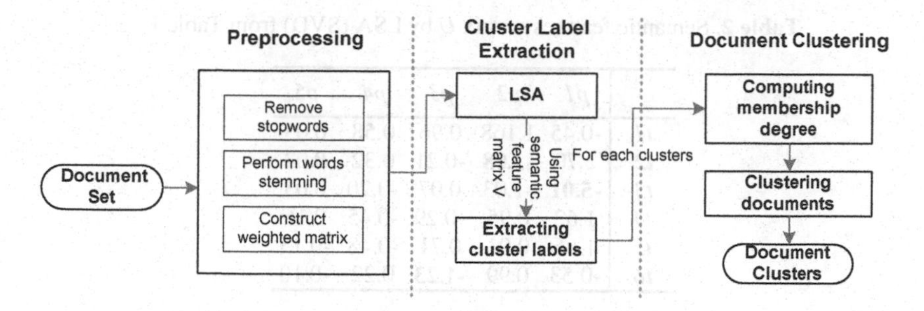

Fig. 1. Document clustering method using semantic features and fuzzy association

4.1 Preprocessing

In the preprocessing phase, we remove all stop-words by using Rijsbergen's stop-words list and perform words stemming by Porter's stemming algorithm [2, 11]. Then we construct the term-frequency vector for each document in the document set [1, 2, 11]. Let $T_i = [t_{1i}, t_{2i} \ldots t_{ni}]^T$ be the term-frequency vector of document i, where elements t_{ji} denotes the frequency in which term j occurs in document i. Let A be $m \times n$ terms by documents matrix, where m is the number of terms and n is the number of documents in a document set.

4.2 Cluster Label Extraction by LSA

In the cluster label terms extraction phase, we use semantic features by LSA [1, 2, 4, 11, 17] to extract cluster label terms. The proposed cluster label terms extraction method is described as follows. First, preprocessing phase is performed, and then the term-document frequency matrix is constructed. Table 1 shows the term-document frequency matrix with respect to 5 documents and 6 terms. The singular value decomposition (SVD) is computed on the correlation matrix in Table 1 by LSA. Table 2 shows the semantic features matrix U by LSA from Table 1. The largest singular values are selected from under $p3$ in the columns of Table 2 for 3 cluster label terms. Terms having the largest singular value are selected in each column for the cluster label terms in Table 2. Table 3 shows the extracted cluster label terms form Table 2.

Table 1. Term-document frequency matrix

term \ document	*d1*	*d2*	*d3*	*d4*	*d5*
t1	2	0	0	1	0
t2	0	4	1	0	0
t3	3	0	1	6	0
t4	0	1	0	0	1
t5	1	2	0	0	1
t6	0	0	0	2	0

Table 2. Semantic features matrix U by LSA (SVD) from Table 1

	p1	*p2*	*p3*	*p4*	*p5*
t1	-0.45	1.168	0.96	0.58	0.03
t2	2.70	**-2.08**	-0.21	0.32	0.03
t3	**-5.01**	-1.03	0.07	-0.20	0.03
t4	1.62	0.95	-0.29	-0.45	0.16
t5	1.66	0.01	0.71	-0.48	-0.15
t6	-0.53	0.99	**-1.23**	0.22	-0.10

Table 3. Result of cluster label terms extraction from Table 2

	cluster label term
p1	*t3*
p2	*t2*
p3	*t6*

4.3 Document Clustering by Fuzzy Association

Document clustering phase is described as follows. First, we construct the term correlation matrix M with respect to relationship between cluster label terms and terms of document set using the fuzzy RT relation by Equation (5). The term correlation matrix is an n by n symmetric matrix whose element, m_{ij}, has the value on the interval [0, 1] with 0 indicates no relationship and 1 indicates full relationship between the terms t_i and t_j. Therefore, m_{ij} is equal to 1 for all $i = j$. since a term has the strongest relationship to itself [3].

Second, a document d_i is clustered into the cluster C^j where the membership degree $\mu_{i,j}$ is the maximum by Equation (6). The term t_a in i is associated to cluster C^j if the terms k_b's in CT^j (for cluster C^j) are related to the term t_a [3]. Where CT^j is cluster label terms.

4.4 The Proposed Document Clustering Algorithm

The proposed document clustering algorithm using both the semantic features and fuzzy association is as follows:

Algorithm. Document Clustering by semantic features and fuzzy association
Input: The term frequency matrix A, the number of cluster k,
the number of whole document n
Output: clustering document C^k
1. Perform the preprocessing.
2. Perform the SVD.
3. *Repeat*
4. Select the c'th left singular vector from the semantic features matrix U
5. Select the term which has the largest index value with the c'th left singular vector for the category label terms

6. *Until* $c = 1, \ldots, k$
7. Calculate the fuzzy RT relation
8. *Repeat*
9. Select $d_j = \underset{1 \le n \le j}{\arg\max}\{\mu_{i,j}\}$ *then* include d_j in cluster C^i
10. *Until* $i = 1, \ldots, k$

In step 2 to 6, the selected cluster label terms phase uses the semantic features by LSA. In step 7 to 10, document clustering phase uses fuzzy association between cluster label terms and the terms of document.

5 Performance Evaluations

We use the Reuters[1] document corpora to evaluate the proposed method. Reuters corpus have 21578 documents which are grouped into 135 clusters [13]. We use the normalized mutual information metric $\overline{MI}$ for measuring the document clustering performance [13, 14, 15].

We have conducted performance evaluation by testing proposed method and comparing it with 5 other representative data clustering method using the same data corpora. We implemented 6 document clustering methods: FLSA, RNMF, KM, NMF, ASI, and CLRG. In figure 2, FLSA denotes our proposed method. RNMF denotes our previous method by using NMF and cluster refinement [10]. KM denotes partitioning method using traditional *k*-means [1, 2, 4, 11]. NMF demotes Xu's method using non-negative matrix factorization [15]. ASI denotes Li's method using adaptive subspace iteration [9]. CLRG denotes Wang's method using local and global regularization [13].

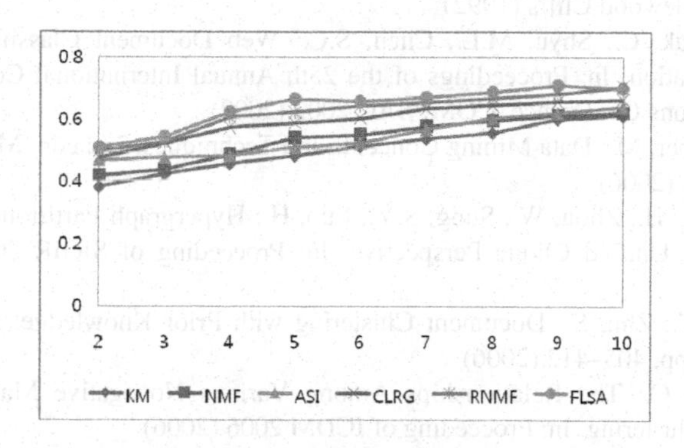

Fig. 2. Evaluation results of performance comparison

The evaluation results are shown in Figure 2. The evaluations were conducted for the cluster numbers ranging from 2 to 10. For each given cluster number *k*, 50 experiment runs were conducted on different randomly chosen clusters, and the final performance values were obtained by averaging the values from the 50 experiments.

[1] http://kdd.ics.uci.edu/databases/reuters21578/reuters21578.html

In Figure 2, the average normalized metric $\overline{MI}$ evaluation results of FLSA is approximately 26.45% higher than that of KM, 18.89% higher than that of NMF, 17.78% higher than that of ASI, 7.39% higher than that of CLRG, and 2.98% higher than that of RNMF. In this experiment, The FLSA showed the best performance because it uses the semantic features for extracting the cluster label terms with respect to repressing the topic of document set and clustering document with respect to cluster label term by fuzzy association that can identify dissimilar documents more successfully than other methods.

6 Conclusions

In this paper, we have presented a document clustering method using the semantic features by LSA and fuzzy association. The proposed method in this paper has the following advantages. First, it can identify dissimilar document between clusters by fuzzy association with respect to cluster label terms and documents. So, it can improve the quality of document clustering. Second, it can easily extract cluster label term are well covered with the major topics of document using the semantic features by LSA. Experimental results show that the proposed method outperforms five different summarization methods.

References

1. Chakrabarti, S.: Mining the Web: Discovering Knowledge from Hypertext Data. Morgan Kaufmann, San Francisco (2003)
2. Frankes, W.B., Ricardo, B.Y.: Information Retrieval, Data Structure & Algorithms. Prentice-Hall, Englewood Cliffs (1992)
3. Haruechaiyasak, C., Shyu, M.L., Chen, S.C.: Web Document Classification Based on Fuzzy Association. In: Proceedings of the 25th Annual International Computer Software and Applications Conference, COMPSAC 2002 (2002)
4. Han, J., Kamber, M.: Data Mining Concepts and Techniques, 2nd edn. Morgan Kaufmann, San Francisco (2006)
5. Hu, T., Xiong, H., Zhou, W., Sung, S.Y., Luo, H.: Hypergraph Partitioning for Document Clustering: A Unified Clique Perspective. In: Proceeding of SIGIR 2008, pp. 871–872 (2008)
6. Ji, X., Xu, W., Zhu, S.: Document Clustering with Prior Knowledge. In: Proceeding of SIGIR 2006, pp. 405–412 (2006)
7. Li, T., Ding, C.: The Relationships Among Various Nonegative Matrix Factorization Method for Clustering. In: Proceeding of ICDM 2006 (2006)
8. Liu, X., Gong, Y., Xu, W., Zhu, S.: Document Clustering with Cluster Refinement and Model Selection Capabilities. In: Proceeding of SIGIR 2002, pp. 191–198 (2002)
9. Li, T., Ma, S., Ogihara, M.: Document Clustering via Adaptive Subspace Iteration. In: Proceeding of SIGIR 2004, pp. 218–225 (2004)
10. Park, S., An, D.U., Char, B.R., Kim, C.W.: Document Clustering with Cluster Refinement and Non-negative Matrix Factorizaion. In: Proceeding of ICONIP 2009 (2009)
11. Ricardo, B.Y., Berthier, R.N.: Moden Information Retrieval. ACM Press, New York (1999)

12. Wang, J., Zeng, H., Chen, Z., Lu, H., Tao, L., Ma, W.Y.: ReCoM: Reinforcement Clustering of Multi-Type Interrelated Data Objects. In: Proceeding of SIGIR 2003 (2003)
13. Wang, F., Zhang, C.: Regularized Clustering for Documents. In: Proceeding of ACM SIGIR 2007, pp. 95–102 (2007)
14. Xu, W., Gong, Y.: Document Clustering by Concept Factorization. In: Proceeding of SIGIR 2004, pp. 202–209 (2004)
15. Xu, W., Liu, X., Gong, Y.: Document clustering based on non-negative matrix factorization. In: Proceeding of ACM SIGIR 2003 (2003)
16. Zadeh, L.A.: Fuzzy Sets. In: Dubois, D., Prade, H., Yager, R.R. (eds.) Readings in Fuzzy Sets for Intelligent Systems. Morgan Kaufmann Publiishers, San Francisco (1993)
17. Zhang, D., Dong, Y.: Semantic, hierarchical, online clustering of web search results. In: Yu, J.X., Lin, X., Lu, H., Zhang, Y. (eds.) APWeb 2004. LNCS, vol. 3007, pp. 69–78. Springer, Heidelberg (2004)
18. Zhang, X., Hu, X., Zhou, X.: A Comparative Evaluation of Different Link Types on Enhancing Document Clustering. In: Proceeding of SIGIR 2008, pp. 555–562 (2008)

Supporting Video Streaming over WiMAX Networks by Enhanced FMIPv6-Based Handover

Phuong-Quyen Huynh, Pat Jangyodsuk, and Melody Moh*

Department of Computer Science
San Jose State University
San Jose, CA 95192-0249, USA
melody@cs.sjsu.edu

Abstract. IEEE 802.16, also widely known as WiMAX (Worldwide Interoperability for Microwave Access), is an emerging broadband wireless standard. The seamless support over an efficient handover (HO) mechanism is vital to its success. This paper proposes a new cross-layer HO scheme, enhanced from FMIPv6 (Fast Mobile IPv6 Handoff Protocol) by adopting two features: reducing control messages and eliminating duplicated care-of-addresses. The resulting HO delay is significantly decreased, which also improves other QoS parameters. The new scheme is evaluated for its support of video streaming traffic while comparing with three other HO schemes: FMIPv6 and two existing enhancements to FMIPv6. We measure HO delay and loss; and PSNR (Peak Signal to Noise Ratio) and MOS (Mean Opinion Score) for video quality. We found that the proposed enhancement has achieved considerable improvement over the existing schemes, especially under heavy network conditions and fast mobility speed.

Keywords: Handover, mobile networks, real-time traffic, video streaming, WiMAX.

1 Introduction

In the world of rapidly changing wireless technologies, WiMAX (Worldwide Interoperability for Microwave Access) has emerged to becoming one of the most promising wireless metropolitan area network technologies [7]. The standard is built on the key premises such as Open architecture IP (Internet Protocol)- based network, high bandwidth equivalent of wire-line xDSL (Extended Digital Subscriber Line) technology, mobility support, QoS (Quality of Service) support, and low cost of deployment and maintenance.

Supporting seamless mobility with QoS is one of the key challenges in offering real-time services. During HO, when a mobile node moves from one place to another, it is essential to provide a satisfactory level of video or voice quality; disruptions or even intermittent disconnections of transmission due to long HO processing time would severely degrade the level of services.

* Corresponding author.

S.K. Prasad et al. (Eds.): ICISTM 2010, CCIS 54, pp. 176–186, 2010.

While the WiMAX technology has enjoyed many advantages such as high data rate, wide transmission range, and flexible scheduling, it however suffers from long-delay HO. This drawback is mainly due to network-layer HO latency. Acknowledging the problem, several major proposals have been presented at the Internet Engineering Task Force (IETF). One of them is the Fast MIPv6 (FMIPv6), which took a cross-layer approach while introducing packet tunneling between current and target base stations to eliminate packet loss [10]. It has improved both delay and loss HO performance [12].

FMIPv6 however requires a sequence of preparation messages. When all these messages have been received in time, it will proceed in the *predictive mode*. If however some messages are delayed or lost, the *reactive mode* will be triggered, this will cause long delay in FMIPv6.

In this paper we present a new algorithm to solve the above mentioned issue. The first idea is to merge the co-functioning data-link layer and network layer messages [2]. This will reduce the number of messages required and avoid message lost during heavy network conditions. Secondly, we adopt the Temporary Care-of-Address (tCoA) generation method [15]. This eliminates the duplicated CoA problem and therefore further reduces HO delay. Finally, by combining the two features, we can further eliminate some other HO messages since they are no longer needed.

2 Related Works

In this section we first discuss FMIPv6 [10, 14, 16]. Next, we briefly describe some WiMAX HO improvements, and for supporting real-time traffic. To address the problem of large HO latency, FMIPv6 has been introduced as a seamless HO scheme [14]. The goal is to reduce the disconnection period so that the time the receiver waits for a packets is minimum. In this version, however, no relationship has been specified between layers 2 and 3 messages. Two later versions have later been proposed: FMIPv6 for IEEE 802.11 [9] and FMIPv6 for IEEE 802.16 [10].

FMIPv6 have its drawbacks. There are two modes in FMIPv6: Predictive and Reactive. Predictive mode will be triggered if every protocol messages reach destination in time. Reactive, on the other hand, will be triggered if some messages such as FBU never reach destination. Typically, Predictive mode gives a much better performance. Fast mobility and heavily congested networks are likely to cause the protocol to run in Reactive mode.

There have been many enhancements proposed for WiMAX HO. Cho et al proposed a HO scheme which requires base station (BS) to actively monitoring signal strength of MNs [4]. Pre-coordinate HO, by Chen et al, improves the process by predicting and associating one target BS beforehand [2].

Many improvements for real time service have also been suggested. Dong and Dai suggested the improvement for QoS during HO [5]. Further enhancement has been achieved by Chellappan et. al [1]. The most obvious enhancement is that layer-3 process will also be changed for each QoS requirement. Furthermore, this scheme support standby mode in case there is no traffic.

While many works focusing on enhancement, there are some focusing on evaluating real time traffic performance during HO. Ivov et al simulated video traffic, including video conferencing, over IEEE 802.11 networks [9]. The result showed that FMIPv6 outperformed MIPv6 in every performance matrices. Kim et al has recently evaluated FMIPv6 for VoIP support over IEEE 802.16 networks [12]. The result confirmed that FMIPv6 is better than MIPv6 in term of supporting real time traffic.

3 Proposed Cross-Layer HO Mechanisms

Although the FMIPv6 HO has been improved in term of HO delay, its overall delay is still considered high [3]. In this section, we first present the two features, followed by the description of the proposed scheme.

3.1 First Feature: Combination of Control Messages

In the original FMIPv6, the numerous number of control messages exchanged during HO can easily congest the air-interface, and may cause the loss of control messages, which will seriously degrade the HO process. Chen et al have proposed to combine correlated layer-2 and layer-3 messages to reduce potential congestion over the air-interface [3], described below.

The first pair of messages to be combined are MOB_NBR-ADV and Proxy Router Advertisement (PrRtAdv) [3], which are sent periodically by the BS to the MN. By combining them, instead of waiting for the MN to send RtSolPr message, the SBS takes the active move to send its router advertisement along with MOB_NBR-ADV of layer 2. Thus, the MN can obtain the needed information to conduct FMIPv6 in advance.

Another pair that can be combined is MOB_HO-IND message of layer 2 and FBU message of layer 3 [3]. The integration of MOB_HO-IND and FBU is named as FBU_MOB_HO-IND. This prevents the HO procedure from waiting for FBU and triggering the reactive Mode, as in the original FMIPv6 [3]. By combining these two messages, SBS would start tunneling packet as soon as it receives FBU. Therefore, some packets will be held and MN will not receive any packet until it reconnects with TBS. Sending FBU right before disconnecting with SBS will keep the connection alive with SBS as much as possible, and will greatly reduce HO delay and HO packet loss.

3.2 Second Feature: New CoA Generation Method

In FMIPv6, the NAR (new access router) needs to execute the Duplicated Address Detection (DAD) process to verify if the CoA of the incoming MN duplicates with any of the existing CoA being used. The time needed to perform the DAD is relatively long compared to other types of delay occur during HO process. According to Lee et al, the average time required for binding update is around 140ms, and the time for layer-2 HO is about 250ms; the DAD process, on the other hand, takes almost 1 second and therefore becomes the major factor of the overall HO delay [15].

To address this, Lee et al suggested an advance method for the MN to generate new CoA [15]. They also verified that their algorithm would guarantee the new CoA, with very high probability, has almost no duplication with any other existing CoA in the network area. Thus, the chance for HO procedure to run in Predictive Mode is significantly increased since little or no time is spent processing DAD. As a result, this procedure can support a much higher mobility speed.

3.3 Description of the Proposed Scheme

Stated above are the two key improvements that can be applied to original FMIPv6 to improve its support of real time applications and greater number of concurrent MN. Our proposed scheme exploits the advantages of both. The message flow is given in figures 2 below. Referring to figure 1, the SBS periodically sends the MOB_NBR-ADV along with the PrRtAdv message to the MN. By receiving these messages, the MN updates necessary information of all BSs in range. When the signal strength between MN and SBS is weak, the MN exchanges MOB_MSHO-REQ and MOB_BSHO-RSP with the SBS. After that, it triggers layer-2 HO by sending the combined FBU_MOB_HO-IND message to SBS. Upon receiving this combined message, the SBS forwards FBU message to PAR. The PAR starts to buffer packets destined to the current CoA, and sends the Handover Idication (HI) message to NAR indicating the proposed nCoA of the terminal. Note that in this scheme, the nCoA is generated at the MN using the new algorithm described above [15], so it is hardly to be duplicated at the new network. Therefore, after the PAR sends HI to NAR, NAR, upon receiving, will return with HAck as usual according to FMIPv6 process. However, PAR does not need to send FBAck to MN since the generated address is unique. The process after that is similar to that of the original FMIPv6.

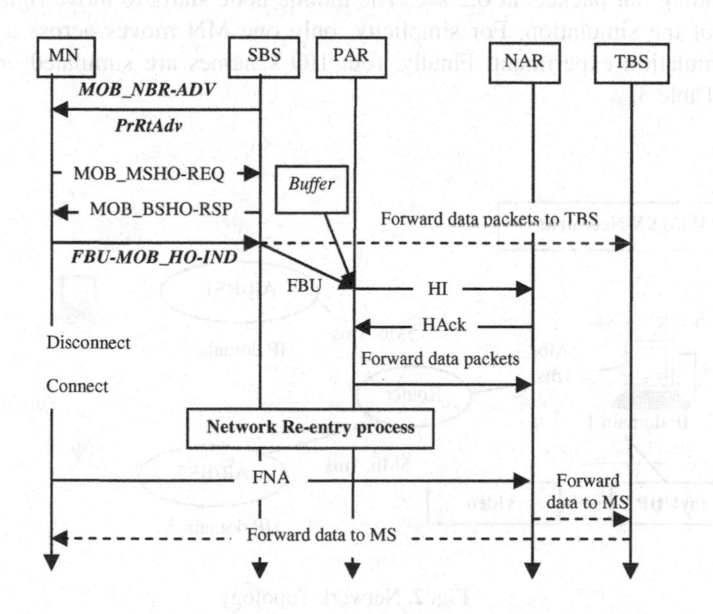

Fig. 1. Message flow of Proposed Scheme with Separated BS and AR

4 Performance Evaluation

4.1 Simulation Settings

As shown in Figure 2, the network has three domains. Domain 1 contains CN which plays as the sender to send video and CBR traffic. The receivers, which are located in the other domain (domain 2), include a mobile node (MN) to receive video traffic, and other fixed nodes for other background CBR traffics. The MN will move from domain 2 to domain 3 with a constant speed throughout a simulation experiment. For the ease of simulation, we assume that each BS has all the functionalities of an AR integrated within.

Network Simulator version 2.28 (ns-2.28) [20] is used for the simulation. Additionally, the NIST WiMAX patch [19] is exploited to simulate the behavior of PHY and MAC layers in WiMAX networks. Finally, Mobiwan patch [6] has been adopted for MIPv6 functionalities. Based on these patches, we then develop a new module which simulates FMIPv6 proposed in RFC 5270 [10], including message sequence, packet tunneling, and cross-layer triggering of messages. We also simulate the two improvements described in section 3. Major simulation parameters are summarized in Table 1. To study video streaming performance, we make use of a video tool set, EvalVid [13], developed by Klaue et al, and has later been extended by Ke et al [11].

Except for the last set of experiments shown in Section 4.2, each simulation experiment runs for 20 seconds; the video used is 12 seconds long. Video application starts sending out packets at 0.2 sec. The mobile node starts to move right at the beginning of the simulation. For simplicity, only one MN moves across a domain in every simulation experiment. Finally, four HO schemes are simulated and summarized in Table 3.

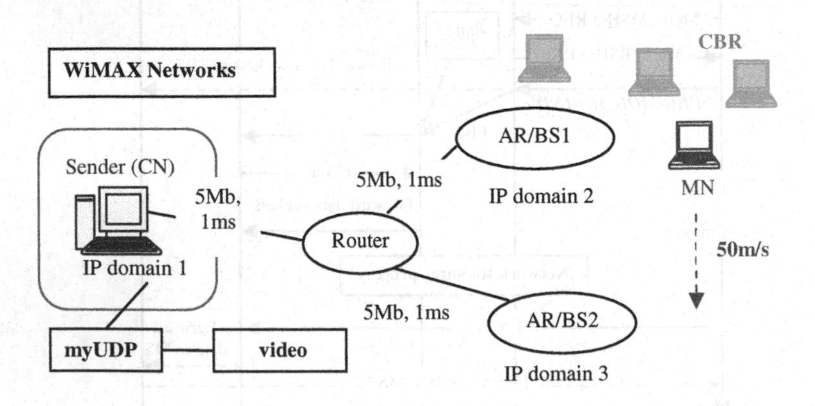

Fig. 2. Network Topology

Table 1. Simulation Parameters

Antenna height	1.5 meters	
Carrier frequency	3.486 GHz	
Coverage radius	750 m	
Distance between BSs	1 km	
Mobility speed	Varied (30m/s, 40m/s, 45m/s, 50m/s) 50 m/s (or 180 km/hr) by default	
Link layer and MAC layer queue size	50 packets each	
Receiver sensitivity	4×10^{-13} dB	
Bandwidth	2 Mbps	
Codec	Video	Background Traffic (CBR)
	- 200 Kb/s - 25 frames/second	- Bit rate: 448 Kb/s
Packet size	Video	CBR
	512 bytes	210 bytes

Table 2. Description of Simulation Schemes

Scheme	Description
FMIPv6	The original cross-layer based FMIPv6 (supports both Predictive and Reactive modes), which incurs 1 second of DAD process.
FMIPv6+tCoA	FMIPv6 plus tCoA created by the new CoA Generation Method (without 1 second of DAD process)
FMIPv6+merge	FMIPv6 plus Message Combination scheme.
Proposed Scheme	FMIPv6, adopts both tCoA and Message Combination scheme.

Performance metrics

There are two groups of metrics measured in this study: general network performance and video quality. For network performance, we measure HO delay and loss. HO delay is the period between last packets received from PAR and the first packets received from NAR. The delay is measured in the period between LINK_SWITCH and LINK_UP event (refer to Figure 5).

For video performance, we measure PSNR (Peak Signal to Noise Ratio) and MOS (Mean Opinion Score). PSNR is evaluated based on the PSNR between the luminance component Y of source image *s* and destination image *d*. Theoretically, PSNR can be calculated as below [13]:

$$\mathrm{PSNR}\,(s,d) = 20 \log [\, V_{peak} / \mathrm{MSE}\,(s,d)]\ \mathrm{dB}\,. \qquad (1)$$

where $V_{peak} = 2^k - 1$, *k*-bit color depth
MSE(s,d) = mean square error of s and d
s = source image, *d* = destination image

MOS is a subjective metric commonly used to judge video quality when displayed at the receiver's application. Usually, MOS measures the human impression of the sound or image at the application layer. MOS can be derived from PSNR, as summarized in Table 3 [13].

Table 3. PSNR and MOS

PSNR	> 37	31-37	25-31	20-25	<20
MOS	5 (Excellent)	4 (Good)	3 (Fair)	2 (Fair)	1 (Bad)

4.2 Simulation Results

In this section, we illustrate the results in terms of network performance and video quality performance. Further results, including those of voice quality, may be found in a detailed technical report [8].

Network Performance

Handover delay: Figures 3 shows the HO delay while varying background traffic load, keeping the default mobility speed of 50m/s. It is clear that the performance of our proposed scheme is superior, especially when the background traffic load is heavy. Note that when the network is in ideal status with no background traffic load, the HO delay of FMIPv6+tCoA and the proposed scheme are all much smaller than that of FMIPv6. However, when the network becomes busier, only the proposed scheme can sustain and maintain a good delay performance. The reason is that in FMIPv6, as well as in FMIPv6+tCoA, the MN needs to wait for PrRtAdv before it sends the HO request to SBS. However, the heavy traffic load in the network may cause that message to be lost during transmission while the MN is waiting, and finally finds out that it has lost the connection with the current BS. This will introduce a long delay in the HO process. Both FMIPv6+merge and the proposed scheme does not have this problem because the PrRtAdv is periodically broadcasted along with the MOB_NBR-ADV, so the MN has many chances to get that message regardless of the traffic load.

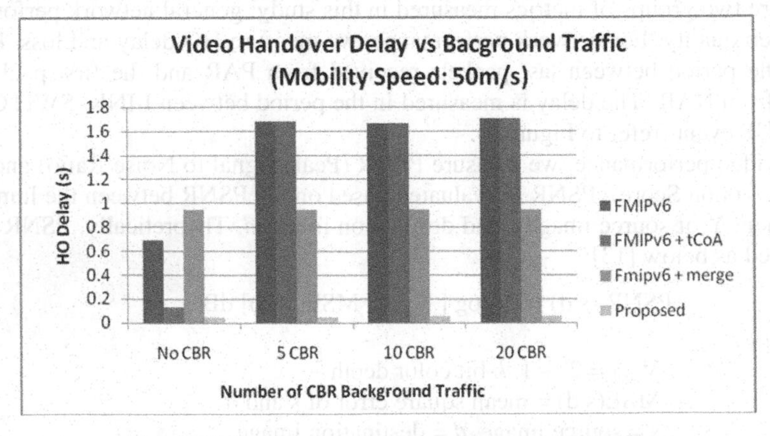

Fig. 3. Video Handover Delay vs Background Traffic

Figure 4 evaluates HO delay performance while varying the mobility speed, keeping 20 CBR background traffic load by default. As the network traffic is very heavy, long delay at the BS scheduler and possible buffer overflow (at either BS or AR) may prevent FBAck message from reaching the MN, both the FMIPv6 and FMIPv6+tCoA scheme will end up in Reactive mode, and thus have the same HO delay, while the proposed scheme yields relatively small value of HO delay regardless of the mobility speed. This happens because it does not require the reception of FBAck. The same applies to FMIPv6+merge, yet, as discussed above, the HO delay of FMIPv6+merge in predictive mode still suffers from DAD delay.

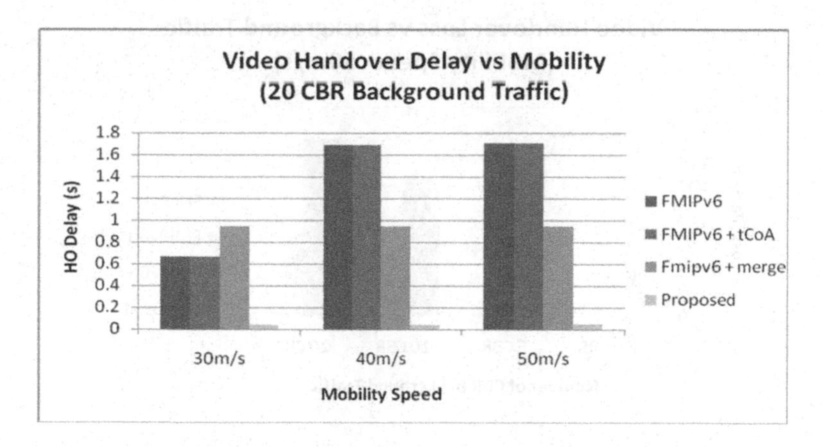

Fig. 4. Video Handover Delay vs Mobility

Handover Packet Loss: Figures 5 is the result of HO Loss of video packets while varying background traffic load. Again the difference between FMIPv6 and FMIpv6+tCoA can only be seen on a network with a few CBR background traffic load. With 5 or more CBR background traffic load, these two schemes tend to have similar poor loss performance (up to 145 packets). Both FMIPv6+merge and the proposed scheme give a better result. They do not suffer from the above problem since PrRtAdv is integrated with MOB_HO_ADV. This message will be sent periodically. However, as the background traffic is increased, the loss in FMIPv6+merge also increases due to buffer overflow because it takes longer to finish HO. The proposed scheme has the smallest number of lost packets during HO.

Video Quality

In this section we evaluate video quality in terms of PSNR and MOS. To clearly demonstrate the differences in MOS, this set of experiments is each run for 10 (instead of 20) seconds. The background traffic load is varied while the mobility speed is kept at 40 m/s. Figures 6 and 7 show the PSNR and the corresponding MOS, respectively. We see that the first two schemes, FMIPv6 and FMIPv6+tCoA, both has begun to suffer lower PSNR and lower MOS when background traffic load increased to 5 CBR, whereas the other two improved schemes, FMIPv6+merge and the proposed, achieve high PSNR and MOS until 20 CBR background traffic load. Again this is due to their ability to maintain smaller HO delay and low packet loss even during heavy network condition.

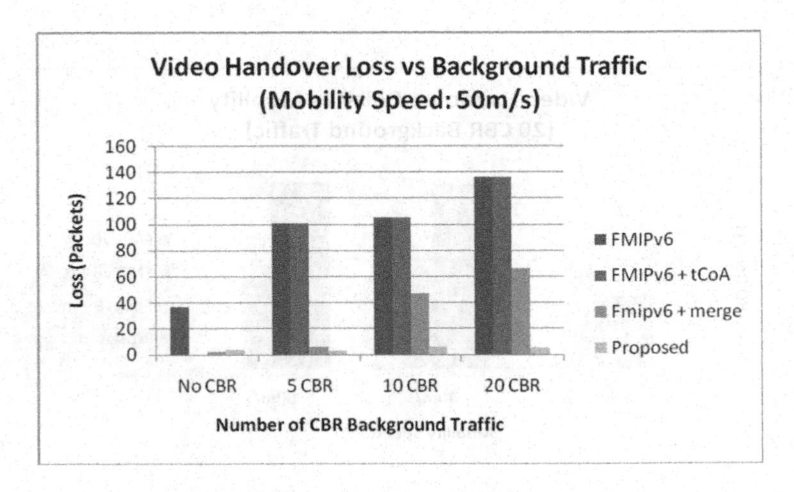

Fig. 5. Video Handover Loss vs Background Traffic

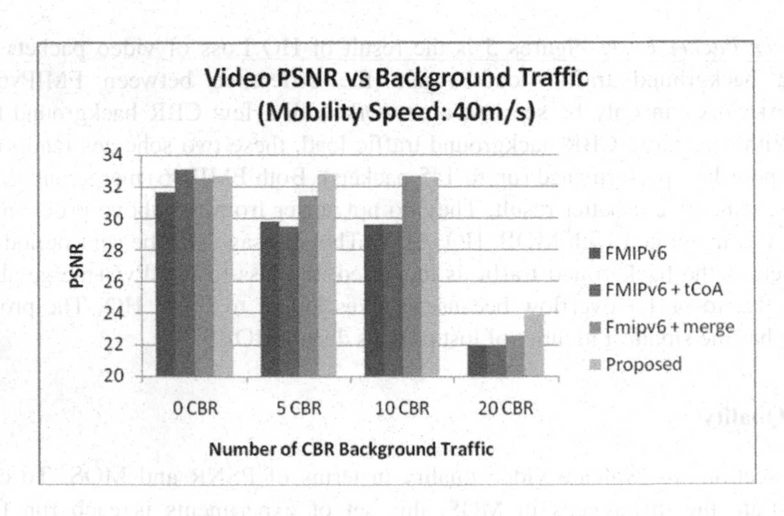

Fig. 6. Video PSNR vs Background Traffic

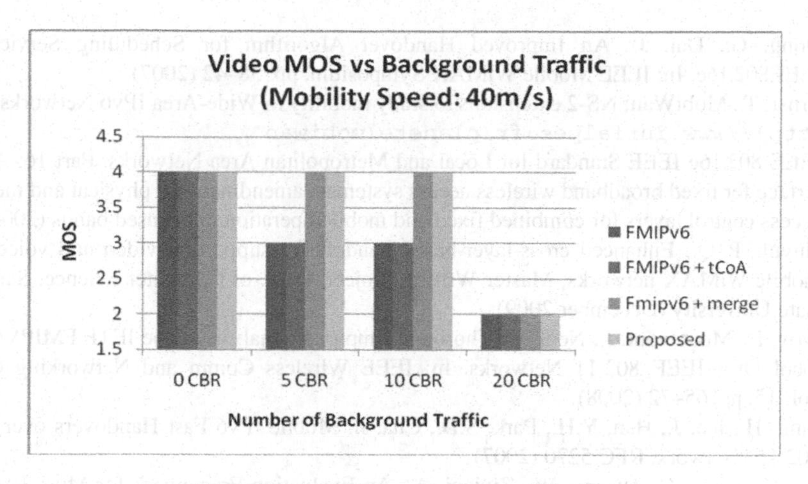

Fig. 7. Video MOS vs Background Traffic

5 Conclusion

As a promising emerging wireless metropolitan-area network technology, it is crucial for WiMAX to support seamless mobility for real-time services. The HO of WiMAX suffers long delay due to both layer-2 and layer-3 HO processes. Cross-layer HO combines and integrates HO in the two layers and improves HO performance. This paper describes a newly proposed enhancement which combines two features that have been previously proposed [3] [15]. Performance evaluation has been conducted using computer simulation, comparing four schemes: FMIPv6, FMIPv6+tCoA [15], FMIPv6+merge [3], and the proposed scheme. We found that the proposed scheme has successfully eliminate several control messages and the DAD delay, therefore performs the best in terms of HO delay, loss, and user perspectives for video. Future works may include extending the proposed scheme for supporting all the five scheduling QoS classes [1], combining the proposed scheme with link-layer retransmission schemes [17], and applying the scheme for high-speed vehicular networks [18].

References

1. Chellappan, B., Moh, T., Moh, M.: Supporting Multiple Quality-of-Service Classes in IEEE 802.16e Handoff. In: Proceedings of the International Conference of Computing, in Engineering, Science and Information, Fullerton, CA (2009)
2. Chen, J., Wang, C.C., Lee, J.D.: Pre-Coordination Mechanism of Fast Handover in WiMAX Networks. In: Wireless Broadband and Ultra Wideband Communications, AusWireless 2007, p. 15 (2007)
3. Chen, W., Hsieh, F.: A Cross Layer Design for Handover in 802.16e Network with IPv6 Mobility. In: IEEE Wireless Comm. and Networking Conf., pp. 3844–3849 (2007)
4. Cho, S., Kwun, J., Park, C., Cheon, J.H., Lee, O.S., Kim, K.: Hard Handoff Scheme Exploiting Uplink and Downlink Signals in IEEE 802.16e Systems. In: 63rd IEEE Vehicular Technology Conference, VTC-Spring, vol. 3, pp. 1236–1240 (2006)

5. Dong, G., Dai, J.: An Improved Handover Algorithm for Scheduling Services in IEEE802.16e. In: IEEE Mobile WiMAX Symposium, pp. 38–42 (2007)
6. Ernst, T.: MobiWan: NS-2 extensions to study mobility in Wide-Area IPv6 Networks, `http://www.inrialpes.fr/planete/mobiwan`
7. IEEE 802.16e IEEE Standard for Local and Metropolitan Area Networks. Part 16: Air interface for fixed broadband wireless access systems – amendment for physical and medium access control layers for combined fixed and mobile operation in licensed bands (2005)
8. Huynh, P.-Q.: Enhanced cross-layer-based handoff for supporting video and voice over mobile WiMAX networks, Master Writing Project, Dept. of Computer Science, San Jose state University (December 2009)
9. Ivov, E., Montavont, J., Noel, T.: Thorough Empirical Analysis of the IETF FMIPV6 Protocol Over IEEE 802.11 Networks. In: IEEE Wireless Comm and Networking Conf., vol. 15, pp. 65–72 (2008)
10. Jang, H., Jee, J., Han, Y.H., Park, S.D., Cha, J.: Mobile IPv6 Fast Handovers over IEEE 802.16e Network. RFC 5270 (2007)
11. Ke, C., Shieh, C., Hwang, W., Ziviani, A.: An Evaluation Framework for More Realistic Simulations of MPEG Video Transmission. J. Inf. Sci. Eng. 24, 425–440 (2008)
12. Kim, H.J., Moh, M.: Performance Study of FMIPv6-based Cross-layer WiMAX Handover Scheme For Supporting VoIP Service. In: IEEE Pacific Rim Conf. on Communications, Computers and Signal Proc. (PACRIM), held in Victoria, B.C (2009)
13. Klaue, J., Rathke, B., Wolisz, A.: EvalVid - A Framework for Video Transmission and Quality Evaluation. In: Kemper, P., Sanders, W.H. (eds.) TOOLS 2003. LNCS, vol. 2794, pp. 255–272. Springer, Heidelberg (2003)
14. Koodli, R.: Fast Handover for Mobile IPv6. RFC 4068 (2005)
15. Lee, J.S., Choi, S.Y., Eom, Y.I.: Fast Handover Scheme Using Temporary CoA in Mobile WiMAX Systems. In: 11th International Conference in Advanced Communication Technology, ICACT 2009, pp. 1772–1776 (2009)
16. McCann, P.: Mobile IPv6 Fast Handovers for 802.11 Networks. RFC 4260 (2005)
17. Moh, M., Moh, T.-S., Shih, Y.: On Enhancing WiMAX HARQ: A Multiple-Copy Approach. In: 5th IEEE Consumer Communications and Networking Conference (CCNC), Las Vegas, NV (2008)
18. Moh, M., Chellappan, B., Moh, T.-S., Venugopal, S.: Handoff mechanisms for IEEE 802.16 networks supporting intelligent transportation systems. In: Zhou, M.-T., Zhang, Y., Yang, L. (eds.) Wireless Technologies for Intelligent Transportation Systems. Nova Science Publishers, Inc. (2009) (accepted to appear)
19. Seamless and Secure Mobility Project. An IEEE 802.16 model for ns-2 by NIST, `http://www.antd.nist.gov/seamlessandsecure/pubtool.shtml#tools`
20. The Network Simulator – ns-2, `http://www.isi.edu/nsnam/ns/`

Two Risk-Aware Resource Brokering Strategies in Grid Computing: Broker-Driven vs. User-Driven Methods

Junseok Hwang, Jihyoun Park*, and Jorn Altmann

Technology Management, Economics and Policy Program,
Seoul National University, Korea
Tel.: +8210-9700-4356
{junhwang,april3}@snu.ac.kr, jorn.altmann@acm.org

Abstract. Grid computing evolves toward an open computing environment, which is characterized by highly diversified resource providers and systems. As the control of each computing resource becomes difficult, the security of users' job is often threatened by various risks occurred at individual resources in the network. This paper proposes two risk-aware resource brokering strategies: self-insurance and risk-performance preference specification. The former is a broker-driven method and the latter a user-driven method. Two mechanisms are analyzed through simulations. The simulation results show that both methods are effective for increasing the market size and reducing risks, but the user-driven technique is more cost-efficient.

Keywords: Grid computing, Risk management, Self-insurance, Risk-performance preference specification.

1 Introduction

Since Grid computing expanded its domain to the business area, demand for hard quality of services (QoS) has been continuously raised [1, 2]. The ideal of Grid computing, which is to share computing resources such as computing power, storage, bandwidth, software and data resources across the administrative boundaries of providers [1-5], has been often restricted by threats of various network security issues [6-8]. Furthermore, using computing resources provided from the network caused management issues. Users are hardly able to control failures and network attacks in remote systems and the dynamic characteristic of the environment makes users more incapable to predict the precise performance of the networked computing environment. As a result of that, although Grid computing promised many benefits with respect to resource utilization and system agility [3, 9], the commercialization of this technology has been slow. In order to expedite the widespread adoption of Grid computing, a safe and risk-aware resource allocation was required. Many technical

---

* Corresponding Author.

S.K. Prasad et al. (Eds.): ICISTM 2010, CCIS 54, pp. 187–197, 2010.

solutions have been developed for addressing uncertainty about security, QoS and privacy, but users are still hesitant to participate in an open computing environment, because technologies cannot eliminate risks in this network computing era, although they can reduce them [10-12].

This paper suggests two risk-aware resource brokering strategies: self-insurance and risk-performance preference specification. Self-insurance is a broker-driven method to supply spare resources when a selected Grid resource fails to complete an assigned task. Risk-performance preference specification is a user-driven method that brokers choose Grid resources based on revealed security requirements of users. Both mechanisms stems from economic concepts rather than technological basis. A simulation method is used to analyze the effects of both mechanisms and compared them with the alternative of not dealing with risk at all. These risk management proposals aim not only to lower the residual risk to an acceptable level, but also to introduce economically feasible solutions for resource brokers to manage their businesses successfully in a Grid computing environment. Thus, a sensitivity analysis to investigate the feasible range of a broker' cost for self-insurance coverage was also investigated. The results showed that the proposed mechanisms could increase the overall market size and reduce risks, but the user-driven method was more cost-efficient.

The remainder of the paper is organized as follows. In Section 2, previous studies of risk management in networked computing environments are discussed. Section 3 describes the structure of a Grid computing market and its risks. Section 4 proposes models of risk-aware resource brokering services for risk management in a Grid computing market. Section 5 explains evaluation methods to measure the effectiveness of the proposed models. Section 6 describes the design of the simulation. After presenting results of the simulation in Section 7, Section 8 concludes with discussions about the results.

2 Literature Review

There have been various approaches to manage risks in network environments. Particularly, a cyber-insurance that offers monetary compensation for system failures has been proposed as a safety net for IT assets [11-16]. However, there are two obstacles for insurance companies to sell cyber-insurances. First, insurance companies do not have sufficient actuarial data to estimate the probability and amount of loss from cyber attacks. Second, interdependency among firms suffering cyber attacks increases the premium price [11, 13, 16]. Bohme and Kataria [11] showed that cyber-insurance is best suited to the class of risk with high risk correlation at the firm-internal level and low risk correlation at the global level. While the firm-internal risk correlation influences a firm's decisions to seek insurance, global risk correlation influences an insurer's decision about setting premiums.

Grid computing resource brokers, who supply computing services to users with risky computing resources, can protect their business by buying cyber-insurance or by investing more money on technological improvement [12]. However, neither is a perfect solution considering that the cyber-insurance market is not yet mature and the technology development cost may not be affordable in some cases [16].

In a decentralized network environment, failures of task execution in any connected node can affect the performance of entire systems. Thus, handling distributed resources prior to job execution and proper job scheduling are another important issues in risk management. Fault tolerance mechanisms [6, 17] and reservation systems [18] are introduced to guarantee the reliability of systems. Correct measurement of risks also helps system operators to control risks [10, 19-21]. Using this method, users can choose a less risky environment so that they can actually reduce the amount of risk that they take. Song et al. [8] allowed users to explicitly tell their security requirements when their job was submitted. Reflecting individual specifications, brokers filter out inappropriate Grid resources.

Based on these previous researches, this paper suggests two resource brokering strategies for brokers. One is that resource brokers embed insurance in their resource brokering service. A self-insurance resource brokering strategy helps brokers to prepare contingencies. The other is that resource brokers exploit users' revealed preference about risk-performance tradeoffs. In this case, users indicate the level of minimum security requirements for their applications.

3 The Structure of a Grid Resource Market

3.1 Characteristics of the Market

A Grid computing environment in this paper is a marketplace in which the right to use computing resources can be traded. Computing resource providers trade the right with users through Grid resource brokers. This Grid computing resource market is considered uncertain, as the supply of resources varies and the performance of resources cannot be precisely characterized [1, 2, 22]. Technologically, this Grid computing environment is based on utility computing or inter-grid [5]. Computing resources are provided by various organizations on the network. Software and hardware are sold by standard units and charged on a pay-per-use basis.

3.2 Risk Cases

The source of risks that cause failures of information systems was probed with two tiers [11]. The two-tier approach classified four risk cases according to the intensity of correlation within a firm and at a global level. Hardware failures of individual computing resources do not affect, either internally or globally, other computers in the network. In contrast, an insider attack that shows high internal correlation causes simultaneous failures in an internal network; spyware and phishing programs are examples of globally correlated failures. Worms and viruses are disastrous, with impacts on a worldwide scale. Internal correlation can be reduced by allowing diversity in systems [13]. In a utility computing structure, the internal correlation is assumed to have little impact; however, hardware failures and global correlations are difficult to handle.

3.3 Entities

There are two entities in this market: users who buy open computing resources to execute their job, and brokers who provide resource services and are responsible for risk management. Resource owners are not considered because they delegate brokers to sell their resources.

Users. Users are entities that are in need of computing resources. Each user, i, has a unique risk tolerance level, σi, and $0 < \sigma i < 1$. A user who has a lower risk tolerance level is less tolerant of risk threats. This risk-tolerance level of users is assumed not to change in the model, and users set their requirements on the quality of service based on this value.

Brokers. Brokers are entities that provide computing resources to users in the Grid computing resource market. It is assumed that one broker operates one resource market. Brokers purchase computing resources, receive users' requests for computing services, select appropriate computing resources for the requests, and execute users' jobs with those resources. If there is excess demand or supply after all interactions with users are completed, brokers adjust the amount of resources to purchase in the subsequent period.

3.4 Resources

Grid computing resources are characterized by performance and fault tolerance. The performance of open computing resources is static information that the broker obtains by scanning the resources. The fault tolerance, δj, of an open computing resource, j, is the capability to counter risks, and is interpreted in the model as the degree of stability in the resource. δj scales from 0 (exclusive) to 1 (exclusive), where $\delta j = 0$ if the computing resource is totally vulnerable, and $\delta j = 1$ if it is perfectly protected. Installation of proper security tools (i.e., antivirus software, firewall, and access control) can improve δj but it is not possible to eliminate risks completely [11].

Meanwhile, service brokers can have their own in-house computing resources. As these computing resources do not suffer uncertain supply, they are called risk-free resources. Risk-free resources are used for contingencies in which brokers must replace troubled Grid computing resources. The cost of this type of resources is much higher than Grid resources. According to [4], the operating cost of risk-free resources is estimated to be 50% higher than that of Grid computing resources.

4 Risk-Aware Resource Brokering Strategies

4.1 Self-insurance

Users of the Grid computing market lose the total value of the computing resources that they have purchased for a job, even by one failure of a distributed task. This is because the users cannot obtain any partial value from an incomplete job. For this reason, the insurance contract for this kind of incident is a heavy burden for insurers, who have to compensate the financial loss of the policyholder. The insurer's burden, however, can be significantly alleviated if the insurance contract is embedded in the broker's resource brokering service. With embedded insurance, the resource broker, who is also an insurer, contracts with users to compensate failures of tasks incurred at allocated open computing resources, for a given period. Instead of reimbursing the

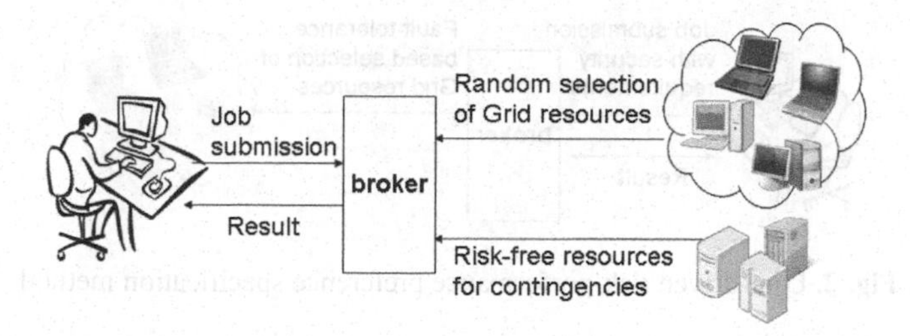

Fig. 1. Broker-driven self-insurance method

financial loss of the entire job, the self-insured resource brokering service actually replaces those troubled Grid computing resources with risk-free resources so that the users' job can be successfully finished (Figure 1). In the point that brokers take an active action to resolve risk issues in the market, this strategy is a broker-driven method.

The costs of the insurance broker for the insurance service is defined as follows [11]:

$$C_{insurance} = E(L) + A + \mu c \,. \tag{1}$$

E(L) is the cost for covering the average claims, A is the administration cost, c is the safety capital for catastrophic contingencies, and μ is the interest rate. As E(L) is predictable and A is negligible, c is the most important variable to determine the insurer's cost [11]. In the model, c will depend on the failures incurred by worms and viruses, which are the most catastrophic events in a Grid computing environment.

4.2 Risk-Performance Preference Specification

The risk of a Grid computing resource can be measured by its fault tolerance level: a smaller value represents a riskier resource. Based on this data, users can request the resource broker to supply only those resources within their risk tolerance level. A problem in composing such resource portfolios is that it is very hard for brokers to determine the exact value of the fault tolerance of each computing resource in a highly heterogeneous computing environment. Instead, the security protection activities that each computing resource adopts (e.g., installation of antivirus software, firewall and access control) is used as a proxy of the fault tolerance level in this paper. As security activities and execution time have tradeoffs [8, 9], low-risk resources, which implement more security-related applications, take more time to return the result with the same performance. The advantage of low fault tolerance cancels out the disadvantage of execution time and vice versa.

The resource broker in the model classifies Grid computing resources according to their security compliance. Users submit a job and specify the minimum security requirements for this job in terms of their risk tolerance level, and request the broker to select resources under this risk level (Figure 2). Since users initiate a risk management process, this strategy is a user-driven method. This method is useful for the broker's decision process on the types and amount of Grid computing resources that they purchase for the business. Some users prefer secure environments and thus accept slow processes. For the same price, others attach importance to prompt

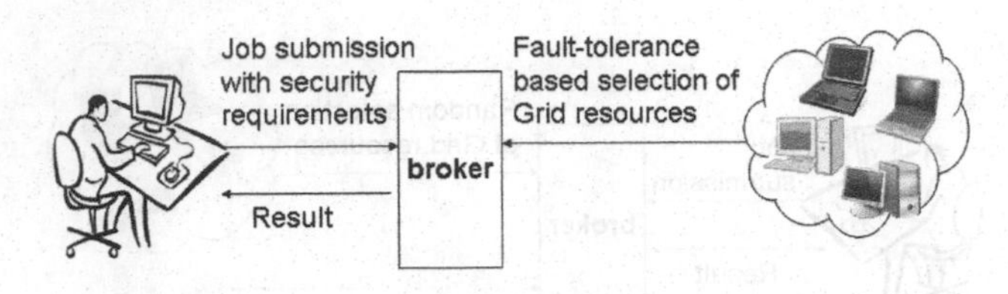

Fig. 2. User-driven risk-performance preference specification method

responses of services [19]. Reflecting users' preferences helps the service broker to predict his or her business requirements more accurately.

As this service only alters the broker's resource selection process, management costs are assumed negligible.

5 Evaluation

The performance of proposed risk-aware resource brokering strategies for risk management is evaluated by two different aspects.

First, users examine a broker's performance during interactions. The performance of a broker is published to users at the end of every service period in terms of the probability of failure about Grid computing resources that the broker has supplied [19]. At the beginning of an interaction, a potential user analyzes this evaluation to decide whether to enter the market. While participating in the market, if the user experiences any failure occurrences of his or her job that are caused by a lack of service capacity of the broker, failing to meet the time of completion, or a privacy violation, the user decides to leave the market. Therefore, the overall market size can be a sign that the risks in the Grid resource market are well managed.

PCR [10] is a comprehensive measurement for a company's losses incurred by information security breaches. The company first calculates the expected loss, expected severe loss, and standard deviation of the loss, respectively, for security breaches. They are then summed with weights derived from the analytic hierarchy process (AHP) matrix. The result of PCR depends on this AHP matrix, which reflects the emphasis of the company's security protection activities. A resource broker calculates a PCR for a certain period (e.g., one year) to evaluate the service quality. The base data are records of miscarried services due to failure occurrences. When the ratio of miscarried services, scaled from 0 (no failure) to 1 (a total failure of requests from a market), exceeds 0.5, it is counted as a severe loss. The service broker in this paper is assumed to weights the stability and predictability of the business, which puts emphasis on variance of the losses. Table 1 shows the AHP matrix for this broker.

Table 1. AHP matrix for the broker who weights business stability and predictability

	Expected loss	Expected severe loss	Standard deviation of loss	Weights
Expected loss	1	1	0.5	0.25
Expected severe loss	1	1	0.5	0.25
Standard deviation of loss	2	2	1	0.5

6 Simulation Design

To analyze the effectiveness of self-insurance contracts and risk-performance preference specifications as risk management mechanisms, a simulation method was adopted.

The simulation program was written using Java and Repast J, which is an agent-based simulation toolkit especially designed for agent-based social science modeling [24].

The simulation allows three parameters to be set. First, the degree of risk in the simulation is determined. The stylized Grid resource market in the model, based on a utility computing technology, is a highly diversified open computing environment. Heterogeneity in systems as well as providers helps to mitigate threats from internally correlated risks [13]. However, performance is not tightly controlled and arbitrary installation of applications downloaded from the Internet increases the exposure to globally correlated risks. Worms and viruses are defined as rarely occurring catastrophic events. In the simulation, such an event happens only once in an entire simulation period. During the simulation, these risk events are generated following Poisson distribution. Figure 3 illustrates the created risk events in the simulation for one year of test period.

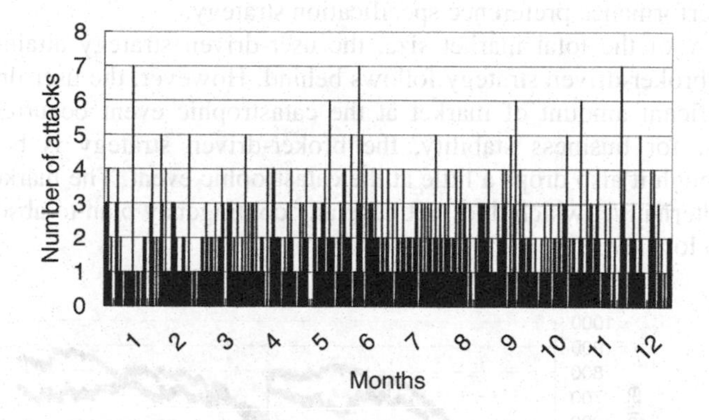

Fig. 3. Risk events simulation

The second parameter is the resource brokering strategy that a broker selects in the market. A broker can have three different resource brokering strategies, do-nothing, broker-driven self-insurance method, and user-driven risk-performance preference specification method, as described in Section 4. An insurance broker is assumed to make an insurance contract for given simulation periods of coverage with users.

Finally, for an insurance service broker, determining the safety capital budget c is an important issue, because c should be determined to assure the feasibility of the broker's service. The upper bound of c, which is the amount of risk-free resources to purchase, can be calculated as 33% of the potential total market demand, if the operating cost of risk-free resources is assumed 1.5 times that of Grid computing resources[1]. However, the lower bound depends on several factors. First, c is only valid

[1] If the market potential of a Grid resource market is M, the maximum number of risk-free resources, x, satisfies this equation: $(M + 1.5x)/M < 1.5$. Therefore, $x < M/3$.

as long as the PCR of the self-insurance strategy is better than that of the user-driven method. If the performance of the user-driven strategy is better, a broker will choose it as his or her risk management strategy because additional risk-free resources do not need to implement this service. The lower bound of c can also be restricted by the broker's acceptable level of residual risk. The residual risk after applying the self-insurance with c should be low enough for the broker to accept [11, 12]. Thus, a sensitivity analysis with regard to c was carried out to find the feasible range of c for the self-insurance strategy.

Each simulation scenario was tested for 365 time periods (iterations), assuming that a service broker assesses one-year performance of risk management proposals. This number does not restrict the result of the simulation, as a steady state, which is considered the equilibrium status in the agent-based simulation [25], was observed within this test period.

7 Results

Figure 4 compares the market size of the do-nothing strategy, the broker-driven self-insurance strategy with its maximum risk-free resource ratio of 33%, and the user-driven risk-performance preference specification strategy.

In the view of the total market size, the user-driven strategy attains the largest volume, and broker-driven strategy follows behind. However, the user-driven method loses a significant amount of market at the catastrophic event occurred at seventh month. Thus, for business stability, the broker-driven strategy is believed more reliable, although it also drops a little at the catastrophic event. The market size of the do-nothing alternative, which does not have any contingency plan to absorb damages, stagnates at a low level.

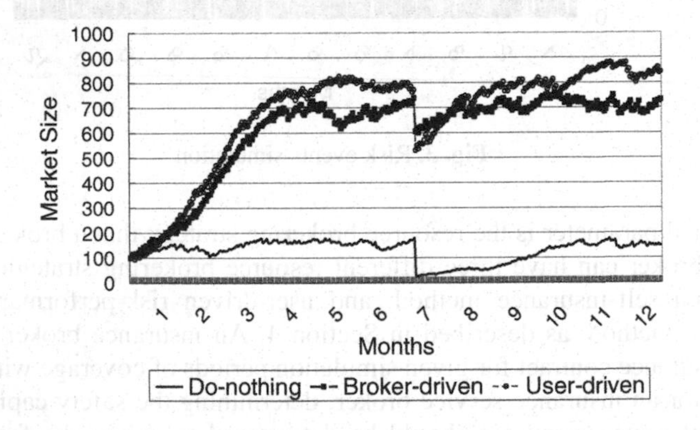

Fig. 4. Market size comparison

Table 2 shows that the broker-driven strategy with the maximum feasible level of risk-free resource ratio (c=33%) is the best alternative at alleviating risks, as assessed by the PCR, in a Grid computing resource market. Furthermore, this combination is superior to any other alternatives in individual loss measurements: the expected loss, expected severe loss, and standard deviation of the loss.

However, it may not be the preferred solution for brokers because this choice is the most expensive alternative. 33% of risk-free resource ratio was calculated assuming that the insurance broker could obtain the entire market potential. However, it turned out that roughly 70% of market potential could be fulfilled with the broker-driven strategy in the simulation. If the upper bound is re-estimated based on this realized market size, the insurance broker may allow at most 23% of risk-free resource ratio for safety capital. It is worth noting that the PCR of the user-driven method falls only marginally behind the broker-driven method in reducing the risk at the realized market size. Further, brokers may want to compromise more on risk and select a low-ratio alternative that spends less money on risk-free resources, or adopt a user-driven strategy that requires no additional resource cost. Such a decision depends on whether the residual risk will be acceptable or not after the alternative method is implemented.

Table 2. PCR of each resource brokering strategy

Resource brokering Service	Expected loss	Expected severe loss	Standard deviation of loss	PCR
Do-nothing	0.0857	0.9403	0.0686	0.2908
Broker-driven (c=20%)	0.0223	0.5195	0.0376	0.1543
Broker-driven (c=21%)	0.0230	0	0.0366	0.0241
Broker-driven (c=22%)	0.0231	0	0.0361	0.0238
Broker-driven (c=23%)	0.0234	0	0.0352	0.0235
Broker-driven (c=33%)	0.0212	0	0.0284	0.0195
User-driven	0.0292	0	0.0330	0.0238

The lower bound of risk-free resource ratio to the market potential has been investigated in two criteria. The lower bound is defined as the minimum amount, with which a broker-driven method can be selected as an effective risk management tool for a broker. First, a broker adopts a broker-driven method as long as its PCR performance is better than that of a user-driven method, and switches to the user-driven one otherwise. 23% is the minimum ratio in this case. The second case is when the broker strongly defends the system from expected server losses. The minimum demand to reserve risk-free resources is 21% of the market potential, but this value will not be invoked because the broker will switch to a user-driven strategy before this point.

8 Conclusions and Discussions

The evolutionary trend of computing systems indicates that a utility computing will be the next-generation computing platform [4, 7]. In order to operate computing jobs in such a diversified and dispersed network environment, developing risk management systems is essential, though the optimal investment on risk management systems differs between brokers in different situations.

In this paper, two risk-aware resource brokering models have been proposed as risk management strategies for Grid computing resource brokers: a broker-driven method and a user-driven method. Brokers could expand their business volume and reduce the overall risks in a Grid computing resource market by adopting those risk-aware resource brokering strategies.

Simulation results demonstrate that the broker-driven method achieves better performance as a risk management mechanism measured by PCR, while the user-driven method could generate slightly higher volume. However, the broker-driven method suffers from the high cost of preparing risk-free resources as self-insurance coverage. To the contrary, the primary advantage of a user-driven method is its cost efficiency. The feasible set of risk-free resource preparation ratios was investigated by sensitivity analysis. This result will help resource brokers, who consider adopting the proposed risk management systems, to decide the most appropriate system within their budgets and acceptable residual risk levels.

The major contribution of this paper is that risk management systems of Grid computing in this paper are designed not at the level of technological system but at the level of resource trading market. Thus, proposed risk management strategies were analyzed with their economic effects such as market size and feasibility as well as with their capability to reduce risks. Technological approaches often ignore those economic effects that they can have when implemented. However, it is very important to check their sustainability in the market.

One limitation of this study is that pricing for these proposed models was not analyzed. Pricing influences users' reactions to service offerings; thus, the final market size that each resource brokering strategy can attain may change. Once pricing is determined, brokers can calculate their profits. This profit would become the third criterion for brokers to decide their risk management schemes, along with the market size and PCR. Implementation of the proposals in real systems is another issue. Subsequent studies are required to address these issues.

Acknowledgments. This research was supported by the KCC(Korea Communications Commission), Korea, under the CPRC(Communications Policy Research Center) support program supervised by the IITA(Institute for Information Technology Advancement) (IITA-2009-(C1091-0901-0003)).

References

1. Kenyon, C., Cheliotis, G.: Architecture requirements for commercializing Grid resources. In: Proceedings of the 11th IEEE International Symposium on High Performance Distributed Computing, pp. 215–224 (2002)
2. Kenyon, C., Cheliotis, G.: Grid resource commercialization: economic engineering and delivery scenarios. In: Grid resource management: state of the art and future trends, pp. 465–478. Kluwer Academic Publishers, Dordrecht (2004)
3. Foster, I., Kesselman, C., Tuecke, S.: The Anatomy of the Grid: Enabling Scalable Virtual Organizations. International Journal of High Performance Computing Applications 15, 200–222 (2001)
4. Minoli, D.: A Networking Approach to Grid Computing. Wiley-Interscience, Hoboken (2004)
5. Thanos, G., Courcoubetis, C., Stamoulis, G.: Adopting the Grid for Business Purposes: The Main Objectives and the Associated Economic Issues. In: Veit, D.J., Altmann, J. (eds.) GECON 2007. LNCS, vol. 4685, pp. 1–15. Springer, Heidelberg (2007)

6. Domingues, P., Sousa, B., Moura Silva, L.: Sabotage-tolerance and trust management in desktop grid computing. Future Generation Computer Systems 23, 904–912 (2007)
7. Plaszczak, P., Wellner, R.: Grid computing: the savvy manager's guide. Elsevier/Morgan Kaufmann, San Francisco (2005)
8. Song, S., Kai, H., Yu-Kwong, K.: Risk-resilient heuristics and genetic algorithms for security-assured grid job scheduling. IEEE Transactions on Computers 55, 703–719 (2006)
9. Boden, T.: The Grid Enterprise — Structuring the Agile Business of the Future. BT Technology Journal 22, 107–117 (2004)
10. Bodin, L.D., Gordon, L.A., Loeb, M.P.: Information security and risk management. Commun. ACM 51, 64–68 (2008)
11. Bohme, R., Kataria, G.: Models and Measures for Correlation in Cyber-Insurance. In: Workshop on the Economics of Information Security (2006)
12. Gordon, L.A., Loeb, M.P., Sohail, T.: A framework for using insurance for cyber-risk management. Commun. ACM 46, 81–85 (2003)
13. Chen, P.-Y., Kataria, G., Krishnan, R.: Software diversity for information security. In: Workshop on the Economics of Information Security (WEIS), Harvard University, Cambridge, MA (2005)
14. Gordon, L.A., Loeb, M.P.: The economics of information security investment. ACM Trans. Inf. Syst. Secur. 5, 438–457 (2002)
15. Kesan, J., Majuca, R., Yurcik, W.: The Economic Case for Cyberinsurance. University of Illinois College of Law uiuclwps-1001 (2004)
16. Ogut, H., Menon, N., Raghunathan, S.: Cyber Insurance and IT Security Investment: Impact of Interdependent Risk. In: Workshop on the Economics of Information Security (2005)
17. Hwang, S., Kesselman, C.: A Flexible Framework for Fault Tolerance in the Grid. Journal of Grid Computing 1, 251–272 (2003)
18. McGough, A.S., Afzal, A., Darlington, J., Furmento, N., Mayer, A., Young, L.: Making the Grid Predictable through Reservations and Performance Modeling. The Computer Journal 48, 358–368 (2005)
19. Battre, D., Djemame, K., Kao, O., Voss, K.: Gaining users' trust by publishing failure probabilities. In: Third International Conference on Security and Privacy in Communications Networks and the Workshops (SecureComm 2007), pp. 193–198 (2007)
20. Kleban, S.D., Clearwater, S.H.: Computation-at-risk: assessing job portfolio management risk on clusters. In: Proceedings of the 18th International Parallel and Distributed Processing Symposium, p. 254 (2004)
21. Yeo, C.S., Buyya, R.: Integrated Risk Analysis for a Commercial Computing Service. In: IEEE International Parallel and Distributed Processing Symposium (IPDPS 2007), pp. 1–10 (2007)
22. Huang, Z., Qiu, Y.: Resource trading using cognitive agents: A hybrid perspective and its simulation. Future Generation Computer Systems 23, 837–845 (2007)
23. Tobias, R., Hofmann, C.: Evaluation of free Java-libraries for social-scientific agent based simulation. Journal of Artificial Societies and Social Simulation 7 (2004)
24. Tesfatsion, L.: Agent-Based Computational Economics: A Constructive Approach to Economic Theory. In: Tesfatsion, L., Judd, K.L. (eds.) Handbook of Computational Economics, vol. 2, ch. 16, pp. 831–880. Elsevier, Amsterdam (2006)

A Socio-Technical Perspective on Computer Mediated Communication: Comparison of Government and Non Government Sectors in India

Payal Mehra

Indian Institute of Management,
Lucknow
payal@iiml.ac.in

Abstract. The Information System discipline, studies the way individuals, groups and organizations use information. Although these methodologies have traditionally focused on the design of hardware, software and data aspects of the IS, newer (the so called 'soft') approaches, involve more consideration of human factors issues These socio-technical methodologies incorporate a higher level of participation by system users and focus on identification of user needs and task satisfaction. This study examines how socio-technical factors (e.g. organizational climate and IT related issues) affect information sharing through their effects on perceived relative satisfaction with the new media (Email, Instant Messaging and Video Conferencing) in the government and non-government sectors of an emerging economy. The findings reveal a digital divide between the non government and the government sector. Though technology has been continuously upgraded in most government organizations, the same cannot be said about communication. The study also reports wide gaps between the use, adoption and application of technology and what people desire socially. The study recommends structuring of the new media to harness data and information from participants.

Keywords: Socio-technical model, media satisfaction, digital divide, GLM regression, CMCS.

1 Introduction

The Information System (hitherto referred to as IS) discipline, studies the way individuals, groups and organizations use information and consists of five aspects: hardware; software; data; people and procedures (Andrew Turk and Kathryn Trees, 1998) Although these methodologies have traditionally focused on the design of hardware, software and data aspects of the IS, newer (the so called 'soft') approaches, involve more consideration of human factors issues (Avison et al., 1993; Crowe et al., 1996). These sociotechnical methodologies incorporate a higher level of participation by system users and focus on identification of user needs and task satisfaction According to the new approach, active participation of the users needs to be maximized at each stage. A thorough examination of the usability of the system and user satisfaction) is

S.K. Prasad et al. (Eds.): ICISTM 2010, CCIS 54, pp. 198–209, 2010.

required for system modification and is a useful step to redesign information systems according to user need.

This study examines how socio-technical factors (e.g. organizational climate and IT related issues) affect information sharing through their effects on perceived relative satisfaction with the new media (Email, Instant Messaging and Video Conferencing).

The study is significant as it examines the complexity of computer and communication issues faced by organizations in an emerging economy. It specifically contrasts the work environments of two sectors in India-the government and the non-government with respect to technology and communication and suggests ways to develop 'a socio-technical mindset'. The study proceeds in three parts. First, it provides an overview of IS and human computer interaction briefly reviewing the major social and technical findings of the field; next, it discusses the findings of a survey that highlights certain socio-technical gaps between the two sectors under study-government and non government in India. It concludes with a discussion of the main findings and potential solutions for reducing the gap.

2 Overview of IS and CMC

2.1 Brief History of the Development of the Socio-Technical Model

In the production model of IS, transaction based IS improved procedures and processes (accounting, payroll etc.) but ignored the user's need to a large extent. While the production model of IS has gradually improved (primarily technologically), it still presumes the technical nature of IS and emphasizes the role of IS in efficient monitoring and control of processes in organizations. According to the interactive model of IS, users are active participants that interact with the system to provide data (their feelings, attitudes and assumptions); to specify requests and get the requisite information. The purpose of these systems was to increase efficiency and effectiveness of decision makers.

Innovations in Artificial Intelligence and advancement of technologies for ES and DSS have enabled more intelligent and more efficient systems to be built. This paved the way for the 'networked model' where attempt was made to understand the information needs of the organization from a holistic perspective. It centered on the belief that one could technically realize the flow of data from specific functions and reuse the information available for organizational purposes. This was in fact a big illusion (Dubravka Cecez-Kecmanovic, 2002), and a socio-technical approach has been suggested as an alternative to the existing approaches. This places communication, coordination, collaboration and shared knowledge and not database at the heart of information management; the IS is projected as a social system which use information technologies to perform certain tasks.

2.2 Literature Review: A Summary

2.2.1 Information System and Communication

The IS presents a rather simplistic model of communication and information management where there is hardly any re-use of the knowledge and experience. One finding is that it is 'sometimes easier and better to augment technical mechanisms with social mechanisms to control, regulate, or encourage behavior' (Sproull, & Kiesler, 1991), rather than the other way round, i.e., social mechanisms to facilitate appropriate technical behavior. Also, systems often assume a shared understanding of information. (Ackerman, 2001). In reality, however, to quote Goffman, (1961, 1971) people exhibit very nuanced behavior as far information-sharing is concerned.

IT can provide access to knowledge, but access is not the same as using or applying knowledge. (Starr Roxane Hiltz, 1988). Research findings by Hsiu-Fen Lin and Gwo-Guang Lee (2005) indicate that social-oriented organizational climate (i.e. top management support, open communication, stimulus to develop new ideas, and reward systems in inducing knowledge sharing) is likely to have positive benefits and compatible beliefs about promoting knowledge sharing. Linking satisfaction with the new media to presence of an open, people oriented and supportive climate in organizations, null hypotheses framed in this context are:

Ho 1: Satisfaction with the new media is independent of the people orientation in the two sectors

Ho 2: Satisfaction with the new media is independent of the communication climate in the two sectors

Media accessibility, especially in an emerging economy, is critical to achieve satisfaction with the media. Previous researches have shown that user apprehension and lack of familiarity with the medium are 'powerful inhibitors of an individual's involvement in communication activities particularly in technology mediated environments' (Brown, Fuller, & Vician, 2004; Easton, Easton, & Belch, 2003; Rao &Dennis, 2000; Scott & Timmerman, 2005).

Ho 3: Satisfaction with the media is independent of the frequency with which the employees use the new media in the two sectors

Ho 4: Satisfaction with the new media is independent of the extent of familiarity of the employees with the new media in the two sectors

Ho 5: Satisfaction with the new media is independent of the extent of technology initiatives undertaken in the two sectors.

2.2.2 Media Choice, Media Satisfaction and Task Perspectives of Computer Mediated Interaction

This section explores the cross linkages between channel choice, channel appropriateness, channel effectiveness, channel availability and accessibility.

The media richness theory developed by Daft and Lengel (1984, 1986) implies that, a sender can (and should) use the richest possible medium to communicate the desired message. Research on media appropriateness by Zmud, Lind and Young (1990) found three factors that influenced the choice of media in the actual or desired use of media: capacity for feedback, accessibility and quality where it was found that appropriateness was highest in face to face communication followed by phone, meeting, voice mail and email (William and Rice (1983) Detractors of the 'cues filtered out theory' espouse that in reality participants rarely have the choice to select as media for a task; it is in essentially a trade off between what is available and accessible to serve a particular need. Walther (1992), in particular, recommends the social information processing perspective as an alternative to the cues filtered out approach. This perspective accepts that CMC has a cue limitation which is why the medium is incapable of conveying all task and social related information as fast as a face to face communication. In other words a manager's choice of a particular media for a certain task is governed by organizational constraints rather than the optimal fit (Fulk, Scmitz & Steinfield 1990). This leads to the next hypothesis:

Ho 6: Satisfaction with the new media is independent of the level of social interaction induced by the medium.

2.3 The India Perspective

Internet and computers have heralded not only a technological but also a social revolution in India. This 'social construction of technology' (the process by which people give meaning to a new technology by discussing it amongst them) has redefined user needs, expectations and satisfactions from technology. An emerging economy, India is fast catching up with the world in terms of technological advancements. Currently witnessing a massive computerization boom, plans are afoot to improve the telecom density further.

The increased use of CMC, particularly in emerging economies such as India, raises concerns relating to its efficacy particularly when compared to the more natural face to face communication. For Indians (and for many Asian countries) the key purpose of communication is often to preserve harmony and avoid offence. A collectivist mentality still drives the social fabric of Indian life; therefore, the strengthening of relationships between members is of critical importance.

Studies by Thompson and Feldman (1998) suggest that the increased use of CMC might subsequently result in the drop or decline in other forms of communication such as face to face (FTF) meetings and one to one conversations. With advances in communication technologies in most Indian companies (read software, financial services, call centers educational concerns and even some government concerns) this is already happening; at present, there is little evidence to show the social effects of technology though.

In the current scenario, a wide 'digital divide' appears to have segregated the Indian industry into the technology intensive information rich companies (typically the private sector) and the information poor government concerns (at least some organizations where IT is confined to the top management and its effect not percolated down to the lower management and administrative rungs.

3 Research Methods

The research study is empirical in nature. A questionnaire was developed aimed to collect responses relating to the demographic details, organizational culture, frequency of use of different media, familiarity with different media, task-use frequency, choice of media for communication tasks and extent of satisfaction with the electronic media. For continuous data the five point Likert scale was utilized. The study was institutionally funded and was conducted in three phases. Initially the questionnaire was pre-tested using 100 managers. The questionnaire was finally administered to 1000 respondents out of which 729 questionnaires were found usable for analysis. 33.4% of the sample represented the public sector and the remaining represented the private sector. The subjects for the study were the knowledge workers in a computer enabled organization having regular access to the email/other facilities in India. The subjects were junior, middle and senior level employees (17.6%; 75.8% and 6.6% respectively) working in listed companies.

The data were analyzed using the statistical package SPSS 16.0. Descriptive statistics, chi-square analysis, ANOVA, reliability analysis, and multivariate analysis (Factor analysis and General Linear Model of regression) were utilized for the findings. The Cronbach alpha was used to measure the internal consistency of the variables. The resulting indexes are .709 for the questionnaire comprising 63 variables; 730 for familiarity with various media (four variables); .740 for frequency of use of media with respect to tasks (eight variables) and, .829 for measuring satisfaction with the media on the socio-economic dimensions of task performance (20 variables). To understand the relationship of the sample with the media, the participants were asked to indicate the level of familiarity with face to face communication, email mode of communication, videoconferencing and instant messaging on Likert scales ranging from 1-low familiarity to 5-high familiarity Responses indicated that participants had the highest levels of familiarity with face-to-face communication, followed by email, videoconferencing and instant messaging.

The research model is presented (see Figure 1).the dependent variable is the satisfaction with the new media (email, IM, Video-Conferencing) in public and private enterprises. The independent variables comprise the effects of an open communication climate in organizations, people orientation, technology intensiveness in organizations, extent of media usage and media familiarity by the respondents.

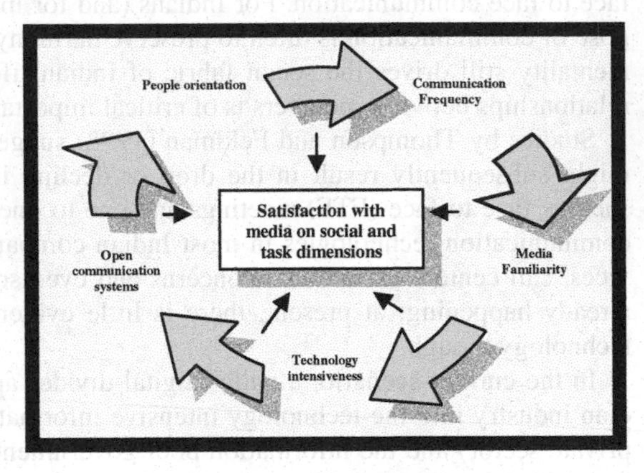

Fig. 1. Research Model

4 Results

4.1 Satisfaction with the New Media on Social and Economic Dimensions of Task Performance

Two tests indicate the suitability of the data for structure detection in factor analysis. The Kaiser-Meyer-Olkin Measure of Sampling Adequacy is.891 (close to 1.0) indicates that a factor analysis may be useful with the data. Bartlett's test of sphericity is significant at .000. The rotated component matrix below shows the four extracted components (See Table 1 below). The first component is highly correlated with sharing personal views, quality of social interaction, shared understanding of the situation and original meaning conveyed as intended (social dimensions of task performance). The second component is highly correlated with task coordination, group coordination

and quality of judgment and analysis (task and group coordination for decision making). The third component is highly correlated with reliability and credibility of information (quality of the shared knowledge in group work); and the fourth component with speed and cost factors (speed, cost and time effectiveness).
(Extraction Method: Principal Component Analysis. Rotation Method: Varimax with Kaiser Normalization. Rotation converged in 5 iterations. Variance explained: 66%)

Table 1. Rotated Component Matrix

Rotated Component Matrix(a)					
		Component			
No.	Constructs	1	2	3	4
Q9.1	Sharing job information with group	-.177	.165	.483	.165
Q9.2	Quick response	.070	**.539**	.083	.526
Q9.3	Task coordination	.196	**.685**	.176	.370
Q9.4	Group coordination	.258	**.681**	.225	.261
Q9.5	Quality of judgment and analysis	.206	**.575**	.003	-.036
Q9.6	Information generated	.072	.369	.161	.110
Q9.7	Reduced impact of hierarchy	.020	.476	.115	-.038
Q9.8	Speed and cost effectiveness	.175	.302	-.107	**.698**
Q9.9	High participation and commitmer	.296	.129	.082	.308
Q9.10	Lends to global operations	.026	.110	.445	**.569**
Q9.11	Handling time constraints	.134	-.070	.121	**.640**
Q9.12	Anonymity in Expressing views	.402	.023	.477	.278
Q9.13	Reliable	.318	.318	**.501**	.195
Q9.14	Amenable to small Group size	.084	.021	**.599**	-.042
Q9.15	Credibility of information	.206	.322	**.534**	.139
Q9.16	Resolving miscommunication/ and conflicts	.414	.274	**.506**	-.101
Q9.17	Sharing personal views on official matters	**.727**	.069	.126	.080
Q9.18	Quality of social interaction	**.764**	.193	-.050	.140
Q9.19	Original meaning Conveyed as inte	**.613**	.364	.012	.152
Q9.20	Shared understanding Of situation	**.634**	.101	.304	.123

Table 2 below indicates the differences in satisfaction with the new media on all the four dimensions of task performance between the two sectors under study. The table reveals that satisfaction levels between the two sectors- non government and government-is not significant for social dimensions of task performance (the respondents in both the sectors show reduced satisfaction on this dimension).

However, there are significant differences in the satisfaction score for the other three dimensions between the sectors. The respondents in the non government sectors are more satisfied with the new media than their counterparts in the government sectors on task and group coordination for decision making, reliability and credibility of information, generation of trust in groups and speed, cost and time effectiveness and less satisfied on the social dimensions.

Table 2. Sector-wise Differences in Satisfaction with the New Media on Task Dimensions

S.N	Satisfaction Dimension	Non-Govt SECTOR	Govt. SECTOR	Significance (ANOVA)	Comment on Sector-wise Differences
1	**Social dimensions of task performance**	-.0119115 1.02103130	0250029 95622714	.659	Not significant Both record low satisfaction
2	**Task and group coordination on decision making**	.0752965 .99932154	-.1580516 .98515881	.005	Significant Private sector records greater satisfaction
3	**Reliability and credibility of information to generate trust in groups**	.0764697 .87270207	-.1605142 1.21167566	.004	Significant Private sector records greater satisfaction
4	**Speed cost and time effectiveness**	.1556308 .96942909	-.3266779 .98647097	.000	Significant Private sector records greater satisfaction

Table 3. Sector-wise Differences in Effects of Frequent use and Familiarity with the new media

Sector	Statistic	E mail Freq	Video Conf Freq	Instant Messaging Freq	Familiarity Email	Familiarity Video conf	Familiarity Instant Messaging
Non-Govt.	Mean	4.2144	2.3492	3.1508	4.7629	3.6021	4.1839
	SD	.48690	1.05539	1.19349	.55991	1.34395	2.0188 7
Govt.	Mean	4.1481	2.2593	2.8264	3.4691	3.5267	1.9835
	SD	.92916	.97206	1.12444	.79388	1.40333	.37446
Difference between the two sectors	ANOVA Sig.	.273 N.S	.266 N.S	.000 N.S	.000 Signif	.483 Not Significant	.000 Significant

The table below (Table 3) reveals that sector wise differences significantly exist as far as the frequent use of Instant messaging is concerned (IM as a communication tool is hardly used in the traditional government led enterprises); familiarity with email (most government sector respondents profess to be low on familiarity with the medium) and familiarity with instant messaging.

The survey revealed considerable differences of perception the on various cultural parameters such as people orientation, technology intensity and presence of an open communication system measured on a 7-point semantic differential scale (See table 4). The non government or the private enterprises exhibit a higher mean than their government counterparts. In other words the government sector respondents believe that their organizations are not that open in terms of communication environment (due to the presence of a bureaucratic structure); technology intensiveness is there but largely used for data processing rather than facilitating interaction and people orientation is rather low as compared to the private sector.

Table 4. Sector-wise Differences in Type of Organization Culture

SECTOR	**Mean SD**	**Highly people oriented**	**Highly technology intensive**	**Open Communication system**
Non-Govt.	**Mean**	5.5103	5.8663	5.0247
	SD	1.13922	1.10323	1.31110
Govt.	**Mean**	4.7037	4.7613	3.2181
	SD	1.23761	1.24674	1.19129
Difference between the two sectors	**ANOVA Significance**	**.000** Significant	**.000** Significant	**.000** Significant

Further analysis of culture and its association with satisfaction with the new media on select dimensions revealed that for dimension on sharing job information with group (1), Task coordination (2) Group coordination (3) and quality of social interaction (6), both the sectors believe that a people oriented culture is helpful. Analysis of a technology intensive culture and its association with perceived satisfaction with the new media reveals that there is no difference between the sectors on dimensions relating to satisfaction with sharing job information with group and Information generated(both the sectors report poor association), task coordination and group coordination (both sectors report good association). In other words while the private sector believes that a technology intensive culture perpetrates social interaction and facilitates better decision making via the new media, the government sector reports quite the opposite. Analysis of an open communication culture and its association with perceived satisfaction with the new media reveals that the private sector believes that an open communication climate fosters better decision making and sharing of information via the new media, the government sector reports quite the opposite.

The section concludes with result of the GLM Multivariate. Proceeding with the function that satisfaction with the new media in the two sectors is dependent upon the frequency of email, video conferencing and instant messaging (Q3), and familiarity with email, video conferencing and instant messaging (Q4) as well as organizational culture (Q6) and value placed on communication outcomes (Q7), the following model is created:

Design: Intercept+Q3.1+Q3.3+Q4.1+Q4.3+Q4.4+Q6.1+Q6.2+Q6.4+Q7.1+Q7.2+Q7.3+Q3.4+Q1.4

The regression model evaluates the impact of the frequency of and familiarity with the new media as well as organizational culture, technology intensiveness people orientation and an open communication system on the satisfaction with the media on social dimensions, task and group coordination, generating trust and speed and cost effectiveness. The GLM interprets that the effects of frequency of email, frequency of video conferencing, familiarity with email and video conferencing, people orientation and value placed on communication outcome as a decision making enabler contribute significantly to the model. The null hypotheses one three and four and six (Ho1; Ho3, Ho4 and Ho6) have been rejected as significant interaction effects of these variables are visible. Thus people orientation, familiarity with email and frequency of email contribute significantly to the model, i.e., these variables have a positive impact on satisfaction scores.

Of particular interest is the data shown in Table 5 below. The effect of non- government enterprises is compared to the effects of government enterprises where it is found that respondents belonging to the non-government organizations express reduced satisfaction with the social dimensions of task performance when using computer mediated tools of communication and greater satisfaction than the government

Table 5. Contrast Results (K Matrix)

Level 1 Vs. Level 2	**Dependent Variable**	**Social dimensions of task performance**	**Task and group coord**	**Reliability and credibility of information**	**Speed cost and time effectiveness**
(Non-Govt sector Vs Govt sector)	Contrast Estimate	-.216	.070	.318	.535
	Hypothesized Value	0	0	0	0
	Difference (Estimate – Hypothesized)	-.216	.070	.318	.535
	Std. Error	.122	.122	.122	.123
	Sig.	**.078**	**.569**	**.009**	**.000**
	90% Confidence Interval for Difference — Lower Bound	-.456	-.170	.079	.294
category = 2	Upper Bound	.024	.309	.557	.776

sector on the other three dimensions of task performance. The significance value of .078 for social dimensions of task performance, of .009 for reliability and credibility of information to generate trust in groups and .000 for speed cost and time effectiveness indicates that this difference is not due to a chance variation but significant. For the dimension on task and group coordination on decision making, the significant value is more than .10 indicating that the difference may be due to a chance variation.

5 Discussion and Conclusion

For the technology specialist, installing a satisfactory computer communication network is the ultimate payoff but interactive communication systems are not just about technology with predictable cause and effect equations. These systems are implemented in a social context and form part of a socio-technical system. Application of new technology may not result in a desired outcome if the users do not have positive expectations, experiences and attitudes. Communication or Interaction can only be fully understood through detailed analysis of the social context in which it occurs: "... at this point we should no longer see people simply as 'users' of given systems, but as social 'actors.' In other words, whether expert computer users or not, people act independently and have their own reasons for what they do, and it is computers and systems that has to adapt to people, not vice versa (Riva, 2002).

The study revealed that wide gaps exist in the government sector as far as the use adoption and application of technology is concerned and what people desire socially. The gap is there because there is a fundamental difference between what is technically possible and what can be accomplished socially. The gap or the 'digital divide' between the non government and the government sector is also wide. There are several reasons for this. Firstly, there is decreased people orientation in the public sector. The employees differ in their skills sets due to lack of training and awareness. Familiarity with email and instant messaging- the communication tools of today-is weak. There is seemingly lack of computer support in most organizations. Where present, computers are reduced to mere machines used for data entry and taking printouts (as stated in most of the qualitative comments recorded in the questionnaires) rather than a communication tool to facilitate task efficiency. Secondly, the organization culture in most government organizations is not only technology deficient but also communication deficient. There appears to be a lack of empowerment because the bureaucratic or the so called 'babu' culture does not permit independent and autonomous style of working. Open channels of communication across hierarchies do not exist and if they do, they are not used to the extent these should have been. There are few government firms that have scaled up processes to include systems which address individual employee concerns and there are yet other government organizations where computer literacy, adoption and application is rather abysmal. While the technology has been continuously upgraded, the same cannot be said about communication. This is why government concerns continue to report decrease in satisfaction with the new media since communication issues have not been adequately addressed.

Respondents in both sectors respond poorly to media's effect on affective dimensions involving emotions and positively to cognitive dimension based on experiences (time and cost efficiency, for example). Interestingly even the high technology software and financial concerns have recorded decrease in satisfaction on social interaction with the new media and comparatively higher satisfaction with respect to the task related economic dimensions (coordination, information-sharing for decision making, time and cost effectiveness and the like). The question is should the new media be structured to facilitate this kind of interaction and if so then to what extent?

Knowledge incubates in a person's mind (Purani, Nair, 2006). For growth and development, data and information must be harnessed efficiently to create knowledge. This needs to happen in the government sector which otherwise records a robust growth in terms of productivity. This can only occur when a conducive, people oriented, flexible and open communication climate is provided to the employees. Installing a CMCS (Computer mediated communication systems) has the potential to reduce communication delays and other costs. Social and communication barriers created by hierarchy, bureaucratic norms and poor professional standards can be sufficiently reduced by providing access to CMC components such as chat, conferencing support and email.. It would herald a techno-social revolution that can change the work pattern of most government firms. The other major initiative would be to sensitize the top management of the pitfalls of ignoring communication issues in facilitating decision making and information sharing via technology. Even if there a lack of media choice for a task (participants have to make-do with the media available and accessible to them), then also a trade off can be made to fit in social arrangements with specific task and media settings. This requires a considerable overhaul in terms of integrating IT, IS and socio-communication issues in organizations.

References

1. Turk, A., Trees, K.: Culture and participation in the development of CMC. In: Ess, Sudweeks, F. (ed.) Proceedings Cultural Attitudes towards Communication and Technology, University of Sydney, Australia, pp. 263–267 (1998)
2. Avison, D., Kendall, J.E., DeGross, J.I. (eds.): Human, Organizational, and Social Dimensions of Information Systems Development. Elsevier Science, Amsterdam (1993)
3. Brown, S.A., Vician, F.: Who's afraid of the virtual world? Anxiety and computer-mediated communication. Journal of the Association for Information Systems 5(2), 79–107 (2004)
4. Burkes, et al.: A study of partially distributed work groups: The impact of media location and time on perceptions and performance. Small Group Research 30, 453–490 (1999)
5. Daft, R.L., Lengel, R.H.: Information richness: A new approach to managerial behavior and organizational design. Research in Organizational Behavior, 191–233 (1984)
6. Daft, R.L., Lengel, R.H.: Organizational information requirements, media richness, and structural design. Management Science 5, 554–571 (1986)
7. Dennis, A.R., Valacich, J.S.: Computer brainstorms: More heads are better than one. Journal of Applied Psychology 78, 531–537 (1993)
8. Cecez-Kecmanovic, D.: The discipline of information systems: Issues and challenges. In: Eighth Americas Conference on information systems (2002)
9. Easton, G., Easton, A., Belch, M.: An experimental investigation of electronic focus groups. Information and Management 40, 717–727 (2003)

10. Goffman, E.: The Presentation of Self in Everyday Life. Anchor- Doubleday in Ackerman, New York (1961)
11. Goffman, E.: Relations in Public. Basic Books in Ackerman, New York (1971)
12. Hiemstra, G.: Teleconferencing, concern for face, and organizational culture. In: Burgoon, M. (ed.) Communication year 6 Sage. in Walther IEEE (1992)
13. Lin, H.-F., Lee, G.-G.: Effects of socio-technical factors on organizational intention to encourage knowledge sharing. Management Decision 44(1), 74–88 (2005)
14. Campbell, J.: Media Richness, Communication apprehension and participation in group videoconferencing. Journal of Information, Information Technology, and Organizations 1 (2006)
15. Kock, N.: Can communication medium limitations foster better group outcomes? An action research study. Information & Management 34, 295–305 (1998)
16. Kock, N.: The psychobiological model: Towards a new theory of computer-mediated communication based on Darwinian evolution. Organization Science 15, 327–348 (2004)
17. Kock, N.: Evolution and media naturalness: A look at e-communication through a Darwinian theoretical lens. In: Applegate, L., Galliers, R., DeGross, J.L. (eds.) Proceedings of the 23rd International Conference on Information Systems, pp. 373–382. Association for Information Systems, Atlanta (2002)
18. Ackerman, M.S.: The Intellectual Challenge of CSCW: The gap between social requirements and technical feasibility. To be published in Human-Computer Interaction (2001)
19. Purani, Nair: Knowledge community: integrating ICT into social development in developing economies. AI and Society 21(3) (2006)
20. Rao, V.S., Dennis, A.R.: Equality of reticence in groups and idea generation: An empirical study. Journal of Information Technology Management 11, 1–20 (2006)
21. Riva, G.T.: The Socio-cognitive psychology of computer-mediated communication: The present and future of technology-based interactions. Cyber Psychology & Behavior 5(6) (2002)
22. Scott, C.R., Timmerman, C.E.: Relating computer, communication, and computer-mediated communication apprehensions to new communication technology use in the workplace. Communication Research 32, 683–725 (2005)
23. Short, J., Williams, E., Christie, B.: The Social Psychology of Telecommunications. John Wiley and Sons, Chichester (1976)
24. Sproull, L., Kiesler, S.: Connections: new ways of working in the networked organizations. MIT Press in Riva. G, Cambridge
25. Hiltz, S.R.: Productivity enhancement from Computer-mediated communication: A systems contingency approach. Communication of the ACM 31, 12 (1998)
26. Templeton, G.F., Lewis, B.R., Snyder, C.A.: Development of a measure for the organizational learning construct. Journal of Management Information Systems 19(2), 175–218 (2002)
27. Walther, J.B.: Interpersonal effects in computer-mediated interaction: A relational perspective. Communications Research 19(1), 52–90 (1992)
28. Walther, J.B.: Anticipated on-going interaction versus channel effects of relational Communication in computer-mediated interaction. Human Communication Research 20, 473–501 (1994)
29. Zmud, W., Lind, M., Young, F.: An Attribute Space for Organizational. Information Systems Research 1(4), 440–457 (1990)

Can You Judge a Man by His Friends? - Enhancing Spammer Detection on the Twitter Microblogging Platform Using Friends and Followers

Teng-Sheng Moh and Alexander J. Murmann

San José State University
Dept. of Computer Science
San José, CA 95192-0249 USA
tsmoh@cs.sjsu.edu, ajmurmann@gmail.com

Abstract. As online social networks acquire a larger user base, they also become more interesting targets for spammers. Spam can take very different forms on social web sites and can not always be detected by analyzing textual content. However, the platform's social nature also offers new ways of approaching the spam problem. In this work we analyze a user's friends and followers to gain information on him. Next, we evaluate them using different metrics to determine the amount of trust his peers give him. We use the Twitter microblogging platform for this case study.

Keywords: Spammer Detection, Twitter, Microblogging, Online Social Networks, Learning Process, Classification Algorithm, Machine Learning, Online Trust.

1 Introduction

Unsolicited email is a problem every Internet user is familiar with. The amount of spam emails increases from year to year and the costs of coping with this problem increase with it. The huge amount of spam is one of the most often heard complaints about the email service [1].

The increasing popularity of social web sites makes them more attractive to spammers. We also face Spam in new forms which poses new challenges for us. However, the social nature of the platforms also offer new approaches to handling these challenges.

Unsolicited email is mainly distinguished from normal email by its content. But on social web sites, the content may only play a minor role. The spammer's behavior is what distinguishes him from a normal user. An example of this would be subscribing to many user's feeds on the Twitter platform in order to get their attention. Oftentimes, this is not driven by immediate commercial interests and the content posted is not necessarily different from that of legitimate users. Furthermore, on some platforms, checking the content may not be a viable option,

S.K. Prasad et al. (Eds.): ICISTM 2010, CCIS 54, pp. 210–220, 2010.

due to the the content's nature. An example for this case would be online video platforms.

In this case the service might be able to cope with spammers by utilizing the social structure underlying the network. This is independent of the form spam takes on the platform. In this paper, we evaluate ways to use information on a user's friends and followers on the Twitter platform to determine if the user is a spammer or legitimate user. We also propose a learning process that consists of two steps. In the first step, a categorization algorithm is trained to distinguish between spammers and legitimate users on a set of basic user features. In the following step we use this trained classifier to generate new features for a user which depend on a user's followers being spammers or legitimate users. These new features also try to capture the implicit trust given to a user by his or her followers. Our goal is to evaluate the information that can be gained by analyzing a user's direct neighbors in the social graph and not to develop the best spam detector possible for the Twitter platform.

The article is organized as follows. In Section 2, we give an overview of related work in the field of spam fighting methods. Section 3 gives a brief introduction to the functionality of the Twitter platform and terminology that will be used throughout this work. Section 4 describes the architecture of our framework and Section 5 describes the test data in detail. In Section 6, we describe the experimental setup that was chosen to evaluate our framework, and the results. The conclusion and potential future work are described in Section 7.

2 Related Work

Heymann et al. [2] categorize existing anti-spam strategies into three categories.

- Detection-based
- Prevention-based
- Demotion-based

Detection-based approaches identify spam or spammers and then either delete them or display them as likely spammers. A common example from this field is spam filters for email.

Prevention-based systems try to keep spam from getting into the system. They do this by authenticating users before they are allowed on the platform or by putting up obstacles that prevent malicious behavior. For example, using CAPTCHAs to keep bots out, or creating costs for the contribution of messages.

Demotion-based approaches rank spam lower than non-spam. This is a strategy often encountered in web-search, where potential spam appears at the bottom of the list of search results.

Using information contained in social graphs to detect spam has already been proposed outside of online social networking sites. Boykin and Roychowdhury [3] proposed an algorithm that used the social graph created by e-mails sent and received by users. Addresses that occurred together in e-mail headers were connected with an edge in the graph. In this graph Boykin and Roychowdhury found strongly connected components which were then analyzed. Because spammers sent e-mails to huge numbers of recipients at the same time, but never sent e-mails to each other, spammers ended up in different components than legitimate users. Based on a component's size, maximum degree and clustering coefficient, these components were categorized as components consisting of spammers or non-spammers.

Krause et al. [4] used a detection-based approach for online bookmarking systems. They collected attributes about users, the tags they used, the locations they accessed the system from, and the way the interacted with the system. The resulting data sets were then categorized by a Support Vector Machine (SVM) into legitimate users and spammers.

Benevenuto et al. [5,6] used a similar approach to categorize users on the YouTube video platform into spammers, promoters and legitimate users based on their attributes. They used 60 attributes that were either based on the user himself, his videos, or his social network metrics. The categorization here was done by a SVM as well.

3 Definitions

The Twitter microblogging platform allows users to write short notes and messages, not exceeding 140 characters. The sum of all messages of a user is referred to as a user's *feed*. Other users can subscribe to such a feed. This process is called *following* and creates a directed relation between two users. If user A follows user B, user B is called a *friend* of A. User A will be a *follower* of B. The process of unsubscribing from another user's feed is commonly referred to as *unfollow*. Throughout this work we will also refer to the sum of friends and followers a user has, as his *peers*. Peers are a user's direct neighbors in the social graph.

4 Spam Categorization Framework

We enriched the basic feature set that describes a user, in two different ways. The first feature set was based on an aggregation of a user's peer's attributes. The second feature set included metrics that predicted whether a user's followers were spammers or legitimate users. The latter feature set also tried to grasp the implicit trust given to a user by his followers.

The first feature set is based on aggregating friend's and follower's attributes. We calculated the average values of certain attributes for all friends and for all followers the user had.

Some versions of the trust metrics had to predict whether a user's followers were spammers. Therefore, we needed an additional step to train a classifier to make those predictions. Thus, in the first step, a preexisting categorization algorithm was trained to categorize users as spammers or non-spammers. For this, we used a training set that contained manually categorized users and their account attributes retrieved from the Twitter platform. Once the categorization algorithm was trained, it was used to predict whether a certain user was a spammer or a legitimate user. So far, the process is similar to existing approaches as described by Krause and Benevenuto [4,5,6]. However, in a second step, an extended attribute set was generated for each user. These additional attributes were based on the predictions provided by the first learner and the user's position in the social network. A second categorization algorithm was then trained on this extended attribute set. The outline of the algorithm used is as follows:

Enhanced Attribute Learner(*trainingSet*, *testSet*)
1 *model*1 ← Train-Basic-Learner(*trainingSet*)
2 *trainSetEx* ← Extend-Data(*trainingSet*, *model*1)
3 *model*2 ← Train-Basic-Learner(*trainSetEx*)
4 **for** *e* ∈ *testSet*
5 **do** *extended* ← Extend-Data(*e*, *model*1)
6 Evaluate(*extended*, *model*2)

5 Test Data

We chose the Twitter microblogging platform as our test platform. The reason for this is that information about users on this platform can be acquired easily.

Most of the account names of spammers were acquired using the web page twitspam.org, where users can submit the names of suspected spammers. A few users were added as spammers which got our attention while we were collecting data. To obtain trusted users, we added Twitter users whom we were following on our own. This resulted in a data set containing 77 spammers and 155 legitimate users. In addition, for each of these users, information on up to 200 of their followers was acquired. Because many users share followers, information on more than 200 of their followers was available for some users in the data set.

5.1 Basic Attribute Set

The attributes in the basic attribute set were intended to be the basic attributes as they were available via the Twitter API. However, preliminary studies showed that an account's age was the attribute associated with the highest information gain. This might be because spammer's accounts were much more likely to be closed by the platform owner. However, since most of the spammers in our data set come from twitspam.org, the spammers in our data set might have been at a higher risk of closure than spammers not listed on twitspam.org. To prevent this from making our results less meaningful, the account age was taken out of the attribute set and other attributes were calculated on a daily basis, so that the account age would not be reflected in other attributes. Thus, for each user the following user features were calculated and resemble the basic feature set:

- follower-friend ratio
- number of posts marked as favorites

- friends added per day
- followers added per day
- account is protected?
- updates per day
- has url?
- number of digits in account name
- reciprocity

Two attributes were added to these basic attributes. The number of digits in the account name was shown to be a useful metric in spammer detection on social tagging services by Krause et al. [4]. We wanted to see how this attribute performs on other platforms. We also added the reciprocity. The reciprocity is the rate with which a user follows his followers. This was added because it used to be a common practice on Twitter to use a script to follow everyone who follows you. Therefore, spammers might accumulate many followers who have this kind of script set up. This might result in valuable attributes based on peers.

5.2 Aggregating Friend and Follower Attributes

These features were based on attributes of a user's peers. The average value for these features was calculated separately for all followers and friends of a user. In the case of the boolean attribute *protected?* the percentage of all protected friend/follower accounts was calculated. This results in 12 new features, since features were aggregated for friends and followers separately. Those values will be referred to as peer values and were calculated for the following attributes:

- follower-friend ratio
- updates per day
- friends added per day
- followers added per day
- reciprocity
- account is protected?

5.3 Trust Propagation

We assumed that legitimate users were less likely to actively seek interaction with spammers or the content they provided. This might result in less users following spammers. We tried to utilize this expected behavior by interpreting a user's interaction with another user as an implicit sign of trust in the user with whom interaction was sought. The follower expected the friend to provide valuable information. Therefore, the follower most likely did not expect for the friend to be a spammer. We tried to capture that implicit trust. However, this was not always the case since some users followed all their followers or were spammers.

We tried different ways to accumulate the trust given to a user by other users and extend a user's attribute set with it. For most added attributes, we made the assumption that as there were more users one followed, less attention was given to each single friend and less trust could be inferred. This assumption was shared with established approaches like the PageRank algorithm [7]. In the basic version we therefore used the following formula:

$$\text{trust metric} = \sum_{followers} \frac{1}{\#\text{users followed}}$$

We also applied the following modifiers to this formula:

legit accumulate only the values coming from users who are predicted to be legitimate users

capped accumulate only values coming from up to 200 users

squared use $\frac{1}{\#\text{users followed} * \#\text{users followed}}$ instead of $\frac{1}{\#\text{users followed}}$

All these modifications were tried in all possible combinations. As an additional attribute, the ratio between predicted spammers and predicted legitimate users following a user was calculated.

6 Experiments

In preliminary tests we evaluated different categorization algorithms for the first and second step. An extended data set was created using the first algorithm to which then the second algorithm was applied with 10 fold cross validation. We used implementations of these algorithms as provided by the WEKA package (http://www.cs.waikato.ac.nz/ml/weka/). Table 1 shows the accuracy achieved in preliminary tests with each combination of algorithms. It can be seen that for this purpose JRIP, which is an implementation of the RIPPER rule learner, turned out to perform the best, with J48, which is an implementation of the C4.5 algorithm, as a close second.

Table 1. Preliminary accuracies for different learner combinations

1st\2nd step	JRIP	J48	SOM	Naïve Bayse
JRIP	93.0%	93.0%	71.5%	79.9%
J48	89.2%	91.9%	73.9%	84.8%
SOM	89.8%	87.6%	76.8%	85.1%
Naïve Bayse	88.2%	86.0%	75.9%	78.3%

To compare the performance between the basic attribute set and our two enriched feature sets, we ran a ten fold cross validation 10 times which resulted in 100 different result sets. This was done for the basic feature set, the basic features enriched with aggregated peer values, the basic feature set enriched with the values based on implicit trust and predictions on follower being spammers and a final test involving the basic feature set with both additional feature sets added.

6.1 Evaluation Criteria

To evaluate if our extended attribute set was able to improve the performance of a classification algorithm, we used different established metrics on test results acquired with and without the extended attribute sets. We calculated *accuracy*, *precision*, *recall*, *F1*, and finally draw a Receiver Operating Characteristic Curve (ROC curve) to evaluate the test results [8].

The most basic metric used is *accuracy*, which is the percentage of all instances that are classified as belonging to their actual class. The *accuracy*, however does not give any information on instances where a class tends to be misclassified or in which way it is misclassified. In spammer detection, it can be worse if a legitimate user is being falsely classified as a spammer, than if a spammer is misclassified as a legitimate user. Falsely accusing legitimate users of being spammers can drive them off the platform. Therefore, we also calculated the *precision* (p), which determines the fraction of actual positives in the group of instances classified as positives. The higher the number of correctly classified spammers and the lower the number of false positives, the higher the *precision* will be. *Recall* (r), on the other hand, measures how many elements from a class (usually the positive class) are correctly classified. In our case, we used it to measure how many of all actual spammers are detected. In addition, we used the *F1* measure, which is the harmonic mean between *precision* and *recall*. Thus, *F1* takes both measurements into consideration but penalizes a big difference between both values.

The ROC curve plots the false positive rate against the true positive rate. Thus, we can see the trade-off between catching more spammers and falsely classifying more legitimate users as spammers. This way the curve shows which results could be achieved by using appropriate probability cut-offs [9].

6.2 Results

Table 2 and 3 show that our extended attribute sets were able to generally improve the results. It is important to note that the *precision* was improved in

Table 2. Evaluation metrics for RIPPER algorithms with the different extended feature sets

Metric	basic	basic + peer values	peer values	basic + trust	all features
Precision	0.79	0.80	0.75	0.88	0.84
Recall	0.84	0.83	0.71	0.85	0.85
F1	0.81	0.81	0.73	0.87	0.84
Accuracy	0.87	0.87	0.82	0.91	0.90

Table 3. Evaluation metrics for C4.5 algorithms with the different extended feature sets

Metric	basic	basic + peer values	peer values	basic + trust	all features
Precision	0.80	0.81	0.72	0.85	0.86
Recall	0.85	0.79	0.67	0.85	0.86
F1	0.83	0.80	0.69	0.85	0.86
Accuracy	0.88	0.87	0.80	0.90	0.90

all combinations, as it shows that fewer legitimate users are being classified as spammers. This compensates for the loss in general accuracy that C4.5 takes when using the average values for peers in addition to the basic attribute set.

Figure 1 shows the ROC curves generated with the C4.5 leaner and the RIPPER learner. Both charts show that our additional attribute sets outperform the basic attribute set. However, the curve using the peer value sin combination with the basic attribute set has some intersection which the curve using the basic attribute set only. Nevertheless, the curves for the trust metrics and the curve for the combination of all attributes confirm what was indicated by the other metrics. For both RIPPER and C4.5 both attribute sets yield very similar results. Again, C4.5 seems to be doing slightly better using all attributes since this curve is far above the trust metric's curve for higher false positive rates.

To measure the value provided to the result by each attribute we calculated the information gain and the chi square values for the attributes included in the extended attribute set. The values were calculated on a data set, that was constructed by adding up the extended data sets created using the Ripper algorithm with 10 fold cross validation for follower classification. The top 10 ranked attributes and their values can be seen in the two tables in table 4 and table 5. The ratio between legitimate followers and spammer followers in both rankings turned out to be the highest rated value. The friend-follower ratio was rated very high by both metrics as well. This might change in the future. Notification emails about new followers from Twitter now include the number of friends and followers a user has. Given that a high number of friends is a good indicator of spammer status, users might stop responding to those spammers as hoped for. However, spammers can easily adjust to this changed situation by unfollowing users and keeping their friend count low. We have already encountered some spammers with reasonably low friend numbers during our studies, which might be an indicator that this is already happening. We expect that the number of friends added daily will thus lose its importance.

Different versions of our trust metric also consistently ranked pretty high. Their usefulness is confirmed by the very good results that they achieved in the other tests. It is interesting to notice that in both measurements the average friend-follower ratio for a user's friends ranks 7, but the same value for once followers had a information gain and Chi Square value of 0.

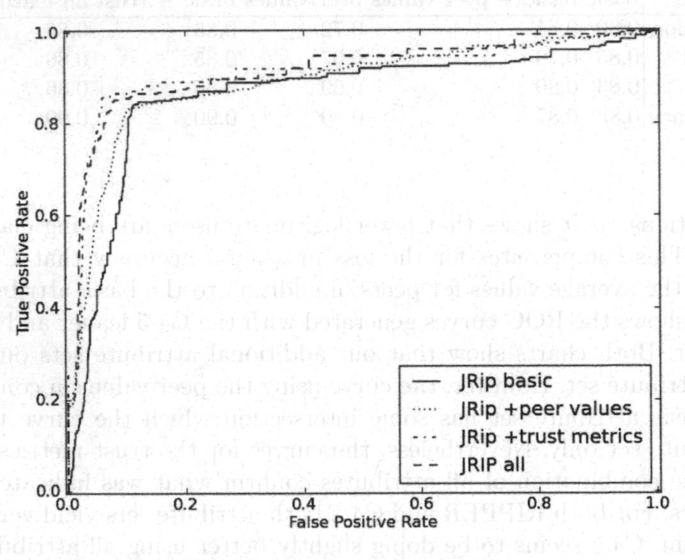

(a) ROC Curve using RIPPER

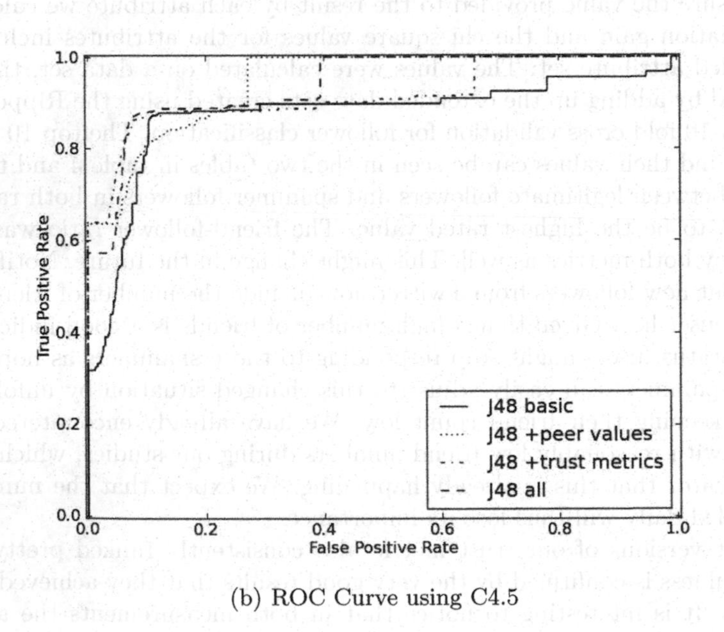

(b) ROC Curve using C4.5

Fig. 1. (a) and (b) show the ROC curves for combinations of our different feature sets

Table 4. Top 10 Chi Square values for data set extended with JRip. Added attributes are bold.

Attribute	Chi square value
spammers to legit followers	128.68
friends per day	106.72
trust metric legit.	105.49
friend-follower ratio	101.23
trust metric legit. capped	94.8697
trust metric	88.78
friend-follower average for friends	81.54
average protected for followers	80.57
trust metric legit. square	79.93
trust metric legit. square capped	74.99

Table 5. Top 10 Information gain values for data set extended with JRip. Added attributes are bold.

Attribute	Information gain
spammers to legit followers	0.48
friend-follower ratio	0.35
friends per day	0.34
trust metric legit.	0.34
trust metric legit. capped	0.29
trust metric	0.29
friend-follower average for friends	0.27
average protected for followers	0.25
trust metric legit. square	0.24
average protected for friends	0.24

7 Conclusion and Future Work

The much improved classification results and the high values received by our additional attributes for both the chi-squared statistic and information gain show that a user's peers indeed tell much about the nature of a user. The RIPPER algorithm was able to get consistently better results on the extended attribute sets as compared to the basic attribute set.

Although the additional features already improved spammer detection, we see several ways of improving our framework further. Some users might be less careful in evaluating whom they follow than others. This is supported by the fact that many users just follow everyone who follows them. This will result in a much higher number of spammers in their friend list. As of now, we only use the number of users that are being followed to weight the value added to a friend's trust metric. We could use the ratio of predicted spammers among a user's friends to weight this value even further and make it more meaningful. However, this would require a much larger data set than we already have.

We would also like to conduct future tests with a bigger set of test data, based on more users that were selected in a more representative fashion. In addition, we wish to conduct tests on other social networking platforms to see if you can in general judge a user by his peers or if this is limited to the Twitter platform.

We would also like to explore community belongingness as an indicator. We expect that legitimate users are more often part of a closely connected subgroup, in contrast to spammers who just follow everyone and therefore will be connected to many otherwise divided groups.

References

1. Carpinter, J., Hunt, R.: Tightening the net: A review of current and next generation spam filtering tools. Computers and Security 25(8), 566–578 (2006)
2. Heymann, P., Koutrika, G., Garcia-Molina, H.: Fighting spam on social web sites. IEEE Internet Computing 11(6), 36–45 (2007)
3. Boykin, P.O., Roychowdhury, V.P.: Leveraging social networks to fight spam. Computer 38(4), 61–68 (2005)
4. Krause, B., Schmitz, C., Hotho, A., Stumme, G.: The anti-social tagger - detecting spam in social bookmarking systems. In: AIRWeb 2008 (April 2008)
5. Benevenuto, F., Rodrigues, T., Almeida, V., Almeida, J., Zhang, C., Ross, K.: Identifying video spammers in online social networks. In: AIRWeb 2008 (April 2008)
6. Benevenuto, F., Rodrigues, T., Almeida, V., Almeida, J., Concalves, M.: Detecting spammers and control promoters in online video social networks. In: SIGIR 2009 (July 2009)
7. Page, L., Brin, S., Motwani, R., Winograd, T.: The pagerank citation ranking: Bringing order to the web (1999)
8. Tan, P.N., Steinbach, M., Kumar, V.: Introduction to Data Mining. Addison-Wesley, Reading (2005)
9. Witten, I.H., Eibe, F.: Data Mining - Practical Machine Learning Tools and Techniques. Morgan Kaufmann, San Francisco (2005)

Encouraging Cooperation in Ad-Hoc Mobile-Phone Mesh Networks for Rural Connectivity

Kavitha Ranganathan[1] and Vikramaditya Shekhar[2]

[1] Indian Institute of Management, Ahmedabad, India
[2] Birla Institute of Technology, Mesra, India

Abstract. This paper proposes a rating based incentive scheme for encouraging user participation in ad-hoc mobile phone mesh networks. These peer-to-peer networks are particularly attractive for remote/rural areas in developing countries as they do not depend on costly infrastructure and telecom operators. Active user participation is however critical for the success of such a network. We evaluate our incentive scheme using extensive simulations and find that our proposed scheme is successful in enhancing the network throughput.

1 Introduction

Peer-to-peer mesh networks for mobile phones have recently been proposed as an alternative to the traditional base-station cellular model [1]. In the proposed system there are no base stations or telecom operators or centralized control of any sort. Specially designed mobile phones can start communicating directly with each other, when they are within range of one another[1]. Since the range is quite limited, intermediary phones can also relay communication between two devices that are out of range from each other. Thus, mobile units form ad-hoc mesh networks and the more the number of units the larger the network can scale[2].

Such ad-hoc peer-to-peer phone networks have tremendous value for applications where the base-station model is not feasible for a variety of reasons. In rural areas, where user density might be low, scale economies might not justify expensive base-station towers. Another application where such networks can be very valuable is in disaster-hit regions. When the traditional communication networks are down, the mesh mobile phones can still communicate with each other in a totally decentralized manner. In this study we will concentrate on the first application – providing cheap and quick connectivity to rural regions.

While the technological challenges (routing, admission control etc) of allowing mobile handsets to form such local networks seem to have been solved to an extent

[1] Within a kilometer of each other if they use TerraNet technology [1] for example.

[2] Such networks cannot scale indefinitely. In TerraNet's proposal for example, there can be 7 intermediaries who relay the voice data to the destination, before the latency gets too high for reasonable quality.

S.K. Prasad et al. (Eds.): ICISTM 2010, CCIS 54, pp. 221–231, 2010.

[1], this study focuses on yet another aspect – incentives for participation -- which is very vital for such a network to work. Since most calls and data have to be routed through intermediaries, it is essential that users collaborate with each other. Such issues do not arise in a conventional cellular network, where all communication is channeled through a base-station. In the proposed network however, individual users have to be willing to route traffic for the benefit of other users. Given that routing such traffic for others will mean consumption of one's own limited resources (like power), there might be a rational tendency to behave selfishly. Especially in rural settings in developing regions, where mobile phone users have to deal with erratic power supply and in some cases travel a kilometer or more to charge their phones, draining away their battery to route other people's traffic might not sound very appealing. In such a situation, it might be tempting for the user to conserve his/her device's power by either switching it off when not in use, or if the user is more skilled -- tamper with it so that the device does not forward other's traffic. If enough users are selfish instead of altruistic, the entire network can come crashing down.

This paper studies the incentives for user participation in such a network. We propose a rating based scheme that encourages users to participate fully in the network. Using simulations, we evaluate the effectiveness of the proposed scheme. To the best of our knowledge, this is the first paper to study incentives for participation in ad-hoc peer-to-peer mobile-phone networks. We find that our rating based scheme is effective in encouraging users to keep their phone switched ON more than they normally would have. This results in a reasonably good network throughput of around 70%, where-as in the absence of such an incentive scheme the network performance suffers dramatically.

The rest of the paper is structured as follows: In section 2 we compare P2P mobile-phone networks to two related networks, MANETS and multi-hop cellular networks. Section 3 contains a discussion of related work. We describe our system model and simulations in sections 4 and 5 respectively and results are presented in section 6. We conclude in section 7.

2 P2P Mobile-Phone Networks

While traditional cellular networks operate on a centralized model where all communication between any two hand-sets travels through a base-station, the recently proposed P2P phone networks are totally decentralized. With no central authority or entity, the phones form ad-hoc P2P mesh networks. Two phones within range of each other can directly talk to each other. If they are out of range from each other, then one or more phones in-between them can act as relays and transmit the data between them. Such a network can be connected to the outside world via a computer with an internet connection.

There are of-course advantages and disadvantages to this model. The obvious advantage for rural regions is that locals need not wait for service providers to set-up costly infrastructure and related services, before they have a functioning phone network. If there are enough mobile phones in the region, a viable network will spring up quite automatically. Assuming a handset range of around 1 KM and a maximum hop-count of 7 [1], a network area of approximately 8 sq kilometers can be covered, provided there are enough users in the area to act as relays. However with a totally decentralized model, well designed routing protocols and solutions for security and privacy issues are crucial for its success. Further, as with almost any peer-to-peer network, incentives for the users to participate in the network must be carefully considered.

Although phone mesh networks are a fairly new phenomenon they are related to two other kinds of networks – MANETS and multi-hop cellular networks.

2.1 MANETS vs. Phone Mesh Networks

Traditional mesh networks (also known as MANETS - Mobile Ad-hoc NETworkS) are a popular research topic and have been studied significantly [2]. Initially conceived for military purposes, these networks comprise of mobile devices (computers, PDAs and sensors) which form ad-hoc data networks without centralized control or infrastructure. Most deployed MANETS use the ad-hoc mode of IEEE 802.11 standard (Wi-Fi). While MANETS are used primarily for data packets, phone mesh networks can be used for voice as well as data. Our study is inspired by the TerraNet project [1] whose peer-to-peer phones are expected to support many types of radio transmission like WiMax, GSM and UTMS.

Applications for MANETS are typically limited to military settings or disaster relief with day-to-day applications still a rarity. P2P mobile-phone networks on the other hand have a very promising application for the common man -- they could potentially be the driving force for bridging the digital divide in rural areas. Telecom operators are usually hesitant to venture out to rural areas, because the demand is spread out and the costly infrastructure needed makes their venture less profitable. In India for example, rural mobile penetration is only at around 15% of the population whereas urban penetration is above 70 % [10].

2.2 Multi-hop Cellular Networks vs. Phone Mesh Networks

Multi-hop cellular networks (MCNs) were originally conceived as a hybrid between the traditional single-hop cellular networks and ad-hoc mobile multi-hop networks (MANETs). In MCNs mobile nodes can communicate directly with each other if they are within range, or they can communicate with a base-station. An intermediary mobile node can act as a relay, so that a node out of range from a base-station (BS) can still access the network. The BS is in-charge of routing and keeping track of the various mobile nodes within its range. While MCNs still rely on base-stations for the backbone network and centralized routing decisions, phone mesh networks are totally decentralized with no central intelligence or coordination.

Figure 1 provides examples of multi-hop cellular networks and phone mesh networks.

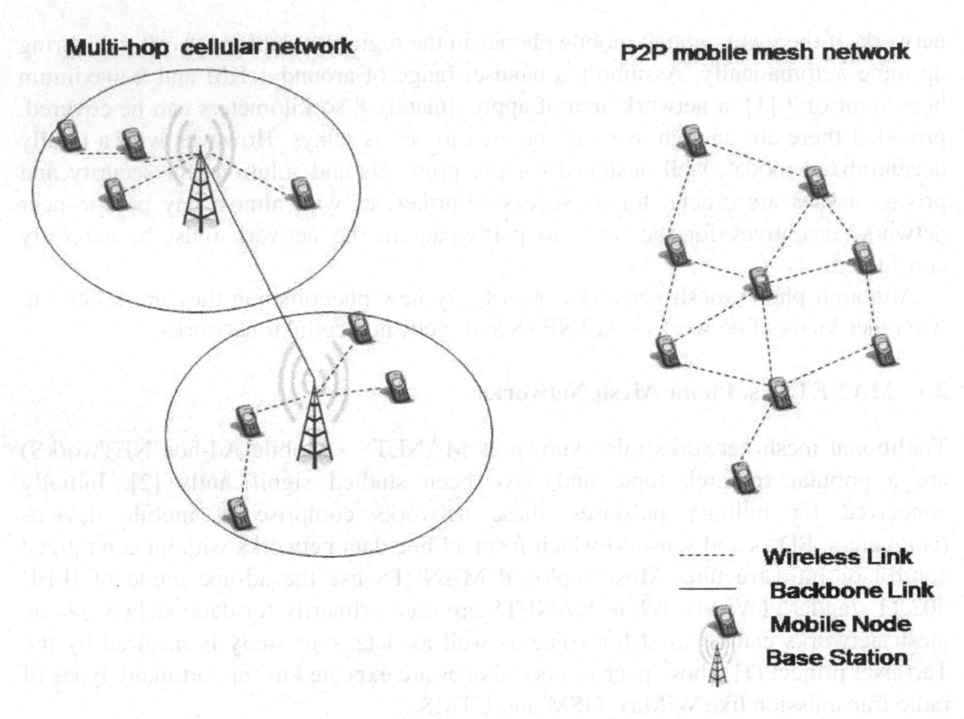

Fig. 1. Multi-hop cellular network and P2P mobile mesh network

3 Related Work

Considerable work has been done in incentives for cooperation in the related fields of mobile ad-hoc networks (MANETs) and multi-hop cellular networks.

Srinivasan et.al [3] uses a game-theoretical approach to study cooperation among nodes in an ad-hoc network. They consider different classes of nodes with different power constraints, and propose a mathematical framework for studying user behavior in this scenario. Their proposed strategies however require each node in the system to know the number of users in each class and the energy constraint in each class. Given that a MANET system is totally decentralized, such information may be difficult to collect – especially if nodes have the incentive to relay incorrect information about themselves.

The Sprite project [4] proposes a credit based system for MANETs where a Credit Clearance Service (CCS) determines the charge and credit to each node in the system. Since such a system revolves around a central authority, it is not suitable for a totally decentralized system like ours.

The Terminodes project [5, 6] comes closest to our work in terms of a pure decentralized design philosophy and use of tamper proof hardware. They assume that a tamper-proof hardware at each node keeps track of a virtual currency called nuggets. A sender would load a packet with nuggets before sending it and each relay node would be paid a nugget for participating. The proposed scheme works under the assumptions that each node generates packets continuously and that generated packets cannot be buffered – they have to either be sent immediately or dropped.

Both these assumptions are invalid in our system. Firstly, users cannot be expected to use their phones continuously – there will be peak times and lag times and high-volume users and low-volume users of the system. Secondly, some amount of buffering will be possible – a user may decide to delay a non-urgent call till the time its phone has acquired a better reputation and the call has a higher chance to go through successfully. Moreover, in the Terminodes project, the billing is done on a per packet basis where-as in our scheme, we look at calculations on a per session basis which is more suitable for voice sessions.

The CONFIDANT protocol [7] enables nodes to find out about the behavior of other nodes and maintain a reputation system. Nodes broadcast information about selfish nodes and selfish nodes may have their requests ignored. While such a system does not rely on a tamper-proof hardware, the overheads of broadcasting node behavior and maintaining individual information about each node may lead to significant overheads.

Incentive studies in multi-hop cellular networks [8, 9] propose charging and rewarding schemes but assume a central authority like an Internet service provider or a network operator who makes sure that all nodes follow the proposed rules. Again, in our network such schemes are not viable due to the lack of a central authority.

4 System Model

We assume a network of N mobile nodes, where each node may be ON or OFF at a given point in time. Our system is measured in discrete time slots – each node either stays on or off for an entire time slot. We also assume that nodes exchange data-packets (voice can be easily modeled as an extension to this system, since we assume that end-to-end links remain alive for an entire slot).

At the beginning of each time slot, multiple sessions are initiated. For each session, two unique nodes (from the set of ON nodes) are randomly selected as the source and destination. 'L-1' other nodes are randomly selected as the relay nodes which link the source to the destination. Note that these nodes may be ON or OFF. It is important to include OFF nodes also in the random selection, to ensure that in situations where there are not enough ON nodes in the network, the session is unable to go through.

If ON, each relay node either participates in that session or refuses to participate, depending on the embedded relay policy described later. If all the relay nodes are ON and all agree to participate in the session, then the data reaches its destination. If even a single relay node does not agree to participate then that session is aborted. It is to be noted that once an end-to-end link is formed between the source and destination, that link is assumed to remain stable for that entire time slot. One session is limited to one time-slot.

We assume that users have the choice of switching their phones ON or OFF (as in any real world situation), depending on their perceived gains. Given such a system, it is easy to see that users may be tempted to keep their phones OFF when they do not want to communicate with anyone and if they are not expecting any important/urgent incoming data. In a bid to conserve power (which as we mentioned earlier is a scare commodity in many rural settings), users may quickly use their phone and then switch it off. This of-course will prove detrimental to the operation of such a network, as it relies on user-participation to function.

4.1 Incentive Structure

We propose the following rating based scheme to "incentivize" users to participate in the network (that is to keep their nodes ON even if they are not using it for themselves). Since the network is totally decentralized with no controlling authority or coordinating entity, the incentive structure has to work via information stored and decisions taken at individual mobile nodes only.

At the beginning of a node's entry to the network, each node is provided a rating M (Max rating). The rating of a node is assumed to be embedded in the device (in a tamper-proof part of the hardware as proposed by Levin et. Al [11]) and cannot be altered by the user. The user can however check to see what the current rating is.

At each time slot, the rating of a node decays at a fixed rate. The only way a node can prevent the decay of its rating is by acting as a relay for other people's data. Every time, a node acts as a relay its rating is increased by a fixed amount. The policy that decides whether or not a node should participate in a session (that is act as a relay) is assumed to be embedded in the device and again cannot be tampered with by the user.

We propose the relay policy Reciprocative and also experiment with two extreme policies Selfish and Altruistic.

4.2 Relay Policies

Reciprocative Relay

When a node receives a request to take part in a session, it also receives the current rating of the source node – the node where the request originated. The probability that the node will agree to act as a relay for that session depends on the rating of the source node, and is calculated as follows:

Assuming a Max rating M = 1000,

$800 < Rating_{source} <= 1000$ Probability(AGREE) = 1
$600 < Rating_{source} <= 800$ Probability(AGREE) = 0.99
$400 < Rating_{source} <= 600$ Probability(AGREE) = 0.96
$200 < Rating_{source} <= 400$ Probability(AGREE) = 0.90
$0 < Rating_{source} <= 200$ Probability(AGREE) = 0.85
$Rating_{source} = 0$ Probability(AGREE) = 0

The above numbers have been chosen so that enough calls get accepted but at the same time, selfish users get sufficiently punished. For example, a node with a rating of 700 and 7 relays in the session has a $(0.99)^7 = 0.93$ chance of a successful session where as a node with a rating of 400 has only a $(0.90)^7 = 0.47$ chance of a successful session. Moreover, if the session has lesser number of relays then the chances of a successful session increases. Note that we have assumed that all the relay nodes were ON.

The Reciprocative policy can be implemented with minimum overheads by piggybacking the rating number of the source with the request for taking part in a session. Since the extra data is a single number, the size of the request will not change significantly.

We also implemented two extreme relay polices to understand the best-case and worst-case scenarios.

Selfish Relay: The node refuses all requests by others to forward data.

Altruistic Relay: The node agrees to all requests by others to forward data.

4.3 User Behavior

Given the above relay policy (Reciprocative), a rational user who normally might have acted selfishly has to now decide between letting others use its device as a relay at least some of the time or quickly loose its rating. Being purely selfish will quickly eliminates the user from the network, as others will not relay its calls at all (when Reciprocative is being used) once its rating reaches zero. The rational user may however decide to keep the phone switched off (and hence conserve power), till a time that the rating reaches a certain lower threshold (where only a certain percentage of its requests are being met). At this point the user may decide to turn the phone ON to increase its rating to a desired upper threshold. We model this kind of behavior in our simulations and call it the Adaptive user behavior.

In our simulations, apart from the three relay policies, we model three kinds of users: Adaptive, Selfish, and Altruistic.

Adaptive User: The adaptive user keeps the phone OFF till its rating reaches a lower threshold, and then keeps it ON till the rating reaches an upper threshold.

Selfish User: Phone is OFF most of the time, except to occasionally send/receive calls/data.

Altruistic User: Phone is kept ON all the time. Typically such a user may not face a major power constraint, and would not mind keeping the phone ON most of the time.

5 Simulations

We studied multiple scenarios with different permutations of relay policies and user behavior. Table 1 provides the parameters used for the simulations. Each experiment was repeated 5 times and the results presented are the average of the 5 runs. The standard deviation in all cases was observed to be less than 3%.

Table 1. Simulation Parameters

Parameter	Name	Value
Max Rating	R_{Max}	1000
Rate of Rating Decay	D	1/Time slot
Rate of Point (Rating) Accumulation	P	4/successful relay
Number of Nodes	N	100
Number of Relays (links) per Session	L	7
Duration of Simulation (Number of Time slots)	T	10000
Lower Threshold for Rating	R_l	850
Upper Threshold for Rating	R_u	1000
No. of sessions initiated per time slot		5

We measured the percentage of successful sessions, that is how many of the 50,000 sessions that were initiated were successfully completed. We also measured the percentage of nodes that were switched ON and the average rating of the nodes, as the simulation progressed. Evaluated together, these three parameters give a fairly good idea of the performance of the relay policy, for different kinds of user behaviors.

We also experimented with scenarios with multiple user behavior.

6 Evaluation

Figure 2 plots the number of successful sessions for the three kinds of user behavior, when Reciprocative relay was used. As expected, in the best case scenario when all users were Altruistic (that is kept their nodes ON always), all the sessions were successful. The decay in the node rating was more than made up by the points accumulated by successful participation. Hence, in spite of Reciprocative relay, 100% of the sessions were successful. When all users are adaptive (the more realistic scenario), the success rate was around 70%. Some session requests were rejected because of the poor ratings of users. Again as expected, when all users were selfish (only occasionally switched on their phones) the system failed miserable.

The results show that Reciprocative relay (which is designed to incentivize users to behave Adaptively instead of selfishly) can bring the system to a relatively flourishing state where 70% of the sessions are successful.

Figures 3 and 4 plot the average rating of nodes over time and the number of ON nodes over time, for Reciprocative relay. Altruistic and Selfish node behavior is self-explanatory. For the case when all nodes are Adaptive, the number of ON nodes fluctuates, but does not fall below 90%, which ensures a thriving network. Similarly, the average rating of a node, when nodes are adaptive, hovers around 80 %. That is, Adaptive nodes are able to balance their rating to ensure an acceptable level of successful sessions. These results show that for Reciprocative relay and Adaptive user behavior the system is stable over time.

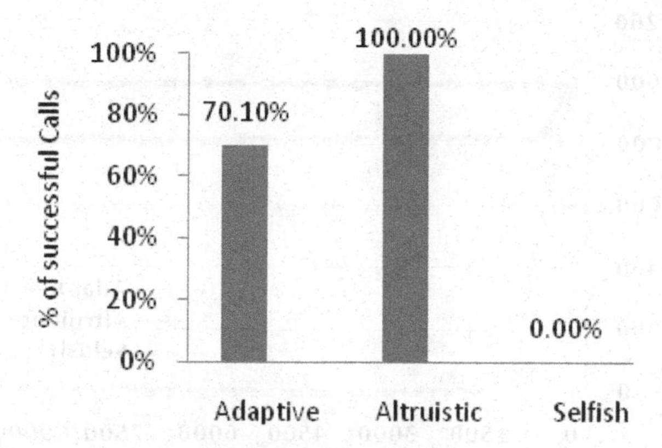

Fig. 2. Percentage of successful sessions for different user behaviors

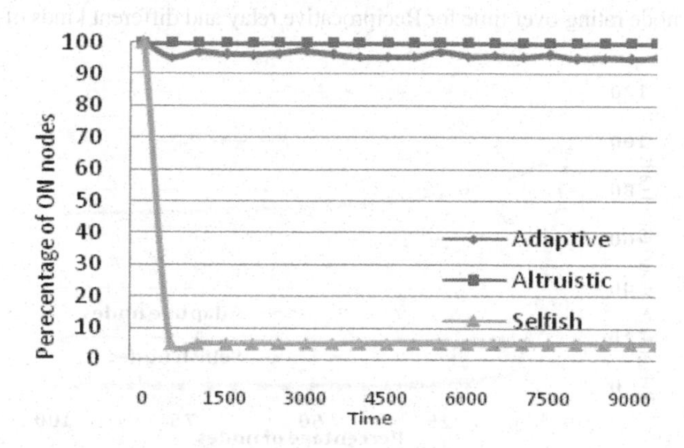

Fig. 3. Percentage of ON nodes over time for reciproactive relay and different kinds of user behavior

We also experimented with various ratios of rating decay (D) to point accumulation (P). A ratio of 1:4 yielded the best results. For lower ratios the average rating of a node decays over time which ultimately brings the network to a halt. For higher ratios, all nodes have very high ratings and hence switch OFF their phones. This obviously adversely affects the throughput of the system. We note that a 1:4 ratios worked well for the current parameters under consideration, like the frequency of sessions initiated. For good results this ratio will have to be adjusted according to the expected load of a particular network.

Figure 5 plots the percentage of successful sessions when there are two kinds of users in the system – Altrusitic users and either Selfish or Adaptive.

As can be seen, as the number of altruistic users decrease, the system performance goes down. However, it goes down much more drastically when users turn Selfish, than when users turn Adaptive. Hence, even when all users turn Adaptive, the system still performs relatively well (70%), where as even with a small percentage of selfish nodes (10%) the throughput of the network drops drastically to 7%.

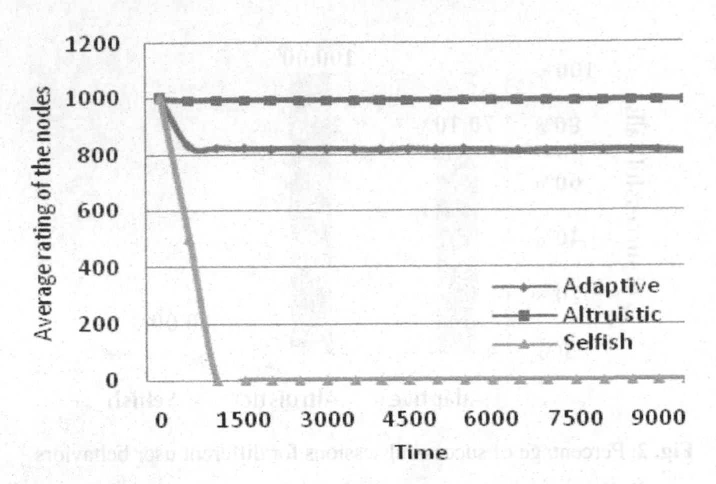

Fig. 4. Average node rating over time for Reciprocative relay and different kinds of user behavior

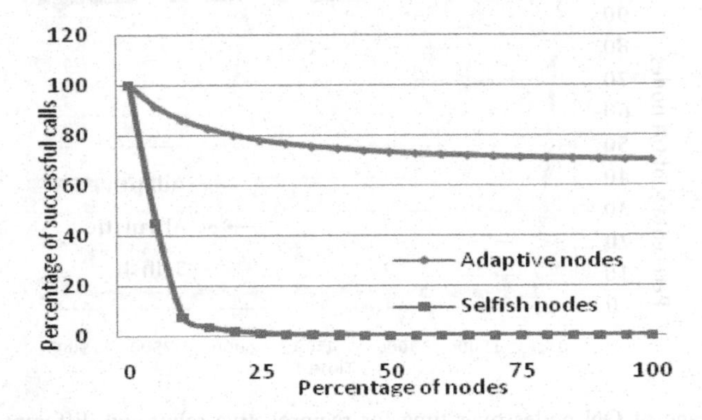

Fig. 5. Percentage of successful calls as composition of network changes from altruistic to adaptive/selfish

7 Conclusions and Future Work

In this paper we have addressed the problem of incentivizing user participation and cooperation in a peer-to-peer phone network. We have proposed a totally decentralized rating-based scheme where nodes gain points for forwarding other people's data and where the treatment received by a node depends directly on its rating. We find that our scheme is successful in stimulating cooperation among nodes and also in maintaining the throughput of the network and an acceptable level.

A current limitation of our study is that we assume uniform upper and lower thresholds for Adaptive user behavior whereas in reality there might be different bands of users. Some may switch off their phones earlier than others, depending on how much they value power conservation vs. the throughput they receive from the network. Others may prefer to keep their phones on to receive calls, even if they lose power in the process. Our future work will look at modeling multiple user behavior patterns.

We also plan to examine how to include users who are at the margin of the network and may not be asked to relay enough calls and thus not get a chance to better their rating.

Acknowledgements

This work was funded by the Idea Telecom Center of Excellence and IIM-Ahmedabad.

References

[1] Frost, Sullivan: Technical Insights, Mobile and Wireless Communications Technology Alert (October 2007)

[2] Kurkowski, S., Camp, T., Colagrosso, M.: MANET Simulation Studies: The Incredibles. Mobile Computing and Communications Review 9(4)

[3] Srinivasan, V., Nuggehalli, P., Chiasserini, C., Rao, R.R.: Cooperation in Wireless Ad Hoc Networks. In: INFOCOM 2003 (2003)

[4] Zhong, S., Chen, J., Yang, Y.R.: Sprite: A Simple, Cheat-Proof, Credit-Based System for Mobile Ad-Hoc Networks. In: INFOCOM 2003 (2003)

[5] Hubaux, J.-P., Gross, T., Le Boudec, J.-Y., Vetterli, M.: Towards Self-Organizing Mobile Ad-Hoc Networks: the Terminodes Project. IEEE Comms. Magazine 39(1), 118–124 (2001)

[6] Buttyan, L., Hubaux, J.: Stimulating Cooperation in Self-Organizing Mobile Ad Hoc Networks. Mobile Networks and Applications 8, 579–592 (2003)

[7] Buchegger, S., Le Boudec, J.-Y.: Performance Analysis of the CONFIDANT Protocol: Cooperation Of Nodes - Fairness In Distributed Ad Hoc Networks. In: Proceedings of MobiHoc (June 2002)

[8] Salem, N.B., Buttyán, L., Hubaux, J., Jakobsson, M.: A charging and rewarding scheme for packet forwarding in multi-hop cellular networks. In: Proceedings of Mobile Ad hoc Networking & Computing, June 1-3 (2003)

[9] Lamparter, B., Plaggemeier, M., Westhoff, D.: Estimating the value of co-operation approaches for multi-hop ad hoc networks. Ad Hoc Networks 3(1), 17–26 (2005); 16EE

[10] Measures to Improve Telecom Penetration in Rural India-The next million subscribers, Telecom Regulatory Authority of India (December 2008)

[11] Levin, D., Douceur, J.R., Lorch, J.R., Moscibroda, T.: TrInc: Small Trusted Hardware for Large Distributed Systems. In: NSDI 2009 (USENIX Symposium on Networked Systems Design and Implementation) (2009)

Fuzzontology: Resolving Information Mining Ambiguity in Economic Intelligent Process

Olufade F.W. Onifade[1,2], Odile Thiéry[1], Adenike O. Osofisan[2], and Gérald Duffing[1,3]

[1] SITE-LORIA, Nancy Université, B.P. 239 54506 Vandoeuvre, France
[2] University of Ibadan, Ibadan, Oyo State, Nigeria
[3] ICN Business School, 13, rue Michel Ney – F5400, Nancy, France
{onifadeo,odile.Thiery,gerald.duffing}@loria.fr,
{mamoshof}@yahoo.co.uk

Abstract. Human beings are seen as a problem solver, and the process follows the simple input-process-output. Thus having identified a decisional problem (input), will apply some rules of inference and problem solving techniques based on experience and cognitive abilities (processing) to deduce or arrive at some conclusion (output). Information is frequently defined as interpreted data. Juxtaposing with infological equation by Langefors, the importance of the interpreter becomes vivid in the process. Earlier on, we proposed an ontological framework for knowledge reconciliation in economic intelligence process. However, we noticed with deep concern that the various mental process shaping and constructing certain knowledge are difficult to comprehend and placed tangibly making it Fuzzy. This research work is based on the horrendous nature of the bi-valued logic (yes/no, true/false) inherently present in human attempt to inform. We therefore propose a new concept tagged Fuzzontology to assist in enhancing the interpretation of ambiguous meaning.

Keywords: Ontology, Decisionability, Knowledge Reconciliation, Translation Credibility, Economic Intelligence, Fuzzontology.

1 Introduction

Knowledge by its very nature has been observed to emerge from people's head [18]. Although the various mental processes shaping and constructing knowledge are very difficult to comprehend, it is still the singular key behind exercising and displaying judgment in human decision making employing intuition, gut feeling, and cognitive abilities amongst others with available information aimed at making logical inferences on the situation. Information has been frequently defined as "interpreted data" consequently, the same data might elicit different interpretation depending on the individual involved [4]. It is generally accepted that human mind i.e., the brain in its bodily context alongside external structures depend to a large extent on its capability to transform complex decision making

S.K. Prasad et al. (Eds.): ICISTM 2010, CCIS 54, pp. 232–243, 2010.

task via information processing into simpler tasks by exploiting structures of the real world and by actively restructuring problems into simpler chunks so that they can adequately fit into peculiarities of human cognition [16].

Constant change of interpretation, and consequently of perceived meaning, (e.g. information) formed the basis for the proposition of the "infological equation" by[17]. The term "meaning" has been defined as the thing one intends to convey especially by language thus pinpointing the ambiguity inherent in huma interpretatio. Deciding not to decide is not the same as indecision goes an old saying. There could be several factors that undermine the process of decision making. One of such factors earlier identified was lack of access to adequate information, but the present story has changed with the advent of the internet and other private and corporate information delivery sources. The impact of human element in the learning process prior to transformation of information into knowledge was properly presented in[8]. Its dynamic and circular relationship was presented by [16]. Burgess, et al., [3] evaluated the impact of information overload in determining quality criteria to assist in information search and decisionability is defined as the performance of the decision maker based on utilization of available information [26].

The above stated elucidate the importance of human sense making and interpretation of messages with focus on the use of languages. Most popular form of communication is via this mean which is grossly inadequate in perfection. Heuristic judgments made in the act of formulating interpretation or delivering decisions can be engulfed with unconscious bias taking into cognizance the imperfect nature of memory over time. With these views in mind, we believed the ontological approach to information quality by [33,34] ontological framework for knowledge reconciliation by [26] amongst other will be grossly inadequate to capture these seemingly intangible factors capable of under mining the quality of information at every stage of its usage.

The rest of this work is structured as follows: section 2 discussed the general misconception amongst the trio of data, information and knowledge with a bid to depict the distinctions, but more importantly presents their interdependencies. In section 3, we present the ontological framework for knowledge reconciliation in economic intelligent process, highlighting the fuzziness in interpretation as the rational to make ontology more robust via the inclusion of fuzzy inference method. Section 4 presents our main contribution for this work based on our earlier proposed framework. We tagged this Fuzzontology. Section 5 concludes the work with future recommendation.

2 The Misconceptions amongst Data, Information and Knowledge

Information was referred to as the knowledge about the decision, the effect of its alternatives, the probability of each alternative amongst others in [14]. It is sometimes referred to as meaning, drawing upon data interpreted (created) by human sense making processes, including application of bias by its producer [1]. Popular knowledge management literatures attempts to make distinction among data, information and knowledge. Tuomi, [32] opined that the general notion is that, data is viewed as a facts that becomes information as they are combined into meaningful structures, which would subsequently result into knowledge through the application of meaningful information in a desired environment towards the goal of making a prediction. Data again was referred to as a set of discrete, objective facts about events [8].

OBrien [23] defines information as data placed in a meaningful and useful context. The impact of human element in the learning process prior to transformation of information into knowledge was properly presented in [9]. Thus knowledge communicated concerning some particular fact, subject, or event, whether be it appraised or told, intelligence, news especially contrasted with data [13]. The above implied, the reader/user is expected to make appropriate referencing from available information.

In the view of Davenport [6], he described data as simple observation of states of real world, while information is data endowed with relevance and purpose, and knowledge is described as valuable information. As quoted in [16], referring to the opinion of [31] where information was described as meaningless, and only becomes meaningful when it is interpreted. There seems to be no end to the inter-usage of these concepts, in another submission two sets of equation was given in an attempt to show the relationship and their interdependency.

$$Data + Information = Knowledge$$
$$Data + Information + Knowledge = Wisdom$$

Lueg, [18] opined that knowledge is often viewed as information with a specific properties. The opinion above elicits some hierarchical transition in the development and understanding of the concepts we have been discussing. The view of [10] is such that: information is data in context. Information is usable, it is the meaning of data, and thus fact becomes understandable. An attempt to translate the above into knowledge appropriate for deriving decisions from the view of [5] can only result from the addition of experts opinion, skills and experience to the combination of data and information.

Harper, [14] argued that information is relevant when it is seen by all the parties involved. Fox, [11] was of the opinion that data are stored in many formats and encoding via increasingly sophisticated conventions and standards. Considering from this perspective, information with meaning was referred to as knowledge [24]. Bringing in again the human dimension into the scene, [6] opined that information and knowledge are characteristically human creations, thus their management will continue to pose a problem unless we give people a primary role.

The issue of defining information is highly controversial, ambiguity,misconception and misrepresentation abound in various existing definitions [22]. Rather, the emphasis on information should be with respect to the relationship to data, information and knowledge with thier management for creating values[16]. The pattern of usage and adoption employed by many researchers on the term information is obviously divergent. This factor thus contributes in no small measure

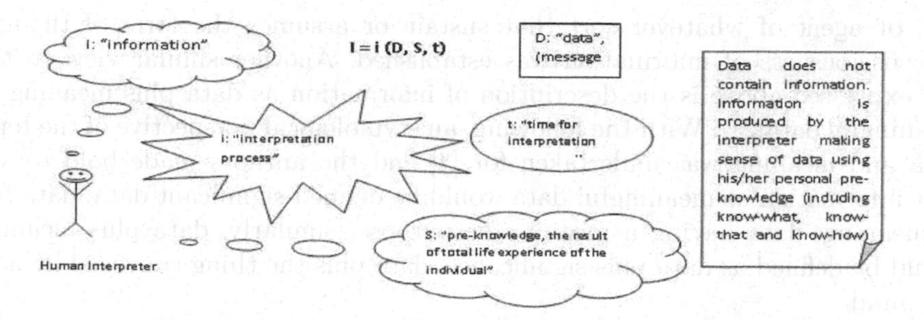

Fig. 1. The Infological Equation (Langefors, 1966)

to present and reflect different perspectives [4]. In the Infological equation proposed by [17] and shown in figure 1, information was paralleled with meaning brought about via interpretation. This was however in contrast with the submission of [30] in the model of communication where the authors reiterated that information must not be confused with meaning. It has however being argued that [30] gave a mathematical definition and not a conceptual one. This stems from the fact that it is wrong to confuse a measure of a thing with the thing measured, let alone confused with metrics, with the thing measured by it [4]. In demonstrating the workability of the infological equation, Langefors [17] was of the opinion that those who are to interpret data in order to inform themselves must be viewed as part of the system. Using the equation I = i(D, S, t), where I is the information (knowledge) produced by a person from the data D alongside with pre-knowledge S through an interpretation process i in interval t. Bednar and Welch, [1] reiterated the impossible nature of communicating meaning between people[17]. This was sequel to the fact that simply transmitting data will not lead to communication of shared understanding knowing fully well that i and S can not be assumed to be common. To this end, communication can only be seen to approach success most closely where individuals interpreting the same data belong to a group with possible vested professional interest.

It is not a common place for a piece of data to generate similar factual meaning when interpreted by different individuals. However, derivable inferences would be likely different more widely in meaning of the data for different individual based on his/her associations, and/or possible consequences depending on the uniqueness of S. Communication and intention is context-dependent. Interpretation of context continually evolves with time thus having great influence on sense-making and communication (by Wittgenstein, (1963) and quoted in [1]).

The infological equation as seen in figure 1 distinctively identify differences between data and information, however, information was synonymously referred to as knowledge brought about via the interpretation and pre-knowledge of the individual in question. Information has been widely defined as interpreted data thus bringing in a reference to the i as depicted in the infological equation. However, if this is anything to follow, then different people viewing same data will come up with different interpretations, i.e. propose different meanings to the same data [4]. The subjective nature, i.e. related to a subject - a mind, ego, or agent of whatever sort that sustain or assumes the form of thought or consciousness of information was established. Another similar view to the one expressed above is the description of information as data plus meaning or meaningful data[21]. With the following, an etymological perspective of the term data and meaning was undertaken for [4] and the authors made bold to say that information as meaningful data would be defined significant data, data full of meaning, data having a meaning or purpose, similarly, data plus meaning would be defined as data plus significance, data plus the thing conveyed by it in the mind.

Relating information and knowledge, [22] emphasized that knowledge representation are not knowledge but rather representation of knowledge. This submission further geared the notion that information is generated inside the mind of a person or a subject. It is dependent on the person where it is generated by the data stimulus, coupled with his/her individual experience. Information in the view of [15] is "decision-relevant data", thus remarking the subjectivity therein embedded. It is no gain saying that the manner by which individual interprets data, information and knowledge will have tremendous impact on what course of action they will undertake in collecting, managing and sharing such information within an organization. The interdependency of the trio again was the vocal point in [16] where several other authors opinions were juxtaposed leading to the formation of what is referred to as the circular relationship between data, information and knowledge. As opposed to the synonymous reference to data and information, [8] disagreed with this stating that information is more than data and thus requires some form of human interaction/involvement relating to interpretation [21]. The element of interpretation is subjective [22] and thus requires human involvement in making sense of something via their unique attribute with which they are endowed. Tuomi [32] argued that there is a reverse hierarchy of data information and knowledge as data emerged last only after knowledge and information are available. Consequently, [16] asserted that data does not exist in isolation but is a result of human intervention by creating data via their knowledge and understanding and this was used to arrive at figure 2 above. In the opinion of [11] information is provided when data answer an implicit or an explicit question made by the data receptor. To this end, for data to

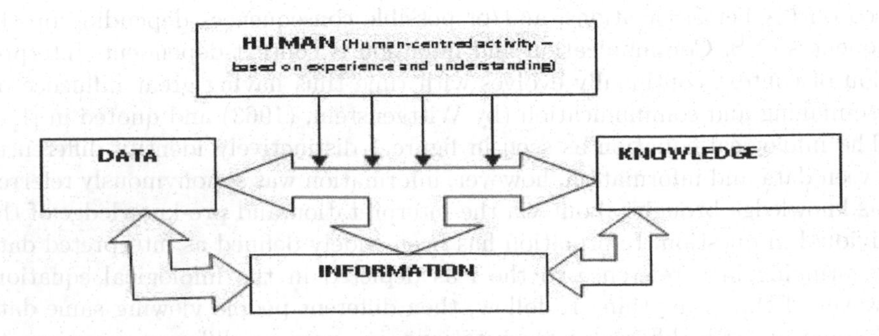

Fig. 2. The Circular relationship amongst data, information and knowledge

be informative, it should be associated with a relevant question i.e. information consists of datum and relevant questions brought about by some operation performed on it. The next section describes our framework based on ontology for minimizing the risk of non-quality data during knowledge reconciliation process.

3 Ontological Framework for Knowledge Reconciliation in EI

EI from its definition is set to present a coordinated action of search and information utilization for timely, effective and strategic decision making [20]. Bouaka and David [2] presented a model that combines both the context of the problem, the representation of the decision-maker and the challenges of the decision-making problem. The objective is to facilitate the identification and the representation of the decision problem on the one hand and preparation of the information research project on the other. It aids the understanding of the problem through identification of the users characteristics and assessment of the level of identified stakes of the problem. Sequel to the above, [2] presented a proposal aimed at assisting the decision maker in explicitly defining his decisional problem. Complexity of problem definition derives from the differences in interpretation of each key actor in a given situation usually results into different level of comprehension of the event and to the explanations of the influences between events. In the formal part, it is important to recognize that personal values play a part in interpretation, and the latter part opined that individuals brings to bear different experiences and wisdom that has created different belief systems [13]. An interpretation is in itself subjective, i.e. related to a subject, prevailing circumstances, a mind, ego, or agent of whatever sort that sustains or assumes the form of thought or consciousness [4]. Wand and Wang [33] are one of the key proponent of ontological dimension to data quality employed a formal model for an information system by considering mapping function from the real

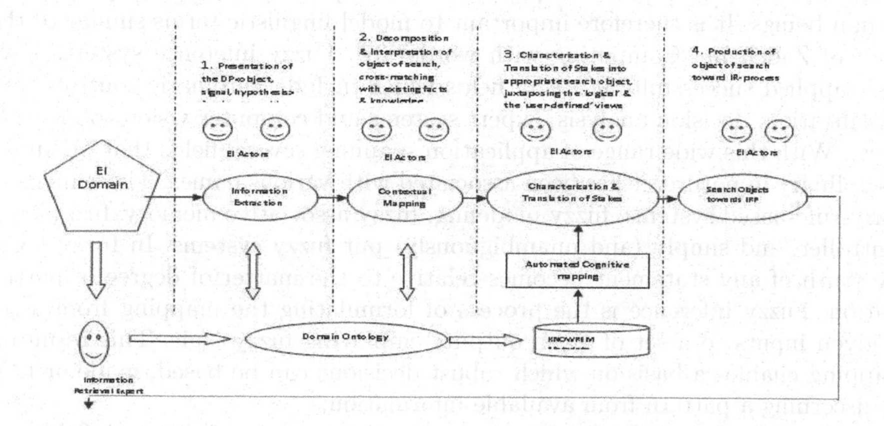

Fig. 3. Ontological Framework for KNOWREM (Onifade, et al., 2008)

world to an information system. Bearing in mind the need to reduce the risk of non-quality data during information search operation towards the delivery of strategic decision making in economic intelligent systems, we proposed ontological framework for knowledge reconciliation to facilitate proper understanding between the decision maker (DM) and the information specialist (Watcher) the duo of which are identified and established actors in economic intelligent process. We define Knowledge reconciliation as the attempt to map the desire for precision in naturally expressed languages employed in describing, sharing of knowledge, interpreting and communication of a need to another person/object in an acceptable degree of accuracy devoid of misinterpretation, disinformation, biases and unnecessary personal preferences to mention a few, which was referred to as risk factors (RFs) in [25]. This becomes imperative because real world is laden with concepts which do not have a sharp boundary e.g. "fine", "useful", "more important than", "old" e.t.c. Consequently, we developed an ontological framework aimed at reducing the risk in the knowledge reconciliations operations. A detail discussion on the KNOWREM framework can be found in [26].

With this background, we present the main contribution of this work in the next section where we employed the same framework in figure 3 to improve the judgment of the actors involved via the inclusion of fuzzy inference system to facilitate robust decision making in "FuzzOntology".

4 Fuzzontology

Information mining is the non-trivial process of identifying valid, novel, potentially useful and understandable patterns in heterogeneous information sources. Considered as an offshoot of knowledge discovery in databases (KDD) or data mining, it is saddled with the exploitation of information for decisional purposes. However, it's without gainsaying to assert that language is human most effective tool to structure his experience and also model his environment. Langefors, [17] in his equation stressed the importance of interpretation which is peculiar to human beings. It is therefore important to model linguistic terms similar to the work of Zadeh in "Computing with words"[37]. Fuzzy inference systems have been applied successfully in many fields which include automatic control, data classification, decision analysis, expert systems and computer vision to mention a few. With this wide range of application spanning several fields that are interdisciplinary in nature, it has been associated with various names. These include: fuzzy-rule-based systems, fuzzy modeling, fuzzy associative memory, fuzzy logic controller, and simply (and unambiguously) put fuzzy systems. In fuzzy logic, the truth of any statement becomes relative to the matter of degree of participation. Fuzzy inference is the process of formulating the mapping from a set of given inputs to a set of given outputs employing fuzzy logic. This resultant mapping enables a basis on which robust decisions can be based, made or help in discerning a pattern from available information.

In the following examples, we present two excerpt from different fields i.e. "Medical Diagnosis" and "Organizational decision process". In the former, the basic notion of ¡o, s h¿ is employed by the medical practitioner to arrive at logical conclusion based on human notion and concepts. Words like "not painful, very painful, and moderately painful" are membership function which allows the fuzzy inference system accommodate the imprecision in human communication

Example 1 (Conversation between a Doctor and an Elderly Person).
Introduction: ***Elderly people (EP)*** *usually have problem describing thier dental* ***pain*** *experience to the* ***doctors*** *(Dr). Classification of the pain is usually of the form "****"very painful"****,* ***"moderately painful"*** *and* ***"not painful"***
Conversation
Dr: what is the problem madam?. **EP:** I have problem with my teeth
Dr: what type of problem do you have and where?
EP: I have **"pain"** at the **"upper part"** on the right **"towards the back"**
Dr: How would you describe the pain? (the Dr begins to use some instruments to touch the suspected areas and asked the EP to describe the pain experienced in the areas)
Dr: How is it here **EP:** No **pain** there
Dr: What about here? **EP:** it's **"moderately painful"** there **Dr:** and here?
EP: oh! yes, it's **"very painful"** there, yes, yes // EP screaming

1. We can represent problems identification in term of $< o, s, h >$. This step is very important because it affords the Dr. the opportunity to understand the problem without which he cannot successfully move on. Here, his competence, sense of judgment, intuition and experience towards interpretation is brought into play. This is what we tagged knowledge reconciliation stage and the result determines the translation credibility
2. The Dr. employs the conviction based on the earlier discussion resulting into translation credibility (TC), and the formulated object of search for the information retrieval process. In the formulation of search object, the signals derived from the dialogue will be employed to formulate and refine his subsequent queries. In all, we belief the inference query-formulation will take an if rule format such that we can have

if ***"pain"*** *is* ***"moderate"*** *and* ***"age"*** *is* ***"elderly"*** *and* ***"located"*** *on the* ***"far right"*** *conner of the mouth, and ..., (some solutions)*

Example 2. It is no gain saying to assert that decisional understanding and information understanding are two distinct phases in decision making/information mining where human involvement is very critical, i.e. its human centred. These phases are responsible for rigorous definition of the decisional problem, estimating its potential benefits, followed by the collection of necessary information. We are convinced that since it involves human beings, then, the imprecision resulting from the use of language of communication alongside other roots of biases and environmental factors are critical to the success of the operation. Unfortunately, human notion and concepts are not easily transformed into computer representation without the loss of originality, hence the need for fuzzy inference systems. Below is the representation of the linguistic variables employed for the research and their corresponding membership function. We attempt to limit the membership functions to 3 for clarity, it can however be extended to accommodate more definitions as the need arises.

- $< ORN >$: what is the need, at stake, (SWOT): well understood, understood, not- understood.
- $< ENF >$: Govt. policies, competitors, customers: Highly favourable, favourable, not- favourable.
- $< INE >$: level of education, cognitive experience, and intuition: Advance, intermediate, low.
- $< BIA >$: Personal preference, misinformation, disinformation: prominent, average, low

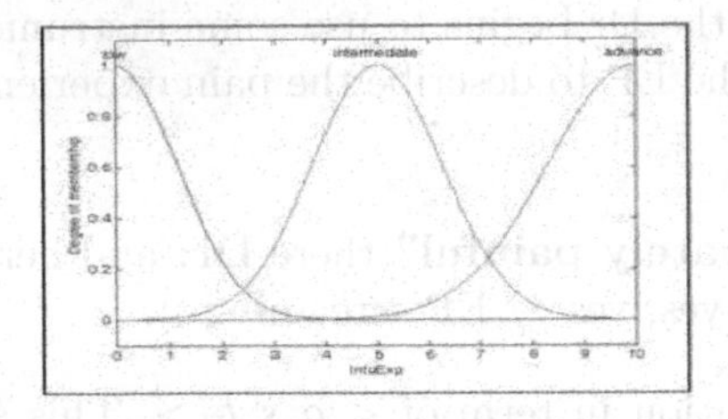

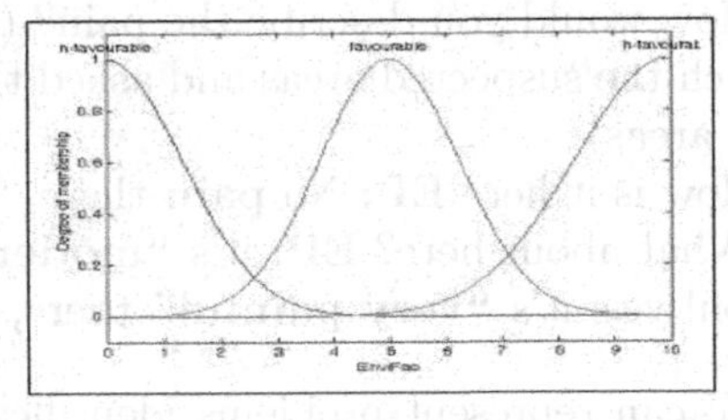

Fig. 4. Membership functions for "Intuition & Experience" and "Environmental factors"

Next we employed the Mamdani fuzzy engine to generate the membership functions for our designs and the examples are given in figure 4 above. The above provides a more flexible way of maintaining consistency when drawing information from both the EI domain and domain ontology as shown in figure 3. While the vagueness of interpretation still exists, the centriod function employed in fuzzy inference system and the membership function accommodates the imprecision during the defuzzification process, thereby reducing the error rate. The output of the de-fuzzified variable can thus be viewed as a chart to show the interdependencies of linguistic variables in consideration.

In figure 5(a) below, we give the result of the effect of two of the variables (Intuition and Experience, and Organizational need). By organizational need, we hope to infer the ability of the decision maker to adequately put into cognizance the need of the organization alongside their strength and weaknesses (SWOT). We believed that inability to adequately define this will spell doom for the organization, and same is for other factors considered in this research. We therefore evaluate the effect pictorially on the factor called translation credibility i.e. ability of the information specialist to adequate map and translate the mind of the decision maker into his information retrieval problem. Elsewhere the focus has been on the understanding of organizational competitive advantage in the production of goods and services. However, knowledge of such factors are not as realistic as the situation itself. While government comes with various policies that can be described as being favourable, highly favourable or unfavourable,

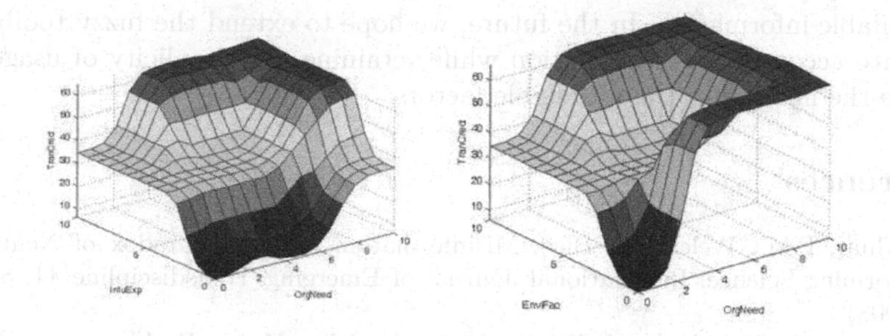

Fig. 5. (a): "organization's need" and "intuition & experience" (b): "organization's need" and "environmental factors" Figure 5: Fuzzification sample results

other factors like the strength of business competitors are not as easily determined by the opponents, it therefore behooves that employing intangible factors in decision making can only be handle the fuzzy way. Another important factor like the type, desire, and expectation of the customers cannot be appropriately determined with precision. Although these factors can be sometimes intangible, their omission hitherto has contributed no small risk to information use for decision making. Figure 5(b) presents the fuzzified result from the combination of the "organizational need" and the "environmental factors". Expressions like favourable is not tangible, and therefore, precise classification is almost unattainable. With the fuzzy inference system, it is no longer an issue as the intangibility can be expressed with appropriate membership function causing us to have results even when both factors are negligible. The result depict the importance of the factors as shown in the manner with which the graph grow equally at inception, but however formed a steeply growth at the organizational need axis thus implying that adequate and appropriate definition of what needs be done is sine-qua-non to the success of the overall operation. This section has showcased the focal point of our research where we combined the strength of fuzzy systems in resolving ambiguity which characterises human communication. Fuzzy solution is not only judge for its accuracy, but also the simplicity and readability thus providing a robust and scalable method for accommodating imprecision.

5 Conclusion and Future Work

It cannot be over emphasized that language is humans most effective tool to structure his intuition, experience and judgmental abilities aimed at employing these to model his environment. Consequently, the ambiguity, vagueness, and imprecision can be modeled only with technology that has such provision - fuzzy. Whenever ontology is faced with "interpretation" there is always possibility of ambiguity which can highly impair the information search operation. With this backgroud, we proposed a new concept tagged "fuzzontology" to assist in knowledge reconciliation process towards delivering strategic decision based on available information. In the future, we hope to extend the fuzzy toolbox to adequate accommodate this notion while retaining the simplicity of usage and enlarge the number of the intangible factors.

References

1. Bednar, P.M., Welch, C.: Bias, Misinformation and the Paradox of Neutrality. Informing Science. International Journal of Emerging Transdiscipline 11, 85–106 (2008)
2. Bouaka, N., David, A.: A Proposal of a Decision Maker Problem for a Better Understanding of Information Needs. IEEE Explore, 551–552 (2004)
3. Burgess, M., et al.: Using Quality Criteria to Assist in Information Search. International Journal of Information Quality 1, 83–99 (2007)
4. Callaos, N., Callaos, B.: Toward a systemic notion of information: Practical consequences. International Journal of an Emerging Transdiscipline 5(1), 1–11 (2002), `http://inform.nu/Articles/Vol5/v5n1p001-011.pdf` (retrieved June 11, 2009)
5. Chaffey, D., Wood, S.: Business information management: Improving performance using information systems. Harlow, Prentice Hall, Pearson Education Limited (2005)
6. Davenport, T.H.: Information ecology. Oxford University Press, Oxford (1997)
7. Davenport, T.H., Prusak, L.: Working Knowledge: How organizations manage what they know. Harvard Business School Press, Boston (1998)
8. Davies, L., Ledington, P.: Information in action. Macmillan Press, London (1991)
9. Dervin, B.: An overview of Sense-Making Research: Concepts, Methods and Results to Date. Presented at the Int'l Communication Association Annual Meeting, Dallas, School of Communications, University of Washington, Seattle (1983)
10. English, L.: Improving data warehouse and Business information quality. Methods for reducing cost and increasing profits. Wiley and Sons, New York (1991)
11. Fox, R.: Moving from data to Information: OCLC system and series. Int'l Digital Library Representatives 20(3), 96–101 (2004)
12. Floridi, L.: Philosophy and Computing, An Introduction. Routledge, Taylor and Francis Group, London (1999)
13. Gackowski, Z.J.: Operations quality of information: Teleological operations research-based approach, call for discussion. In: Proceedings of the 10th Anniversary of ICIQ 2005, November 4-6. MIT, Cambridge (2005)
14. Harper, R.: Information that counts: A sociological view of information navigation. In: Munro, A.J., Hook, K., Benyon, D. (eds.) Social Navigation of Information Space, pp. 81–89. Springer, London (1999)
15. Harris, R.: Introduction to Decision Making, pp. 1–15 (July 1998), `www.virtualsalt.com/crebook5.htm`
16. Knox, T.K.: An Investigation into the Information strategy formulation process within a higher education institution: A case study approach. Nottingham Buz. School. Nottingham, Nottingham Trent University: 256 (2007)
17. Langefors, B.: Essays on Infology - Summing up the planning for the future. Studentlitteratur, Lund (1966)
18. Lueg, C.: Information, Knowledge and Networked minds. Journal of Knowledge Management 5(2), 151–159 (2001)
19. Maqsood, T., Finegan, A.D., Walker, D.H.T.: Biases and Heuristics in Judgment and Decision Making: The Dark Side of Tacit Knowledge. In Issues in Informing Science and Information Technology (2004)

20. Martre, H.: Intelligence Economique et stratégie des enterprises, Rapport du Commissariat General au Plan, Paris, La Documentation Francaise, pp. 17–18 (1994)
21. Mingers, J.: The Nature of Information and Its Relationships to Meaning. In: Winder, R.L., Probert, S.K., Besson, I.A. (eds.) Philosophical Aspects of Information Systems, pp. 73–84. Taylor and Francis, London (1997)
22. Neill, S.D.: Dilemmas in the Study of Information: Exploring the Boundaries of Information Science. Greenwood Press, New York (1992)
23. O'Brien, J.A.: Management Information Systems, 6th edn. Mc Graw-Hill/Irwin (2004)
24. O'Leary, D.E., Selfridge, P.: Knowledge management for best practice. Communications of the ACM 43(11), 12–24 (2000)
25. Onifade, O.F.W.: Cognitive Based Risk Factor Model for Strategic Decision Making in Economic Intelligence Process. In: GDR-IE Workshop, June 16-17 (2008), `http://s244543015.onlinehome.fr/ciworldwide/wp-content/uploads/2008/06/nancy_onifadeofw.pdf`
26. Onifade, O.F.W., Thiery, O., Osofisan, A.O., Duffing, G.: Ontological Framework for Minimizing the Risk of Non Quality Data during Knowledge Reconciliation in Economic Intelligence. In: Proc. of the 13th Int'l ICIQ 2008, pp. 296–309. MIT, Boston (2008)
27. Onifade, O.F.W., Thiery, O., Osofisan, A.O., Duffing, G.: Decisionability: Contending With Information Flow, Information Quality, And Information Overload In Economic Intelligence. In: VSST 2009 (2009)
28. Oxford English Dictionary, OED (1989)
29. Revelli, C.: Intelligence stratégique sur Internet, Paris, Dunod, pp. 18–19 (1998)
30. Shannon, C.F., Weaver, W.A.: Mathematical Model of Communication. University of Illinois Press, Urbana (1949)
31. Sveiby, K.E.: The new organizational wealth: Managing and measuring knowledge-base assets. Berret-Koehler, San Francisco (1997)
32. Tuomi, I.: Data Is More than Knowledge: Implication of the Reversed Knowledge Hierarchy for Knowledge Management and Organizational Memory. Journal of Management Information Systems 16(3), 103–117 (Winter 1999)
33. Wand, Y., Wang, R.Y.: Anchoring Data Quality Dimension in Ontological Foundations. Communications of the ACM 39(11), 86–95 (1996)
34. Wand, Y., Weber, R.: On the Ontological Expressiveness of Information Systems Analysis and Design Grammars. Journal of Information Systems 3, 217–237 (1993)
35. Wand, Y., Weber, R.: On Deep Structure of Information Systems. Journal of Information Systems 5, 203–223 (1995)
36. Wilson, T.D.: Information needs in social services. Research highlights in Social Work 13, 12–24 (1986), `http://information.net/tdw/publ/paper/infoneed85.html`
37. Zadeh, L.A.: Fuzzy Logic= Computing with words (CW). IEEE Transactions on Fuzzy Systems 4(2), 103–111 (1996)

Personalizing Web Recommendations Using Web Usage Mining and Web Semantics with Time Attribute

Teng-Sheng Moh and Neha Sushil Saxena

San Jose State University
Dept. of Computer Science
San Jose, CA 95192-0249, USA
tsmoh@cs.sjsu.edu, neha0929@gmail.com

Abstract. Web personalization is the process of customizing web pages using a user's navigation patterns and interests. With the increase in the number of websites and web pages available on the internet, directing users to the web pages in their areas of interest has become a difficult problem. Various approaches have been proposed over the years and each of them has taken the solution of creating personalized web recommendations a step farther. Yet, owing to the possibilities of additional improvement, the system proposed in this paper takes generating web recommendations one more step ahead. The proposed system uses information from web usage mining, web semantics and the time spent on web pages to improve user recommendations.

Keywords: Web Personalization, Web Usage Mining, Web Semantics, Time Attribute, and Clustering.

1 Introduction

In recent years, the popularity of the Internet has grown to a great extent. Nearly every person, young or old, uses it for a variety of purposes. People use the internet to collect information, do research, save money, get the latest news, etc. With each passing day, more informative web sites, web pages and web documents are added to the already huge collection. Any popular search engine returns thousands of related links to a search query. It has become difficult for users to get the most relevant information due to this plethora. Users often spend considerable time browsing web pages in order to get the right information. If the users' motive for browsing a web page can be identified, it will be easier to make that aspect of information a higher priority.

Identifying user intentions helps personalize their web browsing experience. It also has other advantages. Web pages that a user is likely to browse can be cached, and thus, the retrieval time and load on the network can be considerably reduced. Knowing the user's interests is profitable, too. Only related and targeted marketing advertisements will be shown to the user. This results in an increased number of customers.

S.K. Prasad et al. (Eds.): ICISTM 2010, CCIS 54, pp. 244–254, 2010.

This paper proposes a system which evaluates a user's interest based on a combination of other users' browsing patterns, the page's contents, and the time spent by users on said web pages.

2 Related Research

The basics of web personalization are explained in [1]. The approach described in this paper is comprised of two components: an offline component for data preparation and usage mining, and an online component which is a recommendation engine. In this paper the authors have focused on using only web usage mining and user profiles for web personalization. Research that focuses only on web usage mining in order to create personalization has the short coming of not considering the web pages' contents. Web page content adds to the knowledge of the user's interest area and thus is an important factor in improving web personalization. A few of the other papers that only focus on web usage mining are [2], [3], [4]. In [2], the authors explore the relationship between the queries that the users use for when searching web pages and their navigational patterns. A site-keyword graph is formed based on these two attributes. This is used to generate recommendations for new users. In [3], the authors propose to improve web personalization by considering web usage data along with user preferences that the users specify via websites registration. In [4], too, it aims at improving web personalization using only web usage mining.

Recent papers have explored the possibility of including web semantics to improve web personalization. Papers [5], [6], [7], [8], [9], [10] have targeted this area. While [5], [6], [7] have used both web usage and web semantics mining, the others have solely concentrated on semantic mining. Semantic mining is where the documents are clustered together based on their content and the user is recommended to pages in the cluster the current document belongs to. This method does not take into account the activities of previous users found in the web logs. The approach in [7] is interesting, as it explores the ways in which a web document's semantics can be combined with the navigational pattern obtained from the web log and used to give better recommendations.

The research presented in this paper is taken a step further by considering one more attribute, the time spent on web pages. The time spent on web pages indicates the interest level of the users. There has been some research on the time attribute in [11], but it has not considered web semantics for personalizing the web. Thus, examining the time attribute in addition to web usage and web semantics mining is a new step in improving web personalization.

3 Motivation

A lot of research has been done to improve the quality of web personalization. However, none of the authors have experimented with the time attribute alongside web usage data and web semantics. While web usage data gives insight to the users' browsing patterns and web semantics reveals the web page's content, the time spent on a web page indicates the user's interest level in the area that the web page covers.

To better understand this, consider web pages 1, 3, 4, 7, and 8, which are not semantically similar to each other. A user who has navigated pages 1, 3, 4, and 7, and spent less than 1 minute, 3 minutes, 5 minutes, and 10 minutes respectively on these pages. Another user has visited pages 1, 3, 4, and 8, and spent less than 1 minute, 10 minutes, 3 minutes, and 6 minutes respectively on these pages. Now, suppose we have a new user, who visits pages 1, 3, and 4 for duration less than 1 minute, 10 minutes, and 3 minutes. Which page recommendation would be more relevant, page 7 or page 8? Recommending page 8 would make more sense and be closer to the new user's interests.

The web personalization approaches that rely solely on web usage data would recommend both pages 7 and 8. Approaches that rely on web usage data and web semantics would also recommend both pages 7 and 8 along with other pages semantically similar to these pages. But now that we have considered the time spent on each page, we know that the new user's interests will be closer to those of users who have spent approximately the same time on the pages browsed prior to web page 8. Thus, when we combine web usage data and web semantics along with the time spent on the other web pages, we get the new recommendation as web page 8 and all other pages that are semantically similar to web page 8 only.

Also, instead of considering the time spent as it is, it makes more sense to have time slots and associate the web pages with the corresponding time slots. For example, one could have three time slots where the time spent on a web page was less than 2 minutes, between 2 minutes and 5 minutes, and greater than 5 minutes. The range of time slots can be decided based on the average time spent on the pages of the web site being considered. The advantage of using time slots over individual time duration is that it is more flexible and does not differentiate between pages that have similar amounts of time spent on them by different users.

For huge websites or search engine results, every single browsed page or search query returns large sets of similar pages as recommendations. It is impossible to list all these suggestions as recommendations. Instead, it is important for the recommendations provided to the current user to be extremely focused. Considering the time attribute narrows the suggestions list while making it even more focused towards the current user's interests.

4 Proposed System

As shown in Figure 1, the proposed system consists of four components. The first one generates web logs to create synthetic data (in the case of actual web logs, this component will be replaced by the web log pre-processing component). The second one processes the web usage data to get frequent itemsets which are used to determine users' browsing patterns. The third one processes the web semantics and clusters the documents based on semantic similarity. The fourth one uses the second and third components to generate recommendations.

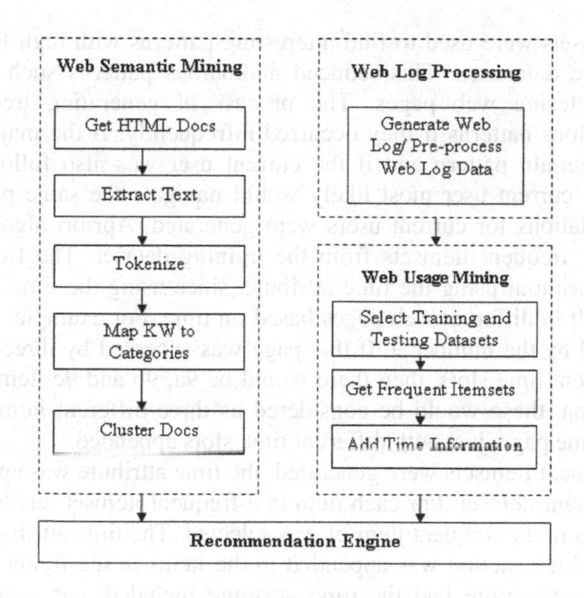

Fig. 1. Architecture of proposed system

4.1 Web Log Processing

This is the process we used to generate synthetic web logs. First, a tree of the web pages to be included in the web log was created. Once the tree was created, all the paths and sub paths (with a minimum length of three web pages) of the tree were found. These paths and sub paths represented the users' navigation patterns. Then, time was added to the navigation paths. Three time slots indicated as 'a', 'b' and 'c' were used. The time slots represented short, average and long amounts of time spent on the web pages. For example, 'a' represented 1 sec to 1 min, 'b' 1 min to 5 min, and 'c' more than 5 min. A random time slot was appended to every web page. The synthetic web log contained web page browsing transactions with access time information appended to each page.

In an actual web log, any single web page browsing pattern is duplicated by many other users. In order to mimic an actual web log, these paths were randomly duplicated. The synthetic web log therefore resembled a real web log containing a large dataset of repeated transactions with a time attribute on each page.

4.2 Web Usage Mining

In this step, the large dataset of transactions (previous users' web browsing information) was searched for those web pages that occur frequently. The dataset was broken into a ratio of 8:2, where 80% of the transactions were used as a training set to generate the frequent itemsets for recommendations, and the other 20% of the transactions were used as a testing set to test the recommendations.

Frequent itemsets were used to find interesting patterns with high frequencies that occurred in large databases. This reduced anomalous patterns such as leaving for lunch without closing web pages. The process of generating frequent itemsets excluded anomalous patterns if they occurred infrequently. If the majority of people navigated in a certain pattern and if the current user was also following a similar pattern, then the current user most likely would navigate the same pages. This was how recommendations for current users were generated. Apriori algorithm [12] was used to generate frequent itemsets from the training dataset. The frequent itemsets were generated without using the time attribute, since using the time slots for a web page would result in different web pages based on time. For example, consider a web page represented by the number 9. If this page was accessed by three different users with three different time slots, then there would be 9a, 9b and 9c items. For frequent itemset generation, these would be considered as three different items, even though they were the same page, just with different time slots appended.

Once the frequent itemsets were generated, the time attribute was appended to each page of the frequent itemset. For each item in a frequent itemset, the transaction with most of the items in the frequent itemset was selected. The time attribute for the items from the selected transaction was appended to the items in the frequent itemset. The frequent itemsets therefore had the time attribute included and were later used to create recommendations for the testing dataset.

4.3 Web Semantic Mining

Web semantic mining aims at mining the content of web pages and finding similarity between web pages based on content. A web page can have text, images, video and/or audio content. For the purpose of mining, only using the text of the web pages helps the most. One of the most popular ways to determine what the text of a document is about is by using term frequency, where a term is a keyword (word of relevance) in the document.

For our web semantic mining, list of categories were formed based on the areas of browsed web pages. The text was extracted and processed to get all the keywords and their frequencies. These keywords were mapped to their categories using the Wordnet thesaurus [13] and the Jiang and Conrath similarity measure [14]. This mapping enabled one to determine which categories a web page belonged to. Once the similarity measure of each document to all the categories had been evaluated, the documents were then clustered, using a modified version of the DBSCAN algorithm to bring together pages with similar content. This process helped to cluster those pages which were contextually similar, but not structurally connected to each other. Even though previous users might not have visited unconnected pages with similar content, semantic clustering brought together such pages and recommends them to the current user.

4.4 Recommendation Engine

This was the final component of the system. It combined the analysis of the usage mining and semantic mining components and produced recommendations for current users. The current user's navigation path was compared with the frequent itemsets generated by the usage mining component. If the user's current path was three items long, then only frequent itemsets of length four items were considered. Each selected frequent itemset produced one recommendation. For each recommended page, it was checked for clustering information and all the documents within that cluster were recommended to the user.

5 Experimental Setup

The proposed system was built using JAVA and Eclipse IDE. We were unable to obtain real web logs due to privacy considerations. Thus, we used synthetic data generated from university websites. Groups of related pages were selected – pages that were interlinked and thus more likely to be navigated to. These groups of interlinked pages were formed into random paths. Extensive experiments were carried out using a synthetic dataset of 10,000 transactions containing 120 URLs. The URLs were selected from four different universities. From each of these universities, the web pages selected belonged to four different areas: science (computer science), art, business and fees/scholarship. About 10 web pages were selected from each of the above mentioned areas for each of the four universities.

To conduct our experiments the original dataset was divided into two parts. One part, called the training dataset, was used to train the system, and the second part, called the testing dataset, was used to test the system. The ratio of transactions was 8:2. The training dataset consisted of 8000 transactions while the testing dataset consisted of 2000 transactions.

For the experiments, each transaction in the testing dataset was divided into halves. The first half of the transaction represented the current user's navigation path while the second half of the transaction represented the web pages that the user would visit in future. The correctness of the recommendations produced by the system was compared with the second half of the transaction. For every first half of the transaction with size w, all frequent itemsets generated by the training dataset that had size w + 1 were selected. All web pages that had not yet been visited by the user (that were not in the first half of the transaction) were candidate recommendations. Since web semantic mining was also being applied, all unvisited web pages that belonged to the cluster, where the candidate recommendation page was, were also included in the recommendations. This procedure was repeated with different support values. Support values narrowed down the number of frequent itemsets generated from the training dataset. A higher support value meant more frequent itemsets.

Precision and coverage (recall) are the two popular metrics that were used to evaluate and compare the performance of the proposed system. Precision measures how well a system generates correct or relevant recommendations time after time. The precision of a transaction is given as the number of web pages correctly predicted divided by the total number of web pages predicted. On the other hand, coverage measures how well the system covers the pages of the area under consideration [15]. The coverage of a transaction is given as the number of web pages correctly predicted divided by the total number of web pages visited by the user. The precision and coverage were evaluated for all the transactions in the testing dataset and their averages were calculated. The average precision and average coverage values helped to evaluate the system.

6 Results and Evaluation

For the evaluation of the proposed system, experiments and comparisons were carried out between the proposed system (the system that considered the time spent on web pages) and a vanilla system (the system that did not consider the time spent on web pages). Two types of experiments were done. One was between the two systems with semantic clustering, and the other was without semantic clustering.

For each of these experiments, the average precision and average coverage were calculated at different support values. The results that were obtained are shown in graphs below.

Precision Comparison with Clustering: The graph shows that the proposed system gave results with better precision than the vanilla system. The proposed system generated recommendations that took the time attribute into consideration. So the recommendations were more precise. The vanilla system produced a large number of recommendations thus reducing the preciseness of the system.

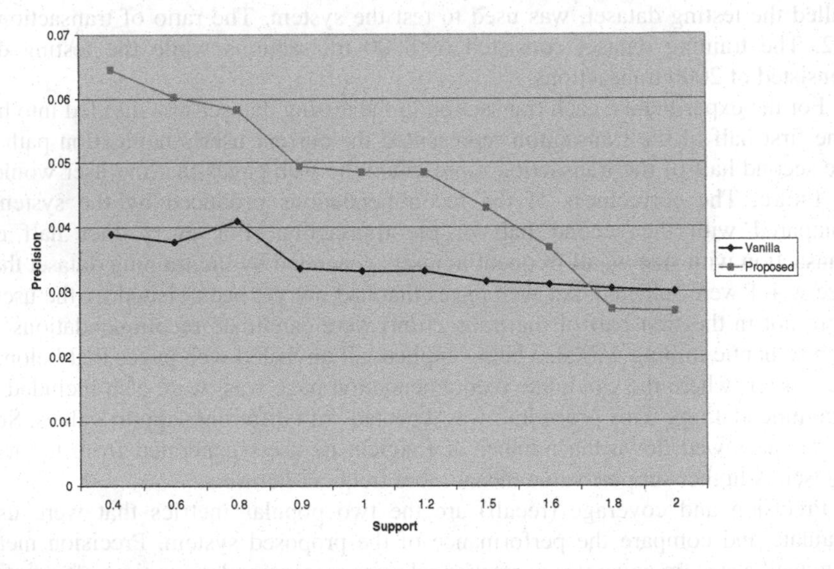

Fig. 2. Precision with clustering

Precision Comparison without Clustering: The graph below shows the precision comparison of the proposed and vanilla systems without semantic clustering. This graph clearly shows that the proposed system was better than the vanilla system. Again, the proposed system produced results with better precision than the vanilla system.

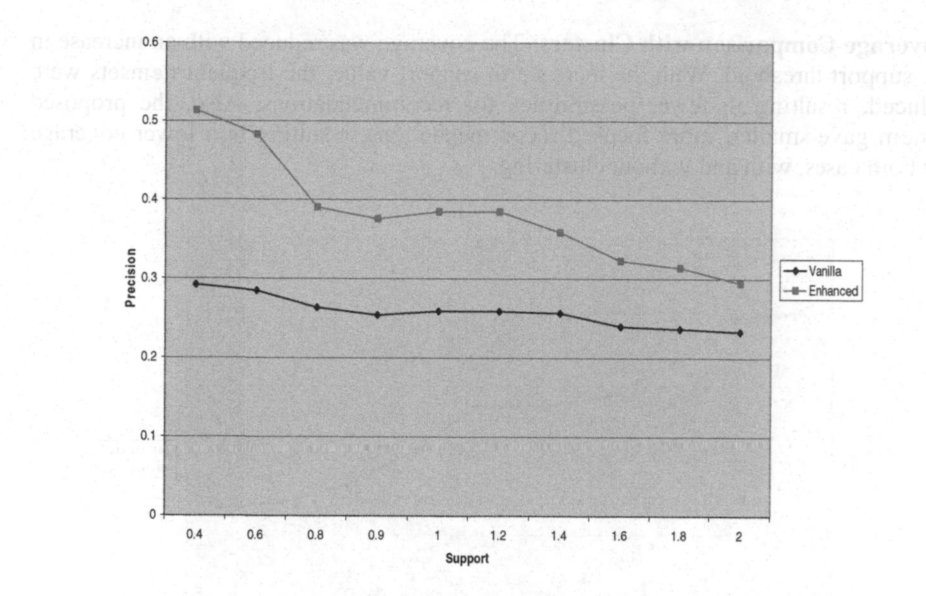

Fig. 3. Precision without clustering

Precision Comparison of All Systems: The graph compares all four systems: vanilla with and without clustering and proposed with and without clustering. In both cases, with and without clustering, the proposed system gave higher precision. The reason why the precision of system without clusters is better is due to the fact that the system with clusters considered and recommended extra pages which might not be visited by other users for lacking of link connectivity. Thus, the recommendations of systems with clusters are larger than that of without clusters, leading to reduced precision.

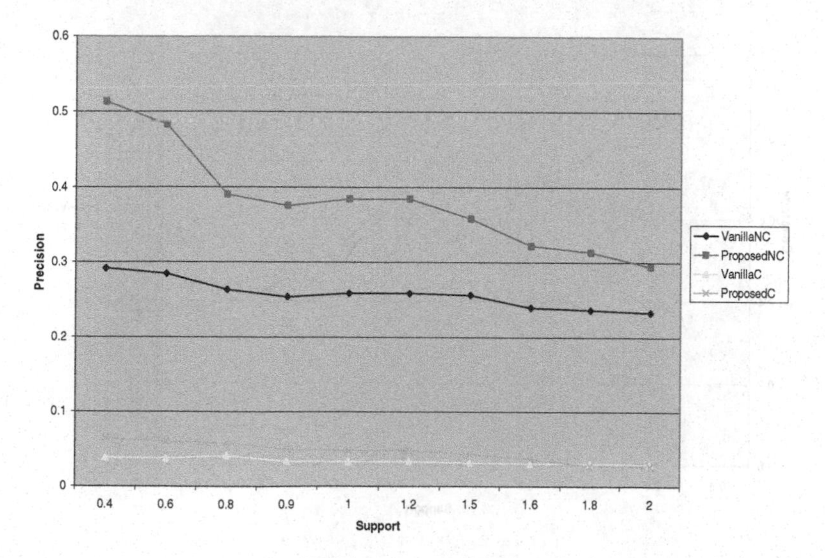

Fig. 4. All precisions

Coverage Comparison with Clusters: The coverage was reduced with an increase in the support threshold. With the increase in support value, the frequent itemsets were reduced, resulting in fewer possibilities for recommendations. Also, the proposed system gave smaller, more focused recommendations, resulting in a lower coverage for both cases, with and without clustering.

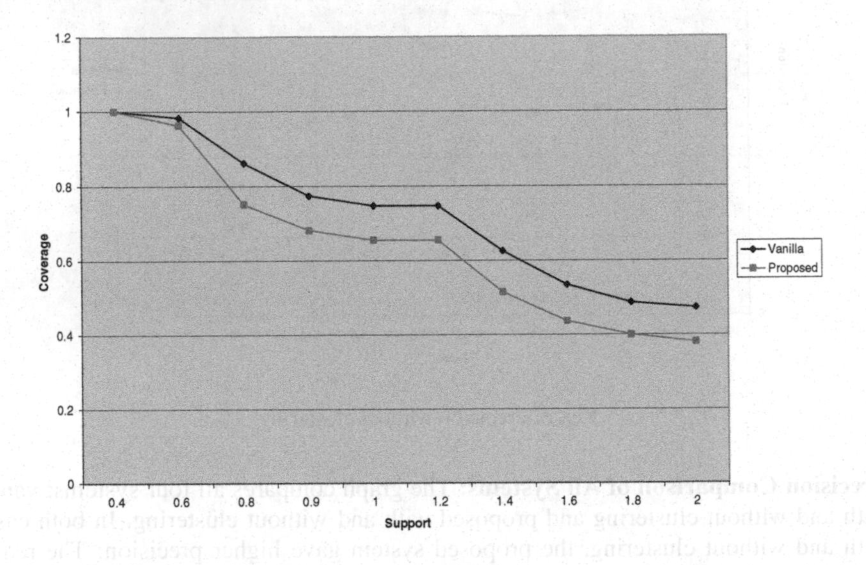

Fig. 5. Coverage with Clusters

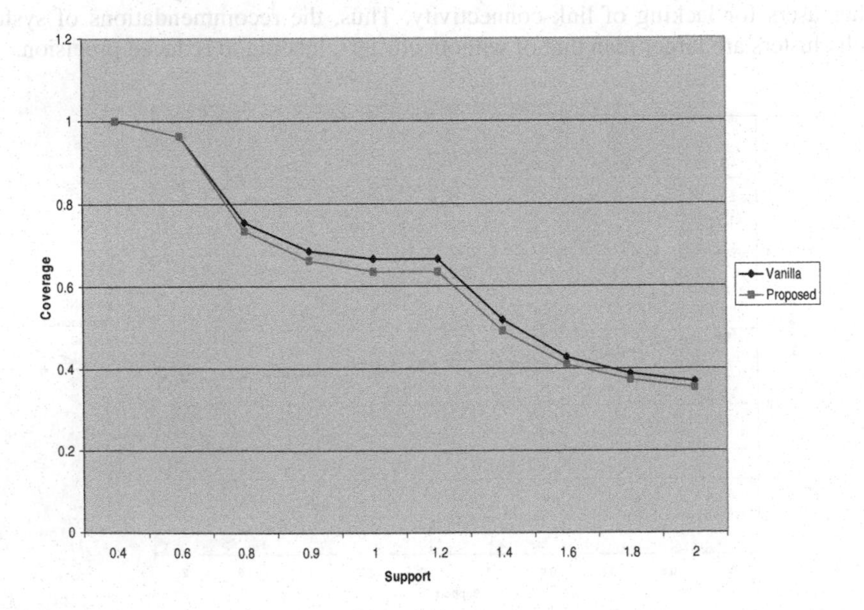

Fig. 6. Coverage without Clusters

Coverage Comparison of All Systems: As shown in the graph, the coverage of systems with clustering was better than that of those without clustering. This is because with clustering even those pages which are not structurally connected were considered, thus giving the users a wider but more targeted recommendation.

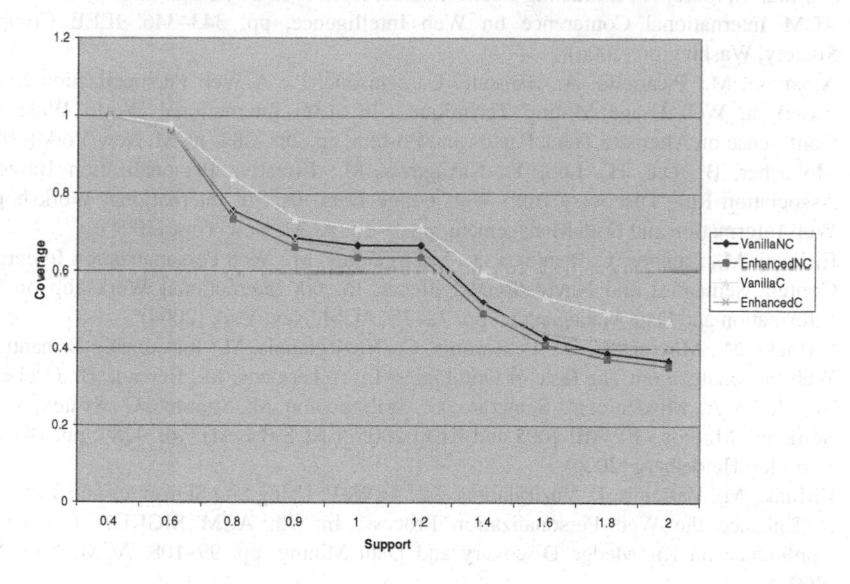

Fig. 7. All Coverage

7 Conclusion and Future Work

This paper proposes a system which aims to improve personalized web recommendations by taking the time spent on web pages (the time attribute) into consideration in addition to using web usage and web semantics mining. During web usage mining, after obtaining frequent itemsets based on the browsing patterns of previous users, the time that users have spent on web pages is also accounted for and appended to the frequent itemsets. This adds more information about the users' interests and makes recommendations more focused and personalized. Web semantics mining finds similar web pages based on page contents and clusters them together. This helps recommend unlinked 'orphan' pages that were not browsed by previous users. Our results support the above argument and demonstrate that taking the time spent on web pages into consideration improves the system recommendations.

This work could be further extended by taking more of the user's attributes into consideration. Attributes like gender, location, age, etc. tell us more about the user, whereas considering the time when the web page was browsed or the number of times that the web page was browsed help increase the knowledge about a web page. Using these attributes to generate web page recommendations will create more personalized and useful recommendations for the users.

References

1. Bamshad, M., Cooley, R., Srivastava, J.: Automatic Personalization Based on Web Usage Mining. Communications of the ACM 43(8), 142–151 (2000)
2. Murata, T., Saito, K.: Extracting Users' Interest from Web Log Data. In: 2006 IEEE/WIC/ACM International Conference on Web Intelligence, pp. 343–346. IEEE Computer Society, Washington (2006)
3. Albanese, M., Picariello, A., Sansone, C., Sansone, L.: A Web Personalization System Based on Web Usage Mining Techniques. In: 13th International World Wide Web Conference on Alternate Track Papers and Posters, pp. 288–289. ACM, New York (2004)
4. Mobasher, B., Dai, H., Luo, T., Nakagawa, M.: Effective Personalization Based on Association Rule Discovery from Web Usage Data. In: 3rd International Workshop on Web Information and Data Management, pp. 9–15. ACM, New York (2001)
5. Eirinaki, M., Lampos, C., Paulakis, S., Vazirgiannis, M.: Web Personalization Integrating Content Semantics and Navigational Patterns. In: 6th International Workshop on Web Information and Data Management, pp. 72–79. ACM, New York (2004)
6. Eirinaki, M., Mavroeidis, D., Tsatsaronis, G., Vazirgiannis, M.: Introducing Semantics in Web Personalization: The Role of Ontologies. In: Ackermann, M., Berendt, B., Grobelnik, M., Hotho, A., Mladenič, D., Semeraro, G., Spiliopoulou, M., Stumme, G., Svátek, V., van Someren, M. (eds.) EWMF 2005 and KDO 2005. LNCS (LNAI), vol. 4289, pp. 147–162. Springer, Heidelberg (2006)
7. Eirinaki, M., Varlamis, I., Vazirgiannis, M.: SEWeP: Using Site Semantics and Taxonomy to Enhance the Web Personalization Process. In: 9th ACM SIGKDD International Conference on Knowledge Discovery and Data Mining, pp. 99–108. ACM, New York (2003)
8. Yu, J., Luo, X., Xu, Z., Liu, F., Li, X.: Representation and Evolution of User Profile in Web Activity. In: IEEE International Workshop on Semantic Computing and Systems, pp. 55–60. IEEE Computer Society, Washington (2008)
9. Luo, X., Xu, Z., Yu, J., Liu, F.: Discovery Of Associated Topics for the Intelligent Browsing. In: 1st IEEE International Conference on Ubi-Media Computing, pp. 119–125. IEEE, New York (2008)
10. Luo, X., Fang, N., Hu, B., Yan, K., Xiao, H.: Semantic Representation of Scientific Documents for the E-Science Knowledge Grid. Concurrency and Computation: Practice and Experience 20(7), 839–862 (2008)
11. Ahmad, A.M., Hijazi, M.H.A., Abdullah, A.H.: Using Normalize Time Spent on a Web Page for Web Personalization. In: 2004 IEEE Region 10 Conference, vol. B, pp. 270–273. IEEE, New York (2004)
12. Agrawal, R., Imielinski, T., Swami, A.: Mining Association Rules Between Sets of Items in Large Databases. In: ACM SIGMOD International Conference on Management of Data, pp. 207–216. ACM, New York (1993)
13. `http://wordnet.princeton.edu/`
14. `http://nlp.shef.ac.uk/result/index.html`
15. Shyu, M., Haruechaiyasak, C., Chen, S., Zhao, N.: Collaborative Filtering by Mining Association Rules from User Access Sequences. In: International Workshop on Challenges in Web Information Retrieval and Integration, pp. 128–135. IEEE Computer Society, Washington (2005)

Towards Analyzing Data Security Risks in Cloud Computing Environments

Amit Sangroya, Saurabh Kumar, Jaideep Dhok, and Vasudeva Varma

International Institute of Information Technology,
Hyderabad, India
{amit_s,saurabh,jaideep}@research.iiit.ac.in, vv@iiit.ac.in

Abstract. There is a growing trend of using cloud environments for ever growing storage and data processing needs. However, adopting a cloud computing paradigm may have positive as well as negative effects on the data security of service consumers. This paper primarily aims to highlight the major security issues existing in current cloud computing environments. We carry out a survey to investigate the security mechanisms that are enforced by major cloud service providers. We also propose a risk analysis approach that can be used by a prospective cloud service for analyzing the data security risks before putting his confidential data into a cloud computing environment.

1 Introduction

In a cloud computing environment, the underlying computing infrastructure is used only when it is needed. For example, in order to process a user request, a service provider can draw the necessary resources on-demand, perform a specific job and then relinquish the unneeded resources and often dispose them after the job is done. Contrary to traditional computing paradigms, in a cloud computing environment, data and the application is controlled by the service provider [1,2,3,4,5]. This leads to a natural concern about data safety and also its protection from internal as well as external threats. Despite of this, advantages such as *On demand infrastructure, pay as you go, reduced cost of maintenance, elastic scaling etc.* are compelling reasons for enterprises to decide on cloud computing environments.

Usually, in a cloud computing paradigm, data storage and computation are performed in a single datacenter. There can be various security related advantages in using a cloud computing environment. However, a single point of failure can not be assumed for any data loss. As shown in Figure 1, the data may be located at several geographically distributed nodes in the cloud. There may be multiple points where a security breach can occur. Compared to a traditional in house computing, it might be difficult to track the security breach in a cloud computing environment.

In this paper, we present the advantages and disadvantages (in the context of data security) of using a cloud environment. We carry out a small survey on

S.K. Prasad et al. (Eds.): ICISTM 2010, CCIS 54, pp. 255–265, 2010.

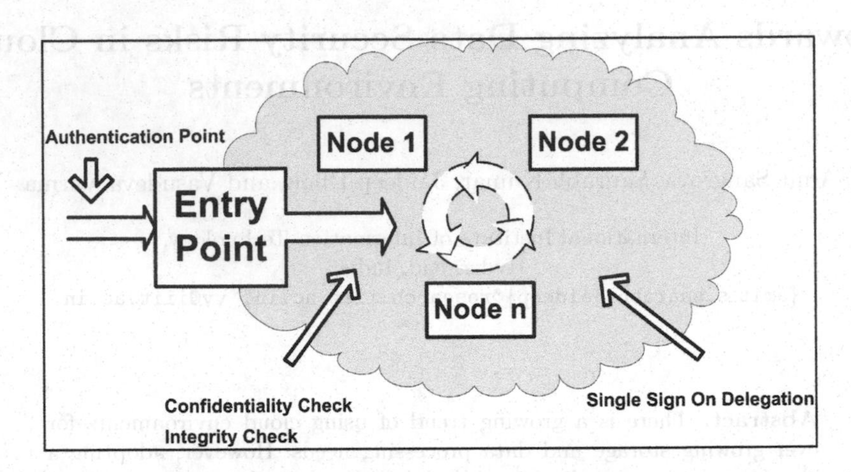

Fig. 1. Typical Data Security Checkpoints in a Cloud Computing Environment

major cloud service providers to investigate the prominent security issues. We investigate the security mechanisms that are used by major service providers. Our study supports that in the context of data security *trust* is a major element which is missing in the currently existing computing models. We believe a lack of trust management mechanism between the cloud service provider and users. Despite the fact that service providers use diverse mechanisms to retain high level of data security, we find a general lack of trust (in the context of confidential data) among the cloud service users.

In order to build a better trust mechanism, we present a risk analysis approach that can be primarily used by the perspective cloud users before putting their confidential data into a cloud. Our approach is based on the idea of trust model, principally used in distributed information systems [6,7]. We extend the general idea of trust management and present its use in analyzing the data security risks in cloud computing.

The contributions of this paper can be summarized as follows:

(a) We investigate the major security issues in cloud computing paradigms.
(b) We also carry out a survey of major cloud service providers to explore the security mechanisms in the context of security issues.
(c) Further, we also present a risk analysis approach that can be used by a prospective cloud service user to evaluate the risk of data security.

The structure of rest of the paper is as follows:

Section 2 provides insight into the security issues in the cloud computing environment from the perspective of service providers and service users. In section 3, we present our risk analysis approach from the perspective of a cloud user. In section 4, we present the related work. We present the limitations of our research and give some direction for future work in section 5.

2 Security Issues and Challenges

IaaS (Infrastructure as a Service), PaaS (Platform as a Service), and SaaS (Software as a Service) are three general models of cloud computing. Each of these models possess a different impact on application security [8]. However, in a typical scenario where an application is hosted in a cloud, two broad security questions that arises are:

- *How secure is the Data?*
- *How secure is the Code?*

Cloud computing environment is generally assumed as a potential cost saver as well as provider of higher service quality. *Security, Availability, and Reliability* are the major quality concerns of cloud service users. Gens et. al. [9], suggests that security in one of the prominent challenge among all other quality challenges.

2.1 Security Advantages in Cloud Environments

Current cloud service providers operate very large systems. They have sophisticated processes and expert personnel for maintaining their systems, which small enterprizes may not have access to. As a result, there are many direct and indirect security advantages for the cloud users. Here we present some of the key security advantages of a cloud computing environment:

Data Centralization: In a cloud environment, the service provider takes care of storage issues and small business need not spend a lot of money on physical storage devices. Also, cloud based storage provides a way to centralize the data faster and potentially cheaper. This is particularly useful for small businesses, which cannot spend additional money on security professionals to monitor the data.

Incident Response: IaaS providers can put up a dedicated forensic server that can be used *on demand* basis. Whenever a security violation takes place, the server can be brought online. In some investigation cases, a backup of the environment can be easily made and put onto the cloud without affecting the normal course of business.

Forensic Image Verification Time: Some cloud storage implementations expose a cryptographic check sum or hash. For example, Amazon S3 generates MD5 (Message-Digest algorithm 5) hash automatically when you store an object [10]. Therefore in theory, the need to generate time consuming MD5 checksums using external tools is eliminated.

Logging: In a traditional computing paradigm by and large, logging is often an afterthought. In general, insufficient disk space is allocated that makes logging either non-existent or minimal. However, in a cloud, storage need for standard logs is automatically solved.

2.2 Security Disadvantages in Cloud Environments

In spite of security advantages, cloud computing paradigm also introduces some key security challenges. Here we discuss some of these key security challenges:
Data Location: In general, cloud users are not aware of the exact location of the datacenter and also they do not have any control over the physical access mechanisms to that data. Most well-known cloud service providers have datacenters around the globe. Some service providers also take advantage of their global datacenters. However, in some cases applications and data might be stored in countries, which can judiciary concerns. For example, if the user data is stored in X country then service providers will be subjected to the security requirements and legal obligations of X country. This may also happen that a user does not have the information of these issues.
Investigation: Investigating an illegitimate activity may be impossible in cloud environments. Cloud services are especially hard to investigate, because data for multiple customers may be co-located and may also be spread across multiple datacenters. Users have little knowledge about the network topology of the underlying environment. Service provider may also impose restrictions on the network security of the service users.
Data Segregation: Data in the cloud is typically in a shared environment together with data from other customers. Encryption cannot be assumed as the single solution for data segregation problems. In some situations, customers may not want to encrypt data because there may be a case when encryption accident can destroy the data.
Long-term Viability: Service providers must ensure the data safety in changing business situations such as mergers and acquisitions. Customers must ensure data availability in these situations. Service provider must also make sure data security in negative business conditions like prolonged outage etc.

Compromised Servers: In a cloud computing environment, users do not even have a choice of using physical acquisition toolkit. In a situation, where a server is compromised; they need to shut their servers down until they get a previous backup of the data. This will further cause availability concerns.

Regulatory Compliance: Traditional service providers are subjected to external audits and security certifications. If a cloud service provider does not adhere to these security audits, then it leads to a obvious decrease in customer trust.

Recovery: Cloud service providers must ensure the data security in natural and man-made disasters. Generally, data is replicated across multiple sites. However, in the case of any such unwanted event, provider must do a complete and quick restoration.

Security Issues in Virtualization
Full Virtualization and Para Virtualization [11,12] are two kinds of virtualization in a cloud computing paradigm. In full virtualization, entire hardware architecture is replicated virtually. However, in para virtualization, an operating system

is modified so that it can be run concurrently with other operating systems. VMM (Virtual Machine Monitor), is a software layer that abstracts the physical resources used by the multiple virtual machines. The VMM provides a virtual processor and other virtualized versions of system devices such as I/O devices, storage, memory, etc.

VMM Instance Isolation ensures that different instances running on the same physical machine are isolated from each other. However, current VMMs do not offer perfect isolation. Many bugs have been found in all popular VMMs that allow escaping from VM (Virtual machine). Vulnerabilities have been found in all virtualization softwares, which can be exploited by malicious users to bypass certain security restrictions or/and gain escalated privileges. Below are few examples for this:

(a) Vulnerability in Microsoft Virtual PC and Microsoft Virtual Server could allow a guest operating system user to run code on the host or another guest operating system [13].
(b) Vulnerability was found in VMware's shared folders mechanism that grants users of a guest system read and write access to any portion of the host's file system including the system folder and other security-sensitive files.
(c) Vulnerability in Xen can be exploited by "root" users of a guest domain to execute arbitrary commands [12].

2.3 Survey

We carry out a small survey of major cloud service providers to investigate the security mechanisms to overcome the security issues discussed in this paper. We consider ten major cloud service providers. These providers provide their services in all major areas of cloud computing, including SaaS, PaaS and IaaS. Table 1 shows the list of service providers that we studied in this survey. In order to analyze the complete state of art of security in cloud computing, the survey needs to be more exhaustive. However, due to the fact that the scope of our work is not just to explore the state of art but to look at the major factors that affect security in cloud computing. Therefore we have intentionally not considered other cloud service providers in this survey.

In table 2, we present the results of the survey that depicts the current state of security mechanisms. Information given in table 2 is based on the information available online at the official websites of these providers.

Table 1. Major Cloud Service Providers

Service Provider Type	Names
IaaS	Amazon EC2, Amazon S3, GoGrid
PaaS	Google App Engine, Microsoft Azure Services, Amazon Elastic Map Reduce
SaaS	Salesforce, Google Docs

Table 2. Summary of Security Mechanisms by Major Cloud Service Providers

Security Issue	Results
Password Recovery	90% are using standard methods like other common services, while 10% are using sophisticated techniques.
Encryption Mechanism	40% are using standard SSL encryption, while 20% are using encryption mechanism but at an extra cost. 40% are using advance methods like HTTPS access also.
Data Location	70% have their datacenters located in more than one country, while 10% are located at a single location. 20% are not open about this issue.
Availability History	In 40% there is a reported downtime alongwith a result in data loss, while in 60% cases data availability is good.
Proprietary/Open	Only 10% providers have open mechanism.
Monitoring Services	70% are providing extra monitoring services, while 10% are using automatic techniques. 20 % are not open about this issue.

3 Risk Analysis Approach

The cloud computing service providers use various security mechanisms to ensure that all the security risks are fully taken care of. However, there are two broad questions:

- *How to estimate the risk to data security before putting a job into the cloud?* and
- *How to ensure customers that their data and programs are safe in provider's premises?*

If a cloud service user is able to estimate the risk of his data security then he can have a level of trust with the service provider. If there is a high risk about the data security then it leads to a decrease in trust and vice-versa.

3.1 Need of a Risk Analysis Approach

The service users need a clear communication about the methods adopted by the service providers to maintain security. Current security technology provides us with some capability to build a certain level of trust in cloud computing. For example, SSL (Secure Socket Layer), digital signatures, and authentication protocols for proving authentication and access control methods for managing authorization. However, these methods cannot manage the more general concept of *Trustworthiness.* SSL, for instance, cannot on its own prove that if a communication between server and multiple hosts is secure. Also, as we discussed in section 1, there are multiple points of failures in a cloud environment.

Current security technology is lacking the complementary tool for managing trust effectively. Gambetta et. al. [14] define trust as *"trust (or, symmetrically, distrust) is a particular level of the subjective probability with which an agent will perform a particular action, both before [we] can monitor such action (or independently of his capacity of ever to be able to monitor it) and in a context in which it affects [our] own action"*. Based on this definition, we can say that trust is a subjective property and is affected by those actions that we cannot monitor.

Three kinds of trust models have been discussed in distributed computing [6]:

- *Direct Trust;*
- *Transitive Trust; and*
- *Assumptive Trust*

In a cloud computing paradigm, where the data and programs cross the organizational boundaries, transitive trust and assumptive trust can be crucial for certain type of applications. A direct trust model in a cloud exists in cloud computing environments, when there is a common trust entity that performs all original entity authentications and the generation of credentials that are bound to specific entities.

A key difference with other models is that the direct trust model does not allow the delegation of original entity authentication. And every relying party must use this entity directly for all validation processes. An example of such type of trust entity is the use of PKI[1] based authentication where a root certification authority (CA) do all kinds of trust relationships. The responsibility of secure data transfer lies in the hands of the certificating authority.

3.2 Risk Assessment Using Trust Matrix

Although no single unit of measure is adequate to the definition of trust, several dependent variables (such as data cost), can be used to describe it. Based upon the prominent security factors discussed in section 2, we build a trust matrix to analyze the data risk. To build the trust matrix, a number of heuristics can be used for selecting the security parameters. However, a simple way to select the security factors is to prioritize them based on subjective opinion and select two most important parameters. We select following two trust variables to build the trust matrix:

(a) *Data Cost*
(b) *Provider's History*

The reason for selecting these trust variables is explained here:

In cloud environment, data can be assigned a cost by the users based on the criticality of the data. The data criticality needs to be computed by the service users. There may be multiple factors that affect the data criticality. For example,

[1] The Public Key Infrastructure (PKI) is a set of hardware, software, people, policies, and procedures needed to create, manage, distribute, use, store, and revoke digital certificates. PKI is an arrangement that binds public keys with respective user identities by means of a certificate authority (CA) [15,16].

confidential business data can be critical and therefore we can assign it a higher cost as compared to less critical data.

Similarly, service provider history can be a possible parameter to estimate the risk. History includes a provider's profile of past services. If users are dissatisfied with a particular service, they can record their experience. If a service provider do not possess a good history of data security (e.g. there is a past record of security failures), then it may also decrease the trust factor. However, other variables can also be used for building the trust matrix. Some of these variables can be Encryption support, Service Cost, Monitoring support etc.

Variable Parameters

Along with trust variables, few parameters used in measuring trust can be applied to fine-tune these trust variables. The parameters which we choose in this category are:

(a) *Data Location*
(b) *Regulatory Compliance*

As we have explained in section 2, data located at the sites which are geographically or politically sensitive would likely to have lower trust than other locations. Similarly, if a service provider is assuring the customers using a centralized regulating authority, it will lead to an increase in trust level among the service users.

We make use of variable parameters as a support mechanism in the trust matrix. It is used as a validation factor that provides a support in the risk analysis.

3.3 Risk Analysis

We capture the relationship using a trust matrix where the axes represents the variables used. The variables used should be meaningfully related to each other. Figure 2, represents an example trust matrix with area representing the Low Risk/High Trust zone and, High Risk/Low Trust zone. This can be explained as:

- x axis represents the data cost.
- y axis represents the service provider's history. and
- z axis represents the data location.

Now, it is obvious that a high data cost with poor service provider history combining with a very sensitive location will result in a higher risk/lower trust.

High trust zone signifies the region of high trust. It can specify the security risk for the current transactions and also for future transactions with that service provider. Similarly, *low trust zone* signifies the region of low trust.

As a risk preventive approach, we also define here a *trust action*, which can be taken as part of a preventive or reactive measure. For example, an added level of authentication and/or verification can be used for the activities which are related to the low trust zone. We have used these variables in a common

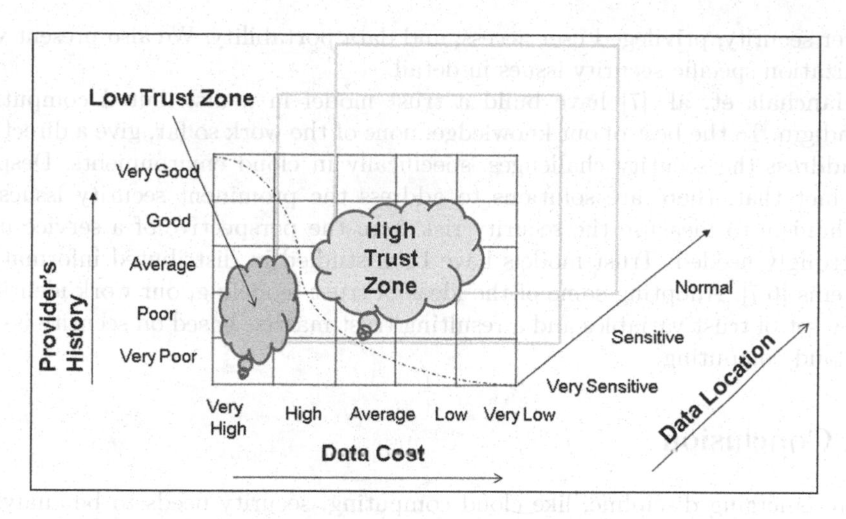

Fig. 2. A Trust Matrix for Risk Analysis

cloud computing scenario, where we have some past statistics about the service provider. The method has been used to measure the trust and will be used for all future transactions. Based on this method, we were able to define the trust actions, for all future transactions with the service provider.

4 Related Work

A survey conducted by IDC (International Data Corporation) suggests that cloud services are still in the early adoption phase. There is a long list of issues cloud service providers need to address. The survey has rated security as the most prominent concern [9]. Buyya [17], provide a survey on current state of the art in cloud computing and identify key challenges that must be addressed in order to make cloud computing a reality.

Cachin et. al. [18], in their survey, give insight into the well known cryptographic tools for providing integrity and consistency for data stored in clouds. The security solutions explored and discussed by them are keeping a local copy of the data, use of hash tree, protocols such as Proofs of Retrievability (POR), and Proofs of Data Possessions (PDP), Digital Signatures etc. These solutions still require a testing on some live data to validate their suitability and ease of use. A whitepaper by AWS (Amazon Web Services) discusses physical security, backups, and certifications in their context [10]. Similarly, other providers such as Google, Microsoft etc. have discussed the security issues in cloud computing [19,20].

Heiser and Nicolett [21] have identified seven prominent risks that customers must assess in order to utilize cloud computing infrastructure. In addition to these seven issues, we have also identified several other major issues that must be addressed by the cloud service providers. These issues include data storage, server security, privileged user access, and data portability. We also present virtualization specific security issues in detail.

Manchala et. al. [7] have build a trust model in a distributed computing paradigm. To the best of our knowledge, none of the work so far, give a direction to address the security challenges, specifically in cloud environments. Despite the fact that, there are solutions to address the prominent security issues, a mechanism to measure the security risk from the perspective of a service user is strongly needed. Trust models have been studied in distributed information systems [6,7]. Adopting some of the ideas of trust modeling, our work identifies a key set of trust variables and a resulting trust matrix, based on security issues in cloud computing.

5 Conclusion

In an emerging discipline, like cloud computing, security needs to be analyzed more frequently. With advancement in cloud technologies and increasing number of cloud users, data security dimensions will continuously increase. In this paper, we have analyzed the data security risks and vulnerabilities which are present in current cloud computing environments.

The most obvious finding to emerge from this study is that, there is a need of better trust management. We have built a risk analysis approach based on the prominent security issues. The security analysis and risk analysis approach will help service providers to ensure their customers about the data security. Similarly, the approach can also be used by cloud service users to perform risk analysis before putting their critical data in a security sensitive cloud.

At present, there is a lack of structured analysis approaches that can be used for risk analysis in cloud computing environments. The approach suggested in this paper is a first step towards analyzing data security risks. This approach is easily adaptable for automation of risk analysis.

References

1. Hayes, B.: Cloud Computing. Communications ACM 51, 9–11 (2008)
2. Amazon elastic compute cloud (2008), http://aws.amazon.com/ec2/
3. Twenty Experts Define Cloud Computing (2008), http://cloudcomputing.syscon.com/read/612375_p.htm
4. Llanos, D.R.: Review of Grid Computing Security by Anirban Chakrabarti. Queue 5, 45 (2007)
5. Weiss, A.: Computing in the Clouds. NetWorker 11, 16–25 (2007)
6. Andert, D., Wakefield, R., Weise, J.: Trust Modeling for Security Architecture Development (2002), http://www.sun.com/blueprints
7. Manchala, D.W.: E-Commerce Trust Matrix and Models (2000)
8. John, H.: Security Guidance for Critical Areas of Focus in Cloud Computing (2009), http://www.cloudsecurityalliance.org/guidance/ (Accessed 2 July 2009)
9. Gens, F.: IT Cloud Services User Survey, part 2: Top Benefits and Challenges (2008)

10. Overview of Security Processes (2008)
11. Rose, R.: Survey of System Virtualization Techniques (2004), http://www.robertwrose.com/vita/rose-virtualization.pdf
12. Xen Multiple Vulnerabilities (2007), http://secunia.com/advisories/26986/
13. Microsoft Security Bulletin MS07-049 (2007), http://www.microsoft.com/technet/security/bulletin/ms07-049.mspx
14. Diego, G.: Can we trust Trust? Oxford:Trust Making and Breaking Cooperative Relations (1990)
15. Two Factor Authentication, http://en.wikipedia.org/wiki/
16. Public Key, http://en.wikipedia.org/wiki/Public_key_certificate
17. Buyya, R., Yeo, C.S., Venugopal, S., Broberg, J., Brandic, I.: Cloud Computing and Emerging IT Platforms: Vision, Hype, and Reality for delivering Computing as the 5th Utility. Future Generation Computer Systems 25, 599–616 (2009)
18. Cachin, C., Keider, I., Shraer, A.: Trusting The Cloud. IBM Research, Zurich Research laboratory (2009)
19. Google App Engine (2008), http://appengine.google.com
20. Microsoft Live Mesh (2008), http://www.mesh.com
21. Brodkin, J.: Seven Cloud Computing Security Risks (2008), http://www.gartner.com/DisplayDocument?id=685308

Time-Sequential Emotion Diagnosis Based on the Eye Movement of the User in Support of Web-Based e-Learning

Kiyoshi Nosu, Takeshi Koike, and Ayuko Shigeta

School of High Technology for Human Welfare, Tokai University,
Nishino 317, Numazu-shi, Shizuoka 410-0395, Japan
nosu@wing.ncc.u-tokai.ac.jp,
8afcm002@mail.u-tokai.ac.jp, 9atad006@mail.u-tokai.ac.jp

Abstract. Learning processes are becoming increasingly mediated by computers and the internet. E-learning, especially Web-based learning, is penetrating into various fields. An ordinary e-Learning system usually evaluates learners' states from their inputs to the pre-programmed questions. The time-sequentially changing emotions of learners are important for the management of the learning environment, just like face to-face classes. The present paper describes an emotion diagnosis system that can judge the emotions of an e-Learning user based on his/her eye movement speed and fixation duration time. The criteria for classifying four pairs of semantically different emotions (eight emotions in total) were established through a time-sequential subjective evaluation of the emotions of the subject and a time-sequential analysis of the eye movements of the subject. The achieved coincidence ratios between the discriminated emotions based on the criteria of emotion diagnosis and the time-sequential subjective evaluation for "tired" and "concentrating" were 77% and 60%, respectively.

Keywords: emotion diagnosis, e-Learning, eyeball movement.

1 Introduction

This instruction file for Word users (there is a separate instruction file for LaTeX users) may be used as a template. Kindly send the final and checked Word and PDF files of your paper to the Contact Volume Editor. This is usually one of the organizers of the conference. You should make sure that the Word and the PDF files are identical and correct and that only one version of your paper is sent. It is not possible to update files at a later stage. Please note that we do not need the printed paper.

We would like to draw your attention to the fact that it is not possible to modify a paper in any way, once it has been published. This applies to both the printed book and the online version of the publication. Every detail, including the order of the names of the authors, should be checked before the paper is sent to the Volume Editors.

Learning processes are becoming increasingly mediated by computers and the internet [1]. E-learning, especially Web-based learning, is penetrating into various

S.K. Prasad et al. (Eds.): ICISTM 2010, CCIS 54, pp. 266–274, 2010.

fields enabling learners to develop essential skills for knowledge-based workers by the use of information and communications technologies within the curriculum stored in the Web servers. For instance, in business fields, a company network delivers training courses to employees, while, at Universities, e-learning is used for a on-line courses or program where the students rarely or never meet face-to-face, nor access on-campus educational facilities, because they study online.

An ordinary face-to face university class, a teacher provides a lecture by monitoring students responses for the teacher's questions as well as their emotions from their voices, facial expressions and body actions.

On the other hand, an ordinary e-Learning system usually evaluates learners' states from their inputs to the pre-programmed questions. Therefore, the e-Learning system does not care of the learners' emotions, such as whether learners are interested in the subject, whether they are tired and so on. These emotional factors are important for the management of the learning environment, just like face to-face classes. Therefore, the recognizing individual learners emotional states are crucial for the advanced e-Learning system so that the system can provide appropriate learning material suitable for the emotional states of the individual learners.

From this reason, the importance of affective states or emotions in influencing learning processes has been noted in previous studies [2]-[4]. For example, Kurokawa and Nosu developed a system that can characterize an emotion of an e-Learning user by analyzing his/her facial expression and biometric signals [3],[4]. The criteria used to classify the eight emotions were based on a time-sequential subjective evaluation of emotions, as well as a time-sequential analysis of facial expressions and biometric signals. However, monitoring the heart rate, respiratory rate, and other biometric signals requires sensors to be attached to the body of the user.

Eye movement is affected by both the real-world environment and the mind of the user [5]-[9]. Eye movement is monitored by a camera that detects the reflected infrared (IR) wave of an eye tracking system. Therefore, the relationship between eye movement and visual information has been investigated extensively. Moreover, research on the human-computer interface has investigated eye movement in order to clarify display-based information processes and discover the factors affecting interface usability. For example, eye movement data provides detailed and specific information on the cognitive processes of a user visiting a web site. Recently, eye tracking methods have employed for usability testing of web sites by measuring the user's total fixation duration time, total number of fixations, average fixation time, scan path, and distribution of fixation time on the screen.

The present paper describes an investigation into an emotion-diagnosis system for a Web-based e-Learning user that can discriminate the emotions of a user from his/her eye movements. This system was designed based on the following three strategies: (A) estimations were limited to a specific usage, e-Learning and a specific emotion classification for an e-Learning user group; (B) emotion discrimination criteria (template) consisted of eye movement signals; and (C) a simple diagnosis algorithm was used for real-time processing using the emotion discrimination criteria and Mahalanobis distance.

2 System Configuration

Viewing web pages involves various eye movements, including saccades and fixation [7]. Eye tracking is a useful tool for obtaining immediate information regarding the distribution of a learner's attention among the elements of a video/image. Fig. 1 shows the system configuration. A video camera in the eye movement measurement equipment monitors eye movement data of the user. Emotion diagnosis software provides the estimated emotional state based on emotion diagnosis criteria. Following emotion diagnosis, the computer displays an appropriate Web page.

Emotion discrimination criteria were obtained by a time-sequential subjective self-evaluation of the emotions and a time-sequential analysis of the eye movement speed and fixation duration times of an e-Learning user.

The emotion having the minimum Mahalanobis distance between the criteria and the measured data for the facial expression was selected as the estimated emotion. A Web page corresponding to a discriminated emotion is generated to appropriately direct an e-Learning user.

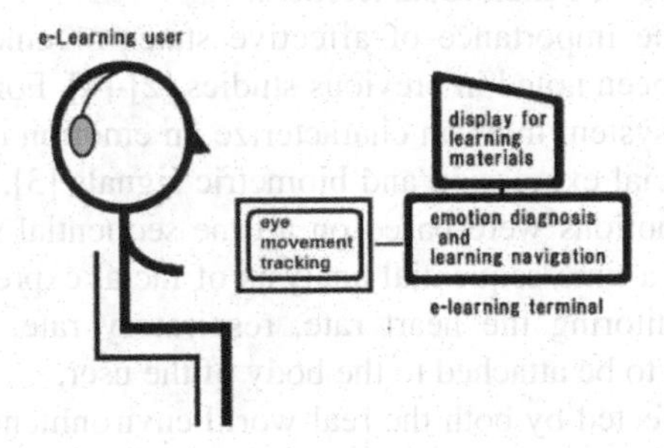

Fig. 1. Configuration of an eye-movement-signal-based real-time emotion diagnosis system for an e-Learning user

A Free View Model 2920B (Tekei Kiki Kogyou Co. Ltd., Japan) was used for the eye tracking measurement. The measurement sampling rate was set to 30 Hz, and the spatial resolution was set to 0.1 deg. Fixation was defined as an eye movement speed of less than 5 deg/s. The eye tracking principle is described as follows:

1) A weak infrared light-emitting diode is located in an eye sensor box, which illuminates the eye of the subject.
2) An infrared camera in the eye sensor box monitors the reflected infrared waves.
3) The fixation direction of the subject was obtained from the relative positions of the pupil and the first Purkinje reflections.

The configuration of the measurement system is shown in Fig. 2. The web pages are accessed from an e-Learning material server and are presented through the PC-based controller of eye movement measurement on a display in front of the subject. Eye movement was detected by the eye sensor, which was located between the display and the subject. The controller converts the eye movement measurements to determine the fixation direction of the subject so that the viewpoint of the subject on the display can be estimated.

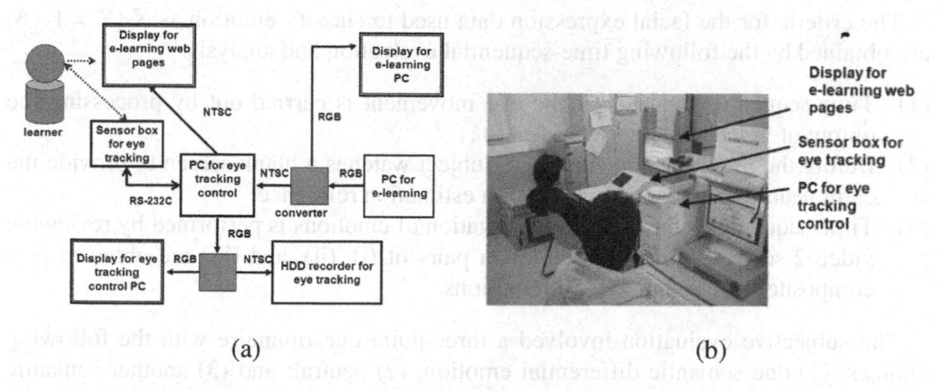

Fig. 2. (a) Schematic diagram of the eye movement measurement equipment, (b) photograph of the measurement equipment

3 Procedure for Obtaining the Criteria for Classifying the Emotions of e-Learning Users

3.1 Emotion Classification

In order to diagnose the temporal change in the emotion of an e-Learning user, the following semantic differential sets of emotions were used for the criteria of facial expressions to classify the emotions of e-Learning users.

i)	easy	difficult
ii)	boring	interesting
iii)	not understand	understand
iv)	tired	concentrating

This categorization differs from the set of general, basic emotions described by Ekman [10].

3.2 Emotion Procedure for Deriving the Criteria for Eye Movement Signals

The internal measurement from the time-sequential subjective evaluation of emotion data is selected so that a specific strong representative emotion of a user is detected. The criteria used to classify the emotion of an e-Learning user as a representative emotion were obtained from a time-sequential analysis of his/her eye movements.

Fig. 3 illustrates the configuration of the measurement system used to determine the criteria of eye movement signals. Fig. 4 summarizes the procedure for deriving the criteria for facial expressions used to classify the emotions of e-Learning users. As the subjects used e-Learning materials, two video cameras monitored the following:

(a) The eye movements of the subject, which is used for the time-sequential analysis of eye tracking trajectory (Video-1, taken by video camera #1).
(b) The e-Learning display screen, which is used for time-sequential subjective evaluation of emotions after the eye movement measurement (Video-2, taken by video camera #2).

The criteria for the facial expression data used to classify emotion as X (X = 1 - 8) are obtained by the following time-sequential evaluation and analysis:

(1) Time-sequential analysis of the eye movement is carried out by processing the output of video camera #1 (Video-1) ,
(2) Before the measurement starts, the subject watches a blank screen to provide the calm neutral state data as an emotion estimation reference,
(3) Time-sequential subjective self-evaluation of emotions is performed by reviewing video-2 so that stronger emotions in pairs of (i), (ii), and (iii) are obtained as a composite representative set of emotions.

The subjective evaluation involved a three-point questionnaire with the following choices: (1) one semantic differential emotion; (2) neutral; and (3) another semantic differential emotion. If a subject selects (2) neutral, the facial expression data for the corresponding period (30 seconds) is neglected for the analysis.

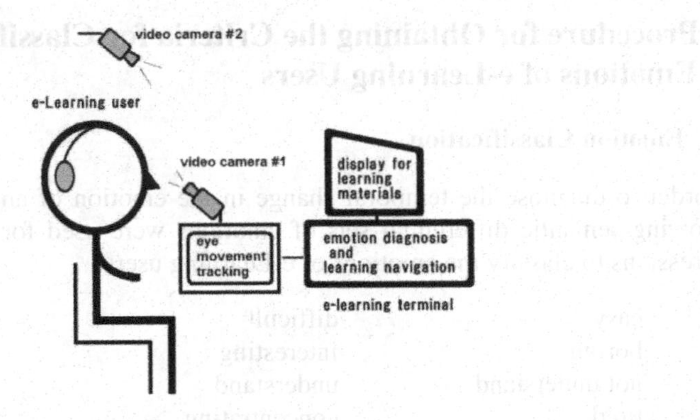

Fig. 3. Configuration of the measurement system used to determine the criteria of eye movement signals

3.3 Measurements to Determine the Diagnostic Criteria

Digital video cameras #1 and #2 and the eye measurement equipment shown in Fig. 3 were used to derive the criteria of facial expressions. The subject operates an e-Learning controller and watches a screen. Video camera #1 monitors the eye movement so that time-sequential analysis of the fixation point movement speed and the fixation duration time can be performed. Video camera #2 records the terminal screen that the subject operates, as shown in Fig. 1.

The measurements to obtain the diagnostic criteria were carried out using introductory communication engineering course material as the e-Learning material. The subjects were 10 male university students from the technology department. Finally, the measurement time was five minutes per subject.

Following the procedure shown in Fig. 4, the eye movement data was obtained for the eight emotions. Fig. 5 shows an example of time-sequential changes of a web page inspected by a learner and the self-evaluated emotions. The samples with self-evaluated emotions of (2) neutral were omitted from the 100 samples (10 samples/subjects x 10 subjects). The number of samples having stronger emotions that remain are as follows:

i) difficult-easy pair 78
ii) boring-interesting pair 82
iii) confused-comprehending pair 51
iv) tired-concentrating pair 65

From these data sets, the criteria for the emotion diagnosis were obtained as shown in Table 1.

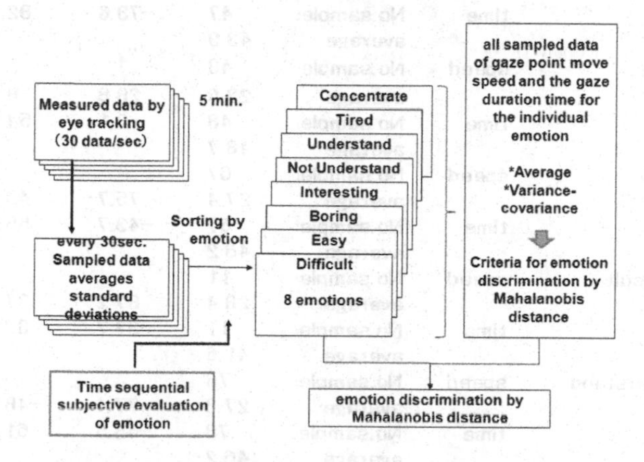

Fig. 4. Procedure for deriving the criteria for eye movement used to classify the emotions of e-Learning users

	2sec/cell																													
time (sec.)	0sec					10sec										30sec														60sec
wabe page Page1																														
Page2																														
Page2-1																														
Page3																														
Page4																														
Page4-1																														
Page4-2																														
Page5																														
Page6																														
emotion Interesting	4	4	4	4	4	4	4	4	4	4	4	4	4	4	4	4	4	4	4	4	4	4	4	4	4	4	4	4	4	4
Boring																														
Concentrate	4	4	4	4	4	4	4	4	4	4	4	4	4	4	4	5	5	5	5	5	5	5	5	5	5	5	5	5	5	5
Tired																														
Easy	5	5	5	5	5	5	5	5	5	5	5	5	5	5	5															
Difficult																2	2	2	2	2	2	2	2	2	2	2	2	2	2	2
Understand	5	5	5	5	5	5	5	5	5	5	5	5	5	5	5	4	4	4	4	4	4	4	4	4	4	4	4	4	4	4
Not Understand																														

Fig. 5. Example of time-sequential changes of a web page inspected by a learner and self-evaluated emotions

Table 1. Criteria for Emotion Diagnosis

emotion	items		sample & avarage	variance-covariance matrix	
interesting	speed	No.sample:	33		
		average	32.2	83.7	-64.8
	time	No.sample:	33	-64.8	66.0
		average	42.9		
boring	speed	No.sample:	57		
		average	23.7	37.7	-26.3
	time	No.sample:	57	-26.3	72.6
		average	48.5		
concentrating	speed	No.sample:	47		
		average	31.2	97.9	-73.6
	time	No.sample:	47	-73.6	82.7
		average	43.9		
tired	speed	No.sample:	48		
		average	23.5	28.8	-8.1
	time	No.sample:	48	-8.1	54.3
		average	48.7		
easy	speed	No.sample:	67		
		average	27.4	75.7	-43.7
	time	No.sample:	67	-43.7	55.6
		average	46.2		
difficult	speed	No.sample:	11		
		average	28.4	87.2	-37.7
	time	No.sample:	11	-37.7	32.2
		average	41.6		
understand	speed	No.sample:	78		
		average	27.3	77.4	-48.5
	time	No.sample:	78	-48.5	61.8
		average	46.2		
not understand	speed	No.sample:	9		
		average	26.0	81.5	-30.1
	time	No.sample:	9	-30.1	27.6
		average	41.6		

4 Algorithm for Emotion Diagnosis and Its Performance

Fig. 6 shows the procedure used to diagnose the emotions of the e-Learning users based upon the criteria given in Table 1. The emotion that had the shorter Mahalanobis distance between the measured facial data and the criteria for emotions was selected as the estimated emotion. The Mahalanobis distance, D, is obtained as follows:

$$D^2 = \left(x-\bar{x}, y-\bar{y}\right)\begin{pmatrix} Sx^2 & Sxy \\ Sxy & Sy^2 \end{pmatrix}^{-1}\begin{pmatrix} x-\bar{x} \\ y-\bar{y} \end{pmatrix}$$

where and are the average values of the variables x and y, respectively, S_x^2 and S_y^2 are variances of x and y, respectively, and S_{xy} is the covariance of x and y.

The effectiveness of the emotion diagnosis system using the procedure shown in Fig. 6 was evaluated based on the coincidence ratio between the emotions discriminated using the criteria of emotion diagnosis and the results of time-sequential subjective evaluation of the emotions of the subjects using the leave-one-out method. This ratio is defined as the ratio of the computer-diagnosed emotions coincident with the subjectively evaluated emotions. Table 2 shows the coincidence ratios.

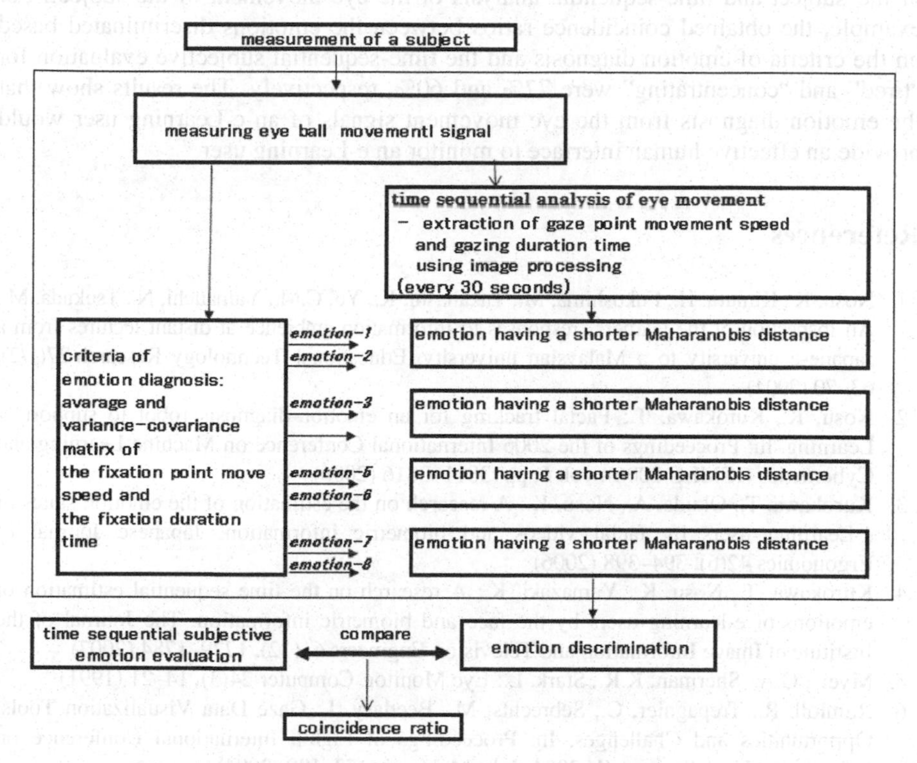

Fig. 6. Procedure used to diagnose the emotions of e-Learning users

Table 2. Emotion diagnosis coincidence ratio

	A	B	C
difficult	11	5	45%
easy	67	47	63%
boring	49	57	86%
interesting	33	18	55%
confused	4	9	44%
comprehending	47	78	60%
tired	37	48	77%
concentrating	38	47	60%

A: number of samples obtained by subjective evaluation
B: number of samples in A coincident with the computer discrimination
C: coincidence ratio

5 Conclusion

The present paper describes an emotion diagnosis system that can judge the emotions of an e-Learning user based on eye movement speed and fixation duration time. The criteria for classifying four pairs of semantically different emotions (eight emotions in total) were established using a time-sequential subjective evaluation of the emotions of the subject and time-sequential analysis of the eye movement of the subject. For example, the obtained coincidence ratios between the emotions discriminated based on the criteria of emotion diagnosis and the time-sequential subjective evaluation for "tired" and "concentrating" were 77% and 60%, respectively. The results show that the emotion diagnosis from the eye movement signals of an e-Learning user would provide an effective human interface to monitor an e-Learning user.

References

1. Nosu, K., Kimura, H., Fukushima, M., Takahashi, K., Yu, C.M., Yamauchi, N., Tsukada, M.: An inspection of the learners' responses to information ambience at distant lectures from a Japanese university to a Malaysian university. Educational Technology Research 27(1/2), 63–70 (2004)
2. Nosu, K., Kurokawa, T.: Facial tracking for an emotion-diagnosis robot to support e-Learning. In: Proceedings of the 2006 International Conference on Machine Learning and Cybernetics (ICMLC 2006), vol. 6, pp. 3811–3816 (2006)
3. Kurokawa, T., Okuda, A., Nosu, K.: A research on the estimation of the emotion states of e-Learning users by facial videos and biometric information. Japanese Journal of Ergonomics 42(6), 394–398 (2006)
4. Kurokawa, T., Nosu, K., Yamazaki, K.: A research on the time sequential estimation of emotions of e-learning users by the face and biometric information. The Journal of the Institute of Image Information and Television Engineers 61(12), 1779–1784 (2007)
5. Myers, G.A., Sherman, K.R., Stark, L.: Eye Monitor. Computer 24(3), 14–21 (1991)
6. Ramloll, R., Trepagnier, C., Sebrechts, M., Beedasy, J.: Gaze Data Visualization Tools: Opportunities and Challenges. In: Proceedings of Eighth International Conference on Information Visualization, IV 2004, July 14-16, pp. 173–180 (2004)
7. Nosu, K., Kanda, A., Koike, T.: Voice navigation in web-based learning materials - an investigation using eye tracking. IEICE Trans. Information and System 90-D(11), 1772–1778 (2007)
8. Cowing, R., et al.: Emotion recognition in human-computer interaction. IEEE Signal Processing Magazine, 33–80 (January 2001)
9. Kim, Y., Bae, J., Joen, B.: Students' Visual Perceptions of Virtual Lectures as Measured by Eye Tracking. In: Jacko, J.A. (ed.) HCII 2009, Part 1. LNCS, vol. 5610, pp. 85–94. Springer, Heidelberg (2009)
10. Ekman, P.: Emotion in the Human Face. Cambridge University Press, Cambridge (1982)

Critical Success Factors in Managing Virtual Teams

Ganesh Vaidyanathan[1] and Carmen Debrot[2]

[1] Indiana University South Bend, South Bend, IN 46634, USA
gvaidyan@iusb.edu
[2] Project Executive, IBM Corporation, Venezuela
cdebrot@ve.ibm.com

Abstract. The corporate world is changing from the boardroom meetings and group gatherings to a virtual team world. Mobile employees who work from home or a remote office are becoming more common. More and more companies are adopting virtual teams because they bring many advantages to organizations and their employees. However, virtual teams carry additional complexity and increase risk of unsuccessful projects in terms of reaching their goals. This paper defines virtual teams, discusses their advantages and challenges, and identifies several success factors to manage virtual teams. These factors are aggregated using a four-dimensional framework. This paper discusses factors from the structural, cognitive, relational, and technological dimensions and the relationships between them. Finally, the relationships between some factors of the previous dimensions and the performance of the virtual team are analyzed.

Keywords: Virtual teams, success factors, structural, cognitive, relational, technical, framework.

1 Introduction

Loughran [1] states that virtual teams are defined as groups of distributed people working together to achieve a common goal. They solve a shared problem through the use of computer-mediated communication technologies, linking them across time, space, and cultural barriers. There are different elements that can make a team virtual, such as geographical separation of team members, skewed working hours, temporary or matrix reporting structures, and multi-corporation or multi-organizational teams [2]. Virtual team can have different shapes and styles using various combinations of whether they belong to same or different organizations and whether they belong to same or cultures and nationalities.

The implementation of virtual teams is a growing trend that has become an integral part of our society. The rapid and substantial growth of information and communication technologies has allowed the fast development of virtual teams. In addition to technology, other factors that have contributed to the prevalence of virtual teams are mergers, acquisitions, downsizing, outsourcing, technology, and technical specialization [2]. At present, more and more employees are opting for tele-work alternatives. Teleworkers currently make up the fastest growing segment of the workplace [3]. Haywood [2] has identified different advantages of virtual teams. She presented the

S.K. Prasad et al. (Eds.): ICISTM 2010, CCIS 54, pp. 275–281, 2010.

advantages from both the managers' perspective and team members' perspective. The managers' perspectives are generally based on the factors that include access to a less expensive labor pool, reduced office space, greater utilization of employees, round-the-clock work force, greater access to technical experts, and larger pool of possible job experience. However, the perspectives of team members includes factors such as increased independence, less micromanagement, larger pool of jobs to choose from, greater flexibility, and opportunity for travel.

While some of these factors may seem opposing in thought processes, it is necessary that both managers and team members understand how virtual teams benefit all participants, in order that the implementation of virtual teams be successful. Telecommuters show an increase of productivity between 15% and 80% with an average increase of 25% to 35%. This advantage is not limited to work-at-home workers; also employees working on projects headquartered at geographically remote locations have improved their productivity because they have a more structured and less-interrupt-driven communication style. Virtual teams can be implemented to reduce susceptibility to environmental disaster, such as what the Oracle Corporation's customer support division has done. A study of telecommuting in Silicon Valley showed a 24% improvement in employee satisfaction. IBM implemented a hoteling for its marketing and services personnel that reduced by 60% the real estate costs per location. Commuters produce millions of gallons of exhaust fumes per year. Studies in the United Kingdom showed that it is possible to eliminate 1.2 tons of carbon dioxide by taking one commuter off the road. Xerox sent their sales and support personnel to work at customer sites in one location, due to the necessity of employees to spend more time at a client's site because of an increase in product complexity.

However, there are different perspectives of the challenges of implementing virtual teams in these two groups. According to Haywood [2], 70% of the time the concerns of the managers have to do with control, including monitoring performance, and training and mentoring of new employees. The second area of concern is communication. In addition, they are also worried about team building, cultural issues, cost and complexity of technology, and finally process and workflow. The areas of most concerns in team members are communication and support. Team members are worried about being excluded from key meetings and decision. Technical support is an overwhelming concern for non-technical team members. In addition, they are worried about recognition, inclusion vs. isolation and management resistance. Also, they are concerned that all promotions go to team members working at the same site as the manager.

In this paper, we plan to propose a framework that may be used to measure success in managing virtual teams. Such a framework is not available in the current literature. This framework may be used by managers to analyze the need for virtual teams as well as measure success through the employment of such teams. The next section illustrates the formation of such a framework along with discussions of impacts of the framework factors on each other.

2 A Framework to Analyze Success Factor to Manage Virtual Teams

Different authors have identified several success factors to manage virtual teams and such factors can be aggregated in different dimensions. Vaidyanathan [4] developed a four-dimensional framework of distributed project knowledge management. These dimensions are identified in figure 1. The structural dimension refers to the pattern of relationships between the project team members [5, 6]. This dimension consists of the following factors that include ties, configuration, or the pattern of linkages among team members, stability, informality, and management of project teams.

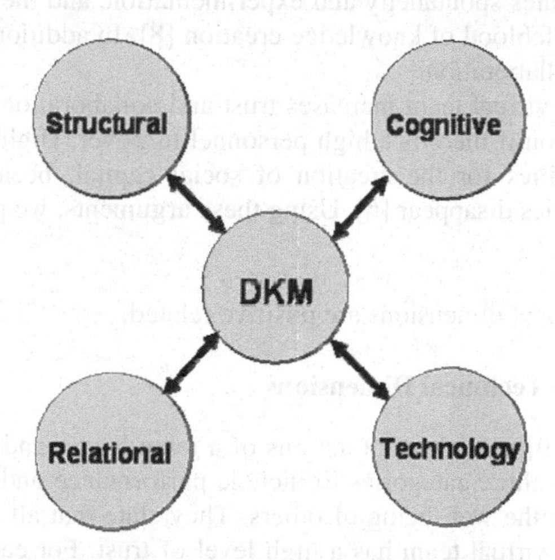

Fig. 1. Four dimensions of Distributed Project Knowledge Management

The cognitive dimension involves resources that provide shared meaning and understanding between project team members. This dimension contains factors that include shared goals, shared culture, and learning. Shared goals include collective goals, aspiration, and a shared vision for project team members. If project team members are culturally diverse, the different cultures need to be understood and accommodated. The project team should have an integrated learning culture that includes education, training, and mentoring.

The relational dimension is centered on the role of the direct ties between team members. This dimension factors focus on trust and collaboration among distributed project teams. Trust in a firm focuses on the role of direct ties between actors and the relational, as opposed to structural, outcomes of interactions [6]. Trust is a key element in the willingness of team members to share knowledge. Collaboration is defined as the degree to which team members actively help one another in their work [7]. Trust plays an important role in successful collaboration.

The technical dimension refers to IT tools, support, and maintenance. IT tools include system designs, technology acceptance and adoption, compatibility, ontology, and security. IT facilitates collection, storage, and exchange distributed project knowledge, and foster all modes of knowledge creation [8].

2.1 Relational and Structural Dimensions

Strong ties promote trust and collaboration. As trust develops over time, opportunities for knowledge transfer between members should increase [6]. Configuration of a team should encourage collaboration as configuration influences the flexibility and ease of communication and knowledge exchange between team members. A decentralization of authority to members such as the development of lateral ties improves communication and collaboration [6]. Strong ties encourage trust. Moreover, the concentration of decision-making authority inevitably reduces creative solutions, whereas the dispersion of power facilitates spontaneity and experimentation, and the freedom of expression which are the lifeblood of knowledge creation [8]. In addition, leadership should promote trust and collaboration.

The stability of a virtual team increases trust and collaboration. There is more difficulty in collaboration if there is a high personnel turnover. Highly unstable network may limit opportunities for the creation of social capital, because when an actor leaves the network, ties disappear [6]. Using these arguments, we propose

Preposition 1:
Structural and relational dimensions are positive related.

2.2 Relational and Technical Dimensions

Duarte and Snyder [9] consider that actions of a team leader and team members that impact trust fall into three categories to include performance and competence, integrity, and concern for the well-being of others. They state that all three factors should exist in order that a virtual team has a high level of trust. For each category, Duarte and Snyder [9] specify many factors that are included in Table 1.

Category	Factors
Performance and competence	Developing and displaying competence Following through on commitments and showing results
Integrity	Ensuring that your actions are consistent with your words Standing up for your convictions; displaying integrity Standing behind the team and its members Communicating and keeping everyone informed about progress Showing both sides of an issue
Concern for the well-being of others	Helping team members with transitions Being aware of your impact on others Integrating team needs with other teams, department, and organizational needs

In order to select the right technology for a global virtual team, it is necessary to consider the nature of the task and the context where it occurs. Context has an impact on the reliability of the communication technology on which members depend [10]. In addition, local habits are developed in a determined context and can have big variations across team members. Gibson and Cohen [10] identify aspects to consider choosing the right technology that include infrastructure, differences in local power, telephone or cable infrastructure, culture and language, accessibility of information, time zone gap, team size, technology maturity, and task complexity.

Technology can send subtle messages about which members of the team are considered high performer or not. The communication between high performers tends to be through one-on-one electronic messages. Thus, being left out of one-on-one communication patterns could indicate that a team member is perceived as less competent than others [9]. Regarding the second factor, the integrity of team processes and decision making can be facilitated through technology. For example, groupware with anonymity features allows sharing opinions and ideas, especially when there is a disagreement or when only a few members of the team do not share an idea. Using technology, members can express their opinion without fear of recrimination. In addition, electronic distribution lists make it easy to get the same information to everyone in a timely manner [9]. In relation to the third factor, the fact that virtual teams operate in an isolated environment can generate less need for social posturing than in traditional teams. This situation can produce a tendency to display less concern. One study showed that computer-mediated groups communicate more negative messages than face-to-face groups do [9]. To avoid this situation, members of virtual teams should avoid waiting for face-to-face meetings or at least tele-conference to express criticism in order to avoid misunderstanding.

In summary, we propose that

Proposition 2a:
Use of technology can have a positive impact on performance and competence as well as integrity factors of trust

Proposition 2b:
Use of technology has a negative impact on concern for the well-being of others.

3 Conclusion and Future Research

Different factors of the dimensions explained above have a diverse impact on the performance of the virtual team. In particular, the relationship between trust and virtual team performance has been analyzed by various authors. Also, the relationship between conflict management and virtual team performance has been studied extensively. Regarding the relationship between trust and virtual team performance, different studies have shown contradictory results. A number of studies show the positive effects that trust has on performance [11, 12]. However, more recent studies have failed to find a positive relationship between trust and performance [13, 14, and 15].

Aubert and Kelsey [15] use the distinction between efficiency and effectiveness to explain the lack of the relationship between trust and virtual team performance. They use the notion of process loss or gain as an explanation for this lack of relationships between trust and performance. Based on this study, trust does not necessarily augment the quality of the task performance. Two additional studies are consistent with the study of Aubert and Kelsey [15]. Dirks and Ferrin [14] show that trust does not necessarily have a strong relationship with actual performance. Jarvenpaa, Shaw and Staples [13] found no relationship between trust and task performance. Moreover, these authors suggest that trust can have unpleasant consequences including that high levels of trust may not always be justified because of the risk that others will take advantage of the situation [14].

Montoya-Weiss, Massey and Song [16] explore the relationship between conflict management behavior and virtual team performance. They distinguish five conflict handling modes to conflict management in an organization: avoidance, accommodation, competition, collaboration, and compromise. They found that avoidance conflict management behavior has a significantly negative effect on performance. However, contrary to the authors' expectation, accommodation had no significant effect on performance. This situation is different from traditional non-virtual teams where past research has found a negative relationship between accommodation and performance. The explanation of the authors is that because of the use of asynchronous communication environment, it is possible that, no matter how much an individual may express accommodation, the team does not experience it [16]. In addition, these authors found positive relationships between collaboration conflict management of behavior and performance. This finding is consistent with the study of Paul et al. [17] who focus their study on the effectiveness rather than efficiency of virtual teams. Also, Paul et al. [17] found that the effect of collaboration is more important for heterogeneous than for homogeneous groups. Finally, Montoya-Weiss, Massey and Song [16] have established a negative relationship between compromised conflict management and performance.

Virtual teams have being adopted by many corporations. They have many advantages but also additional challenges. These challenges can be aggregated in structural, cognitive, relational, and technological dimensions. We are investigating the relationships between relational and structural dimensions. We believe that strong ties promote trust and collaboration, the configuration of the team affects collaboration, and the stability of the team increases trust and collaboration. We are also investigating the relationship between relational and technical dimensions. We believe that technology factors have an impact in building trust in virtual teams. In addition, technology has a role in development virtual teams. Finally, the performance of virtual teams is impacted by different factors of these dimensions. In particular trust and conflict management has an impact in the performance of virtual teams.

The implications of the findings from this study can be classified into three categories— contribution to literature, insights about various dimensions affecting virtual teams, and insights about the impacts of these factors on virtual team performance. From a practical standpoint, we present a framework to examine virtual team. Future studies can use this framework to examine the performance of virtual teams and measure various factors on how they impact the performance. This framework offers a promising avenue for future research.

References

1. Loughran, J.: Working Together Virtually: The Care and Feeding of Global Virtual Teams (2005), `http://www.thoughtlink.com/files/html-docs/TLI-ICCRTS00.htm` (Retrieved February 15, 2005)

2. Haywood, M.: Managing Virtual Teams: Practical Techniques for High-Tecnology Project Managers. Artech House, Boston (1998)
3. Johnson, P., Heimann, V., O'Neil, K.: The wonderland of virtual teams. Journal of Workplace Learning 13(1), 24–30 (2001)
4. Vaidyanathan, G.: Networked Knowledge Management Dimensions in Distributed Projects. International Journal of e-Collaboration 2(4), 19–36 (2005)
5. Nahapiet, J., Ghoshal, S.: Social capital, intellectual capital, and the organizational advantage. Academy of Management Review 23(2), 242–266 (2001)
6. Inkpen, A.C., Tsang, E.W.K.: Social capital, networks, and knowledge transfer. Academy of Management Review 30(1), 146–166 (2005)
7. Hurley, R., Hult, T.: Innovation, market orientation, and organization learning: An integration and empirical examination. Journal of Marketing 62(3), 42–54 (1998)
8. Lee, H., Choi, B.: Knowledge management enablers, processes, and organizational performance: An integrative view and empirical examination. Journal of Management Information Systems 20(1), 179–228 (2003)
9. Duarte, D., Snyder, N.: Mastering Virtual Teams. Jossey-Bass Publishers, San Francisco (1999)
10. Gibson, C., Cohen, S. (eds.): Virtual Teams that Work. Jossey-Bass Publishers, San Francisco (2003)
11. Jarvenpaa, S., Knoll, K., Leidner, D.: Is anybody out there? The antecedents of trust in global virtual teams. Journal Management Information Systems 14(4), 29–64 (1998)
12. Jarvenpaa, S., Leidner, D.: Communication and Trust in Global Virtual Teams. Organizational Science 10(6), 791–815 (1999)
13. Jarvenpaa, S., Shaw, T., Staples, D.: Toward Contextualized Theories of Trust: The Role of Trust in Global Virtual Teams. Information System Research 15(3), 250–267 (2004)
14. Dirks, K., Ferrin, D.: The role of trust in organizational settings. Organization Science 12(4), 450–467 (2001)
15. Aubert, B., Kelsey, B.: Further understanding of trust and performance in Virtual Teams. Small Group Research 34(5), 575–618 (2003)
16. Montoya-Weiss, M., Massey, A., Song, M.: Getting it Together: Temporal Coordination and Conflict Management in Global Virtual Teams. Academy of Management Journal 44(6), 1251–1262 (2001)
17. Paul, S., Samarah, I., Seetharaman, P., Mykytyn, P.: An Empirical Investigation of Collaborative Conflict Management Style in Group Support System-Based Global Virtual Teams. Journal of Management Information Systems 21(3), 185–222 (2005)

3D Reconstruction and Camera Pose from Video Sequence Using Multi-dimensional Descent

Varin Chouvatut[1], Suthep Madarasmi[1], and Mihran Tuceryan[2]

[1] Department of Computer Engineering
King Mongkut's University of Technology Thonburi
126 Pracha-uthit Rd., Bang-mot, Thung-khru
Bangkok 10140, Thailand
varin@cpe.kmutt.ac.th, suthep@kmutt.ac.th

[2] Department of Computer and Information Science
Indiana University-Purdue University Indianapolis
723 W. Michigan St., Indianapolis
IN 46202-5132, USA
tuceryan@cs.iupui.edu

Abstract. This paper aims to propose a novel and simple method for estimating 3D-point reconstruction and camera motion. Given a video sequence of a target object with a few feature-points tracked, the points' 3D-coordinates can be reconstructed along with the estimation of the camera's position and orientation in each frame. The proposed method is based on combining Powell's method using parabolic graph with the well-known Gradient Descent to guess the direction to estimate the unknown variables. The unknowns include six components for camera pose in each frame, one focal length, and three values for each point. Using this proposed method, the problem of missing points due to self-occlusion can be eliminated without using any other special strategies. A synthetic experiment shows accuracy of computing 3D-points and the camera pose in each frame. A real-world experiment from only one off-the-shelf digital camera is also shown to demonstrate the robustness of our approach.

Keywords: Camera Pose, 3D Reconstruction, Structure from Motion, Active-Appearance Model, Feature-Point Tracking, Gradient Descent, Powell's Multi-dimensional Minimization.

1 Introduction

There are many known techniques for estimating camera pose and 3D-point structure from multiple video image frames. Some methods add some specific textured objects known as markers whose 3D positions are known to the scene while capturing images [10][11][12][13][14]. Throwing some referenced markers, such as a checkered paper or some fiducial or color markers into the scene, may not be suitable or even possible in some cases. Even in cases where markers can be added, the manual measurement of 3D relative position of these markers to get precise distance is still difficult.

S.K. Prasad et al. (Eds.): ICISTM 2010, CCIS 54, pp. 282–292, 2010.

The auto- or self- calibration technique is widely used as a robust method for estimating structure from motion without using markers. Unfortunately, there are so many steps and details to work with and an optimization approach is needed after every level of reconstruction. Also, the higher level of reconstruction may suffer from the noisy result obtained from the lower level of the reconstruction. In auto-calibration, the number of available input images is used as the main criterion in selecting the appropriate method to estimate camera motion and 3D point reconstruction. For example, the fundamental matrix [15], trifocal tensor [16], quadrifocal tensor [1], and factorization [17] are used when given two, three, four, and multiple views, respectively. Furthermore, the auto-calibration needs to solve the correspondence problem for the missing point's first, which is even harder in the case of self-occlusion. Thus, this paper proposes a simpler and more practical, but yet robust method where a point's self-occlusion can also be ignored.

Gradient descent is known as a robust minimization method and has been used for various applications. In 1999, Karayiannis [3] used gradient descent for training radial basis function (RBF) neural networks. In 2000, Chen [4] used 1D gradient descent for predicting point motion in images. In 2001, Guerrero and Sagues [5] estimated camera's 3D motion from constraints on brightness gradient on lines. In 2002, Smolic [7] presented an algorithm for generating 360-degree panoramic views from video sequences moved with the planar perspective motion model. In 2009, Po *et al* [6] proposed a fast method for searching point motion using gradient descent. However, one main drawback of the original gradient descent method is the calculation of the derivative from system equations or the conjugate direction as used in [4] so that it will later be used to define the step-size for parameter updating. This calculation can add more complexity to the system due to the equations. It may also be prone to errors in the formula.

Powell's method in multi-dimensions [2][18][19] is another approach for finding the minimization or maximization of system functions [2]. Its concept can be illustratively explained as climbing down the valley shaped like a 3D paraboloid. Although the many unknown variables define a multi-dimensional direction, a line minimization along one dimension at a time is considered. This helps simplify the problem from high dimensionality. This key concept is adapted to our estimation of many parameters. But, for each dimension of climbing down the parabola, the lowest point or the minimization of the graph may not be the correct result in terms of a global solution; thus, we chose not to climb deep down the valley until line minimization is obtained. Instead, we climbed down the valley by one small step in the correct direction at a time.

Our proposed method is adapted from combining line minimization used in Powell's method for multi-dimensional search and simple Gradient Descent optimization. The proposed method is based on back-projecting a feature point from its 2D points seen in several images extracted from a video sequence to the same 3D point in the real-world environment [1]. Given M frames and N feature-points, the constraints from the video sequence with matched feature points are sufficient to solve the problem given 2-D correspondences for N feature points in M frames, considering there are more equations than unknowns. For estimating camera motion, this method uses a well-known feature of video recordings; namely, each pair of the consecutive frames has a small difference in motion. To estimate the 3D reconstruction of feature points, world coordinate in the Z-axis of feature points cannot be positive so that all feature points are only allowed to be in front of camera's lens. Finally, the focal length is well-known to be non-zero and positive. For point correspondence, the popular Active-Appearance Model (AAM) introduced in [8][9] is used to track the natural feature-points of target objects without adding any fiducial markers to the scene. Although some points are missing due to self-occlusion, this does not pose a problem in our proposed method, as long as the occluded points are visible in a sufficient number of frames since there are enough equations to serve as constraints. Given sufficient parallax, each point's 3D structure can still be obtained by this proposed method.

The obtained 3D reconstruction of the target object and camera pose can then be used for applications such as Augmented Reality (AR), face reconstruction, architecture, medical treatment, etc. The real-world (Euclidean) size of the target object can be recovered by rescaling the reconstructed object by using a known actual distance between any two contiguous points.

2 Methodology

Let the camera's extrinsic parameters including rotation and translation parts of each video frame be in the format of 4×4 model-view matrix as:

$$M = \begin{bmatrix} R_{3\times3} & \boldsymbol{T}_{3\times1} \\ 0_{1\times3} & 1 \end{bmatrix} = \begin{bmatrix} m_{11} & m_{12} & m_{13} & m_{14} \\ m_{21} & m_{22} & m_{23} & m_{24} \\ m_{31} & m_{32} & m_{33} & m_{34} \\ m_{41} & m_{42} & m_{43} & m_{44} \end{bmatrix}. \quad (1)$$

And let three components of each 3D feature point (X, Y, Z) be a homogeneous coordinate represented by a 4×1 vector as $\boldsymbol{W} = [X_w \quad Y_w \quad Z_w \quad 1]^T$, 2D points are projected into video images from their 3D points by projection equations. 3D-to-2D point transformations can be shown in the form of matrices as shown in the following equations. First, the world-coordinates of feature points are transformed into camera-coordinates. Then, the camera-coordinates will be transformed into image-coordinates using the focal length.

Transforming world coordinates, $\boldsymbol{W}$, to camera coordinates, $\boldsymbol{C}$, is formed by matrix multiplication, $M_{4\times4}^{w\to c}$, of the camera's rotation and translation, R and T respectively:

$$\boldsymbol{C}_{4\times1} = M_{4\times4}^{w\to c} \boldsymbol{W}_{4\times1} \quad (2)$$

$$\begin{bmatrix} X_c \\ Y_c \\ Z_c \\ 1 \end{bmatrix} = \left(R_x^{-1} R_y^{-1} R_z^{-1} T^{-1}\right) \begin{bmatrix} X_w \\ Y_w \\ Z_w \\ 1 \end{bmatrix};$$

$$R_x^{-1} = \begin{bmatrix} 1 & 0 & 0 & 0 \\ 0 & \cos\theta_x & \sin\theta_x & 0 \\ 0 & -\sin\theta_x & \cos\theta_x & 0 \\ 0 & 0 & 0 & 1 \end{bmatrix}, \quad R_y^{-1} = \begin{bmatrix} \cos\theta_y & 0 & -\sin\theta_y & 0 \\ 0 & 1 & 0 & 0 \\ \sin\theta_y & 0 & \cos\theta_y & 0 \\ 0 & 0 & 0 & 1 \end{bmatrix},$$

$$R_z^{-1} = \begin{bmatrix} \cos\theta_z & \sin\theta_z & 0 & 0 \\ -\sin\theta_z & \cos\theta_z & 0 & 0 \\ 0 & 0 & 1 & 0 \\ 0 & 0 & 0 & 1 \end{bmatrix}, \quad T^{-1} = \begin{bmatrix} 1 & 0 & 0 & -t_x \\ 0 & 1 & 0 & -t_y \\ 0 & 0 & 1 & -t_z \\ 0 & 0 & 0 & 1 \end{bmatrix}.$$

Projecting the obtained camera coordinates to image coordinates, $\boldsymbol{S}$, the focal length, f, will be used as a scaling factor using the following equation:

$$\boldsymbol{S}_{3\times1} = \begin{bmatrix} u \\ v \\ 1 \end{bmatrix} = \begin{bmatrix} f & 0 & 0 \\ 0 & f & 0 \\ 0 & 0 & 1 \end{bmatrix} \begin{bmatrix} X_c/Z_c \\ Y_c/Z_c \\ 1 \end{bmatrix}. \tag{3}$$

From all the above-mentioned equations, the unknown variables include camera's rotation $(\theta_x, \theta_y, \theta_z)$ and translation (t_x, t_y, t_z) for each frame, a single focal length (f), and the 3D reconstructed points (X_w, Y_w, Z_w) of target objects. Consequently, given N 2D-points with three variables (X_w, Y_w, Z_w) each and M video-frames with six variables $(\theta_x, \theta_y, \theta_z, t_x, t_y, t_z)$ each, there will be $6M + 3N + 1$ variables (including the focal length f) in total to be estimated.

The only input of this proposed method is a video sequence recording images of the target object. Having image points tracked, two components (u, v) of each point's image coordinate, $\boldsymbol{S}$, are acquired. Hence, two equations per point are added as the input used to solve for all variables. Thus, given N image-points which are composed of two components (u, v) each and visible in M frames, there will be a system of $2MN$ available equations. As mentioned before, there must be more equations than unknowns so the main constraint is satisfied.

The main methodology of this proposed method is based on gradient descent optimization so global and local energies used as conditions for loop-ending are calculated from average pixel-distances between the observed and re-projected image-points. The global energy is calculated from all video frames and points but the local energy is calculated from only frames and points relevant to the currently considered variable. The following is a brief description of the main algorithm proposed:

1. Given 2D feature-points of the target object, not necessarily visible in all frames, the (u, v) coordinates are defined. Here in this paper, 2D points are tracked with sub-pixel precision by the AAM approach to obtain point correspondences.
2. Initialize the components of the camera pose and 3D points to zero. Note that Z_w of the 3D points must be negative so one may initialize it as -1.0. The focal length can be initialized to a sensibly positive value. One may choose a value between 500.0 and 1000.0. The value 600 is chosen in this paper. All step-sizes used for updating variables (every variable has its own step-size) are initialized with a positive small value; here we use 2.0.
3. Repeat for a specific number of iterations. One may check the condition for loop-ending from the convergence of the global energy.
4. Repeat for the number of variables.
5. Repeat until local energy of the currently considered variable is decreased or the step size is too small, say until smaller than 1.0e-8, or until the maximum number of iterations defined is reached.

5.1. Try adding step-size to the current variable and then calculate local energy of the current variable. If the calculated energy is not decreased, recover the variable's old value and reduce its step-size by half (or by a ratio smaller than 1.0). Note that, for each step-size, we try to update using both a positive and a negative step.

From the described algorithm, swapping the signs of each step-size is used instead of the derivative-calculation used in typical Gradient Descent methods. Compared to original Gradient Descent, the value obtained from derivative calculation at the current iteration is used as the step-size for updating the next iteration. The derivative calculation is based on the current data set (all relevant variables used in the derivative equations) although the data set may not yet be the correct/final results. So the obtained value for defining the step size in Gradient Descent may cause the overall system to jump into a worse direction. This bad data set may then end up being used in the next iteration and so on. This may cause the system to take rather long to converge.

Although the local energy is decreased and hence the variable updated, some criteria for double-checking over constraints on each variable type are also considered to change the unacceptable value to a better direction for the next main iteration (i.e., step 3). These constraints on each variable type are:

1. **Camera Pose:** difference of R and T between any pair of consecutive frames is small. If the variable is too much different from the previous frame, reset its step-size and copy the variable from the previous frame.
2. **3D points:** the Z_w component of 3D points cannot be positive. If Z_w becomes positive, swap its sign to negative.
3. **Focal length:** f must be non-zero and positive. If f becomes negative, swap its sign to positive, if f becomes zero, re-initialize it to a positive value such as 1.0.

3 Experimental Results

After the estimation of all variables, the real-world or Euclidean size of target objects used in the experiments can be recovered by rescaling the reconstructed object using a known real distance between any two contiguous points. The camera's translation, too, can be rescaled the same way. So, in this paper, 3D accuracy will be proved by experiments on synthetic data. 3D graphic-results of real and synthetic experiments are displayed in OpenGL's 3D environment to show camera motion of all video frames and the tracked feature-points. For all OpenGL 3D-results, 3D cones are used as icons to represent cameras in all frames where the first camera is illustrated in the darkest pink and the last in the darkest yellow. The first two points of the target object(s) are distinguished from the others by rendering the first two in red and the others in green. For 2D graphic-results, average pixel-errors are calculated from Root Mean Square (RMS) error of point re-projection and the errors are illustrated by two sets of color dots: one set shown in pink represents the observed 2D-points of the target object(s) and one shown in green represents the re-projected 2D-points. The 2D numerical-errors can be calculated for both experiments on real and synthetic data from RMS error of point re-projection.

3.1 Experiment in a Synthetic Environment

The results obtained from the experiment on synthetic data can be illustrated by both the OpenGL 3D-environment display and 2D pixel-error. The experiment is tested on a synthetic video sequence generated from 3DS Max with 101 frames in length, frame rate of 15 frames per second, and frame size of 640×480 pixels. Target objects including a house and a hexagonal prism with ten and 24 feature-points tracked, respectively, are recorded in a perfect 360-degree circular motion. The 2D graphic-results are illustrated by pixel-errors of image re-projection in Fig. 1.

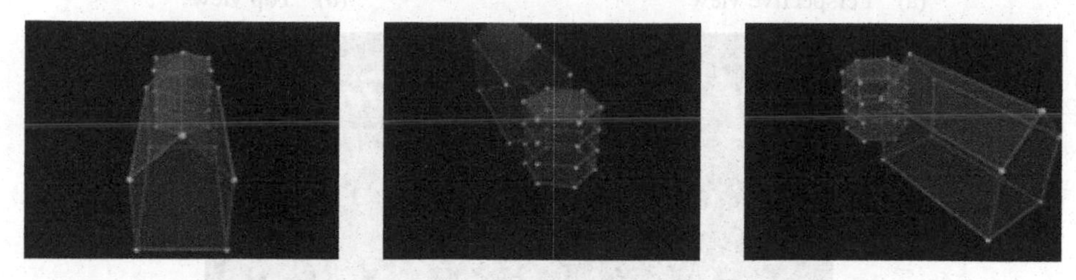

(a) Frame 1, 45, and 90 respectively of an input video with 34 tracked corners.

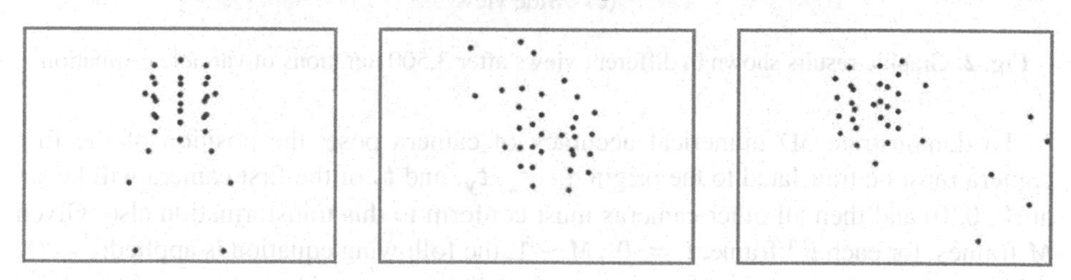

(b) Graphic pixel-errors after 3,500 iterations.

Fig. 1. Pixel-errors shown as displacement between pink (observed) and blue (re-projected) dot-sets of 34 tracked corners of a house and a hexagonal prism

The accuracy in terms of numerical pixel-errors is shown in Table 1 as followed.

Table 1. RMS pixel-errors of point re-projection

#Points	#Frames	#Equations	#Variables	Iterations	Time (sec.)	RMS (pixels)
34	101	6,868	703	3,500	120	1.96e-4

The 3D graphic-results after 3,500 iterations are displayed in the OpenGL environment as shown in Fig. 2. The first two corner points of the house are distinguished from the others by showing them in red and the others in green.

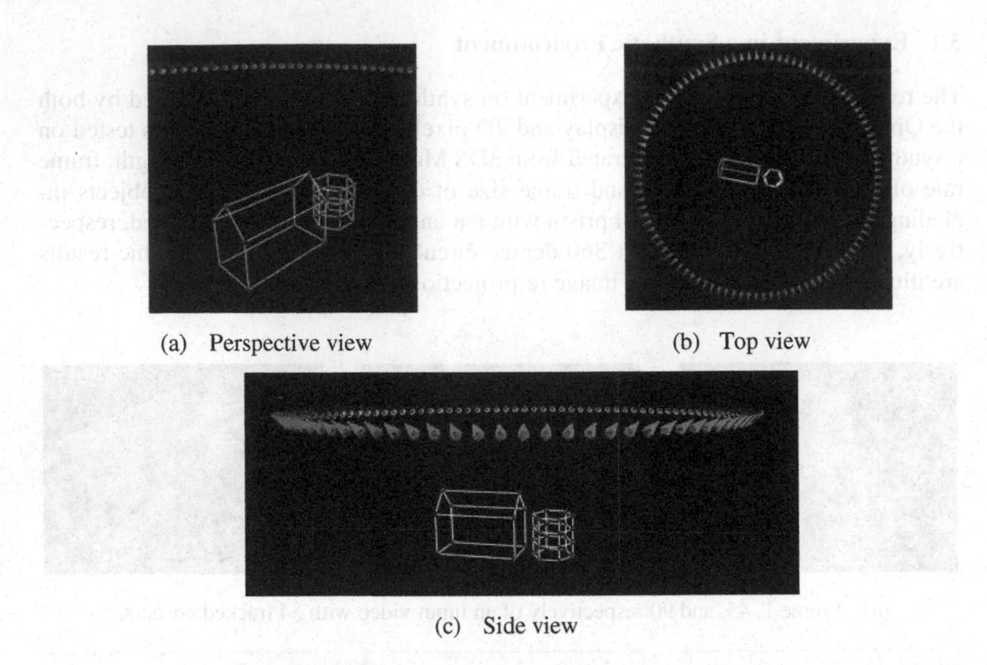

(a) Perspective view (b) Top view

(c) Side view

Fig. 2. Graphic results shown in different views after 3,500 iterations of variable estimation

To demonstrate 3D numerical accuracy of camera pose, the position of the first camera must be translated to the origin, i.e. t_x, t_y, and t_z of the first camera will be set at (0, 0, 0) and then all other cameras must conform to this transformation also. Given M frames, for each ith frame, $i = 0..M-1$, the following equation is applied:

$$\begin{bmatrix} t_x^i \\ t_y^i \\ t_z^i \end{bmatrix} = \begin{bmatrix} t_x^i \\ t_y^i \\ t_z^i \end{bmatrix} - \begin{bmatrix} t_x^0 \\ t_y^0 \\ t_z^0 \end{bmatrix} \quad (4)$$

The 3D numerical errors calculated from camera's 3D pose and objects' 3D structures are as shown in Table 2. The rotation parameters are in the unit of degrees whereas the camera's translation and object's position have user-specified unit.

Table 2. RMS errors of camera pose from 101 frames and 3D structure from 34 points

Rotation (degrees)			Translation (specified unit)			Objects (specified unit)		
θ_x	θ_y	θ_z	t_x	t_y	t_z	X_w	Y_w	Z_w
0.448256	0.007447	0.006824	0.05601	0.030735	0.014168	0.019161	0.002412	0.002895

3.2 Experiment in a Real-World Environment

To show the robustness of this proposed method, a video sequence of the real-world environment is recorded by a digital camera carried by human hand without using any dolly to move the camera around the target object. Consequently, the camera is allowed to move more realistically; i.e., the camera motion will vary with high frequency in almost every dimension (X, Y, Z) and in terms of both rotation and translation. For this experiment, a video sequence with 136 frames in length is recorded with a frame rate of 15 frames per second and a frame size of 320×240 pixels. A target object recorded to the video file is a paper file box with ten corners tracked.

To demonstrate the accuracy of camera motion and 3D reconstructed points for the experiment on real data, the 3D reconstructed points together with the estimated camera pose of each video frame are used to re-project back to all image frames in order to compare the re-projected 2D points with the observed ones. Some example results of 2D-point re-projection are graphically shown in Fig. 3. Pixel-errors can be observed from the displacement between the pink and the blue dots representing the observed and the re-projected 2D-points, respectively.

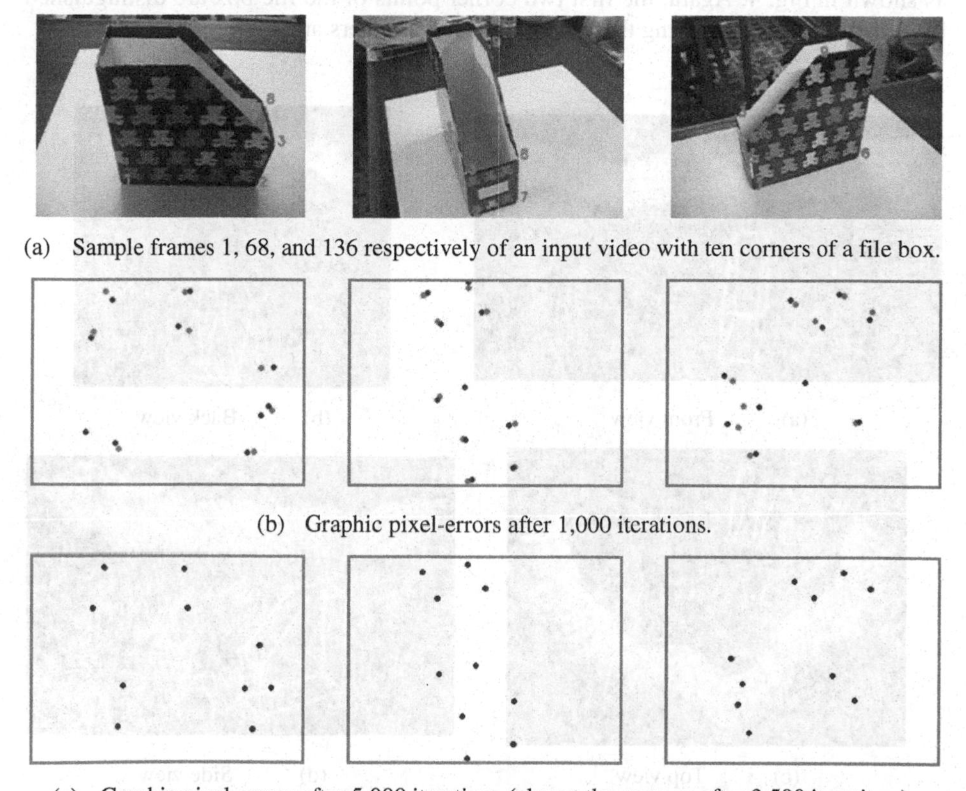

(a) Sample frames 1, 68, and 136 respectively of an input video with ten corners of a file box.

(b) Graphic pixel-errors after 1,000 iterations.

(c) Graphic pixel-errors after 5,000 iterations (almost the same as after 3,500 iterations).

Fig. 3. Pixel-errors shown as the displacement of ten tracked corners of a file box

From Fig. 3, one may observed that, after 5,000 iterations, all pink dots representing the observed 2D points almost disappear, leaving only the blue dots visible. This shows that the pixel displacement between the pink and the blue sets is very small. And this means the re-projection error is very low, implying that the estimation of 3D-point reconstruction and camera pose is good.

The accuracy in terms of pixel-errors calculated from RMS errors is shown in Table 3:

Table 3. RMS pixel-errors of point re-projection

#Points	#Frames	#Equations	#Variables	Iterations	Time (sec.)	RMS (pixels)
10	136	2,268	841	1,000	10	1.52e-1
10	136	2,268	841	3,500	35	2.36e-2
10	136	2,268	841	5,000	50	2.22e-2

3D graphic-results after 5,000 iterations are displayed in the OpenGL environment as shown in Fig. 4. Again, the first two corner points of the file box are distinguished from the others by rendering them in red, while the others are rendered in green.

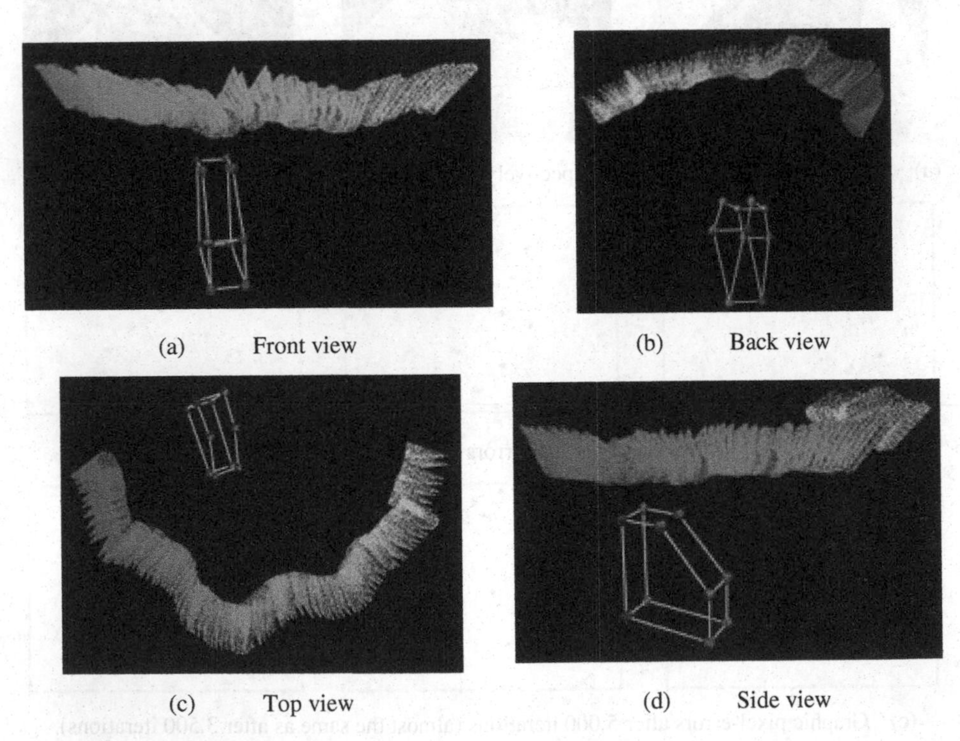

(a) Front view (b) Back view

(c) Top view (d) Side view

Fig. 4. Graphic results shown in different views after 5,000 iterations of variable estimation

From Fig. 4, the camera was moved around the file box side to side with very high frequency change in motion, in accordance to human's walking style while recording this video file. Thus, the camera path of several cones representing camera poses of all video frames has a snake-like shape.

4 Conclusion

Given point correspondences of the target object tracked from a video sequence, the 3D structure of these points can be reconstructed and camera motion can also be estimated at the same time by this proposed method. With our proposed method, the derivative calculation of Gradient Descent is not needed, the missing image-points do not pose any problems, and the obtained results from both real and synthetic experiments show that this method is robust enough while still simple and practical.

References

1. Hartley, R., Zisserman, A.: Multiple View Geometry in Computer Vision, 2nd edn., Cambridge (2006)
2. Press, W.H., Teukolsky, S.A., Vetterling, W.T., Flannery, B.P.: Numerical Recipes: The art of Scientific Computing, 3rd edn., Cambridge (2007)
3. Karayiannis, N.B.: Reformulated Radial Basis Neural Networks Trained by Gradient Descent. IEEE Transactions on Neural Networks 10(3), 657–671 (2000)
4. Chen, O.T.-C.: Motion Estimation Using a One-Dimensional Gradient Descent Search. IEEE Transactions on Circuits and Systems for Video Technology 10(4), 608–616 (2000)
5. Guerrero, J.J., Sagues, C.: Estimating the Motion Direction from Brightness Gradient on Lines. IEEE Transactions on Systems, Man, and Cybernetics – Part C: Applications and Reviews 31(3), 419–426 (2001)
6. Po, L.M., Ng, K.H., Cheung, K.W., Wong, K.M., Uddin, Y., Ting, C.W.: Novel Directional Gradient Descent Searches for Fast Block Motion Estimation. IEEE Transactions on Circuits and Systems for Video Technology 19(8), 1189–1195 (2009)
7. Smolic, A.: Robust Generation of 360-Degree Panoramic Views from Consumer Video Sequences. In: 4th EURASIP-IEEE Region 8 International Symposium on Video/Image Processing and Multimedia Communications (VIPromCom), June 16-19, pp. 431–435 (2002)
8. Edwards, G.J., Taylor, C.J., Cootes, T.F.: Interpreting Face Images using Active Appearance Models. In: 3rd IEEE International Conference on Automatic Face and Gesture Recognition, April 14-16, pp. 300–305 (1998)
9. Cootes, T.F., Edwards, G.J., Taylor, C.J.: Active Appearance Models. In: Burkhardt, H., Neumann, B. (eds.) ECCV 1998. LNCS, vol. 1407, pp. 484–498. Springer, Heidelberg (1998)
10. Chouvatut, V., Madarasmi, S.: A Comparison of Two Camera Pose Methods for Augmented Reality. In: 7th IASTED International Conference on Signal and Image Processing (SIP), August 15-17, pp. 554–559 (2005)
11. Chouvatut, V., Madarasmi, S.: Estimation of Camera Pose for Use in Augmented Reality System. In: 20th International Technical Conference on Circuits/Systems, Computers, and Communications (ITC-CSCC), July 4-7, vol. 3, pp. 979–980 (2005)
12. Tsai, R.Y.: A Versatile Camera Calibration Technique for High-Accuracy 3D Machine Vision Metrology Using Off-the-Shelf TV Camera and Lenses. IEEE Journal of Robotics and Automation RA-3(4), 323–344 (1987)
13. Kato, H., Billinghurst, M.: Marker Tracking and HMD Calibration for a Video-Based Augmented Reality Conferencing System. In: Proceeding 2nd IEEE and ACM International Workshop on Augmented Reality, October 1999, pp. 85–94 (1999)

14. Okuma, T., Sakaue, K., Takemura, H., Yokoya, N.: Real-Time Camera Parameter Estimation from images for a Mixed Reality System. In: IEEE Proceeding 15th International Conference on Pattern Recognition, September 3-7, vol. 4, pp. 482–486 (2000)
15. Hartley, R.I.: Projective Reconstruction and Invariants from Multiple Images. IEEE Transactions on Pattern Analysis and Machine Intelligence 16(10), 1036–1041 (1994)
16. Avidan, S., Shashua, A.: Novel View Synthesis by Cascading Trilinear Tensors. IEEE Transactions on Visualization and Computer Graphics 4(4), 293–306 (1998)
17. Li, J., Chellappa, R.: A Factorization Method for Structure from Planar Motion. In: IEEE Workshop on Motion and Video Computing (WACV/MOTIONS), January 2005, vol. 2, pp. 154–159 (2005)
18. Sasson, A.M.: Combined Use of the Powell and Fletcher – Powell Nonlinear Programming Methods for Optimal Load Flows. IEEE Transactions on Power Apparatus and Systems PAS-88(10), 1530–1537 (1969)
19. Xu, X., Dony, R.D.: Differential Evolution with Powell's Direction Set Method in Medical Image Registration. In: IEEE International Symposium on Biomedical Imaging: Nano to Micro, April 15-18, vol. 1, pp. 732–735 (2004)

Adapting Content Presentation for Mobile Web Browsing

Boonlit Adipat, Dongsong Zhang, and Lina Zhou

University of Maryland Baltimore County,
Maryland, USA

Abstract. As the use of mobile handheld devices for Web access becomes increasingly popular, ease of use has become an important issue for mobile devices. However, due to their small screen size, users often suffer from the poor content presentation that seriously hinders the usability of Mobile Web applications. Therefore, adapting Web content presentation for more effective user browsing on handheld devices becomes very important. In this paper, we propose a new hybrid approach to presentation adaptation that integrates three techniques, namely tree-view adaptation, hierarchical text summarization, and keyword highlighting. A prototype system called M-Web is implemented based on the proposed approach. We conducted a preliminary user study to evaluate the effect of presentation adaptation on mobile Web browsing. The results show that presentation adaptation improves users' performance and perception of mobile Web browsing.

Keywords: mobile computing, content adaptation, mobile Web, human-computer interaction.

1 Introduction

With mobile handheld devices, users can benefit from information access at any time and at any location. Within this trend, mobile Web, which refers to enabling Web access via mobile handheld devices, has received more and more attention [19]. However, since many Web pages are originally designed for desktop computers, directly presenting them on the small screen of mobile handheld devices makes the content look aesthetically unpleasant and difficult to read. Due to the small display screen, it is difficult for mobile users to obtain clear organization of a Web page [1]. As users navigate through information, they need to memorize different parts of the information that they have just viewed and relate the parts together in order to understand it. Finally, they have to decide whether the information matches their interests or not. This process is tedious and increases users' cognitive load, which in turn makes users tend to generate more errors.

To address the above problems, the presentation of Web content should be adapted to better fit the limited screen size of handheld devices for effective mobile Web browsing. Presentation adaptation is the process of automatically re-authoring a Web page (e.g., changing a page layout and format configurations) for easy Web browsing [4, 7, 11]. Although the importance of presentation adaptation in mobile information

S.K. Prasad et al. (Eds.): ICISTM 2010, CCIS 54, pp. 293–303, 2010.

systems has been increasingly recognized, research in this area is still at its early stage. In this study, we propose a new hybrid presentation adaptation approach that combines three individual adaptation techniques and implement it in a prototype system called *M-Web*.

The rest of the paper is organized as follows. In the next section, we introduce related work on presentation adaptation on mobile handheld devices. Then, we describe our proposed approach in detail in Section 3. In Sections 4 and 5, we describe our preliminary user evaluation of the M-Web system and its findings. The final closing section discusses the conclusion and future direction of this research.

2 Related Work

Presentation adaptation mainly involves the use of graphical presentation to help users better understand the information. Without an effective strategy to manage the presentation of such information, users will experience problems in locating specific information, finding relationships among information, and understanding the information [17]. To date, some presentation techniques have been introduced reduce usability problems of mobile Web browsing. We classify them into the following categories:

- The *page splitting approach* divides an original Web page into several smaller sub-pages that fit the screen size of a specific mobile device [10]. Users can perform page-by-page navigation by clicking the *next* or *previous* button or indexed page numbers. The problem with this approach is that as the size of a Web page increases, the number of split pages will increase accordingly. As a result, locating specific information will become more difficult and time consuming.
- The *content outlining approach* converts the structure of original Web content into the form of table of content (TOC). Using heuristic approaches, some predefined rules are applied to parse HTML tags and extract content headers, which later are converted into a list of hyperlinks. Users can click one of those links to switch to a new page and view the full content [4]. This approach is intuitive as readers can navigate content through TOC. However, predefined rules in the heuristic approach are not flexible enough to be applicable to any Web documents.
- The *summary-based approach* presents only a portion or a summary (or summaries of separate sections) of a Web page (e.g., [12, 15]). It enables users to read those snippets or summaries before deciding if they want to see further details.
- The *visualization based approach* uses visualization techniques, such as focus & context approach [5] and zooming technique [3], to view the content of a Web page on a handheld device. For example, Collapse-to-Zoom [3] utilizes a zooming approach, a progressive process, for expanding detailed information upon a user's request. The system first presents a thumbnail overview of a Web page. Based on the overview, users can quickly identify certain content sections in which they are interested. Then, they can zoom in a specific section and a new page containing detailed information of the selected part is presented. However, multiple zooming levels may cause users to lose the context easily.

3 A Hybrid Presentation Adaptation Approach

Existing presentation adaptation approaches contains both pros and cons. There is great potential that those techniques could be deployed together to improve usability of mobile Web applications. Therefore, we propose a novel hybrid presentation adaptation technique, which integrates three effective information presentation approaches (i.e., tree-view adaptation, hierarchical text summarization, and keyword highlighting).

3.1 Tree-View Adaptation

Locating desirable information on a small screen is difficult and time consuming. Users cannot quickly scan through a web page to locate specific information as they usually do on desktops. According to the visual information seeking mantra [16], it is helpful to provide users with an overview of a document and enable them to drill down to particular sub-topics upon their preferences later. Therefore, we propose a DOM (Document Object Model) tree-view adaptation, an approach that automatically generates hierarchical (tree-view) presentation for any Web pages.

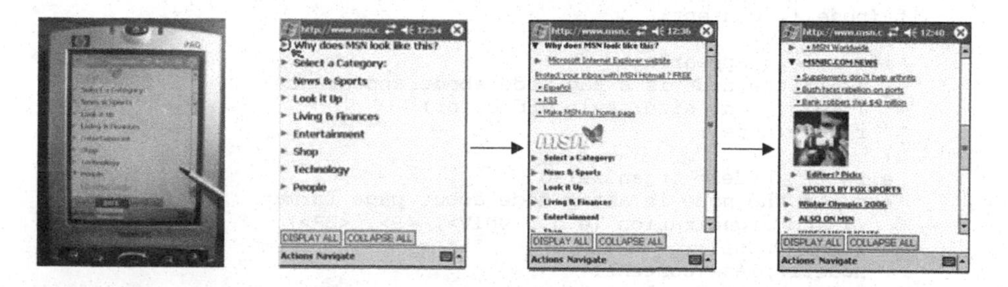

Fig. 1. A generated DOM tree-view adaptation interface of the MSN index page

As shown in Figure 1, the M-Web system displays Web content in the form of tree-view presentation. When a user opens a Web page, he first sees an overview (a list of section topics) of the page. If he is interested in a specific section, he can click at the bullet in front of the section topic to drill down to the next level of the tree, which will be either sub-sections or content of the selected section.

We generate the tree-view presentation based upon the Document Object Model, a standard interface that creates a logical structure of HTML and XML documents in the form of tree-like representation [18]. Each node in a DOM tree represents an object embedded in a Web document. Examples of objects are HTML tags, images, and text. These objects are related to each other as parent and child nodes. These nodes can be either tag nodes (i.e., html/xml tags) or content nodes (i.e., image and text) (Figure 2).

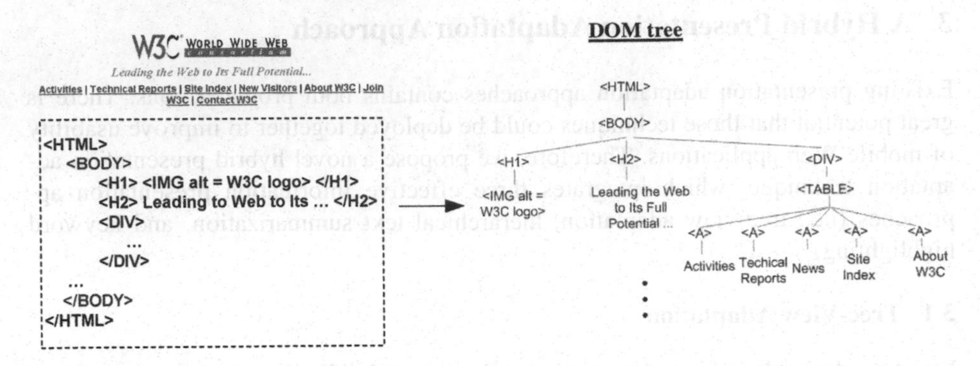

Fig. 2. An example of a DOM tree of the W3C Web page (http://www.w3.org)

Generating tree-view presentation generally involves analyzing the original source code of a Web page, removing tags, and replacing them with the nearest children content nodes. We use the following algorithm to build the tree-view interface.

```
While (node != NULL)
{
  if (node != content)
  {
    if (node = appearTag)
    { // if the node is a tag node about appearance
      //(e.g.,font size/style and color)
      remove(node);
    }
    else if (node = organizeTag)
    { // if the node is a tag node about page format
      // and organization (e.g., <DIV>, <P>, <BR>)
      remove(node);
      node.replace(content)
      tree.shiftUpLevel( );
    }
    AssignTreeLevel( ); // assign tree level numbers to
                        // nodes
  }
  else
  { // if the node is a content node, go to its child
    // node
    node = node.nextChild( );
  }
}
```

First, an HTML parser is used to generate an HTML DOM tree, which gives the position and information of each node in the DOM tree. Then, the M-Web system first removes tag nodes that represent text/image appearance such as font and color. After removing the appearance tags, the tree will remain only content nodes or tag nodes (e.g., <P>, <DIV>) that represent the format and organization of a Web page. After that, the system traverses from the top node. If the node is a tag node, the system replaces it with its child node, which is a content node at the nearest lower level, and shifts the entire lower level of the tree up. If the node is a content node, the system traverses down to its children. The process is repeated until the tree contains only content nodes. Figure 3 illustrates the process of generating a tree-view interface on a handheld device based on the DOM tree structure.

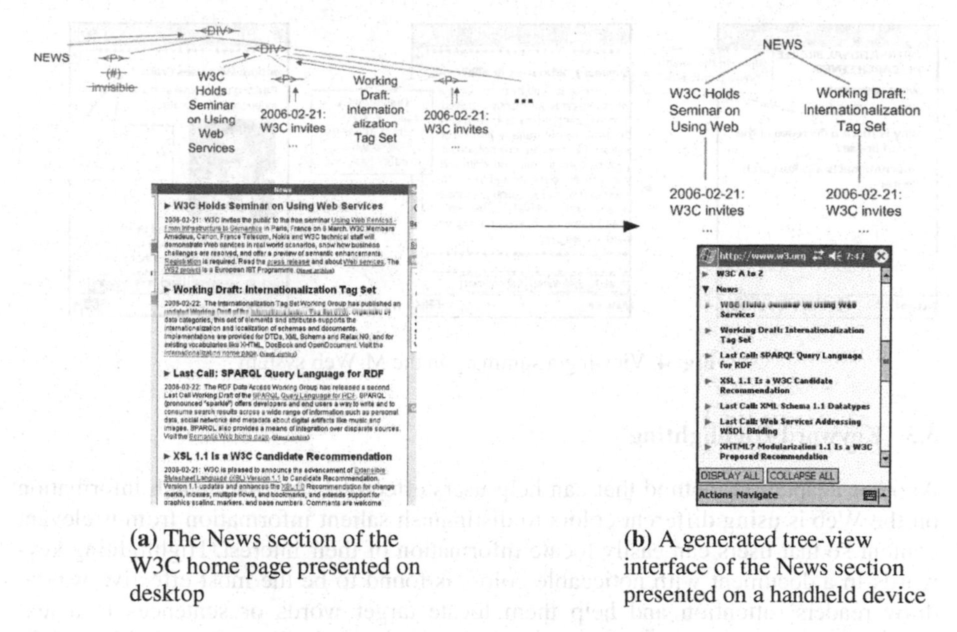

(a) The News section of the W3C home page presented on desktop

(b) A generated tree-view interface of the News section presented on a handheld device

Fig. 3. Generating a DOM tree-view adaptation interface

3.2 Hierarchical Text Summarization

In a mobile handheld device, if original content is displayed as it is, users may have a difficult time in finding information related to their interests due to the large amount of information displayed and the frequent scrolling. It becomes even worse if they find out later that the content does not match their needs after spending a lot of time on reading it. Some studies [6, 12, 20] have shown that summarization can help mobile users quickly identify content sections of their interest. Therefore, it is beneficial to apply summarization to mobile applications.

On the basis of the tree-view adaptation, we generate hierarchical summaries for the content that users are reading. By presenting summaries as a glimpse of large-sized content, users are provided with a useful cue to determine the relevance of information to their interests, potentially resulting in saving time by not to peruse the full content of irrelevant sections in a Web page.

We employ a heuristic approach to summarization, which has been used in other studies [2, 8]. In our research, a summary is generated by aggregating the first sentence of each paragraph in a Web page together. Harper and Patel [8] found that summaries generated by this approach were effective and able to help users quickly understand the content.

In the tree-view adaptation interface, users can view the summary of a content section by clicking its section title. Then, the M-Web system switches to the summary page that presents the summary of the selected section. If a user is interested in reading the full content after viewing its summary, he can click the "full content" button at the bottom of the interface (Figure 4).

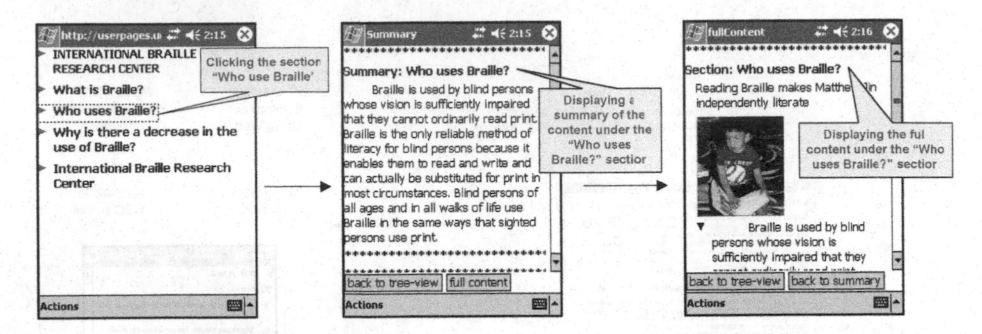

Fig. 4. Viewing a summary in the M-Web system

3.3 Keyword Highlighting

Another adaptation method that can help users effectively browse/search information on the Web is using different colors to distinguish salient information from irrelevant content so that users can easily locate information of their interest. Highlighting keywords in a document with noticeable colors is found to be the most effective way to draw readers' attention and help them locate target words or sentences in a text document [9]. We adapt Web content presentation to facilitate users with explicit visual cues in the information search task. Users' keywords of interest specified by users are used to adapt the content display of a Web page by highlighting those keywords appeared in that page (Figure 5).

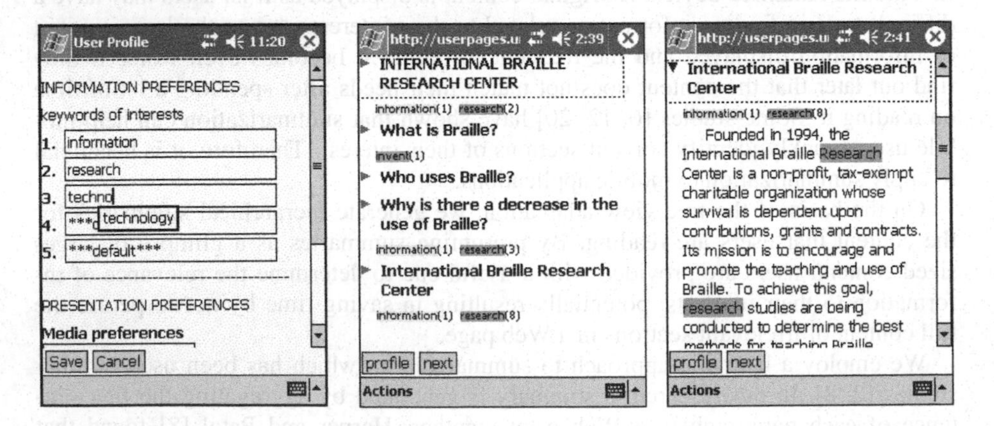

Fig. 5. Entering keywords

Fig. 6. A tree-view interface with a visual cue (results of the keyword search)

Fig. 7. When expanding a section, keywords are highlighted

The M-Web system automatically searches the content of a Web page for the keywords of interest, and presents the number of keywords that appear in each content section. The number of keywords is appended to each section title in the tree view (Figure 6). The number inside parentheses (Figure 6) represents the number of occurrences of each keyword in that Web page, providing cues to help users identify the location of specific information they are looking for. When users expand a section, those keywords in the content are highlighted with different color codes for easy keyword spotting (Figure 7).

4 Evaluation

We conducted a user empirical study to evaluate the effectiveness of the proposed presentation adaptation approach. Five different mobile Web systems used in the experiment were as follows:

1. O – direct display without any adaptation. A Web page was directly presented with a variable horizontal width and length for up/down and left/right scrolling. In this study, O was used as a benchmark.
2. T – The M-Web system with tree-view adaptation only.
3. T+S – The M-Web with tree-view and hierarchical text summarization.
4. T+H – The M-Web with tree-view and keyword highlighting.
5. All – The M-Web with all three presentation adaptation approaches: tree-view presentation, hierarchical text summarization, and keyword highlighting.

Subjects: A total of 35 undergraduate and graduate students at a university located on the east coast of the U.S. participated in this study. Every participant had prior experience with Web access and using a handheld device (e.g., cell phone and pocket PC).

Mobile device: We used an HP iPAQ h4355 Pocket PC in the experiment. We installed the four versions of M-Web systems and IE on this device. This Pocket PC operates with the 400 MHz Intel technology processor, a 3.5” transflective screen with 64K colors and 64 Mega bytes memory. It connects to a local wireless network via the integrated Wi-Fi (802.11b) adapter with the speed up to 10 Mbps and WLAN.

Task: Our experiment followed a within-subject design. We asked every participant to search information on five different Web sites, each site per one mobile Web system. Those Web sites, although the contents were different, were very similar in terms of content structure and length. The order of the five mobile Web systems used by each participant was randomized based on Latin square design (of size five) [14]. The Web sites were shuffled to pair with every mobile Web system. The experiment was conducted in the laboratory setting.

For each Web site, a participant was required to find two answers. All questions were fact-based. The maximum score a participant could receive was ten points, with one point for each correct answer. There were no partial points. After finding the answer to a question from the designated Web site, Participants were asked to write down the answer instead of choosing an answer from a list of choices, preventing them from randomly guessing the answer. We set up a cut-off time limit (3 minutes) on finding each answer. If the participant could not find an answer within the time limit, his/her answer was considered incorrect. This time limit was determined based on the results of a pilot study.

Measurement: We measured 1) search time – time that a participant took to find an answer and 2) accuracy - the percentage of correctly identified answers to the given factual questions. In addition, after participants completed all tasks using all five systems, we asked them to rank all five systems based on their preference to use for mobile Web browsing.

Experiment Procedure: Before the experiment, each participant was asked to go through a training session. In the training session, we introduced all versions of the M-Web systems. Then, he/she was permitted to practice a few search exercises to familiarize himself/herself with all five systems. The formal experiment started once the participants felt comfortable with using the systems. At the end of the experiment, an informal short interview was conducted. Each participant was asked to provide feedback about the M-Web systems and experience of using them to browse the Web in general.

Result: Figure 8 shows the search time of users when they used each individual system. The results show that all four versions of M-Web systems performed better than O version (the system without presentation adaptation). Among the four M-Web systems, T+H performed the best, All the second, T+S the third, and T the fourth.

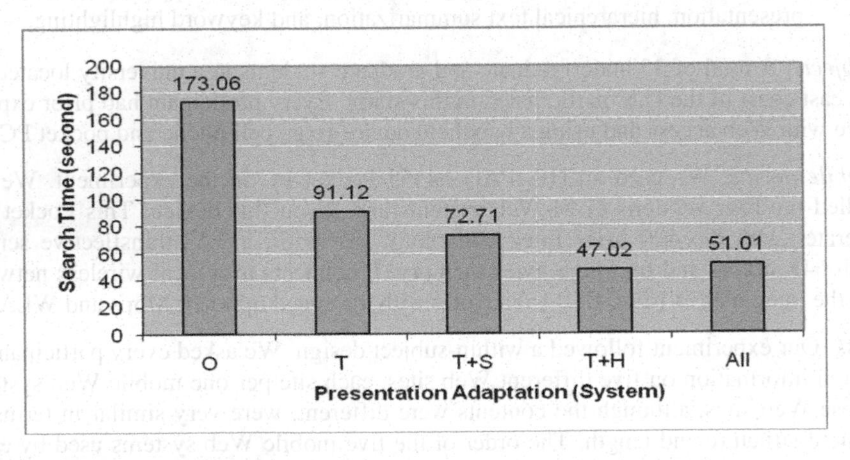

Fig. 8. Average search time (second) for each of the five mobile Web systems

Figure 9 shows the accuracy results from the experiment. The results show that participants achieved higher accuracy when using systems featured with presentation adaptation (i.e., T, T+S, T+H, and All) than when using systems without presentation adaptation (i.e., O). The accuracy values of the four M-Web systems were comparable to each other.

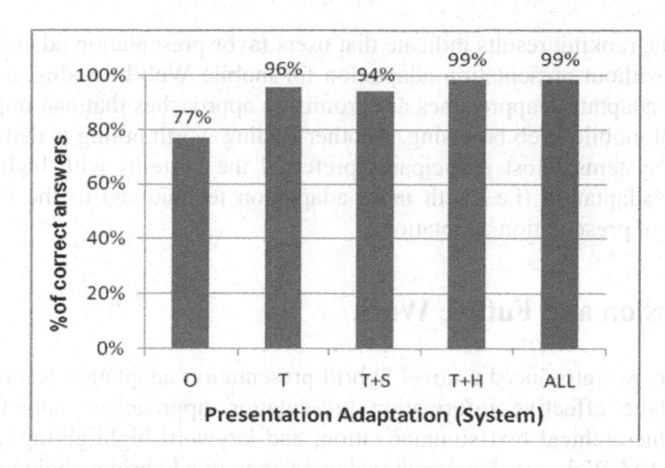

Fig. 9. Percentage of correct answers for each of the five mobile Web systems

In addition, the post-study user interview results show that the order of the four M-Web systems with presentation adaptation, from the most favorable to the least favorable, is 1) All, 2) T+H, 3) T+S, 4) T, 5) O.

5 Discussion

The results of this study provide some insights into how presentation adaptation may affect users' performance and users' perception of mobile Web browsing.

First, we found that presentation adaptation had a positive impact on users' performance of mobile Web browsing. When compared to mobile Web systems without presentation adaptation, mobile Web systems with presentation adaptation reduced users' searching time and increased users' searching accuracy.

Second, Web structure information is vital to improving information search via browsing process on mobile handheld devices. In this study, tree-view adaptation reorganizes original Web content and presents it in the form of a hierarchical structure, thereby enabling users to gain a general overview of the entire content at a glance. With such a presentation format, users can see a global view and quickly access particular content sections that match their interests, reducing the need to scroll through content that is irrelevant to their interests.

Third, keyword highlighting has proved to be a very effective approach in information search via mobile Web browsing. With keywords of users' interest highlighted, users can quickly shift attention to specific portions of content and easily find information they want. Participants using the M-Web system with keyword highlighting (T+H) took the shortest time to find answers.

Fourth, text summarization is beneficial to mobile Web browsing. It saves users time to peruse the entire content and help them quickly identify documents or content sections that are interesting to them. In the experiment, T+H outperformed T+S in search time. This may come from a reason that with text summarization, users still need to spend time reading a summary. However, with keyword highlighting, they can promptly spot an answer if a sentence contains a keyword.

Finally, the ranking results indicate that users favor presentation adaptation significantly over without presentation adaptation for mobile Web browsing, and our three presentation adaptation approaches are promising approaches that can improve users' perception of mobile Web browsing. Another finding worth noting is that when asked to rank the systems, most participants preferred the systems with higher levels of presentation adaptation (i.e., with more adaptation techniques) to the systems with lower levels of presentation adaptation.

6 Conclusion and Future Work

In this paper, we introduced a novel hybrid presentation adaptation technique, which integrates three effective information presentation approaches, namely tree-view adaptation, hierarchical text summarization, and keyword highlighting. A prototype system called M-Web was developed to demonstrate this hybrid technique.

To understand the importance and benefits of presentation adaptation in mobile environments, we conducted a preliminary user study to evaluate the M-Web system. The findings of the study suggest that presentation adaptation is important to mobile Web browsing, and our approach is shown to improve users' performance (i.e., reduces search time and increases search accuracy) and users' perception of mobile Web browsing.

As future work, we plan to test the generalizability of our results by evaluating the effectiveness of presentation adaptation with other information seeking tasks such as information search tasks with no specific goals [13]. In addition, we plan to examine the effect of presentation adaptation on different levels of users' skills and experiences with mobile Web browsing.

References

1. Albers, M.J., Kim, L.: User Web Browsing Characteristics Using Palm Handhelds for Information Retrieval. In: Proceedings of IEEE professional communication society international professional communication conference, pp. 125–135. IEEE, Cambridge (2000)
2. Amato, G., Straccia, U.: User Profile Modeling and Applications to Digital Libraries. In: Abiteboul, S., Vercoustre, A.-M. (eds.) ECDL 1999. LNCS, vol. 1696, pp. 184–197. Springer, Heidelberg (1999)
3. Baudisch, P., Xie, X., Wang, C., Ma, W.-Y.: Collapse-to-Zoom: Viewing Web Pages on Small Screen Devices by Interactively Removing Irrelevant Content. In: Proceedings of the 17th annual ACM symposium on User interface software and technology, pp. 91–94. ACM Press, Santa Fe (2004)
4. Bickmore, T.B., Schilit, B.N.: Digestor: Device-independent Access to the World Wide Web. In: Proceedings of the 6th World Wide Web Conference, Santa Clara, CA, April 7-11, pp. 665–663 (1997)
5. Björk, S., Holmquist, L.E., Redström, J., Bretan, I., Danielsson, R., Karlgren, J., et al.: WEST: a Web browser for small terminals. In: Proceedings of the 12th annual ACM symposium on User interface software and technology, Asheville, North Carolina, pp. 187–196 (1999)

6. Buyukkokten, O., Kaljuvee, O., Garcia-Molina, H., Paepcke, A., Winograd, T.: Efficient Web Browsing on Handheld Devices Using Page and Form Summarization. ACM Transactions on Information Systems 20(1), 82–115 (2002)
7. Chen, Y., Xie, X., Ma, W.-Y., Zhang, H.-J.: Adapting Web Pages for Small-Screen Devices. IEEE Internet Computing 9, 50–56 (2005)
8. Harper, S., Patel, N.: Gist Summaries for Visually Impaired Surfers. In: Proceedings of the 7th international ACM SIGACCESS conference on computers and accessibility, Baltimore, MD, October 9-12, pp. 90–97 (2005)
9. Hoadley, E.D., Simmons, L.P., Gilroy, F.D.: Investigating the effects of color, font, and bold highlighting in text for the end user. Journal of Business and Economic Perspectives 26(2), 44–64 (2000)
10. Hong, Y.-S., Park, I.-S., Ryu, J.-T., Hur, H.-S.: Pocket News: news contents adaptation for mobile user. In: Proceedings of the 14th ACM Conference on Hypertext and Hypermedia (HYPERTEXT 2003), Nottingham, UK, August 26-30, pp. 79–80 (2003)
11. Laakko, T., Hiltunen, T.: Adapting Web Content to Mobile User Agents. IEEE Internet Computing 9, 46–53 (2005)
12. Lam, H., Baudisch, P.: Summary Thumbnails: Readable Overviews for Small Screen Web Browsers. In: Proceedings of the SIGCHI conference on Human factors in computing systems, Portland, Oregon, USA, April 2-7, pp. 681–690 (2005)
13. Marchinonini, G.: Information Seeking in electronic environments. University Press, New York (1995)
14. Maxwell, S.E., Delaney, H.D.: Designing Experiments and Analyzing Data. Lawrence Erlbaum Associates, Mahwah (2000)
15. Otterbacher, J., Radev, D., Kareem, O.: News to Go: Hierarchical Text Summarization for Mobile Devices. In: Proceedings of the 29th annual international ACM SIGIR conference on Research and development in information retrieval (SIGIR 2006), pp. 589–596. ACM Press, Seattle (2006)
16. Shneiderman, B.: The Eyes Have It: A Task by Data Type Taxonomy for Information Visualizations. In: Proceedings of the IEEE Symposium on Visual Languages, Washington, DC, September 3-6, pp. 336–343 (1996)
17. Spence, R.: Information Visualization. Addison-Wesley, Reading (2001)
18. W3C. Document Object Model (DOM) (2005), `http://www.w3.org/DOM/` (Retrieved September 24, 2009)
19. W3C. Mobile Web Best Practices 1.0 (2006), `http://www.w3.org/TR/mobile-bp/` (Retrieved September 24, 2009)
20. Yang, C.C., Wang, F.L.: Fractal Summarization for Mobile Devices to Access Large Documents on the Web. In: Proceedings of the Twelfth International Conference on World Wide Web (WWW2003), pp. 215–224. ACM, Budapest (2003)

Application of Decision Tree Technique to Analyze Construction Project Data

Vijaya S. Desai[1] and Sharad Joshi[2]

[1] Associate Professor, Head of IT, NICMAR, India
vdesai@nicmar.ac.in
[2] Director of Institute of Management Education Research and Training, India

Abstract. Data mining is the analysis of data from data-warehouse using series of mathematical and statistical methods to detect patterns. Such analysis derives important information which has proved the basis of accurate decision making in retail, banks, fraud detection, customer analysis etc. Construction Industry has benefited a little from these analytical tools mainly due to unavailability of work showing the application of these techniques in analysis of data in construction. A methodology to apply decision tree to analyze the construction labor productivity is presented in the paper. The methodology addresses three areas in decision tree construction process. These include selecting more influential attributes, combining multiple attributes and defining the threshold to ignore irrelevant attributes. The methodology gives more realistic results than the traditional method of decision tree.

Keywords: User Intervention, Unknown Characteristics, Data Warehouse, Data Mining, Decision Tree, Entropy, Gain, Uncertainty Coefficient.

1 Introduction

Increasing use of computers in every business, leads to accumulation of valuable historical data of an organization. This phenomenon is demanding the need of sophisticated data handling tools at all levels of business organization. Data warehousing technology comprises a set of new concepts and tools which support the knowledge workers (executive, manager, and analyst) with information material for decision making (Gatziu and Vavouras, 1999).

Data warehouse is a database created by combining data from multiple databases for the purposes of analysis (AHMAD and NUNOO). Data Mining is the analysis of (often large) observational data sets to find unsuspected relationships and to summarize the data in novel ways that both understandable and useful to the data owner (David, et al. 2004). These techniques typically address four basic applications such as data classification, data clustering, Association between data and finding sequential patterns between the data. Decision tree is one of the classification techniques which generates a tree and set of rules, representing the model of different classes, from a given data set. Various algorithms used to develop a decision tree are CART, ID3, C4.5, and C5, CHAID, QUEST, OC1, SAS, SLIQ, SPRINT, RainForest, Approximation Method, CLOUDS, BOAT (Pujari, 2001).

S.K. Prasad et al. (Eds.): ICISTM 2010, CCIS 54, pp. 304–313, 2010.

2 Related Work

Some amount of research work is found in construction business. Data mining techniques were applied to a state transportation agency's (STA) database to reveal unknown patterns in the database of paving projects using classification technique (Nassar Khaled, 2007). A neural network model was developed for investigating the applicability of data mining in the prediction of tunnel support stability using artificial neural networks (ANN) algorithm (Leu et al 2000). Soibelman et al. presented a process for data preparation for construction knowledge generation through knowledge discovery in databases (Soibelman 2002), and a process for knowledge generation and dissemination (Soibelman 2000). Soibelman et al discussed a data fusion methodology, which was designed to provide timely and consistent access to historical data for efficient and effective management of knowledge discovery (Soibelman et al, 2004). Another study by Lucio et.al presented a prototype of a KDD application to support property valuation (Soibelman and Gonzalez, 2002). Lee et.al presented an application of data mining to analyze the historical maintenance data from the company's residential building project records to discover the hidden causes of problems existing in planning, design, procurement, and construction processes. (Lee et. al. 2008).

2.1 Decision Tree

The construction of decision tree involves three main phases (Pujari 2001). They are 1) Construction phase where the initial tree is constructed based on the entire training data set. 2) Pruning phase where the lower branches and nodes are removed to improve its performance. 3) Pruned tree is processed to improve understandability.

Generic algorithms for decision tree construction use information theory to measure the expected amount of information in a data set. Given a sample S, the average amount of *information needed (entropy)* to find the class of a case in S is estimated by the function,

$$\text{info}(S) = -\sum_{i=1}^{k} \frac{|S^i|}{|S|} \times \log_2 \frac{|S^i|}{|S|}$$

where $S^i \subseteq S$ is the set of examples S of class i and k is the number of classes. If the subset S is further partitioned than suppose t is a test that partitions S into S1, S2, ….., Sr; then the *weighted average entropy* over these subsets is computed by,

$$\sum_{i=1}^{r} \frac{|S_i|}{|S|} \times \text{info}(S_i).$$

The information gain represents the difference between the information needed to identify an element of test t and the information needed to identify an element of test t after the value of attribute X is obtained. The *information gain* due to a split on the attribute is computed as,

$$\text{gain}(t) = \text{info}(S) - \sum_{i=1}^{r} \frac{|S_i|}{|S|} \times \text{info}(S_i)$$

To select the most informative test, the information gain for all the available test attributes is computed and the test with the maximum information gain is then selected. Information gain of a test by the entropy of the test outcomes themselves, which measures the extent of splitting done by the test,

$$\text{split}(t) = -\sum_{i=1}^{r} \frac{|S_i|}{|S|} \log \frac{|S_i|}{|S|},$$

Giving the *gain-ratio* measure,

$$\text{gain-ratio}(t) = \frac{\text{gain}(t)}{\text{split}(t)}.$$

3 Methodology

The methodology included a *"User Intervention"* approach at different stages of construction of decision tree. The user was defined as a *"person having sound knowledge and experience of the domain area on which analysis is to be carried out"*. User intervened to; 1) select the dependent and independent attributes that would participate in tree construction, 2) select the attributes to be combined, 3) define the threshold to stop the growth of the tree.

The information gain for the selected attribute was calculated to measure the influence. Uncertainty Coefficient (UC) was used as threshold to remove irrelevant attributes and stop growing of the tree. The UC for each attribute was carried out by,

$$UC(X) = \frac{gain(X, T)}{Info(T)}.$$

The average UC was defined as the specified threshold. Attributes with UC >= the specified threshold were considered for the classification.

3.1 Method to Combine Multiple Variables

The user intervened to select the numeric attributes to be combined. If user selected two numeric attributes, A1 and A2 then each attribute was divided into two data sets one to the left and second to the right of the median. A1 has L1, R1 and A2 has L2, R2 sides. Let M1 and M2 be the medians of A1 and A2 respectively as shown below. L1⟵ M1⟶ R1 --- for attribute A1 and L2 ⟵ M2⟶ R2 ---for attribute A2. Four data sets formed were, A1<=M1, A1>M1, A2<=M2 and A2>M2. If n number of attributes were selected then 2n data sets were formed. Two data sets were combined to form one class. The combinations formed were $2nC_2$. Four combinations formed in the above example were, C1 = (A1<=M1 and A2<=M2), C2 = (A1>M1 and A2>M2), C3 = (A1<=M1 and A2>M2) and C4 = (A1>M1 and A2<=M2).

A class formed by combining two data sets of same attribute such as A1<=M1 and A1>M1 and A2<=M2 and A2>M2 had no meaning and value. Hence if n attributes were selected then n classes were deducted from the number of combinations formed. The combinations therefore were calculated by the formula $(2nC_2) - n$ or $2n(n-1)$.

4 Application of the Method

The methodology was used to analyze the labor productivity in construction projects. Laborer's work data of 27 projects from various locations of a construction company in India was collected. The data was mainly included details of projects, activities and laborers. Project type included building, road, bridge, jetty, power, plant, railway, etc. Activity type included brick work, shuttering, plaster, concrete, fabrication, etc. Surrounding area of the project included: metro, rural, urban. Location of the project included Karnataka, Maharashtra (states). Minimum and maximum temperatures were the temperatures during the working day. Age groups of the labor were 18-25, 26-35, 36-50 and above 50. Data was standardized, summarized, cleaned and organized in a multidimensional model. Microsoft Access was used to create the database. Labor data of Masonry Brick Work was used for the study. First 5 records out of 80 are given in Table 1.

Table 1. Number of Outcomes for Activity Type - Masonry Brick

Case ID	Location	Age Group	Productivity	Min Temp	Max Temp	Climatic Condition	Physical Mental Overburden	Surrounding Area
1	Karnataka	18-25	0.300	20	35	normal	medium	metro
2	Karnataka	18-25	0.500	18	36	normal	less	urban
3	Maharashtra	18-25	1.375	15	42	good	more	Rural
4	AP	18-25	1.083	37	42	normal	more	Rural
5	Maharashtra	18-25	3.330	40	6	normal	medium	urban

4.1 Selecting the Attribute at Root Node

The independent attributes such as Location, AgeGroup, MinTemp, MaxTemp, ClimaticCondition, PhysicalMentalOverburden, SorroundingArea, and the dependent attribute such as LabourProductivity were selected. Let the *S* be the data set of Masonry Brick Work. Two classes formed were LabourProductivity >=1.44 and LabourProductivity <1.44. The number of outcomes in data set S for LabourProductivity >=1.44 was 25 and for LabourProductivity < 1.44 was 55. The entropy of S was calculated as,

$$= -\frac{25}{80}\log_2\frac{25}{80} - \frac{55}{80}\log_2\frac{55}{80} = 0.896$$

The sub sets of *S* are *S1, S2, S3, S4* for Maximum and Minimum Temperature, Surrounding Area, Physical Mental Overburden, Climatic Condition, Age Group, Location respectively. The sub sets of *S1, S2, S3, S4, S5, and S6* are given in Table 2.

Table 2. Sub Sets of ***S1, S2, S3, S4, S5, S6***

S1	**S11** - MinTemp<=19 and MaxTemp>39	**S12** - MinTemp >19 and MaxTemp<=39	**S13** - MinTemp<=19 and MaxTemp<=39	**S14** MinTemp>19 and MaxTemp>39
S2	**S21** - metro	**S22** - Rural	**S23** - Urban	
S3	**S31** - more	**S32** - Medium	**S33** - Less	
S4	**S41** - Normal	**S42** - Good	**S43** - Extreme	
S5	**S51** - 18-25	**S52** - 26-36	**S53** - 35-50	**S54**- Above 50
S6	**S61** - AP	**S62** - Haryana	**S63** - Karnataka	**S64** - Maharashtra

Four sub sets formed by combining MinTemp and MaxTemp were S11, S12, S13, S14 (Table 3). Weighted average entropy of S1,

$$= \frac{38}{80} x0.790 + \frac{29}{80} x0.992 + \frac{12}{80} x0.811 + \frac{1}{80} x0 = 0.857.$$

Gain for S1 = entropy (S) – weighted average entropy (S1) = 0.896 – 0.857 = 0.040. Weighted average entropy and gain for S2, S3, S4, S5 and S6 were computed as shown in Table 4. The gain was sorted in descending order.

Table 3. Weighted Average Entropy (WAE) and Gain for S1 (Max and Min Temperature)

Entropy Calculation for Sub Set S11, S12, S13, S14					
Sets	Total Outcomes	Outcomes of Productivity Class		Entropy	WAE
		>=1.44	<1.44		
MinTemp <=19 and MaxTemp >39	38	9	29	0.790	0.375
MinTemp >19 and MaxTemp <=39	29	13	16	0.992	0.360
MinTemp <=19 and MaxTemp<=39	12	3	9	0.811	0.122
MinTemp >19 and MaxTemp >39	1	0	1	0.000	0.000
				Weighted Average Entropy	**0.857**

Table 4. Comparison of Gain

Ranking	Data Set	Attribute	Gain	Entropy	UC
1	S2	Surrounding Area	0.103	0.793	0.12
2	S6	Location	0.070	0.826	0.08
3	S1	Max and Min Temperature	0.040	0.857	0.05
4	S3	Physical Mental Overburden	0.028	0.868	0.03
5	S4	Climatic Condition	0.022	0.874	0.02
6	S5	Age Group	0.005	0.891	0.01

Data set S2 of Surrounding Area with maximum gain of 0.103 was selected and placed at the root node. The tree constructed at this stage is presented in Fig. 1 (a). Second ranked data set S6 (location) was selected and joined at root node as its branches. Gain for locations such as AP and Haryana was zero and therefore were dropped. Data sets such as, Karnataka, Maharashtra, and Jammu Kashmir were selected for construction of the tree. The tree constructed at this stage is presented in

Figure 1 (b). Third highest data set was S1 (Maximum and Minimum Temperature). As shown in Table 5, last two rows show the number of outcomes under each class, T1, T2, T3 for each location under surrounding area. Classes with zero outcomes were dropped. The classes such as T2 of Karnataka of Metro, T1 & T3 of Maharashtra of Rural, T1 of Karnataka of Rural, T1 & T2 of Maharashtra of Urban and T2 & T3 of Karnataka of Urban were selected to evolve a new tree as shown in Fig. 2.

Table 5. Data classification for activity type – Masonry Brick Work

Surrounding Area	Metro						Rural						Urban					
Location	Maharashtra			Karnataka			Maharashtra			Karnataka			Maharashtra			Karnataka		
Produ-ctivity	Maximum and Minimum Temperature																	
	T1	T2	T3	T1	T2	T3	T1	T2	T3	T1	T2	T3	T1	T2	T3	T1	T2	T3
>=1.44	0	0	0	0	4	0	4	0	0	0	0	0	4	0	0	0	9	3
<1.44	0	0	0	0	3	0	14	0	3	4	0	0	3	3	0	0	10	6

4.2 Defining Threshold to Stop Growing the Tree

The uncertainty coefficient (UC) for data sets were calculated (Table 4). The average UC was 0.05. Those attributes with UC < 0.05 were eliminated and the final decision tree was developed as shown in Fig. 2.

4.3 Meanings Derived from the Decision Tree Data Analysis

Decision tree presented in Fig. 1 (a) can be interpreted as, chances of productivity becoming more than the average productivity are more in metro (57.14%) and less in Rural (46.88%) and very low in urban (14.63%). Similar interpretations can be derived from decision tree presented in Fig. 1 (b), Fig. 2.

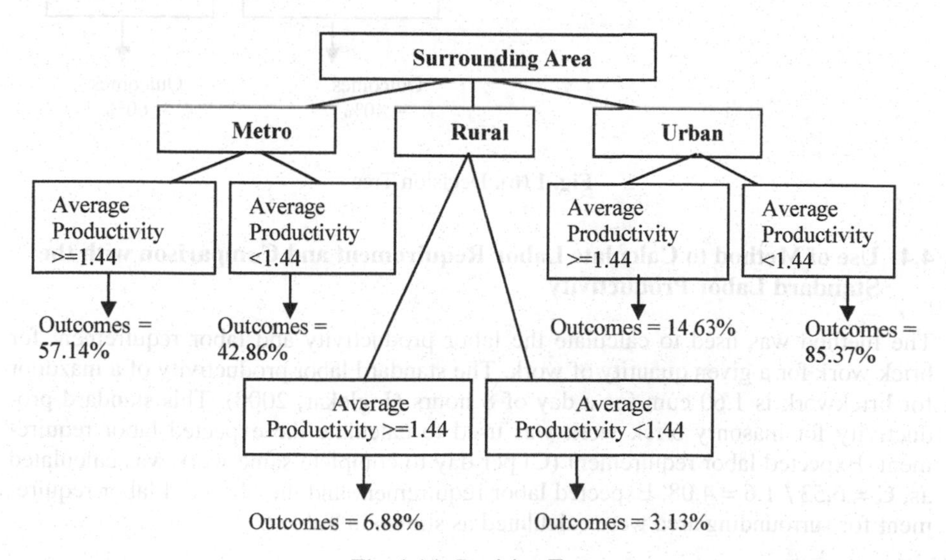

Fig. 1 (a). Decision Tree

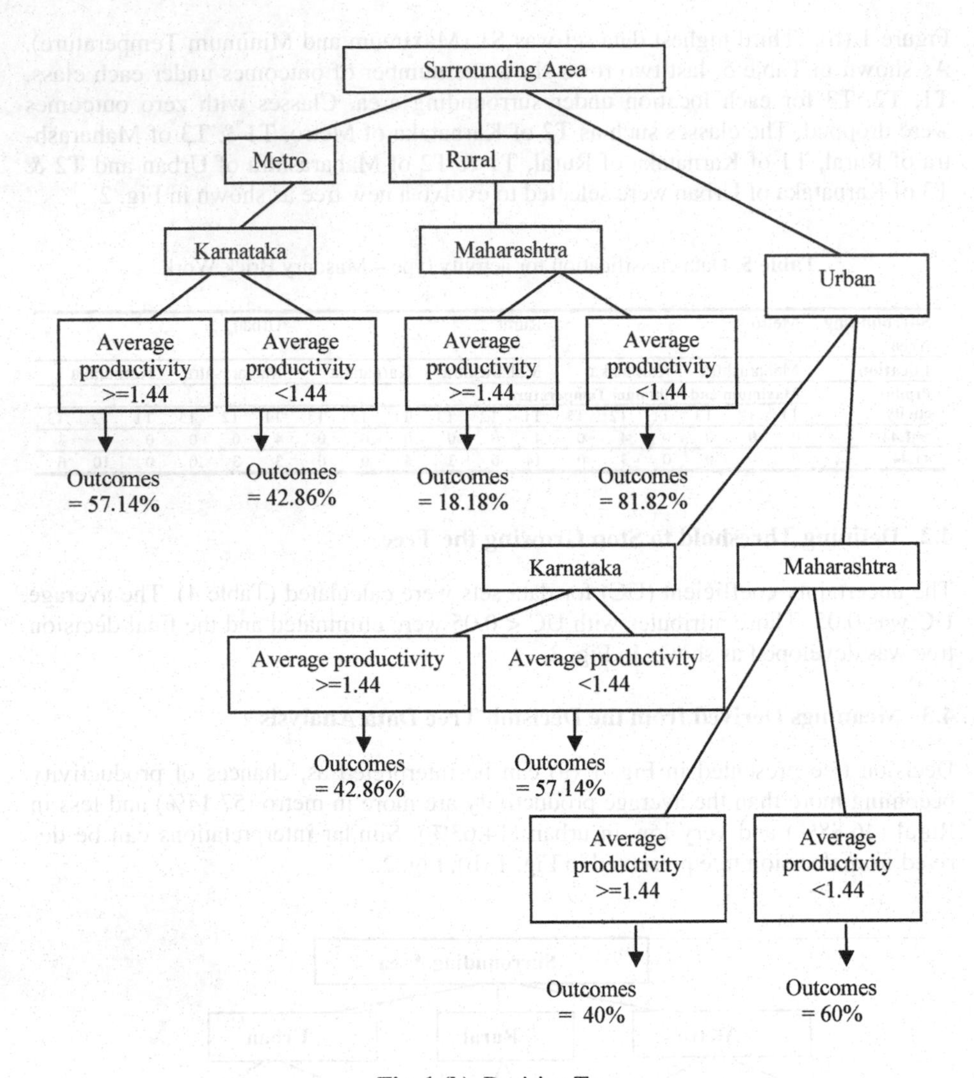

Fig. 1 (b). Decision Tree

4.4 Use of Method to Calculate Labor Requirement and Comparison with the Standard Labor Productivity

The method was used to calculate the labor productivity and labor requirement for brick work for a given quantity of work. The standard labor productivity of a mazdoor for brickwork is 1.60 cum for a day of 8 hours (Joglekar, 2000). This standard productivity for masonry brick work was used to calculate the expected labor requirement. Expected labor requirement (C) per day to complete same work was calculated as, $C = 6.53 / 1.6 = 4.08$. Expected labor requirement and the observed labor requirement for surrounding area were calculated as shown in Table 6.

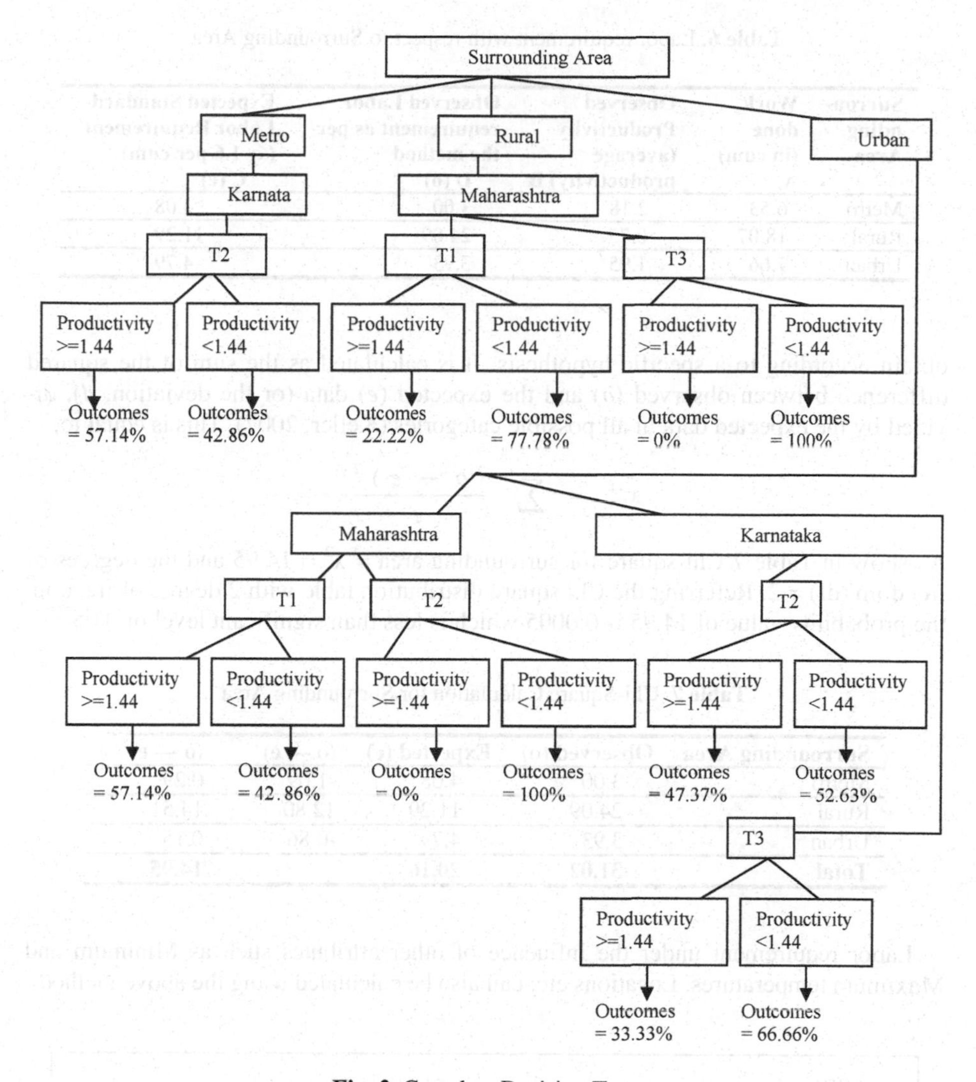

Fig. 2. Complete Decision Tree

The observed productivity for each day's work was calculated as total work done on that day divided by total maydays of laborers used to complete the work. This observed productivity of all work under a data set was averaged to average productivity (B) and was used to calculate the observed labor requirement (D) as, D = Work done (A) / Observed Productivity (B). For example for metro, observed labor requirement per day = D = 6.53/2.08 = 3.00.

Fig. 3, shows the pattern between the standard labor requirements and the labor requirement as per the method with some variation at each data point for surrounding This variation was measured by applying the Chi-square test. Chi-square is a statistical test commonly used to compare observed data with the data we would expect to

Table 6. Labor requirement with respect to Surrounding Area

Surrou-nding Area	Work done (in cum) A	Observed Productivity (average productivity) B	Observed Labor requirement as per the method D (o)	Expected Standard Labor Requirement (@ 1.6 per cum) C (e)
Metro	6.53	2.18	3.00	4.08
Rural	18.07	0.75	24.09	11.29
Urban	7.66	1.95	3.93	4.79

obtain according to a specific hypothesis. It is calculated as the sum of the squared difference between observed (*o*) and the expected (*e*) data (or the deviation, *d*), divided by the expected data in all possible categories (Keller, 2007). This is equal to,

$$x^2 = \sum \frac{(o - e)^2}{e}.$$

As show in Table 7 Chi-square for surrounding area = x^2 = 14.95 and the degrees of freedom (df) = 2. Referring the Chi square distribution table with 2 degree of freedom the probability value of 14.95 is 0.0005 which is less than significant level of 0.05.

Table 7. Chi-Square Calculation for Surrounding Area

Surrounding Area	Observed (o)	Expected (e)	(o — e)	$(o — e)^2/e$
Metro	3.00	4.08	-1.09	0.29
Rural	24.09	11.29	12.80	14.51
Urban	3.93	4.79	-0.86	0.15
Total	**31.02**	**20.16**		**14.95**

Labor requirement under the influence of other attributes such as Minimum and Maximum temperatures, Locations etc. can also be calculated using the above method.

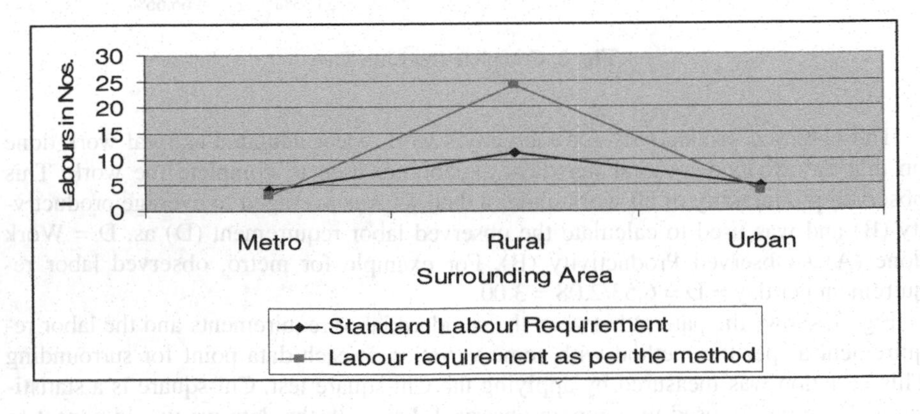

Fig. 3. Labor requirement with respect to Surrounding Areas

5 Conclusion

The methodology presented an improved version of decision tree construction process. It included three major approaches: 1) user intervention, 2) combining multiple variables, 3) defining threshold to ignore irrelevant variables and stop growing the tree. User intervention included participation of domain user to select the most influential variables having more than the average information gain. User also participated in selecting the variables to be combined that had more influence on labor productivity over when it was analyzed alone. The methodology therefore was a combination of statistical methods and experienced human participation to construction a more accurate decision tree predicting accurate meaning.

Labor productivity of brick work was analyzed to predict the influence of other parameters like location, temperatures etc. The patterns derived through the decision tree were used to estimate the labor requirement more accurately. The labor productivity calculated using the methodology was compared with the standard method used by construction professionals in India. The average error level was 0.06.

The approach was applied for Masonry Brick Work. The proposed approach can be applied to any activity as desired by the user and the patterns can be predicted.

References

[1] Ahmad, I., Nunoo, C.: Data Warehousing in the Construction Industry: Organizing and Processing Data for Decision-making. In: 8th International Conference on Durability of Building Materials and Components, Institute for Research in Construction, Vancouver, British Columbia (May-June 1999)

[2] Stella, G., Athanasios, V.: Data Warehousing: Concepts and Mechanisms. Informatique 1, 8–11 (1999)

[3] David, H., Heikki, M., Padhraic, S.: Principles of Data Mining, pp. 141–157. MIT Press, Cambridge (2004)

[4] Lee, J.-R., Hsueh, S.-L., Tseng, H.-P.: Utilizing Data Mining to Discover Knowledge in Construction Enterprise Performance records. Journal of Civil Engineering and Management 14(2), 79–84 (2008)

[5] Leu, S., Chee, N., Shiu-Lin, C.: Data mining for tunnel support: neural network approach. Journal of Automation in Construction 10(4), 429–441 (2000)

[6] Khaled, N.: Application Of Data-Mining To State Transportation Agencies' Projects Databases (March 2007), http://itcon.org/2007/8/

[7] Arun, P.K.: Data Mining Techniques, pp. 7–68. University Press (India) Pvt. Ltd., Hyderabad (2001)

[8] Soibelman, L.: Construction knowledge generation and dissemination. Berkeley-Stanford CE&M workshop: Defining a research agenda for AEC process/product development in 2000 and beyond (2000)

[9] Soibelman, L., Hyunjoo, K.: Data preparation process for construction knowledge generation through knowledge discovery in databases. Journal of Computing in Civil Engineering, ASCE 16(1), 39–47 (2002)

[10] Lucio, S., Stumpf, G., Marco, A.: A Knowledge Discovery in Databases Framework for Property Valuation. Journal of Property Tax Assessment and Administration 7(2), 77–106 (2002)

[11] Lucio, S., Liang, L., Jianfeng, W.: Data Fusion and Modeling for Construction Management Knowledge Discovery. In: International Conference on Computing in Civil and Building Engineering, ICCCBE, 10, Weimar, Bauhaus-Universität (2004)

Effects of Initial Search Bound on the Performance of Self-adaptive Evolutionary Computation Methods

Anjan Kumar Swain

Indian Institute of Management Kozhikode
Kozhikode-673 571
akswain@iimk.ac.in

Abstract. Self-adaptive evolutionary computation methods are widely used for finding global optimum in varieties of problem domains. One of the major demerits of these methods is premature convergence that stuck the search process at one of the local minimum. This paper examines this issue through an exhaustive study on the possible effects of initial search bound on the overall performance of the evolutionary computation methods.

1 Introduction

Evolutionary computing (EC) algorithms broadly consist of genetic algorithms (GAs) [1, 2], evolution strategies (ESs) [3, 4], and evolutionary programming (EP) [5, 6]. The widely used genetic algorithms model evolution on the basis of observed genetic mechanisms with a high dependence on crossover and sparing use of mutation operator. However, EP and ES methods share the majority of the common concepts for continuous parameter optimisation with the exception that EP uses a stochastic tournament selection and ES uses a purely deterministic selection. In addition, ES supports the use of crossover along with the main mutation operator [6-8].

All EC algorithms, can be described mathematically for a population P consisting of individuals p_i , $\forall i \in \{1, \ldots\ldots, \mu\}$, which are subjected to a series of operators such that

$$P(t+1) = s(v(P(t))), \tag{1}$$

where P(t) is a vector of individuals p_i with each p_i representing a vector of object variables p_{ij}: $j \in \{1, \ldots\ldots, n_o\}$ at generation "t" under a particular representation, $v(.)$ is the random variation operator, and $s(.)$ is the selection operator. The action of variation operator on parent individuals generates offspring. Then, the selection operator $s(.)$ operates on both offspring and/or parent individuals to determine which of the individuals will have the opportunity to reproduce. This selection operator usually depends on the fitness or the quality of the individuals determined by some predefined criteria. In EC methods, the mutation rates, mutation variances and recombination probabilities are called as *strategy parameters*. In self-adaptive EC methods the strategy parameters are updated by considering them as a part of the evolution process, i.e., the strategy parameters are allowed to evolve along with the object variables.

S.K. Prasad et al. (Eds.): ICISTM 2010, CCIS 54, pp. 314–324, 2010.

Self-adaptive evolutionary programming (EP) algorithms have been used extensively for global numerical optimization problems [6, 9]. The well-established self-adaptive EP methods work by evolving simultaneously all the object variables and their corresponding strategy parameters. A particular set of strategy parameters associated with an individual survives only when it produces better object variables. The most common variant of self-adaptive EP is the canonical self-adaptive EP (CEP)

$$p_i = \{(p_{ij}, \sigma_{ij}) \mid j = 1, \ldots, n_o\}, \forall i \in \{1, \ldots, \mu\}, \quad (2)$$

where subscript 'i' stands for the ith individual, n_o is the number of object variables and standard deviations; σ_{ij} is the standard deviation or the strategy parameter associated with the j th object variable of the i th individual; μ is the number of individuals in the population pool. Then, the strategy parameters σ_{ij} and the corresponding object variable p_{ij} can be updated as per the following equations:

$$\sigma_{ij}(t+1) = \sigma_{ij}(t)\exp(\tau N_{ij}(0,1) + \tau' N_i(0,1)) \text{, and} \quad (3)$$

$$p_{ij}(t+1) = p_{ij}(t) + \sigma_{ij}(t+1)N_{ij}(0,1) , \quad (4)$$

where $N_i(0, 1)$ and $N_{ij}(0, 1)$ are one-dimensional Gaussian random variates with expectation zero and standard deviation one; and the exogenous parameters τ and τ' are set to $\left(\sqrt{2\sqrt{n_o}}\right)^{-1}$ and $\left(\sqrt{2n_o}\right)^{-1}$, respectively [7].

The Cauchy mutation instead of Gaussian for continuous parameter optimisation has been used to speed up the convergence. Yao and Liu [10, 11], and Yao et al. [12] used a non-isotropic self-adaptation method with the only change made being to replace the normally distributed random variable $N_{ij}(0,1)$ in Eq.(4) by an one-dimensional standard Cauchy random variable $\Delta_{ij}(0,1)$. The probability density function (pdf) of a Cauchy variate can be described as [13]

$$(\pi\varphi)^{-1}\left[1+\left(\frac{x-\theta}{\varphi}\right)^2\right]^{-1}, \quad (5)$$

where $\varphi > 0$ is a scale parameter and θ is a location parameter. The corresponding cumulative distribution function (cdf) is

$$\frac{1}{2}+\pi^{-1}\tan^{-1}\left(\frac{x-\theta}{\varphi}\right). \quad (6)$$

This distribution is symmetrical about $x = \theta$. Also, it does not possess a finite expected value or standard deviation; its finite moments only exist for order less than one. Hence, a standard form for the Cauchy distribution $\Delta(0,1)$ can only be obtained by substituting zero for θ and one for φ. So, the standard pdf and cdf are represented as

$$\text{pdf} = \pi^{-1}\left(1+x^2\right)^{-1} \text{ and} \quad (7)$$

$$\text{cdf} = \frac{1}{2}+\pi^{-1}\tan^{-1}x . \quad (8)$$

Hence, the standard Cauchy variate $\Delta(0,1)$ is nothing but simply a problem independent unimodal, symmetric random variate with heavier tails than the normal variate.

These Cauchy-based modified algorithms are known as fast EP (FEP) or fast ES (FES) in the context of EP and ES research, respectively. FEP which is being discussed in this paper performs much better on multimodal functions with many local minima, while being comparable to CEP in most cases for unimodal and multimodal functions with few local optima. This better performance is attributed to the much flatter tail of the Cauchy variate that helps it to escape local optima very easily. Yao et al. [14] reported that the long jumps that the Cauchy variates perform compared to normal variates accelerate the convergence speed during initial period of search and are detrimental towards the final convergence. This is responsible for the poor performance of the Cauchy variate-based methods on certain optimization problems. Hence, Yao et al. showed that a blend of Cauchy and normal variates might help to find an overall better performance. This motivated subsequent empirical works by Sarvanan and Fogel[9] and Chellapilla [15] on hybrid Cauchy and Gaussian–based algorithms. The initial success of empirical results lead Rudolph [16] to analyse theoretically the relative behaviours of Cauchy and normal mutations. He reported that the normal variates provide faster local convergence on convex functions, whereas Cauchy variates are more helpful in escaping local optima. The FEP method can be described exactly in the same manner as in CEP (Eqns. 3 and 4) with the only exception that here the offspring are generated as per the following expression:

$$p_{ij}(t+1) = p_{ij}(t) + \sigma_{ij}(t+1)\Delta_{ij}(0,1)\,. \tag{9}$$

2 Analysis of Self-adaptive EC Methods

The problems inherent in self-adaptive methods are that all these are normally trapped at some local optimum, thereby leading to premature convergence. This was first reported by Liang et al. [17]. They observed that self-adaptive evolutionary algorithms are not even able to find a global optimum for simple functions and suggested the use of a fixed lower bound on each of the mutated parameters to improve the overall performance of these algorithms. Subsequently, Liang et al. [18] proposed a dynamic lower bound on σ_{ij} that resulted in performance improvement on some test functions. Due to the lack of any concrete method to avoid the premature convergence, researchers normally use a fixed lower bound on the mutation variables [15]. Also, Glickman and Sycara [19] presented three conditions that may be the possible causes of the premature convergence of all self-adaptive evolutionary algorithms. Of course, all these fall within the demerits of self-adaptive methods discussed above. However, their experimentation and assertions are based on training recurrent artificial neural networks (RANN).

These inherent demerits of all the self-adaptive methods [20] are due to one or many of the reasons described below:

i. In all the self-adaptive EC methods, only selection process guides their successive paths toward the optimal solution. Without selection, it is no more than a random search. Hence, this process-blind variation operator cannot know the nature of the fitness landscape. Hence, this may prove fatal in the presence of many local optima. Thus, the effectiveness of this method largely depends on the selection method.
ii. In all these methods, the fitness depends on the object variable values only, two individuals with same object variables and with substantially different standard deviations is treated as equivalent. This may help the solutions to loose diversity and thereby become trapped in prominent local optimum.
iii. In addition, the use of the multiplicative lognormal metaheuristics strategy to mutate the standard deviations, which helps the standard deviations to fall sharply to very low values. These low values of standard deviations effectively provide no modifications of the object variables, which therefore stagnates at a fixed value and behave as if struck at some local minima or wandering on a flat plateau.
iv. Further, this problem may arise due to the use of normal variates with a standard deviation that remains at a constant value of one. Thus, for functions having a global optimum amidst many local optima, it may be very probable that even a small initial search bound may fail to provide adequate search to locate the optimal solution.

Now, two of the following assertions are being derived from point (iv) described above:

Assertion 1: The self-adaptive methods are likely to perform better with large initial search domain only when there are not many local optima at the very close proximity of the global optimum.

Assertion 2: For a large initial search bound(s) the solutions are most likely to be dragged into suboptimal points.

The second assertion states that if the initial search bound is very large and the scale factor of the random variates is one and it uses a lognormal strategy for the variation of the strategy parameters, then all the incremental steps generated during the search process due to mutation will drop rapidly to such a low value that it will ultimately lead to premature convergence. Further, this problem may further aggravate with higher dimensions of the problem and in the presence of many local optima. To verify these assertions, in this paper, the effects of initial search domains with varieties of diverse problems has been studied thoroughly by considering a rich test-function base consisting of 23-benchmark functions.

3 Test-Functions and Simulation

For the experimental verification of our assertions described in the previous section, we have selected to use a large set of 23-benchmark functions that are the same as those originally considered in Yao et al. [14]. This large set of functions ensures that the conclusions drawn from this study are not biased towards a particular class of problems. These functions can be grouped as follows: (i) functions f_1 to f_{13} are high dimensional problems. Out of these, the functions f_1 to f_7 are unimodal, and f_8 to f_{13} are multimodal functions with many local minima. The number of these local minima increases with the dimension of the problem. Further, function f_6 is a discontinuous unimodal step function and f_7 is a unimodal noisy quartic function with an additive uniform random number in the range 0 to 1; (ii) functions f_{14} to f_{23} are all low dimensional. Further, these are all multimodal functions with few local minima. All the 23-benchmark functions are given in Table 1.

All the experiments performed on the CEP and FEP method have been performed under exactly the same conditions with initial standard deviation $\eta_{ij} = 3$, $\forall i \in \{1, ..., \mu\}$ and $\forall j \in \{1, ..., n_o\}$, and the same initial population size $\mu = 100$. The tournament size for all the experiments were fixed at c = 10. The number of generations for each function has been taken as per the work of Yao et al. [14]. For CEP and FEP, a constant lower bound $b_{ij} = 0.0001$, $\forall i \in \{1, ..., \mu\}$ and $\forall j \in \{1, ..., n_o\}$, on the strategy parameters has been included. At this lower bound, Chellapilla [15] reported better performance of FEP and CEP over that of no lower bound. All the results have been averaged over 50 runs. The statistical t-test has been used to study the statistical significance of the obtained results. In the simulation of f_8, the object variables are constrained to stay within the initial search domain. This is because of the reason that its true optimum is at ∞ where the function value is $-\infty$. However, for all other functions, no such restrictions have been imposed, and the initial search domain is only used once while initializing the population pool.

4 Results and Discussion

Initially, this work was started with the observations of the typical behavioral change of FEP on functions f_2, f_7 and f_9, the details of the results on all other function can be found in Swain [21, 22]. The performance of the FEP is poor compared to the CEP, and also the reported results showed some divergence in the convergence characteristics during part of the evolution process. To investigate this typical behavior of FEP, it can be observed that, compared to other similar functions in the function test-bed, the initial search domain or the feasible region is narrow in all these functions. In addition, the fitness value of function f_2 increases very rapidly with increase in the object variable values. Further, we have been observed that the function f_7 yields large objective function values if the randomly generated objective variable value goes far away from the true minimum. This may be true for the methods that use both Cauchy and normal distribution, as in the case of FEP where the Cauchy variates is used to update the object variables and the normal variates to update the strategy

Table 1. The 23-benchmark functions, where n_o is the function dimension, $f_{\min}$ is the inimum function value and SD is the user supplied search domain.

Test Functions	n_o	SD	$f_{\min}$
$f_1(x)=\sum_{i=1}^{n} x_i^2$	30	$[-100,100]^{n_o}$	0
$f_2(x)=\sum_{i=1}^{n} \lvert x_i \rvert+\prod_{i=1}^{n} \lvert x_i \rvert$	30	$[-10,10]^{n_o}$	0
$f_3=\sum_{i=1}^{n}\left(\sum_{j=1}^{i} x_j\right)^2$	30	$[-100,100]^{n_o}$	0
$f_4(x)=\max_i\{\lvert x_i \rvert, 1\le i\le n\}$	30	$[-100,100]^{n_o}$	0
$f_5(x)=\sum_{i=1}^{n-1}\left\{100(x_{i+1}-x_i^2)^2+(x_i-1)^2\right\}$	30	$[-30,30]^{n_o}$	0
$f_6(x)=\sum_{i=1}^{n}(\lfloor x_i+0.5\rfloor)^2$	30	$[-100,100]^{n_o}$	0
$f_7(x)=\sum_{i=1}^{n} ix_i^4+random(0,1)$	30	$[-1.28,1.28]^{n_o}$	0
$f_8(x)=\sum_{i=1}^{n} -x_i\sin\left(\sqrt{\lvert x_i\rvert}\right)$	30	$[-500,500]^{n_o}$	0
$f_9=\sum_{i=1}^{n}\left\{x_i^2-10\cos(2\pi x_i)+10\right\}$	30	$[-5.12,5.12]^{n_o}$	0
$f_{10}(x)=-20\exp\left(-0.2\sqrt{(1/n)\sum_{i=1}^{n} x_i^2}\right)$ $-\exp\left((1/n)\sum_{i=1}^{n}\cos(2\pi x_i)\right)+20+\exp(1)$	30	$[-32,32]^{n_o}$	0
$f_{11}=0.00025\sum_{i=1}^{n} x_i^2-\prod_{i=1}^{n}\cos\left(x_i/\sqrt{i}\right)+1$	30	$[-600,600]^{n_o}$	0
$f_{12}(x)=(\pi/n)\left\{10\sin^2(\pi y_1)+\sum_{i=1}^{n-1}(y_i-1)^2\left[1+10\sin^2(\pi y_{i+1})\right]+(y_n-1)^2\right\}$ $+\sum_{i=1}^{n} u(x_i,10,100,4),\ y_i=1+0.25(x_i+1)\ u(x_i,a,k,m)=k(x_i-a)^m, x_i>a; 0, -a\le x_i\le a; k(-x_i-a)^m, x_i<a;$	30	$[-50,50]^{n_o}$	0
$f_{13}(x)=0.1\left\{\sin^2(3\pi x_1)+\sum_{i=1}^{n-1}(x_i-1)^2\left[1+\sin^2(3\pi x_{i+1})\right]\right.$ $\left.+(x_n-1)^2\left[1+\sin^2(2\pi x_n)\right]\right\}+\sum_{i=1}^{n} u(x_i,5,100,4)$	30	$[-50,50]^{n_o}$	0
$f_{14}(x)=\left[.002+\sum_{j=1}^{25}\left(j+\sum_{i=1}^{n}(x_i-a_{ij})^6\right)^{-1}\right]^{-1}$	2	$[-65.536,65.536]^{n_o}$	0.980004
$f_{15}(x)=\sum_{i=1}^{11}\left[a_i-x_1\left(b_i^2+b_ix_2\right)\left(b_i^2+b_ix_3+x_4\right)^{-1}\right]^2$	4	$[-5,5]^{n_o}$	0.0003075
$f_{16}(x)=4x_1^2-2.1x_1^4+(1/3)x_1^6+x_1x_2-4x_2^2+4x_2^4$	2	$[-5,5]^{n_o}$	-1.0316285
$f_{17}(x)=\left(x_2-(5.1/(4\pi^2))x_1^2+(5/\pi)x_1-6\right)^2+10(1-(1/8\pi))\cos x_1+10$;	2	$[-5,10]^{n_o}$ $[0,15]^{n_o}$	0.398
$f_{18}(x)=\left[1+(x_1+x_2+1)^2\left(19-14x_1+3x_1^2-14x_2+6x_1x_2+3x_2^2\right)\right]\times\left[30+(2x_1-3x_2)^2\left(18-32x_1+12x_1^2+48x_2-36x_1x_2+27x_2^2\right)\right]$	2	$[-2,2]^{n_o}$	3.0
$f_{19}(x)=-\sum_{i=1}^{4} c_i\exp\left[-\sum_{j=1}^{3} a_{ij}(x_j-p_{ij})^2\right]$	3	$[0,1]^{n_o}$	-3.86
$f_{20}(x)=-\sum_{i=1}^{4} c_i\exp\left[-\sum_{j=1}^{6} a_{ij}(x_j-p_{ij})^2\right]$	6	$[0,1]^{n_o}$	-3.32
$f_{21}(x)=-\sum_{i=}^{5}\left[(x-a_i)(x-a_i)^T+c_i\right]^{-1}$	4	$[-10,10]^{n_o}$	-10
$f_{22}(x)=-\sum_{i=}^{5}\left[(x-a_i)(x-a_i)^T+c_i\right]^{-1}$	4	$[-10,10]^{n_o}$	-10
$f_{23}(x)=-\sum_{i=}^{10}\left[(x-a_i)(x-a_i)^T+c_i\right]^{-1}$	4	$[-10,10]^{n_o}$	-10

parameters. These observations encouraged the analysis of the effects of search bound on the solution evolution process. As an attempt to do this, the object variable values during the evolution process were constrained to stay within the prescribed search domain. This was achieved by limiting the values of object variables that cross the upper and lower bounds of the specified search domain to those of the respective boundary values. The results thus obtained have been tabulated in Table 2. This clearly shows that, under these circumstances, it is necessary to constrain the objective variable values during the evolution process to get substantially better results.

Table 2. Comparison of the effect of the search domain on functions f_2 , f_7 and f_9 . Here, SB and NSB represent *with search bound* and *without search bound*, respectively.

FUNC	GEN	NSB		SB	
		Average best	Average mean	Average best	Average mean
f_2	2000	2.057e-01 (8.57e-01)	2.213e-01 (8.99e-01)	7.979e-03 (6.78e-04)	1.118e-02 (8.39e-04)
f_7	3000	3.950e01 (4.88e01)	5.636e06 (1.37e07)	1.798e-02 (5.86e-03)	7.252e-02 (1.12e-02)
f_9	5000	1.227e02 (1.81e02)	5.520e02 (8.88e02)	4.441e-01 (7.26e-01)	4.4996e-01 (7.25e-01)

Encouraged by these results, further tests were conducted on rest of the functions. The results of CEP with search bound (CEPSB) and FEP with search bound (FEPSB) are shown in Tables 3 and 4, respectively. Table 3 shows the average best results averaged over 50 runs for CEP and FEP with no lower bound (NLB), fixed lower bound (FLB), and dynamic lower bound (DSLB) [22]. Table 3 indicates the average mean, averaged over 50 runs for CEP and FEP with NLB, FLB, and DSLB. Table 5 represents the t-test results between with search bound (SB) and no search bound (NSB), with NLB and DSLB for both FEP and CEP.

The results show that, for large initial search bound, constraining the object variables does not provide any remarkable performance change for CEP and FEP. The effects of search bound is profound for the high-dimensional unimodal and multimodal functions with many minima and a narrow search bound, and drastically affects the speed and accuracy of the results. However, for low-dimensional functions with few local minima, constraining the search domain does not affect the results a lot. As shown in Table 5 the results on f_{14} to f_{23} are mostly unaffected or only degraded slightly. Thus, it always seems advisable not to provide hard limiting search bounds on low-dimensional multimodal functions or on high-dimensional functions with a relatively wide search space. In other words, constraining the object variables for high-dimensional functions with a narrow feasible search region is always a good choice.

Table 3. CEP and FEP performance comparison of average best results averaged over 50 runs with NLB, FLB, and DSLB, with hard search bound. Here, NLB, FLB, and DSLB represent no lower bound, fixed lower bound, and differential step lower bound, respectively.

FUN	GEN	NLB		FLB		DSLB	
		CEP	FEP	CEP	FEP	CEP	FEP
f_1	1500	2.678e02 (442.759)	6.292e01 (106.537)	7.114e-02 (0.110595)	1.447e-04 (6.32e-04)	1.295e-04 (8.05e-04)	2.878e-06 (3.72e-06)
f_2	2000	5.296e00 (4.89532)	2.597e00 (2.7224)	2.520e-03 (1.69e-04)	7.979e-03 (6.78e-04)	1.120e-05 (2.76e-05)	1.609e-05 (3.81e-05)
f_3	5000	2.978e03 (1869.66)	1.993e03 (1529.5)	4.778e00 (5.43246)	1.927e00 (2.70533)	6.063e-02 (0.184738)	4.460e-03 (5.44e-03)
f_4	5000	2.796e00 (1.51537)	5.993e00 (4.03235)	1.657e00 (1.17999)	3.127e-01 (0.675225)	9.844e-03 (9.33e-03)	7.028e-05 (6.55e-05)
f_5	20000	2.437e04 (125402)	6.427e02 (2704.95)	2.574e00 (2.85808)	4.233e00 (4.93463)	1.694e00 (2.12981)	2.120e00 (2.01718)
f_6	1500	1.419e03 (1189.33)	1.404e02 (206.966)	1.941e03 (1606.21)	2.800e00 (17.0892)	1.223e02 (457.3)	8.000e-02 (0.274048)
f_7	3000	8.522e-01 (1.54977)	3.926e-01 (1.02354)	9.221e-02 (7.89e-02)	1.780e-02 (5.86e-03)	2.519e-02 (2.62e-02)	1.483e-02 (9.22e-03)
f_8	9000	-7.895e03 (601.779)	-1.098e04 (397.402)	-7.965e03 (702.89)	-1.101e04 (377.739)	-9.331e03 (682.442)	-1.148e04 (340.921)
f_9	5000	8.230e01 (18.3901)	2.821e01 (10.7266)	8.266e01 (29.5006)	4.441e-01 (0.726348)	5.544e01 (16.3822)	6.706e00 (3.7488)
f_{10}	1500	9.467e00 (2.58231)	3.219e00 (1.66619)	8.936e00 (3.03276)	4.355e-03 (6.76e-03)	1.519e00 (1.48235)	6.701e-04 (5.50e-04)
f_{11}	2000	8.890e00 (12.1184)	1.577e00 (2.45007)	9.090e-01 (2.98536)	1.012e-01 (0.153462)	1.455e-01 (0.791963)	1.170e-02 (1.64e-02)
f_{12}	1500	4.850e00 (2.81099)	1.289e00 (1.21332)	2.484e00 (1.94602)	2.554e-02 (6.08e-02)	2.283e-01 (0.368018)	2.073e-02 (6.62e-02)
f_{13}	1500	8.347e01 (309.636)	5.482e00 (4.74588)	5.012e00 (4.51996)	5.035e-03 (2.06e-02)	2.381e-01 (0.680757)	3.023e-04 (1.64e-03)
f_{14}	100	1.899e00 (1.29361)	1.337e00 (0.710715)	1.857e00 (1.42971)	1.408e00 (1.10931)	1.276e00 (0.634871)	1.018e00 (0.140577)
f_{15}	4000	4.794e-04 (3.28e-04)	5.373e-04 (3.31e-04)	3.624e-04 (2.20e-04)	4.906e-04 (3.70e-04)	4.540e-04 (3.39e-04)	5.089e-04 (3.83e-04)
f_{16}	100	-1.031628 (4.49e-16)	-1.031628 (4.49e-16)	-1.031628 (4.49e-16)	-1.031628 (4.49e-16)	-1.031628 (4.49e-16)	-1.031628 (4.49e-16)
f_{17}	100	0.3978874 (2.24e-16)	0.3978874 (2.24e-16)	0.3978874 (2.24e-16)	0.3978874 (2.24e-16)	0.3978874 (2.24e-16)	0.3978876 (1.58e-06)
f_{18}	100	3.0000000 (000000)	3.005668 (4.01e-02)	3.0000000 (000000)	3.095920 (0.478559)	3.0000000 (000000)	3.0000000 (1.41e-06)
f_{19}	100	-3.862782 (1.79e-15)	-3.862782 (1.79e-15)	-3.862782 (1.79e-15)	-3.862782 (1.41e-07)	-3.862782 (1.79e-15)	-3.862782 (1.79e-15)
f_{20}	200	-3.276060 (5.80e-02)	-3.267220 (5.98e-02)	-3.286316 (5.50e-02)	-3.262542 (6.00e-02)	-3.229258 (4.98e-02)	-3.217179 (7.10e-02)
f_{21}	100	-7.062635 (2.85218)	-5.568580 (1.54376)	-7.319284 (2.81869)	-5.836693 (1.91988)	-7.273396 (2.99559)	-5.473270 (1.53511)
f_{22}	100	-7.402268 (3.26269)	-5.703091 (2.29667)	-8.025962 (3.06821)	-6.010586 (2.62509)	-9.104189 (2.62185)	-7.216546 (3.05804)
f_{23}	100	-8.747636 (2.96752)	-6.279178 (3.10888)	-8.954647 (2.76883)	-7.466967 (3.35075)	-9.664067 (2.39478)	-8.450683 (2.79336)

Table 4. CEP and FEP performance comparison of average mean results averaged over 50 runs with NLB, FLB, and DSLB, with hard search bound.

		NLB		FLB		DSLB	
FUNC	GEN	CEP	FEP	CEP	FEP	CEP	FEP
f_1	1500	2.681e02	6.298e01	7.517e-02	1.835e-04	1.318e-04	3.811e-06
		(442.759)	(106.549)	(0.113022)	(7.61e-04)	(8.05e-04)	(4.73e-06)
f_2	2000	5.301e00	2.602e00	3.351e-03	1.118e-02	1.164e-05	2.052e-05
		(4.89293)	(2.72682)	(1.53e-04)	(8.39e-04)	(2.82e-05)	(5.38e-05)
f_3	5000	2.979e03	1.993e03	4.869e00	1.979e00	6.161e-02	5.127e-03
		(1869.35)	(1529.86)	(5.47639)	(2.77576)	(0.185406)	(7.48e-03)
f_4	5000	2.804e00	6.016e00	1.667e00	3.171e-01	1.008e-02	7.380e-05
		(1.51887)	(4.04606)	(1.18718)	(0.680916)	(9.52e-03)	(6.90e-05)
f_5	20000	2.437e04	6.427e02	2.576e00	4.241e00	1.697e00	2.127e00
		(125402)	(2704.95)	(2.85953)	(4.93748)	(2.13322)	(2.02134)
f_6	1500	1.419e03	1.405e02	1.941e03	2.819e00	1.294e02	8.000e-02
		(1189.3)	(206.962)	(1606.26)	(17.0877)	(457.98)	(0.274048)
f_7	3000	9.025e-01	4.470e-01	1.423e-01	7.252e-02	8.528e-02	8.514e-02
		(1.54911)	(1.02291)	(7.89e-02)	(1.12e-02)	(2.98e-02)	(1.76e-02)
f_8	9000	-7.895e03	-1.098e04	-7.965e03	-1.101e04	-9.331e03	-1.148e04
		(601.779)	(397.402)	(702.89)	(377.739)	(682.442)	(340.921)
f_9	5000	8.230e01	2.821e01	8.266e01	4.450e-01	5.544e01	6.706e00
		(18.3901)	(10.7266)	(29.5006)	(0.725103)	(16.3822)	(3.7488)
f_{10}	1500	9.467e00	3.225e00	8.936e00	5.286e-03	1.519e00	7.809e-04
		(2.58224)	(1.66673)	(3.03276)	(6.76e-03)	(1.48235)	(6.59e-04)
f_{11}	2000	8.895e00	1.577e00	9.098e-01	1.012e-01	1.455e-01	1.170e-02
		(12.1212)	(2.45026)	(2.98531)	(0.153462)	(0.791963)	(1.64e-02)
f_{12}	1500	4.857e00	1.291e00	2.514e00	2.603e-02	2.285e-01	2.073e-02
		(2.80717)	(1.21392)	(1.95387)	(6.12e-02)	(0.368055)	(6.62e-02)
f_{13}	1500	8.585e01	5.508e00	5.090e00	6.028e-03	2.381e-01	3.032e-04
		(323.659)	(4.76692)	(4.56177)	(2.48e-02)	(0.680751)	(1.64e-03)
f_{14}	100	2.207e00	1.498e00	2.058e00	1.437e00	1.276e00	1.018e00
		(1.7403)	(0.894568)	(1.661)	(1.13135)	(0.63487)	(0.140577)
f_{15}	4000	4.795e-04	5.377e-04	3.624e-04	4.906e-04	4.540e-04	5.089e-04
		(3.28e-04)	(3.31e-04)	(2.20e-04)	(3.70e-04)	(3.39e-04)	(3.83e-04)
f_{16}	100	-1.031628	-1.031628	-1.031628	-1.031628	-1.031628	-1.031619
		(4.49e-16)	(4.49e-16)	(4.49e-16)	(4.49e-16)	(4.49e-16)	(6.41e-05)
f_{17}	100	0.3978874	0.3978874	0.3978874	0.3978874	0.3978874	0.3979092
		(2.24e-16)	(2.24e-16)	(2.24e-16)	(2.24e-16)	(2.24e-16)	(1.54e-04)
f_{18}	100	3.0000000	3.454556	3.0000001	4.726044	3.0000000	3.0000016
		(000000)	(3.21407)	(2.40e-07)	(9.20109)	(000000)	(1.13e-04)
f_{19}	100	-3.862782	-3.862782	-3.862782	-3.862779	-3.862782	-3.862782
		(1.79e-15)	(1.62e-06)	(1.79e-15)	(8.36e-06)	(1.79e-15)	(1.84e-06)
f_{20}	200	-3.275974	-3.267193	-3.286305	-3.262529	-3.229258	-3.194542
		(5.80e-02)	(5.98e-02)	(5.50e-02)	(6.00e-02)	(4.98e-02)	(0.196917)
f_{21}	100	-7.062191	-5.568230	-7.316297	-5.823788	-7.097563	-5.263785
		(2.85173)	(1.54369)	(2.82229)	(1.93774)	(3.10167)	(1.61585)
f_{22}	100	-7.307106	-5.633761	-7.923894	-5.992247	-9.010666	-7.088381
		(3.33361)	(2.29935)	(3.13801)	(2.64019)	(2.6777)	(3.03909)
f_{23}	100	-8.611583	-6.224645	-8.854568	-7.353902	-9.663200	-7.931150
		(3.02021)	(3.13937)	(2.91609)	(3.44398)	(2.39447)	(3.17695)

Table 5. t-test results between NSB-SB of CEP and FEP. The positive results favour the second method (SB) and vice versa. Dashed lines indicate always with SB.

FUN	CEP		FEP		FUN	CEP		FEP	
	NLB	DSLB	NLB	DSLB		NLB	DSLB	NLB	DSLB
f_1	-0.326248 (0.74563)	-0.703896 (0.48483)	-0.127286 (0.89924)	-0.468675 (0.64138)	f_{13}	1.01158 (0.31671)	-0.816626 (0.41810)	1.00655 (0.31910)	2.11719 (3.93e-02)
f_2	2.06763 (4.40e-02)	1.02503 (0.31038)	1.07777 (0.28642)	1.01138 (0.31680)	f_{14}	-1.000000 (0.32222)	-1.000000 (0.32222)	0.000000 (1.00000)	0.000000 (1.00000)
f_3	0.0534638 (0.95758)	1.259 (0.21399)	0.763773 (0.44867)	-1.3481 (0.18383)	f_{15}	9.04151 (5.11e-12)	7.26028 (2.62e-09)	8.87229 (9.15e-12)	9.20021 (2.97e-12)
f_4	0.591186 (0.55711)	0.399348 (0.69137)	-2.01744 (0.04914)	0.847387 (0.40090)	f_{16}	0.000000 (1.00000)	1.00000 (0.32222)	0.000000 (1.00000)	0.000000 (1.00000)
f_5	1.11384 (0.27078)	1.47239 (0.14731)	1.48265 (0.14457)	1.25798 (0.21436)	f_{17}	0.000000 (1.00000)	0.000000 (1.00000)	0.000000 (1.00000)	-1.00000 (0.32222)
f_6	1.55574 (0.12621)	-0.084165 (0.93327)	0.699824 (0.48735)	0.819288 (0.41659)	f_{18}	0.000000 (1.00000)	0.000000 (1.00000)	-1.00003 (0.32222)	1.00000 (0.32222)
f_7	4.55323 (3.51e-05)	1.26986 (0.21013)	10.037 (1.78e-13)	5.37417 (2.12e-06)	f_{19}	1.000000 (0.32222)	0.000000 (1.00000)	1.5771 (0.12121)	1.43244 (0.15837)
f_8	-----	-----	----	-----	f_{20}	2.50342 (1.57e-02)	-0.650707 (0.51828)	2.55717 (1.37e-02)	-1.38173 (0.17332)
f_9	6.07398 (1.80e-07)	1.92548 (6.00e-02)	5.4532 (1.61e-06)	4.21901 (1.06e-04)	f_{21}	-0.234789 (0.81535)	-1.59828 (0.11641)	-4.4568 (4.84e-05)	-4.83983 (1.34e-05)
f_{10}	-0.454629 (0.65138)	0.0796548 (0.93684)	0.792391 (0.43195)	2.14487 (3.69e-02)	f_{22}	-2.39862 (0.02031)	-0.934669 (0.35454)	-5.44911 (1.63e-06)	-2.73117 (8.75e-03)
f_{11}	0.153643 (0.87852)	-0.33352 (0.74017)	0.60894 (0.54538)	-0.272481 (0.78640)	f_{23}	-0.262911 (0.79372)	-1.37215 (0.17627)	-1.51704 (0.13568)	-2.82662 (6.79e-03)
f_{12}	1.53533 (0.13113)	-0.177354 (0.85997)	1.37004 (0.17692)	-1.66148 (0.10300)					

5 Conclusions

In this paper, the effects of search bounds on the performance of different EP algorithms have been studied. It has been shown that constraining the object variables high dimensional functions with narrow feasible regions to the initial search bounds provides advantages over that of unbounded searching. Further, on low dimensional multimodal functions and high dimensional functions with initial wide search space, the search bound constraints are not advisable.

References

[1] Goldberg, D.E.: Genetic Algorithms in Search, Optimization and Machine Learning. Addison-Wesley Publishing, Reading (1988)

[2] Davis, L. (ed.): Handbook of Genetic Algorithms. Van Nostrand Reinhold, New York (1991)

[3] Schwefel, H.-P.: Evolution and optimum seeking. Wiley, New York (1995)

[4] Schwefel, H.P.: Numerical Optimization of Computing Models. John Wiley, Chichester (1981)

[5] Fogel, L.J., Owens, A.J., Walsh, M.J.: Artificial Intelligence through Simulated Evolution. John Wiley, New York (1966)

[6] Fogel, D.B.: Evolutionary Computation: Towards a New Philosophy of Machine Intelligence. IEEE Press, New York (1995)
[7] Bäck, T., Schwefel, H.-P.: An overview of evolutionary algorithms for parameter optimization. Evolutionary Comput. 1(1), 1–23 (1993)
[8] Bäck, T.: Evolutionary algorithms in theory and practice. Oxford University Press, Inc., NY (1997)
[9] Sarvanan, N., Fogel, D.B., Nelson, K.M.: A comparison of methods for self-adaptation in evolutionary algorithms. BioSystems 36, 157–166 (1995)
[10] Yao, X., Liu, Y.: Fast evolutionary programming. In: Fogel, L.J., Angeline, P.J., Bäck, T. (eds.) Evolutionary Programming: Proc. Fifth Annual Conf. Evolutionary Programming, pp. 451–460. MIT Press, Cambridge (1996)
[11] Yao, X., Liu, Y.: Fast evolution strategies. Control and Cybernatics 26(3), 467–496 (1997)
[12] Yao, X., Liu, Y., Lin, G.: Evolutionary programming made faster. IEEE Trans. Evolutionary Comput. 3(2), 82–102 (1999)
[13] Johnson, N.L., Kotz, S., Balakrishnan, N.: Continuous univariate distributions, vol. 1. John Wiley & sons, Inc., USA (1994)
[14] Yao, X., Liu, Y., Lin, G.: Evolutionary programming made faster. IEEE Trans. Evolutionary Computation 3(2), 82–102 (1999)
[15] Chellapilla, K.: Combining mutation operators in evolutionary programming. IEEE Trans. Evolutionary Comput. 2(3), 91–96 (1998)
[16] Rudolph, G.: Local convergence rates of simple evolutionary algorithms with Cauchy mutations. IEEE Trans. Evolutionary Computation 1(4), 249–258 (1997)
[17] Liang, K.-H., Yao, X., Newton, C., Hoffman, D.: An experimental investigation of self-adaptation in evolutionary programming. In: Porto, V.W., Waagen, D. (eds.) EP 1998. LNCS, vol. 1447, pp. 291–300. Springer, Heidelberg (1998)
[18] Liang, K.-H., Yao, X., Newton, C.: Dynamic control of adaptive parameters in evolutionary programming. In: McKay, B., Yao, X., Newton, C.S., Kim, J.-H., Furuhashi, T. (eds.) SEAL 1998. LNCS (LNAI), vol. 1585, pp. 42–49. Springer, Heidelberg (1999)
[19] Glickman, M.R., Sycara, K.: Reasons for premature convergence of self-adapting mutation rates. In: Proc. Congress on Evolutionary Computation (CEC 2000), San Diego, CA, pp. 62–69 (2000)
[20] Swain, A.K., Morris, A.S.: A novel hybrid evolutionary programming method for function optimization. In: Proc. Congress on Evolutionary Computation (CEC 2000), San Diego, CA, pp. 1369–1376 (2000)
[21] Swain, A.K.: Performance Analysis of Self-Adaptive Evolutionary Computation Methods. In: Proc. National Conference on IT and Soft Computing (ITSC) (November 2006)
[22] Swain, A.K., Morris, A.S.: Performance Improvement of Self-Adaptive Evolutionary Methods with a Dynamic Lower Bound. International Journal of Information Processing Letters 82, 55–63 (2002)

A Computational Study of Margining Portfolios of Options by Two Approaches

Edward G. Coffman, Jr.[1], Dmytro Matsypura[2], and Vadim G. Timkovsky[2]

[1] Columbia University, NY 10027, USA
coffman@cs.columbia.edu
[2] University of Sydney, NSW 2006, Australia
{dmytro.matsypura,vad.timkovsky}@sydney.edu.au

Abstract. This paper presents preliminary results of a computational experiment with the *strategy-based approach* and the *risk-based approach* to portfolio margining with the purpose to clarify which one yields lower margin requirements under different scenarios. There exists a widespread opinion that the risk-based approach is always a winner in this competition, and therefore the strategy-based approach must be disqualified as outdated. However, the results of our experiment with portfolios of stock options show that, in many practical situations, the strategy-based approach yields substantially lower margin requirements in comparison with the risk-based approach.

1 Introduction

Margin payments in *margin accounts* of investors, i.e., *customers* of brokers, are based on established minimum *margin requirements* which depend on a large number of factors, such as current market price, expiry date and other characteristics of securities held in the *positions* of the accounts. Margin rules exist for margining single positions, *position offsets* such as trading *strategies* or *portfolios* comprising all positions with a common *underlying instrument*, and entire accounts which are collections of portfolios.

With adjustments in minimum margin requirements[1], margin regulators keep margin debt of the stock market at a level consistent with the canons of a healthy economy. High margin requirements can reduce investors' activity, lead to underpricing of securities, and cause economic slowdowns. Low margin requirements, in turn, lead to overpricing of securities, high levels of speculation, cash deficits, market crashes, and, again, economic slowdowns. The challenge is

[1] In what follows, we omit the word "minimum" since we mean only minimum margin requirements.

S.K. Prasad et al. (Eds.): ICISTM 2010, CCIS 54, pp. 325–332, 2010.

to find a proper approach to portfolio margining, a "golden mean" that keeps the growth of margin debt within tolerable limits. Such an approach is specified by the definition of the offsets and *margin rates*, i.e., margin requirements per unit of security or offset.

Current margining practice uses two approaches, *strategy-based* and *risk-based.*[2] The strategy-based approach uses offsets that imitate trading strategies while the risk-based approach considers the entire portfolio to be an offset. The strategy-based approach uses margin rates calculated without variations of the current market prices of the underlying instruments. The risk-based approach uses margin rates calculated by artificial variations of the current market prices with the purpose to find a worst-case scenario in their moves. Consequently, the strategy-based approach leads to a computationally complex combinatorial problem while the risk-based approach introduces uncertainty in pricing of derivatives. The goal of this paper is a comparative analysis of these two approaches that, to the best of our knowledge, has never been done before.

2 Strategy-Based Approach

The combinatorial essence of the strategy-based approach arises from the ability to partition a portfolio in many different ways corresponding to different offsets in the margin rule book. Each securities market follows its own margin rule book, for example, NYSE Rule 431 in the U.S. or Reg. 100 in Canada. The strategy-based offsets are of fixed size and imitate trading strategies such as *calendar spreads, straddles, butterfly spreads, box spreads.* In brokerage jargon, offsets of sizes 1, 2, 3 and 4 are *singles, pairs, tripos* and *quads*; every position in an offset is referred to as a *leg.* Complex offsets such as *calendar iron condor spreads* have 12 legs.[3]

Let P be a portfolio, and let $P = x_1\dot{O}_1 + \ldots + x_k\dot{O}_k$ be a partition of P, where $\dot{O}_1, \ldots, \dot{O}_k$ are *prime offsets*, and $x_1, \ldots, x_k$ are their *multiplicities*;[4] see Fig. 1 for an example. If m_j is the margin requirement for the prime offset $\dot{O}_j$ then, in accordance with the strategy-based approach, the margin requirement for P is $m_1x_1 + \ldots + m_kx_k$. Hence, finding a minimum regulatory margin requirement for P is finding multiplicities $x_1, \ldots, x_k$, such that the latter sum achieves the minimum. We will refer to this problem as the *portfolio margin minimization* (PMM) problem. If the maximum offset size in the rule book is d then we will say that the PMM problem is of dimension d. For $d = 2$ it is known as the PMM problem by pairing [1].

[2] The concept also appears in margin regulations and business-related literature under the names "portfolio margining approach" and "risk-based portfolio margining approach".

[3] SEC Release 34-52738, December 14, 2005. We should also mention that it is possible to design *offsets-centipedes* with as many legs as desired [2].

[4] Prime offsets have minimal security quantities; offsets' multiplicities are nonnegative integers.

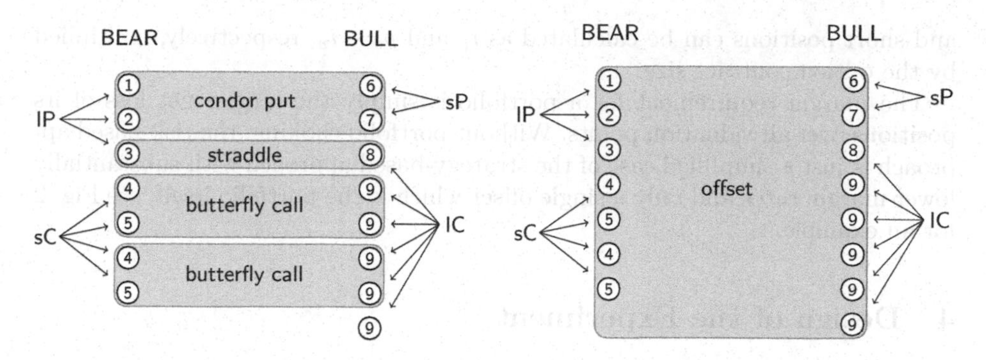

Fig. 1. A portfolio with 9 (5 bearish and 4 bullish) positions in options and 15 option contracts

Fig. 2. The portfolio from Fig. 1. It has only one risk-based offset which is the portfolio itself

Integers represent positions in options. Each option contract is depicted by a circle. The abbreviations IC/IP and sC/sP stand for the set of long and short call/put option contracts, respectively. Figure 1 shows the prime partition $\dot{O}_1 + \dot{O}_2 + 2\dot{O}_3 + \dot{O}_4$ of the portfolio that involves the strategy-based prime offsets $\dot{O}_1, \dot{O}_2, \dot{O}_3, \dot{O}_4$ which are a condor put spread, a long straddle, a butterfly call spread, and a naked long call option contract, respectively.

3 Risk-Based Approach

Similarly to the strategy-based approach, the risk-based approach uses the current market price of a security for margin calculations, and in addition, it considers price variations within certain ranges in an attempt to catch the worst-case price movement for the entire portfolio. This technique is called *portfolio choking.*

According to the risk-based approach, in order to calculate the margin requirement for each unit of a security s whose underlying security u has the market price c, it is necessary to consider in addition to c, ten more *valuation points* $c(1 - ai)$ and $c(1 + bi)$, where $i = 1, 2, 3, 4, 5$, and the constants a and b are chosen such that the lowest/highest valuation points appear to be lower/higher than c by a certain percentage.[5] For stocks, for example, it is 15%.

Let c_v, $1 \leq v \leq 11$, be one of the eleven valuation points including c. If $s = u$, i.e., s is the underlying security, then the difference $o_v = c_v - c$ or $c - c_v$ shows the outcome (*gain* if positive, and *loss* if negative) associated with point c_v for long or short positions in s, respectively, for each security unit.[6] If s is a derivative, then the outcome associated with the valuation point c_v should be calculated in accordance with the mechanism of the derivative. For example, if the security s is an option then, after calculating its in-the-money amount, $i_v = \max\{c_v - e, 0\}$ if s is a call option or $\max\{e - c_v, 0\}$ if s is a put option, its outcome in long

[5] Rule 15c3-1a, section (b)(1)(i)(B), sets the range from 6% to 15%.

[6] These gains and losses are called "theoretical gains and losses" in SEC Release 34-53577.

and short positions can be calculated as i_v and $p_v - i_v$, respectively, multiplied by the option contract size.

The margin requirement for a portfolio is simply the largest net loss of its positions over all valuation points. Without portfolio choking, the risk-based approach is just a simplified case of the strategy-based approach with substantially lower margin rates and only a single offset which is the portfolio itself; see Fig. 2 for an example.

4 Design of the Experiment

The strategy-based approach cannot compete with the risk-based approach in the speed of computations because the latter is free of any combinatorial quantities. Nor does it give an advantage in *initial* margin requirements, which are substantially higher than those obtained by the risk-based approach. But an answer to the question of whether the strategy-based approach always yields higher *maintenance* margin requirements in comparison with the risk-based approach is not obvious. The results of our experiment show that the answer is negative: the strategy-based approach yields substantially lower maintenance margin requirements for large portfolios in certain scenarios. To demonstrate this, we compare the behavior of the portfolio *maintenance* margin requirement as a function of the portfolio size that is calculated by both the strategy-based approach and the risk-based approach in accordance with NYSE Rule 431.

Another goal of our experiment is to contrast different strategy-based algorithms and clarify to what extent offsets with numbers of legs more than two can reduce maintenance margin requirements. Note that the risk-based approach has only one hard-coded algorithm.

We say that a strategy-based algorithm is of *dimension d* if it solves the PMM problem of dimension d, i.e., it takes into consideration all regulatory offsets of sizes $1, 2, \ldots, d$. We consider dimensions two and four only.[7] Strategy-based algorithms of dimensions two and four were the algorithms for solving integer linear programs of the related PMM problems by the CPLEX optimization software package.[8]

The main idea of portfolio variation is to first build a *maximal portfolio* and then randomly remove positions to provide a monotonic reduction of its size. The margin requirement was computed for each generated portfolio by using the strategy-based algorithms of dimensions two and four, and the risk-based algorithm. The portfolios were generated by performing the following steps:

[7] Note that an algorithm of dimension one just calculates the total positions' margin. Dimension three does not deserve a special consideration because option tripos are not permitted by NYSE Rule 431. Offsets of sizes more than four were not permitted up to December 2005 and are currently not being used in margining practice. The popularity of the risk-based approach, which appeared in the U.S. market in July 2005, stalled further developments of strategy-based margining systems.

[8] We used ILOG CPLEX 12.1 on Dell Precision T7400 with two 3.5 GHz Quad-Core Intel Xeon CPUs, 32 GB RAM running Windows XP 64-bit.

Step 1. A group of 16 call options and a group of 16 put options were selected[9] such that exactly 8 options inside each group were in the money; see Fig. 3.

Step 2. The maximal portfolio with 32 positions was built by creating 8 long positions in randomly chosen 8 call options and 8 short positions in the remaining 8 call options; the other 16 positions in put options were created in the same way. This step was repeated 10 times and resulted in 10 unique maximal portfolios. The next steps were repeated for each maximal portfolio.

Step 3. The number of option contracts in each position was randomly generated in the range from 1 to 10. This step was repeated 50 times.

Step 4. A randomly selected position from one side (bearish or bullish) was removed to get a portfolio of the smaller size.

Step 5. The number of option contracts in each remaining position was randomly generated in the range from 1 to 10. This step was repeated 50 times.

Step 6. Steps 4-5 were repeated 29 times to get a total of 30 sets of 50 randomly generated portfolios with sizes monotonically decreasing from 32 to 3. The side from which a position was to be removed in Step 4 was alternated to maintain a balance between the number of bearish and bullish positions.

Step 7. Steps 3-6 were repeated 25 times alternating the starting side in Step 4. Each time the random number generator was restarted to avoid repeated patterns.

Step 8. The margin requirements were averaged for portfolios of the same size. Hence, we calculated 30 averaged margin requirements for each algorithm.

Steps 1 through 8 create a *symmetric scenario* because position quantities chosen at Steps 3 and 5 are distributed uniformly between 1 and 10. The symmetric scenario generates *balanced portfolios* where the numbers of call and put options, in-the-money and out-of-the-money options, long and short positions are approximately the same. Thus, to obtain each point in the graph, we computed and averaged margin requirements for 12,500 portfolios, with the total of $12{,}500 \cdot 30 = 375{,}000$ unique and randomly generated balanced portfolios.

We also performed this experiment in six *asymmetric scenarios* to model *unbalanced portfolios* with different kinds of asymmetry. We performed the same steps as in the above algorithm except for Steps 3 and 5, where the quantities of options in the positions were ranging according to the following three scenarios, where quantities A and B were random integers in the intervals $[7, 10]$ and $[1, 4]$, respectively:

$$\begin{aligned}\textsf{Long Portfolio}: &\ A/B \text{ option contracts for each long/short position,}\\ \textsf{Call Portfolio}: &\ A/B \text{ option contracts for each position in call/put options,}\\ \textsf{Bull Portfolio}: &\ A/B \text{ option contracts for each bullish/bearish position.}\end{aligned}$$

[9] These 32 options were on the IBM stock at the market price of \$84.92 and expired on April 17, 2009. The data was taken as of the end of the day of January 16, 2009, from `http://finance.yahoo.com/`. For pricing options we used interest rate 0.3% and historical volatility 15% (since 2000). Note that we could have chosen any other stock with sufficient number of options.

The other three asymmetric scenarios, Short Portfolio, Put Portfolio, Bear Portfolio, respectively, were obtained by transposing A and B in the above three definitions.[10]

5 Results of the Experiment

The results of the experiment is presented in Figs. 4 through 10, where margin requirements are given in thousands of dollars for portfolio sizes $3, 4, 5, \ldots, 31, 32$. The notation S2, S4, R stands for margin requirements obtained by the strategy-based algorithms of dimensions two, four, and the risk-based algorithm, respectively.

Our main conclusion is that the strategy-based approach does not always yield a higher margin requirement than the risk-based approach.

Comparing the curves S4 and R, we can conclude that the strategy-based approach yields lower margin requirements than the risk-based approach for balanced portfolios (Fig. 4), long portfolios (Fig. 5) and put portfolios (Fig. 8), of sizes more than 21, 15 and 8, respectively; and they are substantially lower for portfolio sizes closer to 32.

In the other four scenarios, i.e., for short portfolios (Fig. 6), call portfolios (Fig. 7), bull portfolios (Fig. 9) and bear portfolios (Fig. 10), the risk-based approach is always more advantageous for all portfolio sizes. It is especially advantageous for investors playing call or bear because the risk-based margining curve is almost flat in Figs. 7 and 10.

With regard to the strategy-based approach, we can conclude that using quads along with pairs, i.e., strategy-based algorithms of dimension four, yields a substantial margin reduction in comparison with using only pairs; it is especially noticeable for balanced portfolios (Fig. 4), long portfolios (Fig. 5), short portfolios (Fig. 6), call portfolios (Fig. 7) and put portfolios (Fig. 8).

Except for scenarios with short portfolios (Fig. 6), bull portfolios (Fig. 9) and bear portfolios (Fig. 10), the trend of the strategy-based requirement is to be *smaller* for *larger* portfolios. Only long portfolios (Fig. 5) reveal the same but weaker trend for the risk-based approach. Thus, all margin requirements are decreasing with the growth of long portfolio size.

It is interesting to notice that the strategy-based approach remains almost insensitive to switching the asymmetry type from call portfolios (Fig. 7) to put portfolios (Fig. 8), while the risk-based approach changes almost flat margin requirement curve into linearly increasing. Thus, the risk-based approach signals about risk associated with put options but remains neutral to call options. Switching the asymmetry type from bull portfolios (Fig. 9) to bear portfolios (Fig. 10) reveals the same behavior of the strategy-based approach but an opposite behavior of the risk-based approach. This phenomenon can be explained only by the portfolio choking technique introduced in the risk-based approach but not used in the strategy-based approach.

[10] Recall that long positions in call options and short positions in put options are bullish, long positions in put options and short positions in call options are bearish.

#	ex price	call opt id	opt price	put opt id	opt price
1	45	IBMDX	39.70	IBMPX	0.45
2	50	IBMDU	35.50	IBMPU	0.67
3	55	IBMDV	31.90	IBMPV	1.00
4	60	IBMDL	25.30	IBMPL	1.45
5	65	IBMDM	21.50	IBMPM	1.90
6	70	IBMDN	17.30	IBMPN	2.70
7	75	IBMDO	13.50	IBMPO	3.90
8	80	IBMDP	10.10	IBMPP	5.34
9	85	IBMDQ	7.10	IBMPQ	7.38
10	90	IBMDR	4.63	IBMPR	10.00
11	95	IBMDS	2.85	IBMPS	14.83
12	100	IBMDT	1.75	IBMPT	17.02
13	105	IBMDA	0.95	IBMPA	21.50
14	110	IBMDB	0.50	IBMPB	26.03
15	115	IBMDC	0.20	IBMPC	28.40
16	120	IBMDD	0.15	IBMPD	32.90

Fig. 3. Selected Options

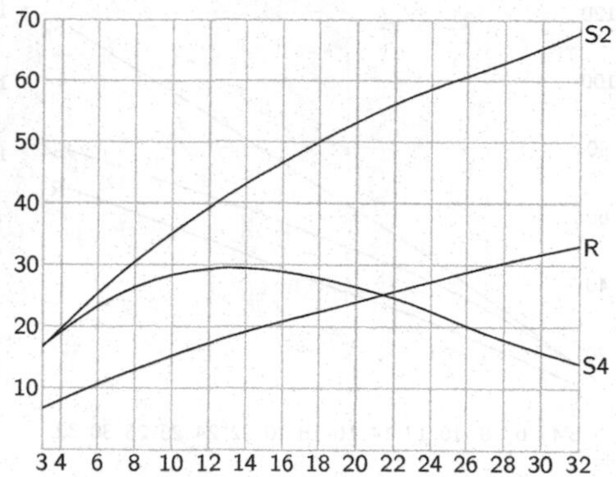

Fig. 4. Balanced Portfolios

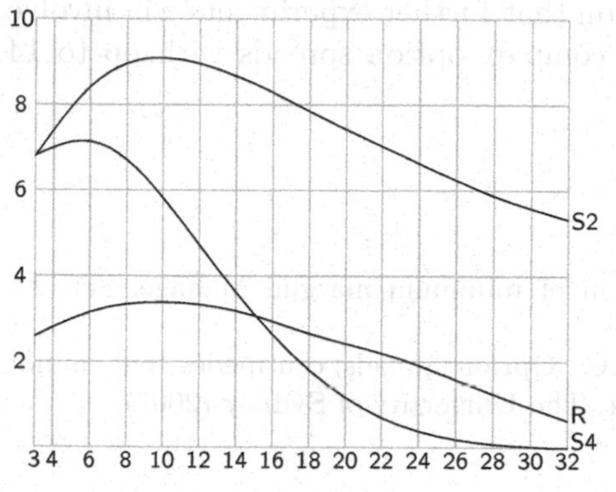

Fig. 5. Long Portfolios

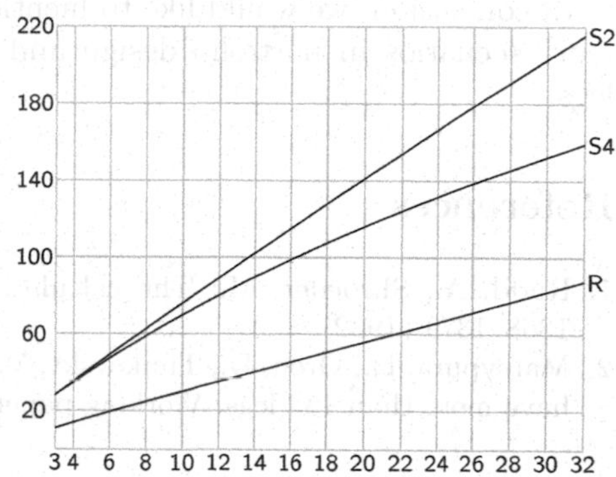

Fig. 6. Short Portfolios

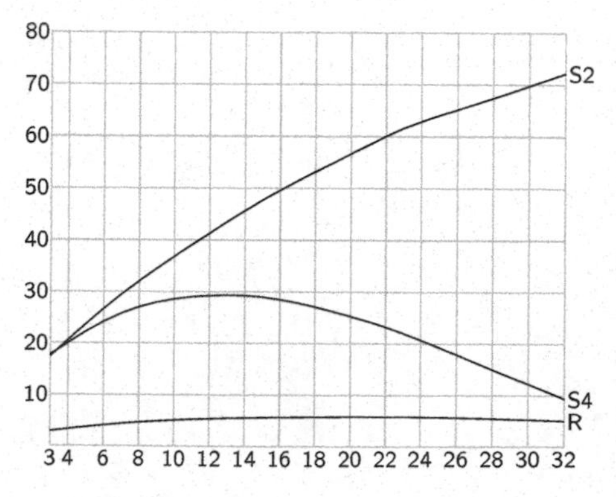

Fig. 7. Call Portfolios

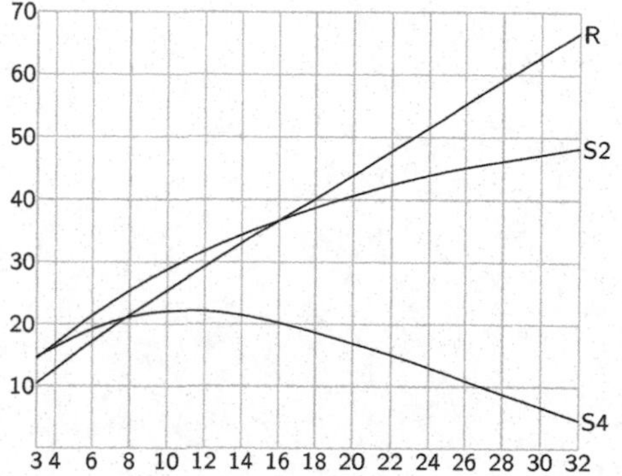

Fig. 8. Put Portfolios

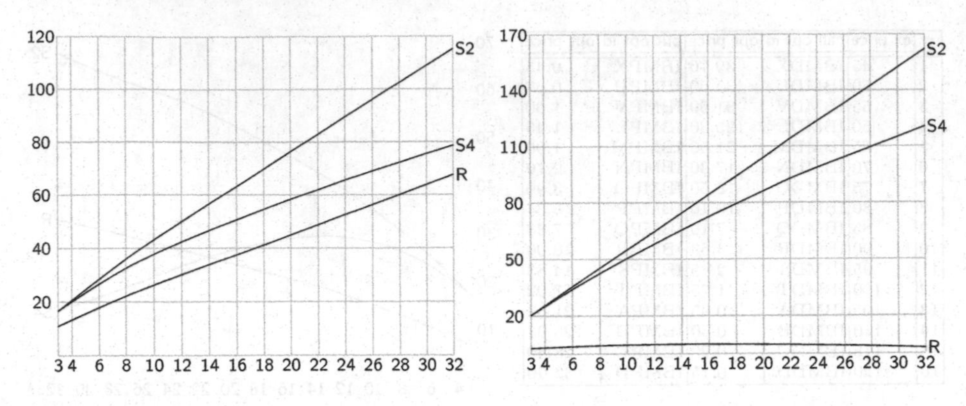

Fig. 9. Bull Portfolios

Fig. 10. Bear Portfolios

In conclusion, we would like to mention that further experiments will involve more scenarios in portfolio design and complex option spreads with up to 14 legs.

References

1. Rudd, A., Shroeder, M.: The calculation of minimum margin. Manage. Sci. 28, 1368–1379 (1982)
2. Matsypura, D., Oron, D., Timkovsky, V.G.: Option spreads: centipedes that cannot have more than 134 legs. Working paper, The University of Sydney (2007)

Improved Substitution-Diffusion Based Image Cipher Using Chaotic Standard Map

Anil Kumar and M.K. Ghose

Computer Science and Engineering Dept.
Sikkim Manipal Institute of Technology, Sikkim, India
dahiyaanil@yahoo.com

Abstract. An improved substitution-diffusion based image cipher is proposed using chaotic standard maps. The first stage consists of row and column rotation, the second stage consists of the mixing the properties of the horizontally and vertically adjacent pixels, respectively, and XORing with an intermediate chaotic key stream (CKS) image which is generated using chaotic standard map, both stages are controlled by combination of secret keys and plain image. Various types of analysis performed such as entropy analysis, difference analysis, and statistical analysis. This proposed technique is tradeoff between security and time. The experimental results illustrate that performance of this is highly secured.

1 Introduction

Security of multimedia data is most important these days due to the widespread transmission over various communication networks. It has been observed that the traditional encryption schemes [1] fails to protect multimedia data due to some special properties and some specific requirements of multimedia processing systems, such as mammoth size and strong redundancy of uncompressed data.

The fundamental features of chaotic systems such as ergodicity, mixing property, sensitivity to initial conditions/system parameters and which can be considered analogous to ideal cryptographic properties such as confusion, diffusion, balance, avalanche properties. Hence, many chaos-based encryption systems also proposed [2-4]. Many chaos-based cryptography schemes have been successfully crypt analyzed [5-6].

In this paper, we have proposed an improved symmetric image encryption by incorporating nonlinearity which is the main limitation of the patidar et al. The proposed algorithm comprises of two stages: first for horizontal, vertical rotation and XORing and second for the vertical, horizontal pixels mixing and XORing the total numbers of rounds in both stages are depending upon the secret keys and plain image.

The rest of the paper is organized as follows: The detailed algorithms of the proposed encryption procedures are discussed in Section 2. In Section 3, analysis of the proposed image cipher and evaluate its performance through various statistical analysis, key sensitivity analysis, differential analysis, key space analysis etc. Finally, Section 4, we conclude the paper.

S.K. Prasad et al. (Eds.): ICISTM 2010, CCIS 54, pp. 333–338, 2010.

2 Proposed Method

In the encryption, it includes the input image, secret keys. The generation of synthetic images explained in section 2.1, pixels rotation and XORing explained in section 2.2, pixels mixing and XORing explained in section 2.3, and decryption discussed in section 2.4.

2.1 Generation of Synthetic Images

Here, the generation of the synthetic images is discussed and also calculating the number of rounds which will be used in rotation and mixing stage.

1. The secret key consists of three floating point numbers and one integer (X_0, Y_0, K, N) where $X_0, Y_0 \in (0, 2\pi)$, K can have any real value greater than 18.0 and N is any integer value, ideally should be greater than 100.

$$\begin{aligned} X_{n+1} &= X_n + K * \sin Y_n \\ Y_{n+1} &= Y_n + X_{n+1} \end{aligned} \quad (1)$$

2. Read the image, let the size of image is $W * H$. XOR all the bytes of the image let it be IXOR.
3. Generate the pseudorandom sequence by iterating the equation (1) by $\mathrm{W} * \mathrm{H} + \mathrm{N}$ times and the pseudorandom sequence generated as *XKey1* and *XKey2* take last $W * H$ values. Scale the key using Algorithm as stated

```
program Synthetic images generation ()
begin
        repeat
              XKey1(i)  =(XKey1(i)/2pi)* 256;
              XKey2(i)  =(XKey2(i)/2pi)* 256;
        until i = W*H
        SImage1=reshape(XKey1,W,H);
        SImage2=reshape(XKey2,W,H);
end.
```

4. Take XOR of the all values of key *XKey1* as *KXOR1* and for *XKey2* as *KXOR2*.

2.2 Rotation and XORing

Here, the concept of the circular rotation [7] is used. The image $[W * H]$ elements are regarded as a set and are fed into an $W * H$ matrix.

1. Transformed each row of the matrix using rotation function which is controlled by key *XKey1*.
2. Transformed each column of the matrix direction using rotation function which is controlled by key *XKey2*.

3. XOR transformed data with synthetic image *SImage1*.
4. Repeat above steps *N1=XOR(IXOR,KXOR1)* times, in order to get completely disordered data.

2.3 Mixing and XORing

In cryptography, diffusion plays most important part especially in images, where the redundancy is large, the process of diffusion is a necessary requirement to develop a secure encryption technique. This removes the possibility of differential attacks by comparing the pair of plain and cipher images.

1. Modify second row by XORing of the first row and second row, modify third row by XORing of the modified second row and third row and the process continues till last row, which in turn gives the vertical diffusion.
2. Modify second column by XORing of the first column and second column, modify third column by XORing of the modified second column and third column and the process continues till last column, which in turn gives the horizontal diffusion.
3. XOR transformed data with synthetic image *SImage2*.
4. The number of rounds *N2=XOR(IXOR,KXOR2)* times, in order to get completely diffusion.

2.4 Decryption

Decryption can be obtained as exact reverse of the encryption as discussed in above sections.

3 Security and Performance Analysis

An robust encryption scheme should resist against all kinds of attacks such as cryptanalytic attacks: cipher text only attack, known plaintext attack, statistical attacks, brute-force attacks, etc. Results of the security and performance analysis performed on the proposed image encryption technique.

3.1 Statistical Analysis

Histograms of encrypted image

Analysis of the histograms of 10 plain images and their corresponding cipher images using various combinations of the secret key. One example of such histogram analysis is shown in (Fig. 1). In the left column, we have shown the plain image *Lena (256 * 256 pixels)* and its histogram right column, cipher image of the image Lena obtained using the secret key, (X_0 = 4.94785238984676, Y_0 = 0.31256128342389, K = 131.7183545564243, and N = 110) and its histogram has been shown. It is clear that the histograms of the cipher image are fairly uniform and significantly different from the respective histograms of the plain image and hence does not provide any type of information to employ statistical attack on the encryption scheme.

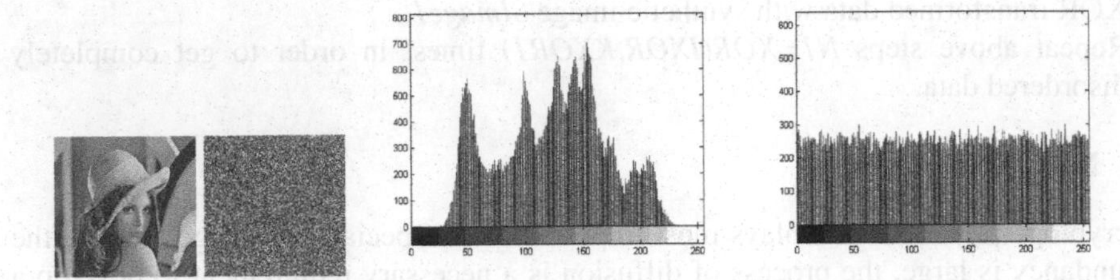

Fig. 1. Histogram Analysis: (a) Lena Image (b) Encrypted Image (c) Histogram of Lena Image (d) Histogram of Encrypted Image

Correlation of two adjacent pixels

For any given image each pixel is highly correlated with its adjacent pixels either in horizontal or vertical direction. An ideal encryption technique should produce the encrypted image with no such correlation in the adjacent pixels. It is an effective tool to assess the measure of encryption-efficiency in terms minimum correlation between corresponding pixels.

Table 1. Correlation coefficients for the two adjacent pixels in the original and encrypted shown in (Fig. 1)

	Original Image	Encrypted Image
Horizontal	0.9187	0.00055344
Vertical	0.9511	-0.0024
Diagonal	0.8934	0.00094164

Correlation between plain and cipher images

We have analyzed the correlation between various pairs of plain and cipher images by computing the 2D correlation coefficients between various colour channels of the plain and encrypted images. We have chosen 10 images for this analysis.

Table 2. Correlation between the plain image 'Lena' and its cipher image

Plain vs Encrypted image	Correlation
	0.0022

3.2 Sensitivity Analysis of the Encryption

A good encryption algorithm must possess the property that a slight change in the key changes the encrypted file almost completely different.

1. Encrypted image should be very sensitive to the secret key. If we use two slightly different keys to encrypt the same plain image then the correlation between two encrypted images should be negligible. Results are shown in (Table 3).
2. The encrypted image should not be decrypted correctly even if there is slight difference between the encryption and decryption keys.

- First by changing least significant bit of X_0 key 4.94785238984676 to 4.94785238984675. The resulting image (Fig 2(b))
- After that changing the least significant bit of Y_0 key 0.31256128342389 to 0.31256128342388. The resulting image (Fig 2(c))

Fig. 2. Sensitivity analysis for the cipher to keys when decrypt with slightly different key (a) Original key (b) Change in x (c) Change in y

Table 3. Sensitivity analysis for the cipher to key

Test item	Test results
Sensitivity for Cipher to Key(key difference(10exp-14)in X_0)	-0.0023
Sensitivity for Cipher to Key(key difference(10exp-14)in Y_0)	0.00066409

Difference attacks – sensitivity analysis for the cipher to plaintext

We have measured the number of pixels change rate by calculating the number of pixel change rate (NPCR) and the unified average changing intensity (UACI) of the two encrypted images (by encrypting two image with same key with one pixel value is changed). In (Table 4) we are showing the average value of the various results.

Table 4. Sensitivity to plaintext

NPCR	0.9959
UACI	0.32156

4 Conclusion

The improved encryption concept is proposed using the rotation, diffusion, and numbers of rounds for both stages depends upon plain image and key combination. This method is immune to various types of cryptographic attacks like known-plain text, chosen plain text attacks. It is lossless also. The proposed scheme paper has high level security, and fast speed.

Acknowledgments

This work is part of the Research Project funded by All India Council of Technical Education (Government of India) vide their office order: F.No:8023/BOR/RID/RPS-236/2008-09.

References

1. Schneier, B.: Applied cryptography: protocols algorithms and source code in C. Wiley, New York (1996)
2. Patidar, V., Pareek, N.K., Sud, K.K.: A new substitution di®usion based image cipher using chaotic standard and logistic maps. Commun. Nonlinear Sci. Numer. Simulat. 14, 3056–3075 (2009)
3. Tong, X., Cui, M.: Image encryption with compound chaotic sequence cipher shifting dynamically. Image and Vision Computing 26, 843–850 (2008)
4. Wang, Y., Wong, K.-W., Liao, X., Xiang, T., Chen, G.: A chaos based image encryption algorithm with variable control parameters. Chaos, Solitons and Fractals 41, 1773–1783 (2009)
5. Li, C., Li, S., Chen, G., Halang, W.A.: Cryptanalysis of an image encryption scheme based on a compound chaotic sequence. Image and Vision Computing 27, 1035–1039 (2009)
6. Rhouma, R., Solak, E., Belghith, S.: Cryptanalysis of a new substitution-diffusion based image cipher. Commun. Nonlinear Sci. Numer. Simulat. (in press, 2009), doi:10.1016/j.cnsns.2009.07.007
7. Kumar, A., Elkhazmi, E.A., Khalifa, O.O., Albagul, A.: Secure Data Communication Using Blind Source Separation. In: Proceedings of the ICCCE, Kuala Lumpur, Malaysia, pp. 1352–1356 (2008)

Linear Multivariable System Reduction Using Particle Swarm Optimization and A Comparative Study Using Relative Integral Square Error

Girish Parmar[1], Mahendra K. Pandey[2], and Vijay Kumar[3]

[1] Department of Electronics Engineering,
Rajasthan Technical University, Kota-324 022, Raj., India
girish_parmar2002@yahoo.com
[2] RJIT, BSF Academy, Tekanpur-475 005, M.P., India
mahendra2003in@yahoo.co.in
[3] Indian Institute of Technology, Roorkee-247 667, U.K., India
vijecfec@iitr.ernet.in

Abstract. An algorithm for order reduction of linear multivariable systems has been presented using the combined advantages of the dominant pole retention method and the error minimization by particle swarm optimization technique. The denominator of the reduced order transfer function matrix is obtained by retaining the dominant poles of the original system while the numerator terms of the lower order transfer matrix are determined by minimizing the integral square error in between the transient responses of original and reduced order models using particle swarm optimization technique. The reduction procedure is simple and computer oriented. The proposed algorithm has been applied successfully to the transfer function matrix of a 10th order two-input two-output linear time invariant model of a practical power system. The performance of the algorithm is tested by comparing the relevant computer simulation results.

Keywords: Dominant Pole, Order Reduction, Particle Swarm Optimization, Power System.

1 Introduction

Every physical system can be translated into mathematical model. The mathematical procedure of system modelling often leads to comprehensive description of a process in the form of high order differential equations which are difficult to use either for analysis or controller synthesis. It is hence useful, and sometimes necessary, to find the possibility of finding some equation of the same type but of lower order that may be considered to adequately reflect the dominant characteristics of the system under consideration.

Numerous methods are available in the literature for order-reduction of linear continuous systems in time domain as well as in frequency domain [1-6]. Further, the extension of single-input single-output (SISO) methods to reduce multi-input multi-output (MIMO) systems has also been carried out in [7-10]. In spite of several methods available, no approach always gives the best results for all systems.

S.K. Prasad et al. (Eds.): ICISTM 2010, CCIS 54, pp. 339–347, 2010.

Further, order reduction of linear time invariant systems using step response matching involves error minimization problem which has been handled with classical techniques so far by different research workers [11-13]. Recently, particle swarm optimization (PSO) technique appeared as a promising algorithm for handling the optimization problems. PSO is a population based stochastic optimization technique, inspired by social behavior of bird flocking or fish schooling [14].

The present attempt is towards evolving a new algorithm for order reduction of linear multivariable systems, which combines the advantages of the dominant pole retention method and the error minimization by PSO. The algorithm has been applied successfully to a 10th order two-input two-output linear time invariant model of a practical power system [15]. The performance of the algorithm is tested by comparing the relevant computer simulation results.

2 Description of the Algorithm

Let, the transfer function matrix of the high order system (HOS) of order 'n' having ‘p’ inputs and ‘m’ outputs is :

$$[G(s)] = \frac{1}{D(s)} \begin{bmatrix} a_{11}(s) & a_{12}(s) & a_{13}(s) & \dots & a_{1p}(s) \\ a_{21}(s) & a_{22}(s) & a_{23}(s) & \dots & a_{2p}(s) \\ \vdots & \vdots & \vdots & \dots & \vdots \\ a_{m1}(s) & a_{m2}(s) & a_{m3}(s) & \dots & a_{mp}(s) \end{bmatrix}$$

or, $[G(s)] = [\ g_{ij}\ (s)\ ],\ i = 1, 2, \ldots\ldots, m;\ j = 1, 2, \ldots\ldots, p$ (1)

is a $m \times p$ transfer matrix.

Let, the transfer function matrix of the LOS of order 'r' having ‘p’ inputs and ‘m’ outputs to be synthesized is:

$$[R(s)] = \frac{1}{\tilde{D}(s)} \begin{bmatrix} b_{11}(s) & b_{12}(s) & b_{13}(s) & \dots & b_{1p}(s) \\ b_{21}(s) & b_{22}(s) & b_{23}(s) & \dots & b_{2p}(s) \\ \vdots & \vdots & \vdots & \dots & \vdots \\ b_{m1}(s) & b_{m2}(s) & b_{m3}(s) & \dots & b_{mp}(s) \end{bmatrix}$$

or, $[R(s)] = [\ r_{ij}\ (s)\ ],\ i = 1, 2, \ldots\ldots, m;\ j = 1, 2, \ldots\ldots, p$ (2)

is a $m \times p$ transfer matrix.

2.1 Retention of Dominant Poles of HOS in LOS [11, 16, 20]

Depending on the order to be reduced to, the poles nearest to the origin are retained. This implies that the over all behavior of the reduced system will be very similar to the original system, since the contribution of the unretained eigen values to the system response are important only at the beginning of the response, where as the eigen values retained are important throughout the whole of the response, and, infact, determine the type of the response of the system.

Therefore, the denominator polynomial in (2) is now known, which is given by:

$$\tilde{D}(s) = d_o + d_1 s + d_2 s^2 + ... + d_{r-1} s^{r-1} + s^r$$

2.2 Determination of the Numerator Coefficients of LOS by Particle Swarm Optimization Technique

The PSO method is a population based search algorithm where each individual is referred to as particle and represents a candidate solution. Each particle flies through the search space with an adaptable velocity that is dynamically modified according to its own flying experience and also the flying experience of the other particles. In PSO, each particle strives to improve itself by imitating traits from their successful peers. Further, each particle has a memory and hence it is capable of remembering the best position in the search space ever visited by it. The position corresponding to the best fitness is known as pbest and the overall best out of all the particles in the population is called gbest [17].

In a d-dimensional search space, the best particle updates its velocity and positions with following equations :

$$v_{id}^{n+1} = wv_{id}^{n} + c_1 r_1^{n} (p_{id}^{n} - x_{id}^{n}) + c_2 r_2^{n} (p_{gd}^{n} - x_{id}^{n}) \tag{3}$$

$$x_{id}^{n+1} = x_{id}^{n} + v_{id}^{n+1} \tag{4}$$

where,

w = inertia weight.
c_1, c_2 = cognitive and social acceleration, respectively.
r_1, r_2 = random numbers uniformly distributed in the range (0, 1).

In PSO, each particle moves in the search space with a velocity according to its own previous best solution and its group's previous best solution. The velocity update in particle swarm consists of three parts; namely momentum, cognitive and social parts. The balance among these parts determines the performance of a PSO algorithm [18]. The parameters c_1 & c_2 determine the relative pull of pbest and gbest and the parameters r_1 & r_2 help in stochastically varying these pulls. In the equations (3) & (4), superscripts denote the iteration number.

In the present study, PSO is employed to minimize the objective function 'E', which is the integral square error in between the transient responses of original ($g_{ij}(s)$) and reduced ($r_{ij}(s)$) order models and is given by [20] :

$$E = \int_0^\infty [g_{ij}(t) - r_{ij}(t)]^2 \, dt \tag{5}$$

where, $i = 1, 2, ..., m; \; j = 1, 2, ..., p$.

and $g_{ij}(t)$, $r_{ij}(t)$ are the unit step responses of original and reduced order models, respectively and the parameters to be determined are the numerator coefficients α_i $(i = 0,1,....,(r-1))$ of reduced order models ($r_{ij}(s)$) of the LOS $[R(s)]$.

In Table 1, the specified parameters for the PSO algorithm used in the present study are given. The computational flow chart of the proposed algorithm is shown in Fig. 1.

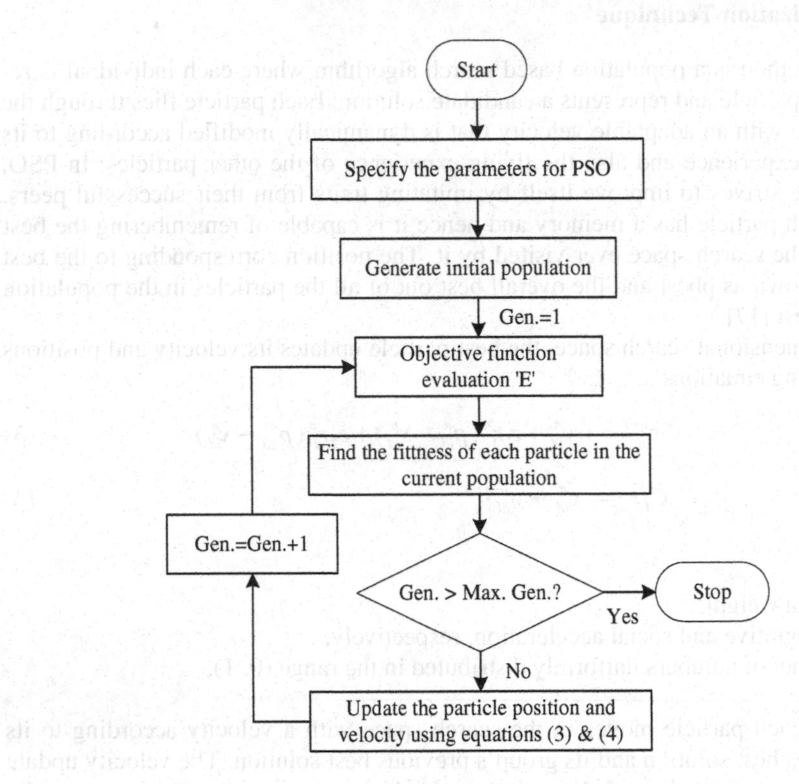

Fig. 1. Flow chart of particle swarm optimization algorithm

Table 1. Parameters used for PSO algorithm

Parameters	Value
Swarm size	20
Max. Generations	200
c_1, c_2	2.0, 2.0
w_{start}, w_{end}	0.9, 0.4

3 Practical Power System under Study

The Phillips-Heffron model of the single-machine infinite-bus (SMIB) power system [15, 20] is considered in this study. The system consists of a three-phase 160-MVA synchronous machine with automatic excitation control system (i.e. a standard IEEE Type-I exciter with rate feedback (RF) and power system stabilizer (PSS)). The numerical values of the parameters, which define the total system as well as its operating point, nomenclature can be found in [19-20].

Based on the state variables and the values of the parameters and the operating point [20], the system (without accounting for the limiters) may be described in state-space form as [15] :

$$\left.\begin{aligned} \dot{x} &= Ax + Bu \\ y &= Cx \end{aligned}\right\} \tag{6}$$

where,

$$x^T = [E_q' \;\; \omega \;\; \delta \;\; v_1 \;\; v_2 \;\; v_3 \;\; v_4 \;\; v_5 \;\; v_R \;\; E_{FD}];\; u^T = [\Delta V_{Ref} \;\; \Delta T_m];\; y^T = [\delta \;\; V_t]$$

and the numerical values of the matrices A, B and C can be found in [20].

4 Application of the Proposed Algorithm to the Power System and Simulation Results

The transfer function matrix (based on the numerical values of the matrices A, B and C) of the 10th order two-input two-output linear time invariant model of practical power system under study is given by [20]:

$$[G(s)] = \frac{1}{D(s)} \begin{bmatrix} a_{11}(s) & a_{12}(s) \\ a_{21}(s) & a_{22}(s) \end{bmatrix} \tag{7}$$

where, the common denominator $D(s)$ is given by:

$$D(s) = s^{10} + 64.21 s^9 + 1596 s^8 + 1.947\times10^4 s^7 + 1.268\times10^5 s^6 + 5.036\times10^5 s^5 + 1.569\times10^6 s^4$$
$$+ 3.24\times10^6 s^3 + 4.061\times10^6 s^2 + 2.905\times10^6 s + 2.531\times10^5$$

The poles of the above system $[G(s)]$ are at $\lambda_1 = -0.1001$, $\lambda_{2,3} = -0.2392 \pm j\,3.2348$,

$\lambda_{4,5} = -0.8977 \pm j\,1.3552$,

$\lambda_6 = -2.1375$, $\lambda_7 = -9.6454$, $\lambda_8 = -11.9632$, $\lambda_{9,10} = -19.0451 \pm j\,2.4859$.

The proposed algorithm is applied to the above multivariable system and the reduced order models ($r_{ij}(s)$) of the LOS $[R(s)]$ are obtained. The general form of 3rd order reduced transfer function matrix is taken as:

$$[R(s)] = \frac{1}{\tilde{D}(s)} \begin{bmatrix} b_{11}(s) & b_{12}(s) \\ b_{21}(s) & b_{22}(s) \end{bmatrix} \tag{8}$$

where, $\tilde{D}(s) = s^3 + 0.5785 s^2 + 10.5690 s + 1.0532$.

and

$b_{11}(s) = 3.5482 s^2 - 16.929 s - 2.30315$,

$b_{12}(s) = -1.03192 s^2 + 30 s + 2.5$,

$b_{21}(s) = -0.63385 s^2 + 7.909 s + 1.03431$,

$b_{22}(s) = 0.515226 s^2 - 1.60344 s - 0.071736$.

The poles of the LOS $[R(s)]$ are at λ_1 = -0.1001, $\lambda_{2,3}$ = -0.2392 ± j 3.2348.

The adequacy of the 3rd order reduced models obtained above is tested by comparing the time responses of the outputs (i.e. δ and V_t) of the original 10th order system (7) and those of the 3rd order reduced system (8), subject to the same input step change. The time responses shown in Fig. 2. (a)-(f) are computed for three distinct input step changes:

(i) with ΔV_{Ref} (s) = 0.05 p.u. and ΔT_m (s) = 0 ;
(ii) with ΔV_{Ref} (s) = 0 and ΔT_m (s) =0.05 p.u. ; and
(iii) with ΔV_{Ref} (s) =0.05 p.u. and ΔT_m (s) =0.05 p.u.

Further, a comparison of the proposed algorithm with some well known existing order reduction techniques (for 3rd order reduced models), is also shown as given in Table 2, by comparing the relative integral square error (R.I.S.E.) 'J' [20-21], in between the transient parts of original ($g_{ij}(s)$) and reduced ($r_{ij}(s)$) order models.

Table 2. Comparison of reduced order models

R.I.S.E. 'J' for $r_{ij}\,(i = 1, 2;\; j = 1, 2)$ $J = \int_0^\infty [\, g_{ij}(t) - r_{ij}(t)]^2\, dt \Big/ \int_0^\infty [g_{ij}(t) - g_{ij}(\infty)\,]^2\, dt$				
Method of Order Reduction	r_{11}	r_{12}	r_{21}	r_{22}
Proposed Algorithm	1.008027	0.037908	1.916702	0.992683
Papadopoulos and Boglou [15]	1.000215	0.683081	2.854987	1.001437
Moore [22]	0.939453	0.568192	3.953609	0.872135
Safonov and Chiang [23]	1.324429	205.8792	79.98665	0.034152
Safonov and Chiang [24]	0.939446	0.563939	3.931755	0.872409
Safonov, Chiang and Limebeer [25]	0.943603	0.047862	17.445711	1.057315

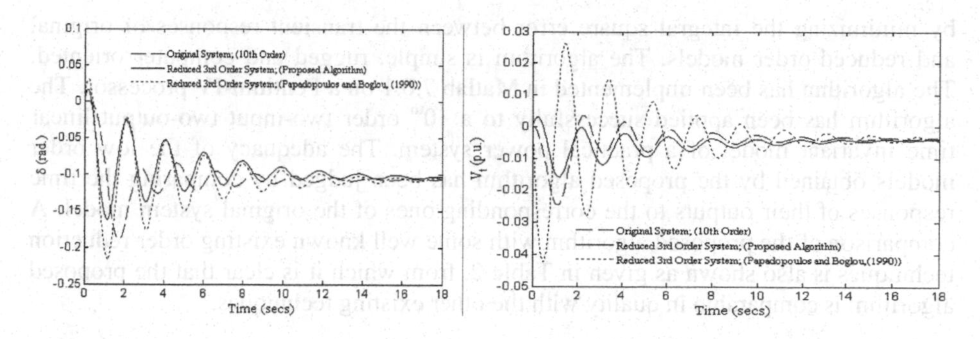

Fig. 2. (a) $\Delta V_{\mathrm{Re}f} = 0.05$ p.u. and $\Delta T_m = 0$. **Fig. 2.** (d) $\Delta V_{\mathrm{Re}f} = 0$ and $\Delta T_m = 0.05$ p.u.

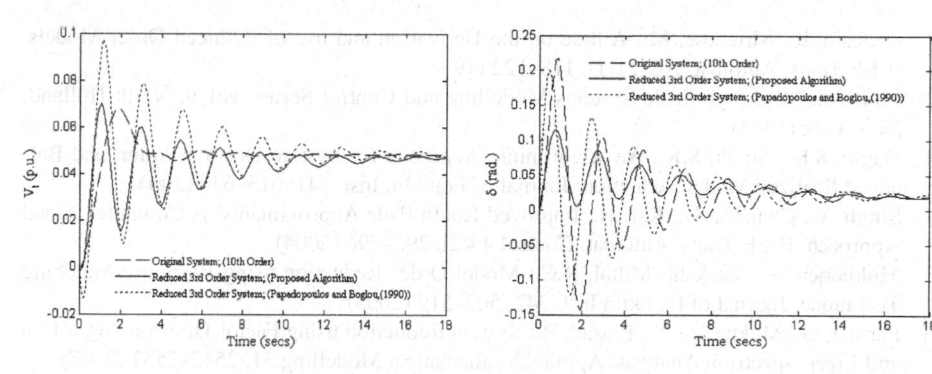

Fig. 2. (b) $\Delta V_{\mathrm{Re}f} = 0.05$ p.u. and $\Delta T_m = 0$. **Fig. 2.** (e) $\Delta V_{\mathrm{Re}f} = 0.05$ p.u. and $\Delta T_m = 0.05$ p.u.

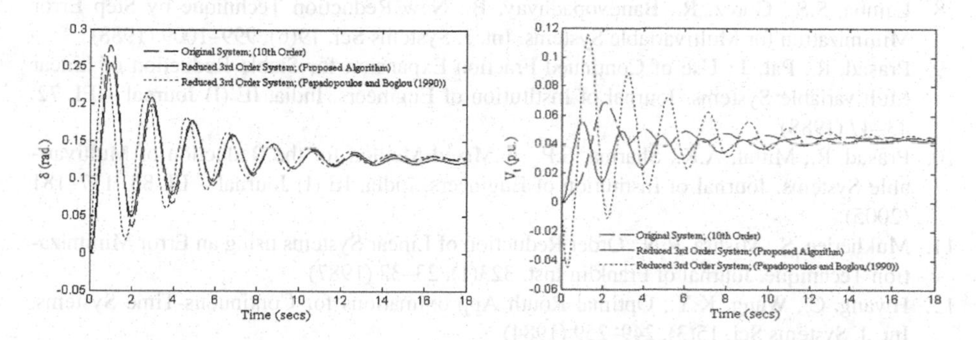

Fig. 2. (c) $\Delta V_{\mathrm{Re}f} = 0$ and $\Delta T_m = 0.05$ p.u. **Fig. 2.** (f) $\Delta V_{\mathrm{Re}f} = 0.05$ p.u. and $\Delta T_m = 0.05$ p.u.

5 Conclusions

An algorithm for order reduction of linear multivariable systems has been presented in which the dominant poles are retained according to the order to be reduced to, while the particle swarm optimization technique has been used to determine the zeros by minimizing the integral square error between the transient responses of original and reduced order models. The algorithm is simple, rugged and computer oriented. The algorithm has been implemented in Matlab 7.0.1 on a Pentium-IV processor. The algorithm has been applied successfully to a 10th order two-input two-output linear time invariant model of a practical power system. The adequacy of the low order models obtained by the proposed algorithm has been judged by comparing the time responses of their outputs to the corresponding ones of the original system model. A comparison of the proposed algorithm with some well known existing order reduction techniques is also shown as given in Table 2, from which it is clear that the proposed algorithm is comparable in quality with the other existing techniques.

References

1. Genesio, R., Milanese, M.: A note on the Derivation and use of Reduced Order Models. IEEE Trans. Automat. AC-21(1), 118–122 (1976)
2. Jamshidi, M.: Large Scale Systems Modelling and Control Series, vol. 9. North Holland, New York (1983)
3. Nagar, S.K., Singh, S.K.: An Algorithmic Approach for System Decomposition and Balanced Realized Model Reduction. Journal of Franklin Inst. 341, 615–630 (2004)
4. Singh, V., Chandra, D., Kar, H.: Improved Routh Pade Approximants: A Computer Aided Approach. IEEE Trans. Automat. Control 49(2), 292–296 (2004)
5. Mukherjee, S., Satakshi, Mittal, R.C.: Model Order Reduction using Response-Matching Technique. Journal of Franklin Inst. 342, 503–519 (2005)
6. Parmar, G., Mukherjee, S., Prasad, R.: System Reduction using Factor Division Algorithm and Eigen Spectrum Analysis. Applied Mathematical Modelling 31, 2542–2552 (2007)
7. Mukherjee, S., Mishra, R.N.: Reduced Order Modelling of Linear Multivariable Systems using an Error Minimization Technique. Journal of Franklin Inst. 325(2), 235–245 (1988)
8. Lamba, S.S., Gorez, R., Bandyopadhyay, B.: New Reduction Technique by Step Error Minimization for Multivariable Systems. Int. J. Systems Sci. 19(6), 999–1009 (1988)
9. Prasad, R., Pal, J.: Use of Continued Fraction Expansion for Stable Reduction of Linear Multivariable Systems. Journal of Institution of Engineers, India, IE (I) Journal – EL 72, 43–47 (1988)
10. Prasad, R., Mittal, A.K., Sharma, S.P.: A Mixed Method for the Reduction of Multivariable Systems. Journal of Institution of Engineers, India, IE (I) Journal – EL 85, 177–181 (2005)
11. Mukherjee, S., Mishra, R.N.: Order Reduction of Linear Systems using an Error Minimization Technique. Journal of Franklin Inst. 323(1), 23–32 (1987)
12. Hwang, C., Wang, K.Y.: Optimal Routh Approximations for Continuous-Time Systems. Int. J. Systems Sci. 15(3), 249–259 (1984)
13. Mittal, A.K., Prasad, R., Sharma, S.P.: Reduction of Linear Dynamic Systems using an Error Minimization Technique. Journal of Institution of Engineers, India, IE (I) Journal – EL 84, 201–206 (2004)
14. Kennedy, J., Eberhart, R.C.: Particle Swarm Optimization. In: IEEE Int. Conf. on Neural Networks, Piscataway, NJ, pp. 1942–1948 (1995)
15. Papadopoulos, D.P., Boglou, A.K.: Reduced-Order Modelling of Linear MIMO Systems with the Pade Approximation Method. Int. J. Systems Sci. 21(4), 693–710 (1990)
16. Davison, E.J.: A Method for Simplifying Linear Dynamic Systems. IEEE Trans. Automat. Control AC-11, 93–101 (1966)

17. Kennedy, J., Eberhart, R.C.: Swarm Intelligence. Morgan Kaufmann Publishers, San Francisco (2001)
18. Eberhart, R.C., Shi, Y.: Particle Swarm Optimization: Developments, Applications and Resources. In: Congress on Evolutionary Computation, Seoul Korea, pp. 81–86 (2001)
19. Papadopoulos, D.P.: Tests and Simulations in Connection with Large Generator Dynamics. In: Proc. of ICEM 1982 Conf. Budapest, Part 3, pp. 881–884 (1982)
20. Parmar, G., Mukherjee, S., Prasad, R.: Reduced Order Modelling of Linear MIMO Systems using Genetic Algorithm. Int. Journal of Simulation Modelling 6(3), 173–184 (2007)
21. Lucas, T.N.: Further Discussion on Impulse Energy Approximation. IEEE Trans. Automat. Control AC-32(2), 189–190 (1987)
22. Moore, B.C.: Principal Component Analysis in Linear Systems: Controllability, Observability and Model Reduction. IEEE Trans. Automat. Control AC-26(1), 17–32 (1981)
23. Safonov, M.G., Chiang, R.Y.: Model Reduction for Robust Control: A Schur Relative Error Method. Int. J. Adaptive Contr. and Signal Process. 2, 259–272 (1988)
24. Safonov, M.G., Chiang, R.Y.: A Schur Method for Balanced-Truncation Model Reduction. IEEE Trans. Automat. Control 34(7), 729–733 (1989)
25. Safonov, M.G., Chiang, R.Y., Limebeer, D.J.N.: Optimal Hankel Model Reduction for Nonminimal Systems. IEEE Trans. Automat. Control 35(4), 496–502 (1990)
26. Varga, A.: Model reduction software in the SLICOT library. In: Datta, B. (ed.) Applied and computational control, signals and circuits, vol. 629. Kluwer Academic Publishers, Boston (2001), http://www.robotic.dlr.de/~varga/

An Anomaly Based Approach for Intrusion Detection by Authorized Users in Database Systems

Bharat Gupta, Deepak Arora, and Vishal Jha

Department of Computer Science Engineering and Information Technology,
Jaypee Institute of Information Technology University, Noida-201307, India

Abstract. This paper is an attempt to introduce a new approach on increasing the security of database systems. Securing databases involves external as well as internal misuse detection and prevention. SQL injection handling and access control mechanism prevents misuse through unauthorized access to the database. This allows only those users to access database contents who are meant to use it. However, if there is an intentional or unintentional misuse by some authorized user, then it becomes very difficult to identify and prevent that misuse then and there only. Such misuse scenarios can be detected later by auditing the transaction log. Therefore the need for a robust query intrusion detection model for database system arises. The model proposed in this paper detects such types of misuses by authorized users and classifies them as legitimate or anomalous by analyzing the nature of queries they fire and tuning itself based on the responses to the alarms raised.

Keywords: Intrusion detection, Query, Database systems.

1 Introduction

Database security involves several tasks such as preventing intentional or unintentional misuse, preserving privacy of a user, data recovery in case of failure and rollback mechanisms. An intrusion detection system focuses on detecting intrusions [12] which intend to breach the security, privacy and integrity of a database. Intrusion detection in databases is in its early stages and most of the IDS prevalent focuses on intrusions in the external network i.e. detecting intrusions through unauthorized users. It is comparatively easy to stop unauthorized access to database as compared to detecting misuse of privileges by an authorized user because we don't know when a person shall do so either intentionally or unintentionally on the specific privileges allocated to him/her for accessing the database.

We propose a new model for Query Intrusion Detection System in Databases. In our approach to the Intrusion detection in database systems [IDDS], we focus on

S.K. Prasad et al. (Eds.): ICISTM 2010, CCIS 54, pp. 348–356, 2010.

detecting anomalous behavior based on the types of queries an authorized user fires and classify them as malicious or legitimate based on the analysis of the query parameters and tuning the IDDS regularly on the basis of responses to the alarms raised by the system. False positive [1,5] and false negative[1,5] alarms generated by the system reduce the accuracy of the system. The tuning itself is a major part because we need to reduce the false alarms generated by the system so as to increase its accuracy and reliability. Initially during the learning of IDDS the role of administrator is vital. As we proceed, we describe how each of the modules of our model works and how the model can be injected into the existing database structure.

We also suggest various approaches to be used for each of the modules in the model. We try to explain these on the basis of related work which has been done in this field. Later in section 4, the experimental setting and results are shown using profile building of an authorized user and by using of data mining for the learning of IDDS.

2 Related Work

Very less number of works is available in the field of Database intrusion Detection. Most of the present intrusion detection mechanism mainly concerned with the intrusion detection in the network or deals with the packet in the network transmission.

Present scenario focuses on intrusion detection in external network. Some work[2] has been done of maintaining security of confidential information in a database while answering as many queries as possible. Some work[3] has been done for finding anomaly in the fired queries. Some of the IDS[9,11] for databases enhance their security through access control, encryption, SQL injection etc which only tackles the issue of unauthorized misuse. However, a fully working mechanism for detection of intended or unintended misuse by privileged users does not exist. Some pitfalls include non considering the case when an authorized user makes some kind of intentional or unintentional anomalous transaction. No robust mechanism to monitor the false alarm rate. There might be cases when a legitimate transaction be treated as a malicious [7] and blocked. There exist different algorithms in data mining genetic algorithms and artificial neural networks but to choose the most efficient one for tuning is a tough task. Each one has its own pros and cons. Since the domain is new, still work is going on to make more improvements in the existing solutions.

3 The Proposed Model

The proposed model is designed to identify malicious query and an anomalous query fired by an authorized user of the database. This model basically does this with the help of user profile. Here the normal user queries are the queries which are generally fired by the user and which are recognized by the administrator as legitimate.

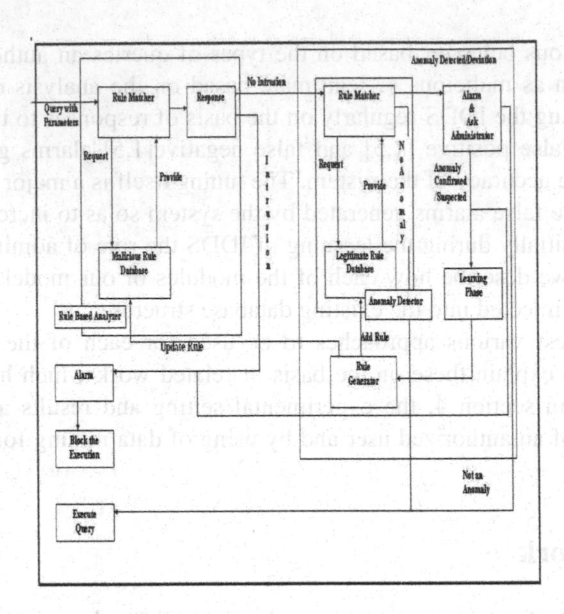

Fig. 1. Proposed model of Query Intrusion Detection in Databases

The Database Intrusion Detection System consists of five modules:

3.1 Query Parser

This module converts the fired query into the desirable format which includes user type, SQL commands used in the query, read set and write set of the query. This is important because intrusion detection system is based mainly on the Role Based Profile and these attributes are necessary for making profile of users.

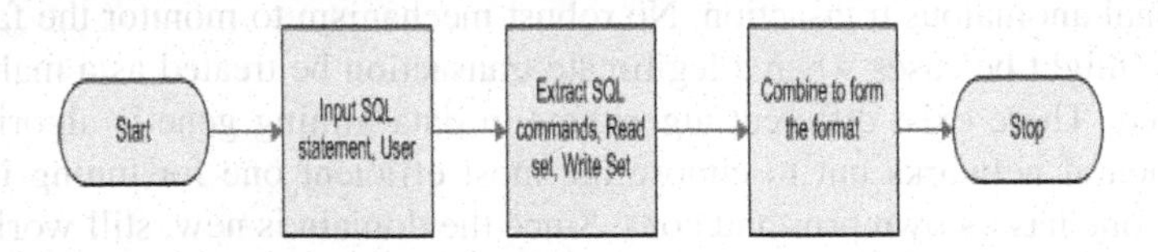

Fig. 2. SQL Parser

3.2 Rule Matcher

The parsed query is then passed to Rule matcher module where it matches exact signature from the Malicious Rule Database. If the signature matches then the query fired is a malicious one and the query is blocked with immediate effect and a message is shown to the user and administrator that the query is blocked by intrusion detection system otherwise it is then passed to deviation detector. The whole process can is shown in the figure no 3.

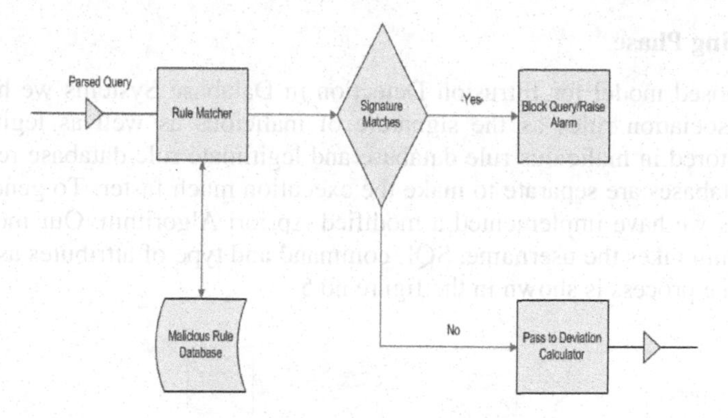

Fig. 3. Working of Rule Matcher

3.3 Deviation Calculator

Deviation detector calculates the deviation of the fired query from the signature present in the Legitimate Rule Database. Deviation is calculated on the basis of priorities and the vales associated to them. In our intrusion detection system since we are focusing on the role based so we have assigned the maximum weightage to the user. The formula is described below:

A: Deviation in User type
If the user name or user type does not match then we add the 3.0 to the deviation.

Dev(A) = 3.0

B. Deviation in SQL commands
Deviation is more when the number of SQL commands in fired query is more than the number of SQL commands in the legitimate rule.

More: Dev(B.1) = 2.0, Less: Dev(B.1) = 0.5

If the relative positions of SQL command is not same then the deviation increases by 0.5.

Dev (B.2) = 0.5

Total Deviation due to SQL command is Dev(B)=Dev(B.1)+Dev(B. 2)

C. Deviation in Attribute set
Deviation is more when the number of attributes in fired query is more than the number of attribute in the legitimate rule.

More: Dev(C.1) = 2.0, Less: Dev(C.1) = 0.5

If the relative positions of attributes are not same then the deviation increases by 0.5.

Dev(C.2) = 0.5

Total Deviation due to attribute is Dev(C) = Dev(C.1) + Dev(C.2)

Total Deviation calculated by the deviation detector is

Dev(Total) =Dev(A) +Dev(B) +Dev(C)

The whole process is described in the figure no. 4.

3.4 Learning Phase

In our proposed model for Intrusion Detection in Database Systems we have generated the association rules as the signature of malicious as well as legitimate rule which are stored in malicious rule database and legitimate rule database respectively. The two databases are separate to make the execution much faster. To generate Association rules we have implemented a modified Apriori Algorithm. Our modified Apriori algorithm takes the username, SQL command and type of attributes associated to the rules. The process is shown in the figure no 5.

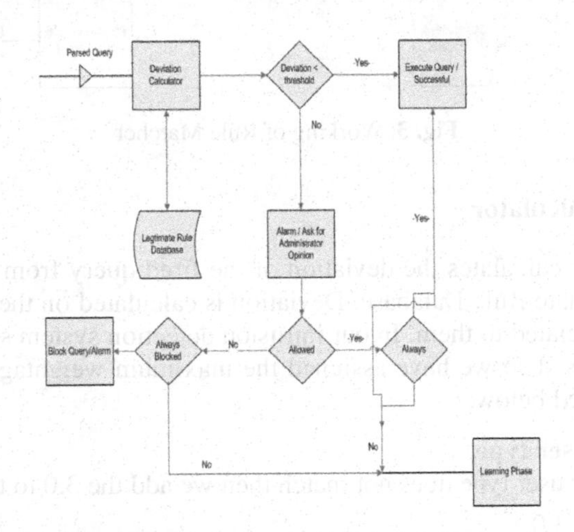

Fig. 4. Working of Deviation Calculator

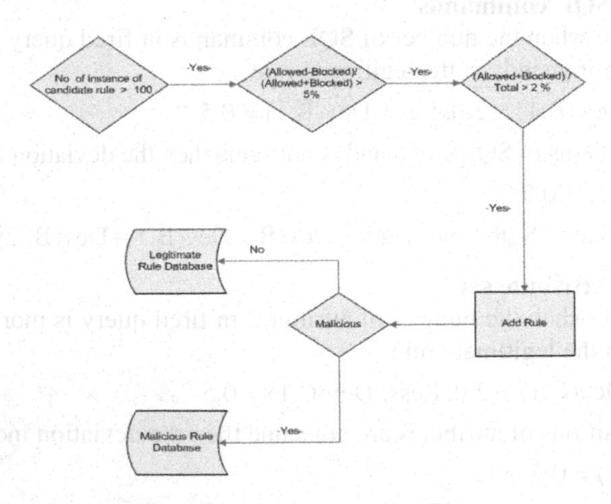

Fig. 5. Working of Learning Phase

Here support is defined as the difference in the number of instance of allowed and blocked of a signature.

Support(allowed of a signature) = No. of allowed(signature) - No. of blocked(signature)
Support(blocked of a signature) = No. of blocked (signature) -No. of allowed(signature)

Confidence here is defined as the ratio of the instance of a query in the whole training dataset.

The minimum number of instances of a signature in the training dataset, minimum support of a signature and minimum confidence of the signature can be varied.

3.5 Alarms

In our IDDS appropriate message is displayed when the query is blocked or allowed to execute. If the query is found to be malicious at the rule based analyzer there is no alarms but we receive a message but when legitimate query is not detected by then an alarm is raised to database administrator.

3.6 Rule Format

We have designed our malicious rule or legitimate rule as follows:

<userclass>; <SQL commands>; <Read_Attributes, Write_Attributes>

Userclass - is required to generate or build user profiles.

SQL commands and Read_attributes, Write Attributes:- are required to generate the signature preciously.

3.7 Constraints

In our SQL parser we have assumed that the fired query is syntactically correct or otherwise it will be handled by database default parser. The system has some limitations in its rule based parser which at present addresses only simple query that to with select, insert delete and update query.

4 Experimental Settings and Results

All the implementations are in C# on .Net Framework version 2.0 on the windows platform. In our case we have used MySQL as the database. Query Intrusion Detection model is developed as a DLL file (Dynamic Linked Library) for its dynamic use any of the C# project.

The model is supposed to work in the various scenarios which include detecting and blocking malicious activities, detecting anomalous query fired by an authenticated user, raising alarms and asking database administrator to react accordingly.

The different cases which are handled by proposed model are discussed below

Case A: The query passes through Rule Based Analyzer and Anomaly Detector without raising any alarm and the query get executed uninterruptedly.

Case B: The query does not get passed through Rule Based Analyzer and the alarm is raised and query is blocked.

Case C: The query gets passed through Rule Based Analyzer but do not pass through Anomaly Detection and alarm is raised and the anomaly is referred to administrator for verification and anomaly is verified the query is blocked and then it is forwarded towards learning phase which in turns pass the anomaly to Rule generator to create rule.

Case D: The query gets passed through Rule Based Analyzer but do not pass through Anomaly Detection and alarm is raised and the anomaly is referred to administrator for verification and anomaly is not verified the query is executed.

The model is verified for various queries and the individual module is checked vigorously to deal with each type of cases that may arise during the real time execution.

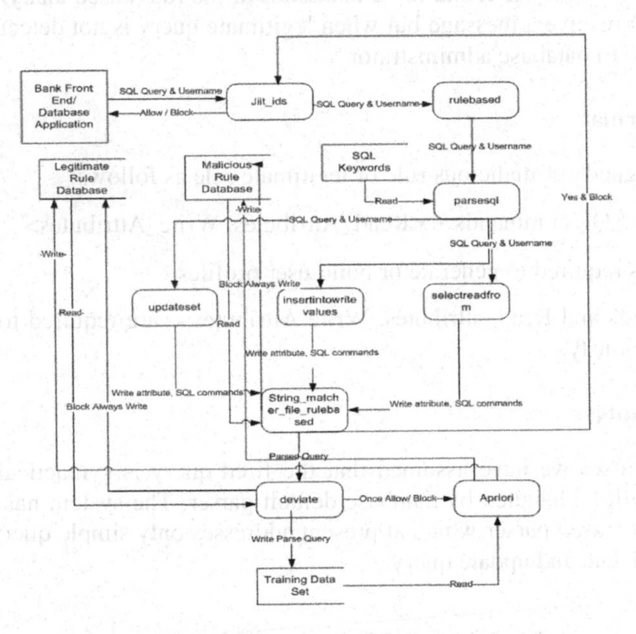

Fig. 6. Data Flow Diagram of Intrusion Detection System

Fig. 7. Memory available during Learning Phase

All the syntactically correct SQL queries are addressed by our system and by changing its various parameters like minimum support, minimum confidence and minimum number of the instance of signature to appear in the training dataset for the learning phase of anomaly detector.

The above figure 7 shows the memory usage and time taken when the number of elements in training dataset was more than 10,000. We can see that when the system starts it consumes maximum memory after which it gradually decreases with time. In the beginning system traverse from beginning to end which requires more memory but after that it has lesser number of loop and the memory uses decreases.

5 Conclusion

In this paper we proposed a framework for the intrusion detection in databases by authorized users. The proposed model can detect anomalies by classifying the anomalous behavior as malicious or legitimate with the help of administrator interference. This framework can be used to merge with any of the scenarios where there is a possibility of malicious transactions. The resource monitoring tool is used to monitor its efficiency and the training tool is used to auto generate training data set for learning phase. The future work can be inclusion of nested and complex queries and the inclusion of time-stamping in rule format.

References

1. Zhong, Y., Zhu, Z., Qin, X.-L.: A clustering method based on data queries and its application in database intrusion detection. In: ICMLC 2005, Guangzhou, August 18-21, pp. 2096–2101 (2005)
2. Gopal, R.D., Goes, P.B., Garfinkel, R.S.: Interval Protection of Confidential Information in a Database. INFORMS Journal on Computing 10(3) (1998)
3. Caulkins, B., Lee, J., Morgan Wang, F.: A Dynamic Data Mining Technique for Intrusion Detection Systems. In: Proceedings of the 43rd annual Southeast regional conference, Kennesaw, Georgia, vol. 2, pp. 2148–2153 (2005)
4. Mokube, I., Adams, M.: Honeypots Concepts, Approaches, and Challenges. In: Proceedings of the 45th annual southeast regional conference, Winston-Salem, North Carolina, pp. 321–326 (2007)
5. Pavlou, K., Snodgrass, R.T.: Forensic Analysis of Database Tampering. In: Proceedings of the 2006 ACM SIGMOD International conference on Management of data, Chicago, IL, USA, pp. 109–120 (2006)
6. Ceri, S., Di Giunta, F., Lanzi, P.L.: Mining Constraint Violations. In: Proceedings of the 43rd annual Southeast regional conference, Kennesaw, Georgia, March 2007, vol. 2, pp. 1–32 (2007)
7. Hu, Y., Panda, B.: Identification of Malicious Transactions in Database Systems. In: Proceedings of the Seventh International Database Engineering and Applications Symposium (IDEAS 2003), pp. 1–7 (2003)
8. Yu, Z., Tsai, J.J.P., Weigert, T.: An Automatically Tuning Intrusion Detection System. IEEE Transactions On Systems, Man and Cybernetics, Cybernetics 37(2), 373–384 (2007)

9. Bertino, E., Terzi, E., Kamra, A., Vakali, A.: Intrusion Detection in RBAC-administered Databases. In: Proceedings of the 21st Annual Computer Security Applications Conference 2005, pp. 170–182. IEEE, Los Alamitos (2005)
10. Lee, S.Y., Low, W.L., Wong, P.Y.: Learning Fingerprints for a Database Intrusion Detection System. In: Gollmann, D., Karjoth, G., Waidner, M. (eds.) ESORICS 2002. LNCS, vol. 2502, pp. 264–279. Springer, Heidelberg (2002)
11. Rietta, F.S.: Application Layer Intrusion Detection for SQL Injection. In: ACM SE 2006, Melbourne, Florida, USA, March 10-12, pp. 531–536 (2006)
12. Li, Z., Dad, A., Zhou, J.: Theoretical Basis for Intrusion Detection. In: Proceedings of the 2005 IEEE, Workshop on Information Assurance and Security United States Military Academy, West Point, NY, pp. 184–192 (2005)

Towards a Knowledge Management Portal for a Local Community

Marcelo Lopez[1], Gustavo Isaza[1], and Luis Joyanes[2]

[1] Caldas University, Street 65 No 26-10, Manizales, Colombia
{mlopez,gustavo.isaza}@ucaldas.edu.co
[2] Pontificia University of Salamanca, Madrid Campus, Paseo Juan XXIII, 3, Madrid, Spain
luis.joyanes@upsam.net

Abstract. This paper discusses how to manage the knowledge associated with dimensions: social, natural, cultural, political, intellectual and physical of local communities. The portal manages communities of practice through collaborative environments, content management, applications on line and intelligence to community, learning environments, relations based on social multi-agent systems and ontologies with semantic representation. Multi-agent systems enable participation and empowerment for e-government, e-business, e-participation and e-learning; the multiple roles aim to reach activity behavior by changing the passive observer becoming a dynamic individual co-constructor of the domain knowledge. The social collaboration process using multi-agent technologies provides distribution, cooperation, intelligence, pro-activity, reactivity, scalability, adaptability among others properties; offering a decentralized model own community interaction; agents perceive and influence aspects of their distributed environment, and learn about it. Multi-Agent Portal systems based on social simulation can improve the recent models incorporating cognitive architectures, as they provide a realistic basis for modeling individual agents and therefore their social interactions.

Keywords: Collective intelligence, collaborative filtering, Community Ontology, Social Agents, Knowledge portal.

1 Introduction

The portal will change the community and influence the lives of people everywhere by applying our locality human collaboration and intelligence to informing and to solving the most important events productive, governmental, education and the community. The portal combines technology, community incentives and human creativity with the open participation and collective intelligence.

The proposed architecture of Knowledge Management Portal for a Local Community is based in dimensions: social, natural, cultural, political, intellectual and physical. As a collection articles, ontology elements and their relations represented the

S.K. Prasad et al. (Eds.): ICISTM 2010, CCIS 54, pp. 357–362, 2010.

dimensions as knowledge capitals of community. Exist the individual level and the collective level, the first linked shared understanding of the personal domain and the second for development understanding of knowledge associated social groups.

This paper presents portal's functions and components, ontology and multi-agent system for enhance learning, collaboration, collective intelligence, participation and content generation in a local community. The ontology's use in the Portal allows to generate technical knowledge of core social development environments, application of social and political analysis to improve development policy management and Social and political analysis.

2 Functions and Components of Portal

The proposed knowledge management portal performs functions as provide to the user a set of Social Web Tools for navigating and finding the necessary information in the portal domain, as: documents edit, audio-image-video editing, recommends system, content filtering system, content syndication and "mash ups" components. Moreover, the Social Portal provides the user with tools for collective intelligence semantic-based searching.

Integrate the collective intelligence resources on portal's subjects as: learning, collaboration, content management, applications on line and contents agents. Provide the user with information support (for example, problems involved with ideas, announcements on various events and actions, chronological and geographical referencing, survey system and voting). Support a flexible user interface that takes into account the user's preferences with respect to local culture and the user's work with the resources and services relevant. Provide access to knowledge and information on various dimensions of local community (domain and main concepts). These are proposed attributes, subject to change by user master (edit or create new levels). In the last level of each dimension, the user enters an article that can add to and edit freely online. Each article has rich interface, references, tags, recommendations, tags, micro formats. With this collection of articles the portal creates models and predictions based on individual reviews of the group or groups of interest. A set of relations among defined concepts as a folksonomy identifying ontologies components (position, rules and roles).

3 Ontology

A community is a social system with a sufficient number of social structures to meet needs of its members through role relations. The local community notion is not just focusing on respect for equality, but it finds its root in claiming the right to difference; new social movements emerge from the knowledge society respecting the uniqueness, different choice, advance social and cultural practices that show a subject in communication with a global and connected world.

For a complete and systematized representation of knowledge and information resources relating to certain culture of community, the ontology of a portal includes domain-independent such as community dimensions and subject domain ontology (positions and roles) [1].

The ontology of community dimensions includes the concepts related to organization of any community activity. For example social group, person, organization, publication, event, problems and solutions, etc. A local community that uses technologies web 2.0 and web 3.0 to share, participates, do, empower its environment, a community prepared for the variety and the different [2]. A community as group of people in a common place (real or virtual), people that share common features, its members are in permanent interaction.

The quality of ontology's collaboration depends on multiple levels of formality that might coexist and have an ontology development in permanent progression.. Ontology represents common understanding of the community involved in social and collaborative processes that depend of learning process of the users. The component education includes: learning objects, learning lessons, collaborative and interactive learning, mentoring, test ware and coaching among others. The process learning is the result of constant compromise and collaborative when applying ontologies (new notion ideas tags, common terminology, formal trivial ontology).

The community members make a start to organize the impression with hierarchical and ad hoc interaction, allows for exploiting relationships for reasoning. The collection articles are associated with the information, issues and contributions of community, may apply dynamics relations between attributes such as disjunction, simple conjunction, complex conjunctive and pattern relations.

In addition, to collaborate and to reach interactive learning process is required consider participation, communication and coordination process, the development of social process and competencies with the social agents. To infer social perceptions, values and norms condition behavior in society; to infer the power relations influence poverty, inequalities, exclusion and vulnerability; to infer the formal and informal institutions are shaped by social and political processes and impact on quality of life of the community.

The collaboration is about work zones, social markets, groupware, task management, form wizard, content notification, evaluation list and analytical case list. The users need to learn to collaborate, common willingness and competencies to interact, learn, compromise, communicate, discuss and recognize rules according a roles and positions. The portal deploys for the community rules, best practices, leadership and sense.

4 Proposed Architecture

The information access layer provides front-end of the portal for user. The information processing layer supports all information (data) flows in the portal from construction of ontology to processing of user query. The base layer maintains data and knowledge management using database and Semantic Web technologies and services as shown in Fig. 1.

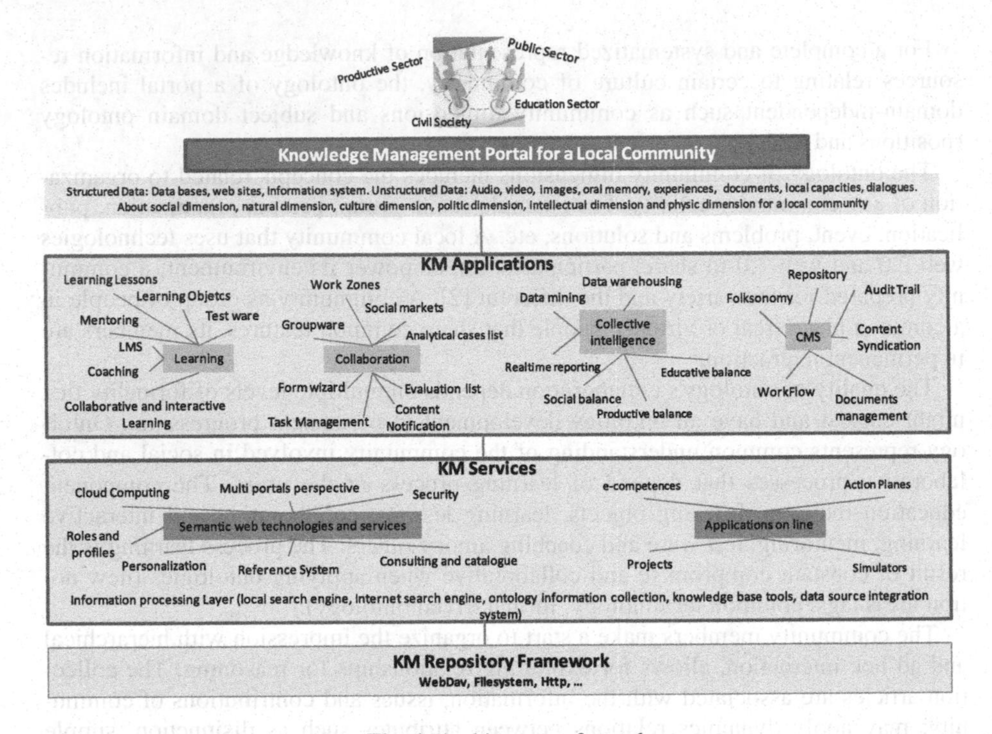

Fig. 1. Architecture of portal

5 Multi-agent System for a Portal Local Community

Based on the system-theoretic and semiotic analysis made in [3], in the multi-agent system context, a digital cities is defined as follows: “An autopoietic organization of social agents communicating by way of computers, such that every social agent is a realization of a semiotics process engendered by navigation taken place in a common (for all agents) environment". In Social communities, people work together systematically toward shared similar goals, sometimes collaboration is stressed, and competition is imminent. Social network analysis is the study of social relationships among individuals in a society.

The multiple roles aim to reach activity behaviour by changing the passive observer becoming a dynamic individual co-constructor of the domain knowledge. The social collaboration process using multi-agent technologies provides distribution, cooperation, intelligence, pro-activity, reactivity, scalability, adaptability among others properties; offering a decentralized model own community interaction; agents perceive and influence aspects of their distributed environment, and learn about it. Multi-Agent based on social simulation can improve the recent models incorporating cognitive architectures, as they provide a realistic basis for modeling individual agents and therefore their social interactions. Delegating some part of knowledge management in agents is a great opportunity to build on the fundamental principles of these technologies, giving a cognitive architecture that may be used for a broad multiple-domain analysis of individual behaviour. An adequate infrastructure could have a strong positive influence on the further development of a real city. The discipline of artificial intelligence gives an interesting perspective to the design of complex systems. Multi-agent systems can change their environment with actuators and perceive the actual state, and thus changes by other agents, with their sensors. They can also directly communicate with other agents. Architecture could roughly be designed from different elements. The explicit information of social relations and dependencies between the agents facilitates the representation of social interactions, and electronic institutions interaction [4], the agent communication distinguishes among private goals, intentions and beliefs, besides public opinions and intentions giving to the digital institutions create a new social reality for agents by creating new powers, normative goals, and in general new social dependencies. [5] Guides of local culture. In our model, agents implementing services need other services according to defined interaction probabilities, in such case, the Multi-Agent will be integrated with the Ontologies described. Deciding whether the current goal has been reached (i.e. decision-making), and choosing and performing the next action (i.e. adjustment of the behavior) Additionally the Agent-based architecture will use Social Phenomena Modeling having the autonomous agents and giving significance to the cognitive architectures [6], [7].The Ontology enables the model to create effective and homogeneous representations and infer new knowledge from axioms, rules and assertions. With the ontological dimension we are increasing the possibilities to understand the semantic in our social system. Therefore, the agent model is based on BDI-agents that express agents' beliefs, desires and intentions as a set of runs. The BDI-model that we present suggests more flexible and scalable behavior having multiple sources of beliefs, goals and intentions. The multi-agent system is a collection of personal agents interacting one another in order to satisfy the requests of their users, for our case each agent uses its locally information stored in the local ontology to suggest new information based on the behavior. The multiple sub-systems describe different models based on the type of information from the specific domain, as well as their properties and communication events; nevertheless for this purpose we are designing a generic interaction model for all the sub-multi-agent architecture.

The main intentions in the agents that participate in the MAS model are described in Table 1.

Table 1. Multi-Agent System roles and intentions

Agent	Intentions
Perception Agent	Capture and read data from the multiple sources/environments and sent them to other agents to be analyzed and processed
BDI Analizer Agent	Receives data from Perception Agent, process the data and integrates intelligent computing techniques to provide linguistic representation for be understood and processed. Additionally, this agent aims to the integration and correlation from multiple events stored in the LCO Ontology. This agent use correlation and classification methods using reasoning tools applied in ontological models
Decision Agent	Manage the events to generate reports based on the information classified and detected using clustered techniques and to create a recommendation model
Management Agent	Manage the events to generate reports based on the information. Allows communicating with the end-users, giving a natural language interface.

6 Conclusions

The proposed portal based on social multi-agent systems and ontologies aims to provide communities' participation and relations, collaborative environments, intelligence to community, learning environments, relations based on social participation, actions and citizen empowerment as the basis of its main concepts and relationships. The concept of a social agent is introduced to create a multi-agent system for the Local Community Portal. Social agents can be individual (BDI) agents or aggregations of agents from groups. The structure of a multi-agent system and that of social agents is characterized in terms of roles and the relationships between them. A the moment, the integrated architecture using the portal, the multi-agent system and the ontology is under development, we have tested the individual components, but with the integration, we provide an approach to social and semantic architecture for the proposed model.

References

1. López, M., Isaza, G., Joyanes, L.: Ontology for Knowledge Management in a Local Community. In: IEEE Latincom, Medellin – Colombia, pp. 1–6 (2009)
2. Lopez Trujillo, M., Joyanes Aguilar, L.: Ecosistema Digital de Desarrollo para la Ciudad-Región de Manizales y Caldas. In: Revista Educación en Ingeniería, Asociación Colombiana de facultades de Ingeniería – ACOFI, pp. 22–32 (2009)
3. Kryssanov, V.V., Okabe, M., Kakusho, K., Minoh, M.: Communication of social agents and the digital city - A semiotic perspective. In: Tanabe, M., van den Besselaar, P., Ishida, T. (eds.) Digital Cities 2001. LNCS, vol. 2362, p. 56–70. Springer, Heidelberg (2002)
4. Aldewereld, H., Dignum, F., Garcia-Camino, A., Noriega, P., Rodriguez-Aguilar, J.A., Sierra, C.: Operationalisation of norms for usage in electronic institutions. In: Nakashima, H., Wellman, M.P., Weiss, G., Stone, P. (eds.) AAMAS, pp. 223–225. ACM, New York (2006)
5. Caire, P.: Designing Convivial Digital Cities. In: Nijholt, A., Stock, O., Nishida, T. (eds.) Proceedings of the 6th Workshop on Social Intelligence Design (SID 2007), Trento, Italy, pp. 25–40 (2007)
6. Zhang, Y., Coleman, P., Pellon, M., Leezer, J.: A Multi-Agent Simulation for Social Agents Department of Computer Science Trinity University San Antonio, TX 78216. SpringerSim (2008)
7. Menges, F., Mishra, B., Narzisi, G.: Modeling and Simulation E-Mail Social Networks: A New Stochastic Agent-Based Approach. In: IEEE Proceedings of the 2008 Winter Simulation Conference Computer Science Department, Courant Institute of Mathematical Sciences, New York University New York, NY 10012, U.S.A, pp. 2792–2800. IEEE, Los Alamitos (2008)

Fusion of Speech and Face by Enhanced Modular Neural Network

Rahul Kala, Harsh Vazirani, Anupam Shukla, and Ritu Tiwari

Department of Information Technology, Indian Institute of Information Technology and Management Gwalior, Gwalior, MP, India
rahulkalaiiitm@yahoo.co.in, harshiiitmg@gmail.com, dranupamshukla@gmail.com, rt_twr@yahoo.co.in

Abstract. Biometric Identification is a very old field where we try to identify people by their biometric identities. The field shifted to bi-modal systems where more than one modality was used for the identification purposes. The bi-modal systems face problem related to high dimensionality that may many times result in problems. The individual modules already have large dimensionality. Their fusion adds up the dimensionality resulting in still larger dimensionality. In this paper we solve these problems by the introduction of modularity at these attributes. Here we divide various attributes among various modules of the modular neural network. This limits their dimensionality without much loss in information. The integrator collects the probabilities of the occurrences of the various classes as outputs from these neural networks. The integrator averages these probabilities from the various modules to get the final probability of the occurrence of each class. This averaging is performed on the basis of the efficiencies of the modules at the time of training. A module that is well trained is hence expected to give a better performance than the one which is not well trained. In this manner the final probability vector may be calculated. Then the integrator selects the class that has the highest probability of occurrence. This class is returned as the output class. We tested this algorithm over the fusion of face and speech. The algorithm gave good recognition of 97.5%. This shows the efficiency of the algorithm.

Keywords: Modular Neural Networks, Artificial Neural Networks, Fusion Methods, Classification, Speaker Recognition, Face Recognition, Ensemble.

1 Introduction

Biometric Identification deals with the identification of people by their biometric identities [1, 2]. There are numerous biometric identities that have been very commonly used. These techniques are usually divided into two categories. These are physiological biometrics that deals with static characteristics. These include face [3], ear, iris, hand geometry, etc. The other category is the behavioral biometrics that includes the active features such as speech, signature, handwriting etc. The limited recognition efficiencies of these systems resulted in a shift towards the multi-modal or

S.K. Prasad et al. (Eds.): ICISTM 2010, CCIS 54, pp. 363–372, 2010.

bi-modal recognition systems [4, 5, 6, 7]. Here the motivation was to mix two or more modalities in order to attain even greater efficiencies. The problems or limitation of one modality could be solved by the other modality. This resulted in fusion of known biometric modalities that include fusion of face and speech [8, 9, 10, 11, 12, 13], fusion of face and ear, fusion of iris and face, face and fingerprint etc. Multi-modal systems with which combine more modalities was also developed. The fused methods employed the mixing up of the attributes from both the biometric identities and giving the same to the recognition systems.

One of the major problems with the biometric identification systems is that of high dimensionality. The extremely large dimensionality of these problems results in the use of good feature extraction systems. In many problems like face recognition, the dimensionality is still large. This may especially have a problem if the size of data in the training data set is too large. This would normally make the system very slow in learning. Also the error at the time of training would be comparatively large. The number of epochs has to be limited. This results in poor performance of these systems.

The bi-modal systems are formed by the fusion of attributes [4, 5, 6, 7]. As a result the dimensionality associated with the fused systems would be still very large. This large dimensionality would result in the same set of problems to an even larger extent. This may make these systems unworkable whenever the size of data set is very large. This induces a big limitation in these systems and restricts their performance.

The Modular Neural Networks (MNN) is the solutions to these types of problems [15, 16]. The basic principle of these networks is to make use of modularity in the system. These divide the entire problem into various modules. Then all the modules calculate the solution to their part of the problem. This happens in parallel between the various modules. The solution so generated is computed to an integrator. The integrator does the task of integration of the results that come from the various modules. It fetches the individual results from each of the modules and then uses these results to calculate the final output that is returned as the output of the system.

There have been numerous ways and models of the ANN. One of the most commonly used models in classification systems is the ensemble [17]. The ensemble architecture is the same as that of the MNN. It divides the problem into various modules and then integrates them by an integrator. In ensemble there is a poling mechanism that is employed to carry out the classification. Each of the module votes for some class. These votes are added and the final class that wins is declared as the winner.

The MNNs developed so far try to employ a hierarchical approach where the input is mapped to some network that performs the task or a group of networks that perform the task which is later of integrated. This would not solve the problems arising out of dimensionality because of the fact that the dimensionality in these systems remains the same. Also it is further not possible to use any dimensionality reduction technique, as this would result in reduced system performance and loss of information. Hence we need better mechanisms for the application of modularity.

In this paper we propose the modularity to be applied at the various attributes. This means that we intend to distribute the attributes among the various modules for the task of identification. This division needs to be done judiciously so that the recognition is good.

Another problem with the ensemble is that the various modules can only return one particular class as their output. This would mean the taking of wrong decisions at various times when a number of modules cannot decide output between some classes. In our approach every module returns a set of probabilities of the occurrence of every class. This gives a lot of information to the ensemble for the combination of the various classes. This system, as against the polling mechanism, may be viewed as a system where a small set of experts sit together and discuss the final output class, in place of just a voting which is not a good solution in case the number of experts are limited.

The novelty of the paper lays four fold. (i) Firstly we modify the present ensemble approach to make it better for the classification problems. This is done by the introduction of probabilities (ii) Secondly we introduce the concept of modularity at the attribute level that would help the system in faster training and dimensionality control along with the performance boost. (iii) Thirdly we suggest the mechanism of weighing of the various modules that may further help in improvisation of the results by eliminating the bad modules from over affecting the decisions (iv) Fourthly this is applied to the problem of fusion of face and speech for the task of person identification for achieving even greater performance. Here we reduce dimensionality of the problem without loss of information.

The innovation lies in the selection of the number of modules and their attributes. This needs to be done in such a way that all attributes get covered and each module is easily able to solve the problem with good efficiencies even if operating alone.

This paper is organized as follows. Section 2 talks about the general recognition systems and various attributes that are extracted for the problem for both the speech as well as the face. Section 3 presents a native fusion algorithm between face and speech. Section 4 would present the entire algorithmic framework. Finally in section 5 we present the results and in section 6 we present the conclusions.

2 Recognition System

Any recognition system involves various stages. The final output is the recognized person or identity. Here the first task is the data collection that acquires the data in the system. In the problem of fusion of face and speech, the camera is used to take the photograph of the person. At the same time the microphone may be used to capture his voice. Ease of the user is a major criterion that needs to be taken care of. Here the system would be very simple to use for the user where the image and speech can be acquired simultaneously.

The next step comes is the image preprocessing. This is needed for the noise removal as well as to highlight the features. In case of the face the input is in the form of image that requires the application of noise removal operators and binarization. In case of speech the input is a signal that may be freed from noise by the application of noise removal filters.

The next task is segmentation. Here we segment the image and the features. In image the task is concerned with application of gradient mask, dialization, filling up of holes, etc. In speech we segment each and every word of the spoken sentence.

Then feature extraction is done. Here we extract the features for dimensionality reduction. The extracted features must be such that they lead to large inter-class distances and small intra-class distances. They must be relatively constant when the same face is clicked numerous time, or the person speaks various times. The features used for the speech and face are discussed below.

For the speech we extract a total of 11 features. These are time duration, number of zero crossing, max cepstral, average PSD, pitch amplitude, pitch frequency, peak PSD and 4 formants (F1-F4). All these features are extracted using the signal processing toolbox of MATLAB and the inbuilt MATLAB functions. These features are all widely used in research for the speech and speaker recognition and verification systems. They are found to remain stable each time the same person is recorded and analyzed.

For the face there were a total of 13 features extracted [18]. These are length of the eye 1, width of the eye 1, center dimension x 1, center dimension y 1, length of the eye 2, width of the eye 2, center dimension x 2, center dimension y 2, length of the mouth, width of the mouth, center dimension x, center dimension y, distance between eye and eye, distance between eye and mouth.

These features as well are extensively used in research related to face recognition and verification. In case of face the problems of the stability of the features is even more important as it is largely dependent on light, expressions and other variations that are possible. These features are relatively more stable.

3 Native Fusion Algorithm

After these features, the first task done was the fusion of the features so collected into a fused system. This is motivated directly from literature where such systems are found to be giving better performances and results as compared to the native methods. We solve the problem in 2 cases. In the first the entire feature vector was used directly as an input to the problem. In the second approach we manually restricted the inputs before giving them as the input.

As the first case we used the entire feature vector of both the systems combined as an input to the system. This included the extracted 11 features from speech and 13 features from face. This made a total of 24 inputs to the classifier for the recognition systems. Naturally the classifier took a lot of time to train. This was the classical approach of fusion of the face and speech that we applied.

In this case we manually selected some features from the speech and some from the face and only these were allowed to be used for the recognition system. This limited the number of features that were used and hence resulted in a better training time and training efficiency. However there was a permanent loss of information as many of the extracted features were not at all used by the system. This may many times lead to reduced efficiencies.

The features that we selected for this problem are Formant Frequency F1 to F4, Peak & Average Power Spectral Density, Length and width of the Mouth, Distance between Center Points of Eye 1 & Eye 2, Distance between Center Points of Eyes & Mouth. These made a total of 7 attributes for the system.

4 Algorithm

The overall algorithm is built over the modular neural network approach. This approach believes in the principles of modularity of the problem where the entire problem is first divided into modules and then solved independently by each module. Once each module returns its results, an integrator is used for the task of integrating the solutions of the various modules and calculating and returning the final output vector.

In this problem of fusion of face and speech, we first divide the problem into modules. Here the modules are divided into modules. Every module gets a set of attributes. It is possible that an attribute is given to more than one module. Similarly it is possible that an attribute is not given to any of the attributes at all. This however must be avoided as it leads to a loss of valuable information.

Then each module independently calculates the output. These use ANNs for the task. At the end each ANN returns the probabilities of the occurrence of each of the class. These probabilities lie in the range of -1 to 1. An occurrence of -1 means that the given class is completely absent and vice versa. The numbers denote the certainty of the ANN in the occurrences of the class as the final output class.

At the end the integrator does the task of combining the individual solutions to give the final output. This is in the form of 1st averaging the various probabilities returned by the individual systems. Here the performance of the systems at the time of training is used as weights. Then the summation takes place to return the final class that is declared as the final output. The basic working of the system is shown in figure 1. We discuss each of the steps in the next section.

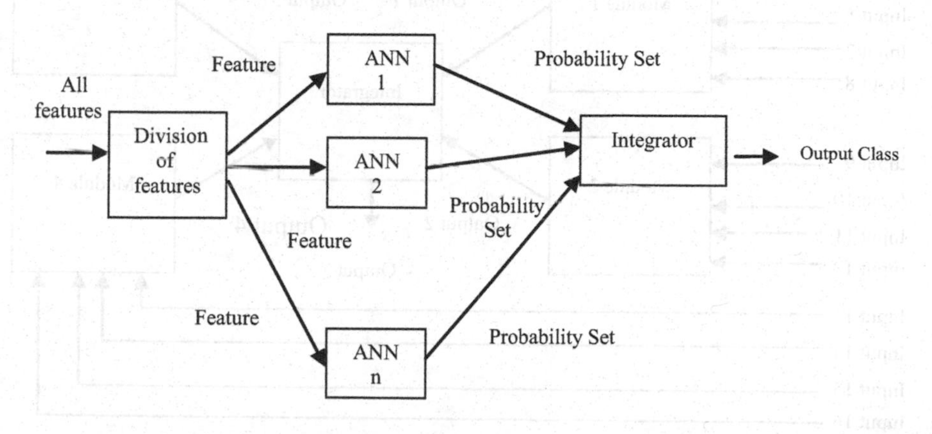

Fig. 1. The general structure of the algorithm

4.1 Modules

Here we are supposed to exploit the modularity in the features. The basic motive is to ensure that each module after getting its feature set must be in a condition to appreciably solve the problem. It must have the related attributes to enable it to do so. Hence the attributes given to any module must be diverse and must collectively supply the entire information. Also loading too many features to an ANN would be not desirable. Here we also try to ensure that all attributes collectively get the complete feature set. This would avoid the loss of information from the system.

In this approach we keep the speech and face features completely different. We then divide the speech attributes into two parts and similarly the face attribute into two parts. In this system we select some speech features to be used as inputs for the first module. These are time duration, max cepstral, pitch amplitude, peak PSD, F2 and F4. The second module contains the rest of the speech features. These are number of zero crossing, average PSD, pitch frequency, F1 and F3.

The last two modules are to cover the facial features. Here also we follow a similar technique. The 3rd module covers the features length of the eye 1, width of the eye 1, center dimension x 1, center dimension y 1, length of the mouth, distance between eye and eye. The rest of the facial features belong to the fourth module. This includes length of the eye 2, width of the eye 2, center dimension x 2, center dimension y 2, width of the mouth, center dimension x, center dimension y, distance between eye and mouth. The structure in general related to the division of the features among modules is shown in figure 2.

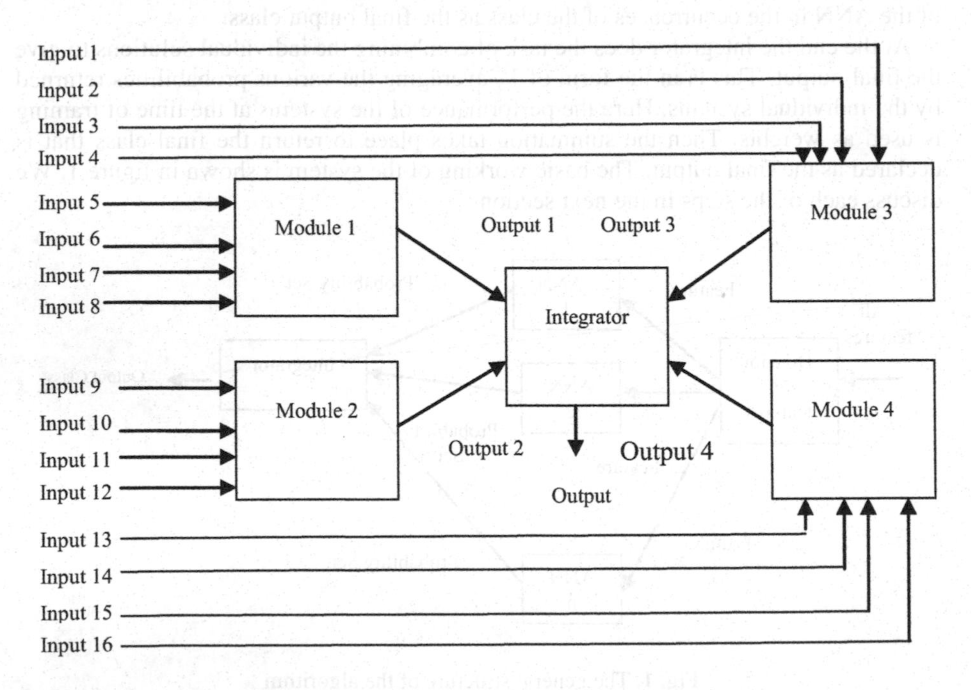

Fig. 2. The division of inputs between modules

4.2 Artificial Neural Networks

The job of classification of the inputs is carried out with the help of ANNs with BPA as the training algorithm. The ANNs are a natural choice because of their ability to learn from the historical data and to generalize the results. The ANNs map any input to some class or person here. We use a classificatory model of ANN here. This has as many output neurons as the number of classes. Each output neuron stands for some person or class. The output at this neuron for any input i is the probability of occurrence of this person or class according to the ANN. Hence the ANN gives as its output c number of probabilities in the output vector. Let this output vector for any input i be represented by $<v_{i1}, v_{i2}, v_{i3}, \ldots v_{ic}>$.

The probabilities here are measured in the range of -1 to 1. A probability of 1 means a certainty of the output class with full confidence. On the other hand a probability of -1 means that according to the ANN this class is not the output with full confidence. Hence the ideal output for any input for the ANN should be a 1 for any one element of the output vector and a -1 for all the other elements. However due to practical reasons, the output lies anywhere in the complete region.

4.3 Integrator

The last part to implement according to the entire algorithm is the integrator. This is the major part of the whole system that does the task of finding the final output class after getting inputs from the individual modules. The integrator analyzes all the outputs by the various modules and then decides the final output class that according to the system is the output.

The input given to the integrator is the solution vector of every module of ANN. Let the vector of the module i be $<v_{i1}, v_{i2}, v_{i3}, \ldots v_{ic}>$ where each v_{ik} is the probability of the occurrence of the class k measured on a scale of -1 to 1. The integrator decides the output by first taking the weighted averages of the probabilities given to it and then selecting the class with the maximum probability. This class is declared as the winner. This is shown in figure 3.

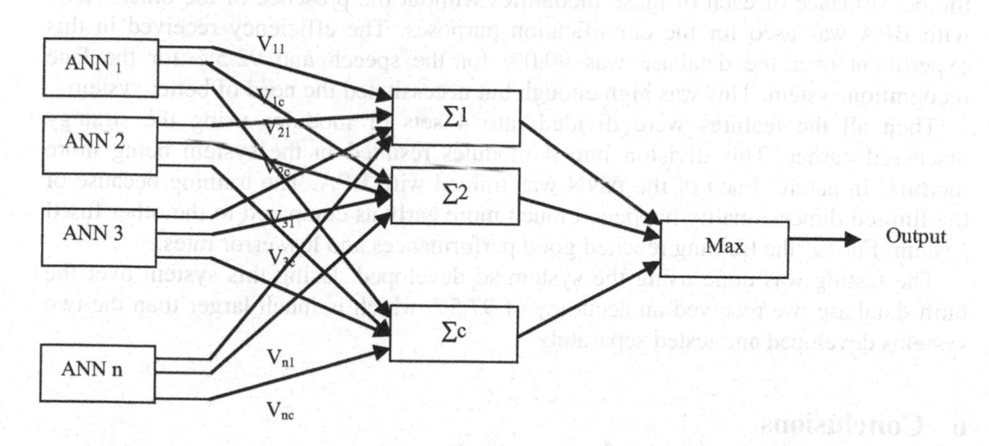

Fig. 3. The Integrator

The weighted average is calculated for each and every class in the system on which the output can map to. The weights of the various modules or ANNs here are their performances on the training data. Each ANN was given the same data set for the training purposes with a different feature set. The performance here is calculated as a ratio between the numbers of elements the ANN correctly classifies in the validation data set by the total number of elements in the validation data set. The higher the performance of the ANN in the validation data set, the more would be its weight and more dominant it would be to decide the final output at the time of integration.

Using these calculated weights, the integrator calculates the weighted average for all the classes. This gives the final probability vector comprising of c probabilities with each probability associated with some class. These probabilities again lie in the range of -1 to 1 with the same meaning of the probabilities.

The next task of the integrator is to find out the final output class. For this the integrator selects the output class with the largest value of the probability out of all the available classes. The class or the person corresponding to this largest probability is declared as the winner and this is the output that the system gives.

5 Results

The testing of the algorithm was done by experimenting and validating the results using a self made database. The database consisted of data of 20 people whole pictures had been taken multiple times. At the same time their voices had been recorded with same words multiple times. This made a system which had to identify the person out of the data of 20 people available in the database.

The features of both face and speech were recorded using MATLAB tools and functions. This transformed the whole data into numerical forms. Normalization was carried out for each of the attributes to make it lie in the range of 0 to 1.

The first experiment conducted was on face and speech separately. We tried to test the performance of each of these modalities without the presence of the other. ANN with BPA was used for the classification purposes. The efficiency received in this experiment over the database was 90.0% for the speech and 92.5% for the face recognition system. This was high enough but necessitated the need of better system.

Then all the features were divided into 4 sets or modules using the strategy discussed earlier. This division into 4 modules resulted in the system being more modular in nature. Each of the ANN was trained with BPA. The training because of the limited dimensionality happened much more early as compared to the other fused system. Further the training reached good performances and low error rates.

The testing was done using the system so developed. Using this system over the built database, we received an accuracy of 97.5% which is much larger than the two systems developed and tested separately.

6 Conclusions

In this paper we took the problem of fusion of face and speech. Using this problem as a means we studied a good method of reducing dimensionality in the problem that was causing effects to the performance of the system. For this we developed an algorithm based on the MNN. The algorithm introduced modularity in the features and divided them into various modules. Each module could be separately used as a system of its own in the classificatory problem with just a little reduced performance.

Then each of these modules was given its share of every input and this in turn returned the probability vector where each output denoted the probability of occurrence of some class. These were combined by using weighted average by the integrator. The integrator then selected the class with the maximum probability of occurrence and this was declared as the final output of the system.

The system so developed has various functionalities that are better than the original fused methods. This solves the problem of high dimensionality that is prevailing in the original methods. This results in better training and training in reduced time and reaching of larger number of epochs. This has a good effect on the system performance and we are able to reach a higher level of accuracies.

The algorithm also proposes a change in the ensemble manner of poling. Here we proposed a probability based poling that can be better than the poling used by the ensemble. Also the weighing of the various modules is carried out which gives a better performance and reduces the influence of bad modules. This further increases the system performance.

Another innovation in this algorithm is the division of attributes between the modules. This division is carried out in a manner to allow maximum efficiencies of each module. This is done by reducing dependencies and giving diverse inputs. This may be generalized to any classificatory problem in general.

The system so obtained gave a good recognition of 97.5% over the self made database and the self extracted features. This is highly encouraging and proves the efficiency of the algorithm as well as the views presented. This was much higher than the performance of the same database with single modalities.

Even though we have proved the working of this algorithm and received good results, a lot may be done in the future. The algorithm needs to be worked upon large databases that would fully exploit and justify the scalability factors of the algorithm. Further we used a set of features for the face and speech. The work over the other possible features may be done. Also different feature combinations may be tried in future. The use of the same algorithm may be done for other multi-modal systems as well. This may even be generalized to any classificatory problem. The weighing factor we discussed was over the validation data set performances. Possibilities of the other weighing strategies may be tried in future and their results may be compared.

Acknowledgments. This work is sponsored and supported by Indian Institute of Information Technology and Management Gwalior.

References

1. Ben-Yacoub, S., Abdeljaoued, Y., Mayoraz, E.: Fusion of Face and Speech Data for Person Identity Verification. IEEE Transactions On Neural Networks 10(5), 1065 (1999)
2. Jain, A., Hong, L., Pankanti, S.: Biometric Identification. Communications of the ACM 43(2), 90–98 (2000)
3. Chen, C.-H., Chu, C.-T.: Combining Multiple Features for High Performance Face Recognition System. In: International Computer Symposium (ICS 2004) Taipei, December 2004, pp. 387–392 (2004)
4. Snelick, R., Indovina, M., Yen, J., Mink, A.: Multimodal Biometrics: Issues in Design and Testing. In: ICMI 2003, Canada, November 5-7, pp. 68–72 (2003)

5. Ross, A., Jain, A.: Information fusion in biometrics. Pattern Recognition Letters (24), 2115–2125 (2003)
6. Rukhin, A.L., Malioutov, I.: Fusion of Biometric Algorithm in the Recognition Problem. Pattern Recogition Letters, 299–314 (2001)
7. Frischholz, R.W., Dieckmann, U.: Bioid: A Multimodal Biometric Identification System. IEEE Computer (33), 64–68 (2000)
8. Bigün, J., Bigün, B., Fischer, S.: Expert conciliation for multi modal person authentication systems by Bayesian statistics. In: Bigün, J., Borgefors, G., Chollet, G. (eds.) AVBPA 1997. LNCS, vol. 1206, pp. 291–300. Springer, Heidelberg (1997)
9. Choudhury, T., Clarkson, B., Jebara, T., Pentland, A.: Multimodal person recognition using unconstrained audio and video. In: Proc. 2ndInt Conf. Audio-Video Based Person Authentication, Washington, DC, March 22-23, pp. 176–180 (1999)
10. Ben-Yacoub, S.: Multimodal data fusion for person authentication using SVM. In: Proc. 2nd Int. Conf. Audio-Video Based Biometric Person Authentication, Washington, DC, March 22–23, pp. 25–30 (1999)
11. Patterson, E.K., Gurbuz, S., Tufekci, Z., Gowdy, J.N.: Noise-based audio-visual fusion for robust speech recognition. In: International Conference on Auditory-Visual Speech Processing, Denmark (2001)
12. Sanderson, C., Paliwal, K.K.: Information Fusion and Person Verification Using Speech & Face Information, IDIAP, Martigny, Research Report, 02-33 (2002)
13. Shukla, A., Tiwari, R.: A Novel Approach of Speaker Authentication by Fusion of Speech and Image Features using ANN. International Journal of Information and Communication Technology (IJICT) (1)(2), 159–170 (2008)
14. Jain, A.K., Hong, L., Kulkarni, Y.: A multimodal biometric system using fingerprints, face and speech. In: Proc 2nd Int Conf Audio-Video Based Biometric Person Authentication, Washington, D.C., March 22-23, pp. 182–187 (1999)
15. Kittler, J., Hatef, M., Duin, R.P.W., Matas, J.: On combining classifiers. IEEE Trans. Pattern Anal. Machine Intell. 20, 226–239 (1998)
16. Fogelman Soulie, F., Viennet, E., Lamy, B.: Multi-modular neural network architectures: applications in optical character and human face recognition. International Journal of Pattern Recognition and Artificial Intelligence 7(4), 721–755 (1993)
17. Perrone, M.P., Cooper, L.N.: When Networks Disagree: Ensemble Methods for Hybird Neural Networks. In: Neural Networks for Speech and Image Processing (1993)
18. Gonzalez, R.C., Wood, R.E.: Digital Image Processing. In: Pearson Education Asia (2002)

Mining Rare Events Data by Sampling and Boosting: A Case Study

Tom Au, Meei-Ling Ivy Chin, and Guangqin Ma

AT&T Labs, Inc.-Research, USA

Abstract. In data mining, popular model ensemble technique like boosting is often used to improve predictive models performance. When mining data with rare events (far less than 5%), though boosting may improve a model's overall prediction power, but the accuracy and efficiency of model estimation is negatively impacted when the simple random sampling procedure is employed. In this study we investigate the performance of applying the boosting technique to an imbalanced sample procedure called case-based sampling. We demonstrate the performance of the combined procedure in predicting customer attrition with an actual telecommunications data. Our results show that the combination of boosting and case-based sampling is very effective at alleviating the problem of rare events.

Keywords: Rare Events, Boosting, Case-Based Sampling and AUC.

1 Introduction

We study the problems in predicting the risk of the occurrence of a rare event. In statistics, the probability of rarity is often defined as less than 5%. Many of the events in real world are rare-binary dependent variables with hundreds of times fewer ones (events) than zeros ("nonevents"). Rare events are often difficult to predict due to the binary response is the relative frequency of events in the data, which, in addition to the number of observations, constitutes the information content of the data set. Most of the popular statistical procedures, such as logistic regression, tree-based method and neural network, will be sensitive to each simple random sample used to train the model.

In order to produce more accurate and efficient estimate of the risk of a rare event of interest, researchers have been using disproportional sampling techniques, such as case-based sample, to do the estimation and bias correction [1]. For rare events study, case-based sampling is a more efficient data collection strategy than using random samples from population. With case-based sample, researchers can collect all (or all available) events and a small random sample of nonevents. In addition to gain accuracy and efficiency, researchers can focus on data collection efforts where they matter. For example, in survival analysis, statisticians have been using the case-based sample in proportional hazards analysis to evaluate hazards rate [2] ; in social sciences, researchers have been using the same sampling technique to evaluate the odds of rare events [1]. Resources available to researchers are different, in clinical trials, the number of subjects

S.K. Prasad et al. (Eds.): ICISTM 2010, CCIS 54, pp. 373–379, 2010.

to study is limited, researchers relies on good quality of collection and sample design; in other disciplines, there may be a vast collection of information, researchers can be overwhelmed by the volume of information available, and the procedures used for assessing rare events can be very inefficient. To the later case, the case-based data sampling can become vital for a successful analysis.

In data mining, boosting is a proven technique for improving the performance of a classification procedure. Boosting works by sequentially applying a classification procedure to re-weighted version of the training sample and then taking a weighted average of the sequence as the final classifier. The seemingly simple strategy does result in improvements in classification performance[11].

In this study, we examine the accuracy, efficiency and performance by applying case-based sampling and booting techniques to an particular data set for predicting customer attrition risk. Customer attrition refers to the phenomenon whereby a customer leaves a service provider. As competition intensifies, preventing customers from leaving is a major challenge to many businesses such as telecom service providers. Research has shown that retaining existing customers is more profitable than acquiring new customers due primarily to savings on acquisition costs, the higher volume of service consumption, and customer referrals [3] [4]. The importance of customer retention has been increasingly recognized by both marketing managers as well as business analysts [3] [5] [6] [4].

In this article, we emphasize the effectiveness of combining case-based sampling and boosting in predicting the risk of a rare event. We evaluated the results based on a dataset from a major telecom service company and the logistic regression.

2 Data Structure for Events Prediction

An event can occur at any time point to a subject of interest. To predict an event occurrence in a future period, we assume the event has not occurred to the subject at the time of prediction. We consider the following structure of events prediction data

$$\{X_{t-m}, X_{t-m+1}, X_{t-m+2}, \ldots, X_t | Y = 0/1 \, in \, (t, t+k)\}, \tag{1}$$

where Xs are explanatory variables measured before the time of prediction t, Y is the event indicator. Y=1 if the event occurred during period (t, t+k); otherwise Y=0.

This data structure for events prediction is not only simple, but also bears real application scenarios. For example, when we score customer database with risk score for retention, we score all currently active customers who are yet to defect, disconnect, or cancel services, then we target those customers with high risk of leaving within a specified period. When we collect data for predictive modeling, we have to bear in mind its application scenario, that is, when we collect data, we need to define a reference time point that all collected individual subjects are 'active' (event not occurred yet), and a period after the time point that events are observed. The length of the window between start point to the reference point of 'active' can be determined by needs and resources available; the length of the period between the 'active' time point to the end of event observation window is specified by retention needs, say if you like to predict event probability for the next quarter or next six-month or next 12-month, etc.

3 Case-Based Sampling

As in most rare event history analyses, the rare event represents only a small percentage of the population (less than 5%). The low rate of event implies that if we draw a random sample to study the event, the sample will be characterized by an abundance of nonevents that far exceeds the number of events, we will need a very large sample to ensure the precision of the model estimation (or we would have to observe a long period of time to collect more cases of event.) Processing and analyzing the random sample will not only impact the efficiency of using resources, but also more importantly reduce the efficiency and accuracy of model estimation. To overcome this problem, practitioners used a type of sampling technique called the case-based sampling, or variations of the ideas like choice-based sampling (in econometrics), case-control designs (in clinical trials) [7], case-cohort designs (in survival analysis) [2]. Here we design a case-based sampling by taking a representative simple random sample of the data in (1) with response 0s, and all or all available subjects in (1) with 1s, such that, the ratio of 0s to 1s is about 3:1, 4:1, 5:1 or 6:1. The combined dataset is called a case-based sample of the data in (1).

More formally, we define a case-based sample as following

Step 1: Take all the observations of $\{(X, Y = 1)\}$ from the population;
Step 2: Take a *representative* random sample of $\{(X, Y = 0)\}$;
Step 3: Combined the data from the above two steps to form a *full case-based sample* $\{(X', Y')\}$.

By *representative*, we mean

$$P\{X'|Y' = 0\} = P\{X|Y = 0\},\ (so\, P\{X'|Y'\} = P\{X|Y\})$$

where $P(\bullet)$ is the density function.

The probability of event based on the case-based sample will be biased and different from population probability of event. We can show that this biasness can be corrected based on population prior information and the bias-corrected estimation is consistent.

Let (X', Y') be from S_n, S_n is a sample of size n from the full case-based sample, then

$$P(Y'|X')\frac{P(Y)P(X')}{P(Y')P(X)} \rightarrow P(Y|X),$$

as n approach the size of the full case-based sample.

Further more,

$$logit(P(Y' = 1|X')) + ln(\frac{\pi(1-\pi')}{\pi'(1-\pi)}) \rightarrow ligit(P(Y = 1|X)), \quad (2)$$

where $\pi = P(Y=1)$ and $\pi' = P(Y'=1)$ are prior information based on the population and the full case-based sample, respectively. The theoretical relationship expressed in (2) shows that the risk ranking based on a simple random sample and a case-based sample are equivalent, but in practice, the parameter estimation based on a case-based sample is more accurate and efficient.

A major task in customer retention is to rank customers by their attrition risks, and to target those most vulnerable customers as a retention effort. The consistency showed in (2) indicates that the ranking of customer attrition risk based on a case-based sample is consistent with the actual ranking, if there is one.

4 Additive LogitBoost

Boosting is one of the important development in classification methodology. Boosting works by sequentially applying a classification algorithm to re-weighted versions of the training data and then taking a weighted average majority vote of the sequence of classifiers thus produced. For logistic regression models, this simple strategy usually carried out by stagewise optimization of the Bernoulli log-likelihood [11]. More precisely, let $\frac{1}{2}logit(P(Y=1|X)) = F(X)$, and $F(X)$ is a linear combination of explanatory variables. Then we start with

Step 1: Assign each individual X with the same weight $w(X) = \frac{1}{N}$, N is the number of observation in the data, $F(X)= 0$ and $P(Y=1|X) = \frac{1}{2}$;

Step 2: Repeat from $m = 1, 2, ..., M$:

– (a) Compute the working response and weights

$$f_m(X) = \frac{Y - P(Y=1|X)}{P(Y=1|X)(1 - P(Y=1|X))},$$

$$w(X) = P(Y=1|X)(1 - P(Y=1|X));$$

– (b)Fit the function $f_m(X)$ by weighted least-squares regression to X using weights $w(X)$;

– (c) Update $F(X) \leftarrow F(X) + \frac{1}{2}f_m(X)$ and $P(Y = 1|X) \leftarrow e^{F(X)}/(e^{F(X)} + e^{-F(X)})$;

Step 3: Output $F(X) = \sum_{m=1}^{M} f_m(X)$ and $P(Y = 1|X) \leftarrow e^{F(X)}/(e^{F(X)} + e^{-F(X)})$.

The case-based sampling is trying to assign more weights to events population $(Y=1)$ by over sampling the event population; the LogitBoot algorithm is trying to assign misclassified individual more weights at each regression stage and force it to be more correctly estimated.

5 The Study Design

There are several data mining methods that may be used to construct models to estimate event probability, such as the logistic regression method, the Cox regression method, the tree-based classification method, and, more recently, the artificial neural network method. Each may be more suitable for a particular application. In this exercise, we set out to use logistic regression to predict customer attrition from a data set of size 71,576, with event rate less than 2%. The data set is from a major telecom service company, which includes a segment of customers with a large number of service lines (in the order of millions) that are active at the beginning of the year. The unit of analysis is the individual subscriber line. For each active service line, information on service termination (yes/no) was collected over the next 3-month period (the first quarter of the chosen year). Customer usage and characteristics were collected over the 3-year period backward from the chosen year. This information includes type of service, service usage, marketing information, customer demographics, and service line transaction history.

We start by randomly splitting (50/50) the data into a training data and a testing data. Then based on the training data set we build the models based on a simple random sample (about 80% of the training data) and a case-based sample (all the events in training data plus a random sample of 79% of the nonevents in the training data). We score the training and the testing data sets by directly applying the estimated models to the data, or by boosting score $P(Y = 1|X) = e^{F(X)}/(e^{F(X)}+e^{-F(X)})$, with M=5. We repeat the above process 1000 times, that is, there are 1000 simple random samples sampled and 1000 case-based samples sampled. Table 1 summarize the study design.

Table 1. The Study Design

	No Boosting	With Boosting
1000 Random Samples	AUCs for Training/Testing	AUCs for Training/Testing
1000 Case-Based Samples	AUCs for Training/Testing	AUCs for Training/Testing

The performance of a predicting model may vary depending on a specific decision threshold used. Thus, an objective evaluation of a risk-predicting model should examine the overall performance of the model under all possible decision thresholds, not only one particular decision threshold. To achieve this, we adopt a useful tool, the receiver-operating characteristic (ROC) curve, to evaluate the performance of a risk predictive model. This curve is the locus of the relative frequencies of false event fraction (FEF) and true event fraction (TEF), which occupy different points on the curve corresponding to different decision thresholds. The area under this curve (on a unit square) is equal to the overall probability that the predictive model will correctly distinguish the "at risk" subjects from the "not at risk" subjects [8] [9]. The area under the curve (AUC) is usually used to summary rise the overall performance of predictive procedure.

6 Results

As mentioned, we repeated each of the analysis described in section 5 one-thousand times. Figure 1 shows the boxplots of the AUCs' distribution. When we use random sampling without boosting, the variation of AUCs are large; when we use both random sampling and boosting, the average performance improved, but the variation is still large. If we use the case-based sampling without boosting, we see the variation of AUCs is dramatically reduced; if we use both the case-based sampling and boosting, then both performance and uncertainty are improved.

The result in this paper will likely apply to some rare event data mining. When mining data with very low event rate is low (far less than 5%), we suggest to use both the case-based sampling and boosting methodology for risk scoring. Since our intention here is to demonstrate the effectiveness of case-based sampling and boosting relative to random sampling in rare event modeling, models discrimination performance is not discussed here. We refer readers to our previous study on the applications applying and evaluating models to predict customer attrition [10].

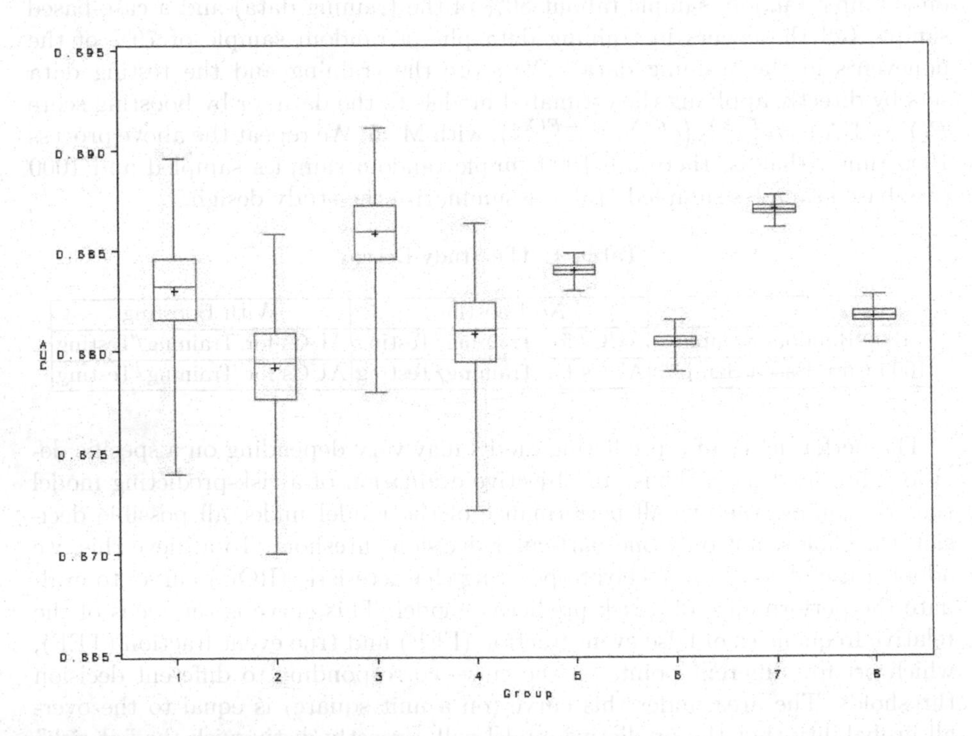

Fig. 1. AUCs distribution: (1)Training Data, Random Sampling, No Boosting; (2) Testing Data, Random Sampling, No Boosting; (3) Training Data, Random Sampling, with Boosting; (4)Testing Data, Random Sampling, with Boosting; (5)Training Data, Case-Based Sampling, No Boosting; (6) Testing Data, Case-Based Sampling, No Boosting; (7) Training Data, Case-Based Sampling, with Boosting; (8)Testing Data, Case-Based Sampling, with Boosting.

References

1. King, G., Zeng, L.: Logistic Regression in Rare Events Data. Society for Political Methodology, 137–163 (February 2001)
2. Prentice, R.L.: A Case-cohort Design for Epidemiologic Cohort Studies and Disease Prevention Trials. Biometrika 73, 1–11 (1986)
3. Jacob, R.: Why Some Customers Are More Equal Than Others. Fortune 19, 200–201 (1994)
4. Walker, O.C., Boyd, H.W., Larreche, J.C.: Marketing Strategy: Planning and Implementation, 3rd edn. Irwin, Boston (1999)
5. Li, S.: Applications of Demographic Techniques in Modeling Customer Retention. In: Rao, K.V., Wicks, J.W. (eds.) Applied Demography, pp. 183–197. Bowling Green State University, Bowling Green (1994)
6. Li, S.: Survival Analysis. Marketing Research, 17–23 (Fall 1995)
7. Breslow, N.E.: Statistics in Epidemiology: The case-Control Study. Journal of the American Statistical Association 91, 14–28 (1996)
8. Hanley, J.A., McNeil, B.J.: The Meaning and Use of the Area under a ROC Curve. Radiology 143, 29–36 (1982)
9. Ma, G., Hall, W.J.: Confidence Bands for ROC Curves. Medical Decision Making 13, 191–197 (1993)
10. Au, T., Li, S., Ma, G.: Applications Applying and Evaluating Models to Predict Customer Attrition Using Data Mining Techniques. J. of Cmparative International Management 6, 10–22 (2003)
11. Friedman, J., Haste, T., Tibshirani, R.: Additive Logistic Regression: A Statistical View of Boosting. The Annals of Statistics 28(2), 337–407 (2000)

Model-Based Regression Test Case Prioritization

Chhabi Rani Panigrahi and Rajib Mall

Department of Computer Science and Engineering
Indian Institute of Technology, Kharagpur, 721 302, India
{chhabi,rajib}@cse.iitkgp.ernet.in

Abstract. We propose a model-based regression test case prioritization technique for object-oriented programs. Our technique involves constructing a graph model of the source code to represent control and data dependence as well as object relations such as inheritance, polymorphism and message passing. This model is further augmented with information, such as message paths and object states, that are available from the UML design models. We perform a forward slice of the constructed model to identify all the statements in the program that may be affected by a change. Information available from the intersection of the forward slice of the model and the relevant slice of each test case is then used to prioritize test cases.

Keywords: Software maintenance, Regression testing, Test case prioritization, Model-based prioritization.

1 Introduction

Regression testing is a crucial activity during the maintenance phase of the software development life cycle. It usually requires large amounts of testing time and resources, and often accounts for almost half the total software maintenance costs [1]. TCP techniques re-order the test cases in the initial test suite according to some criteria such that the higher priority test cases are executed earlier during regression testing. This is advantageous especially in case of unpredicted interruptions like delivery, resource or budget constraints [2],[3] during regression testing. Various empirical studies have shown the effectiveness of TCP techniques even when used to prioritize large test suites [4],[5]. Many existing TCP techniques prioritize test cases essentially based on some code coverage criterion, such as, the number of statements or functions executed by each test case, etc. [4],[6].

This paper is organized as follows: In Section 2, we discuss background concepts required for our paper, and review the existing TCP techniques in Section 3. We present a brief discussion of our TCP technique in Section 4. We present empirical results of our proposed technique in Section 5. In Section 6, we present a comparative study of our technique with the existing approaches available in the literature, and conclude the paper in Section 7.

S.K. Prasad et al. (Eds.): ICISTM 2010, CCIS 54, pp. 380–385, 2010.

2 Background Concepts

In this Section, we review some important concepts that form the basis of our work.

2.1 Graph Model for Object-Oriented Programs

Horwitz *et al.* have introduced the concept of System Dependence Graph (SDG) for procedural programs [7]. Later, Larsen and Harrold have extended the features of the SDG so that it can be used to model object-oriented programs as in [8].

2.2 Object State

An object state is characterized by the values of the state variables that characterize the object. Modification of the values of these state variables may result in state transition of the corresponding object.

2.3 Test Case Prioritization

The TCP problem can be stated as follows [4],[6]: Let T be a test suite designed for validating a program P, and PT be the set of all possible prioritizations (orderings) of T. Let f be a function (called the *objective* function) that maps PT to some award value (according to some ordering criterion) for that particular ordering. Then, the TCP problem can be stated as

$$\textit{Find } T' \in PT \textit{ such that } (\forall T'')\ (T'' \in PT)\ (T'' \neq T'),\quad [f(T') \geq f(T'')].$$

3 Related Work

In this Section, we briefly discuss the different TCP techniques that have been proposed in the literature. Rothermel [9],[6] and Elbaum [4],[5] have proposed several approaches for ordering test cases based on different criteria such as total statement coverage, total function coverage etc.. Srivastava and Thiagarajan [10] presented a technique for prioritizing test cases based on the coverage of impacted basic blocks. Another approach for prioritizing regression test cases based on the concept of a relevant slice of a test case have been presented by Jeffrey and Gupta [12] which is used to assign higher priority to those test cases which execute more number of program statements in the relevant slice. TCP techniques that take into account the design or state models (e.g., EFSM, SDL, UML) are relatively scarce in the literature [14,[15],[16]. The TCP technique proposed by Basanieri *et al.* is based on analysis of UML design models, mainly use case and sequence diagrams [14]. Korel *et al.* have proposed a state model-based TCP technique in which after executing the modified system model on the test suite, the resulting information is used to prioritize the test suite [16].

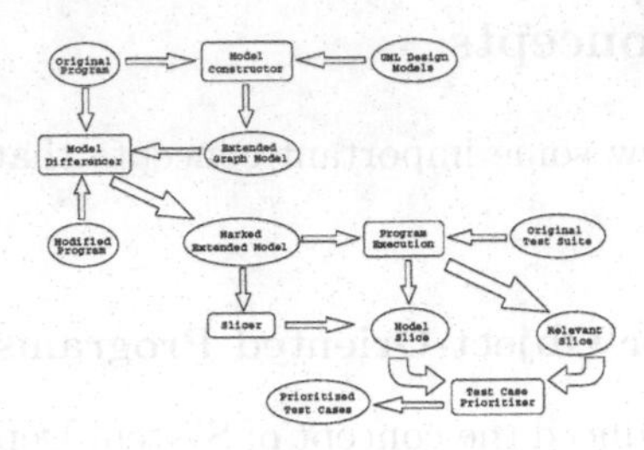

Fig. 1. Schematic representation of our model-based TCP approach

4 MBTPT: Our Proposed Model-Based Approach

We have designed a graph model of the original program as an Extended Graph model as shown in Figure 1. When a change is made to the original program, the Extended Graph model is updated to reflect those changes. We use model slicing on the updated model to identify the set of all marked model elements. The test cases in the regression test suite are executed to identify the relevant slice for each test case. Test cases are assigned a priority in proportion to the number of elements common to both its relevant slice and the model slice. We have named our proposed model as MBTPT (Model-Based Test case Prioritization Technique). A schematic representation of the important components involved in MBTPT has been shown in Figure 1. In Figure 1, ellipses represent artifacts, such as graph models, which act as inputs to the processing units in our technique. The rounded rectangular blocks represent the processing units, such as *Model Constructor*, *Slicer* etc.

We now briefly discuss the different components used in our approach. Model Constructor constructs the graph model by a static analysis of the original program. The constructed model is augmented with information available from the UML design models, such as, statechart and sequence diagrams. This is represented as an *extended graph model.* Model Differencer compares the original and the modified model and identifies the modified model elements. The modified elements are marked on the graph model. In Figure 1, this model is represented as a *marked extended model.* Slicer performs a forward slice [17],[7],[18] on the marked extended model to identify all the directly or indirectly affected model elements. Each modified model element at that point act as the slicing criterion. The set of model elements which are affected by a change, is represented as *Model Slice.* Then we check whether the modified element is present in the relevant slice. This is represented by *Relevant Slice* in the Figure 1. The information from the model slice and the relevant slice is used by *Test Case Prioritizer* to order the test cases.

4.1 Model Slicing

MBTPT use an Extended Graph Model (M_o) of the original program (P) with relevant information from UML design models. The changes made to the original program (P') are identified and M_o is updated (M_u) to reflect those modifications. We perform a forward slice for each marked nodes in M_u. The forward slice of the model gives all those elements which are directly or indirectly affected by the modifications made to the program. Let P be a program under test consisting of m number of statements, and let the forward slice computed from each modified statement $s_j \in P$, $j \in [1, m]$ be denoted by $FS(P_{s_j})$. Then, the set of all potentially affected model elements due to modifications to P, denoted by $MS(P)$, can be represented as follows:

$$MS(P) = \sum_{j=1}^{m} FS(P_{s_j}) \tag{1}$$

Test cases which execute a higher percentage of model elements that can influence the output of a program should be assigned a higher priority [12]. Then we check for the modified model elements which are in the relevant slice. The execution trace helps to determine the model elements in M_o which are executed by each test case.

4.2 Prioritizing Test Cases

The information available from the relevant slice of each test case and the forward slice of the model are utilized to prioritize the test cases. We now give a formal representation of our approach. Let $T = \{t_1, t_2, \ldots, t_n\}$ be a test suite consisting of n test cases used in regression testing of the modified program P'. Let the relevant slice computed for each test case t_i be denoted as $RS(t_i)$. The intersection of the relevant slice of each test case, $RS(t_i)$, and the model slice $MS(P)$ is given as:

$$INT_i = MS \cap RS(t_i) \tag{2}$$

The intersection for a test case actually denotes the number of model elements that are directly or indirectly affected due to modifications and are also included in the relevant slice of the test case. Test cases which have a high value of cardinality of INT_i, i.e., execute a greater number of model elements common to both, are assigned a higher priority during ordering.

5 Empirical Study

In this Section, we present the results obtained from empirical studies of our proposed model-based TCP technique. We have used five different models: ATM model, Library System model, Elevator model, Fuel Pumps model and Vending Machine Model. The sizes of model range from 5 to 12 classes. For each model, the corresponding system prototype is implemented in C++ language. The size of these implementations range from 500 to 1,100 lines of source code. The detail information about the models is shown in Table 1. In order to measure the effectiveness of early fault detection of our test prioritization method, we

Table 1. Results from our empirical studies

Program name	Line of code	No. of classes	No.of test cases	No.of seeded faults	No.of faults detected
ATM	602	8	50	10	8
Library System	1023	12	125	11	7
Elevator	551	5	63	8	5
Fuel Pumps	891	6	97	10	9
Vend.Machine	754	5	112	11	10

seeded faults into implemented systems. For all mentioned programs, we check for the modified elements which are in the relevant slice of the original program , and also calculate the forward slice of each modified model element in the graph model. We observe that the intersection of the relevant slice and the model slice increase the number of faults detected as compared to the approach described in [12]. We have used cardinality value of the set INT_i to prioritize test cases.

6 Comparison with Related Work

The technique proposed in [12] is based on ordering test cases using the concept of a relevant slice of a test case. However, during execution of a test case, there may be other program statements that are potentially affected by a change but do not affect the output of that particular test case. Our technique overcomes this shortcoming by prioritizing test cases based on information from both the relevant slice of a test case and from the forward slice performed on the program model. Furthermore, our technique is based on analysis of a model representation of the program, which is useful for modeling complex programs and quickly identifying the changed element of our model. Our technique differs in the level of abstraction from the TCP techniques proposed by Basanieri *et al.*[14] and Korel *et al.* [15],[16]. Their techniques are defined at a higher level of abstraction since those techniques use UML use case and sequence diagrams, and EFSM system models respectively. Our technique uses an ESDG for modeling the program and also takes into account various dependence relationships (such as control and data) that exist among program elements.

7 Conclusion

It is advantageous to model large and complex programs as graphs because graph algorithms are less expensive than textual analysis of the source code. This has lead to an increased emphasis and focus on developing model-based testing approaches. Our model-based TCP approach is aimed at ordering test cases such that a prioritized subset of test cases is able to detect the maximum number of bugs. Empirical studies show that our proposed slicing-based technique is more effective in terms of fault detection and cost reduction especially in object-oriented programs as compared to code coverage-based approaches.

References

1. Kapfhammer, G.: Software testing. In: The Computer Science Handbook, 2nd edn. CRC Press, Boca Raton (2004)
2. Walcott, K.: Prioritizing regression test suites for time-constrained execution using a genetic algorithm. Technical Report TR-CS05-11, Department of Computer Science, Allegheny College (2005)
3. Walcott, K., Soffa, M., Kapfhammer, G., Roos, R.: Time aware test suite prioritization. In: Proceedings of the Int. Symposium on Software testing and analysis, pp. 1–12 (2006)
4. Elbaum, S., Malishevsky, A., Rothermel, G.: Test case prioritization:A family of empirical studies. IEEE Transactions of Software Engineering 28(2), 159–182 (2002)
5. Elbaum, S., Rothermel, G., Kanduri, S., Malishevsky, A.: Selecting a cost-effective test case prioritization technique. Software Quality Control 12(3), 185–210 (2004)
6. Rothermel, G., Untch, R., Chu, C., Harrold, M.: Prioritizing test cases for regression testing. IEEE Transactions on Software Engineering 27(10), 929–948 (2001)
7. Horwitz, S., Reps, T., Binkley, D.: Interprocedural slicing using dependence graphs. ACM Trans.on Programming Languages and Systems 12(1), 26–61 (1990)
8. Larsen, L., Harrold, M.: Slicing object-oriented software. In: Proceedings of the 18th ICSE, pp. 495–505 (1996)
9. Malishevsky, A., Ruthruff, J., Rothermel, G., Elbaum, S.: Cost-cognizant test case prioritization. Technical Report TR-UNL-CSE-2006-0004, University of Nebraska-Lincoln (2006)
10. Srivastava, A., Thiagarajan, J.: Effectively prioritizing tests in development environment. In: ACM SIGSOFT Software Engineering Notes, pp. 97–106 (2002)
11. Jeffrey, D., Gupta, N.: Test case prioritization using relevant slices. In: COMPSAC 2006: Proceedings of the 30th Annual Inter. Computer Software and Applications Conference, pp. 411–420 (2006)
12. Basanieri, F., Bertolino, A., Marchetti, E.: The cow suite approach to planning and deriving test suites in UML projects. In: Jézéquel, J.-M., Hussmann, H., Cook, S. (eds.) UML 2002. LNCS, vol. 2460, pp. 383–397. Springer, Heidelberg (2002)
13. Korel, B., Koutsogiannakis, G., Tahat, L.: Model-based test prioritization heuristic methods and their evaluation. In: Int. Workshop on Advances in Model-based Testing, pp. 34–43. ACM, New York (2007)
14. Korel, B., Tahat, L., Harman, M.: Test prioritization using system models. In: IEEE Int. Conference on Software Maintenance (ICSM), pp. 559–568. IEEE Computer Society, Washington (2005)
15. Gupta, R., Harrold, M., Soffa, M.: Program slicing-based regression testing techniques. Journal of Software Testing, Verification, and Reliability 6(2), 83–112 (1996)
16. Weiser, M.: Program slicing. In: ICSE 1981: Proceedings of the 5th ICSE, pp. 439–449 (1981)

Software Process Improvement through Experience Management: An Empirical Analysis of Critical Success Factors

Neeraj Sharma[1], Kawaljeet Singh[2], and D.P. Goyal[3]

[1] Department of Computer Science, Punjabi University, Patiala - 147002, Punjab, India
sharma_neeraj@hotmail.com

[2] University Computer Centre, Punjabi University, Patiala - 147002, Punjab, India
director@pbi.ac.in

[3] Management Development Institute, Gurgaon - 122007, Haryana, India
dpgoyal@mdi.ac.in

Abstract. Software product quality is immensely dependent upon the software engineering process. Software engineers have always been experimenting with models like CMM, IDEAL, SPICE and QIP in an endeavor to improve upon their software processes. Recently a new area of software process improvement through experience management has got attention of the software engineering community and developers are experimenting with experience management to improve software process. Though literature emphasizes the various benefits of experience management for software process improvement but there are no empirical studies as to suggest what the critical factors of success are for EM based SPI. This paper explores the critical factors of success for an EM based SPI program and perform an empirical investigation of these factors.

Keywords: Experience management, SPI, Experience Bases, CSF.

1 Introduction

The pious accumulation and use of knowledge in software engineering has gained immense importance. With the introduction of knowledge management (KM) in software engineering environments, a whole new discipline of software process improvement (SPI) within the area of software engineering has emerged with promising future. Software companies are fast realizing the need for sharing experience among its software engineers to improve upon the process with experience bases and help organizations mitigate to a great extent the effects of what is called *experience walking out of the door* syndrome, a phenomenon explaining the crisis faced by software engineering firms for their dependence over software developers for their competence.

Software engineering is a very complex multi-stage process involving teams commonly comprising of many professionals working in different phases and activities. Constant technology changes make the work dynamic: New problems are solved and new knowledge is created every day. The knowledge in software engineering is vast and ever-growing, making it difficult for organizations to keep track of what this

S.K. Prasad et al. (Eds.): ICISTM 2010, CCIS 54, pp. 386–391, 2010.

knowledge is, where it is, and who has it. A structured way of managing the knowledge and treating the knowledge and its owners as valuable assets could help organizations leverage the knowledge they possess. Understanding user requirements, choosing the most appropriate software development methodology, applying suitable tools, storing project related experience etc. are some of the examples which justify the knowledge-intensive nature of the software engineering discipline, and hence make a strong case for KM in the field.

During the last decade SPI through Experience Management (EM) has become an important area of research in software engineering. In the software development context, EM can be viewed as the foundation for continuous improvement of the software process and consequently, the resulting products. Using an EM approach, knowledge created during software processes can be captured, stored, disseminated, and reused, so that better quality and productivity can be achieved.

As software development projects grow larger and the discipline moves from craftsmanship to engineering, it becomes a group activity where individuals need to communicate and coordinate. Individual knowledge has to be shared and leveraged at a project and organization level, and this is exactly what EM proposes - demystify the individual hero and shift the focus to collective creativity, exploiting the emerging behavioral idea - *"none of us is as smart as all of us"* [5].

The present paper identifies the factors responsible for the success of an EM based SPI program and performs an empirical analysis to support the same.

2 Literature Review

A lot of research has been reported about KM in software engineering e.g. [20] and [25]. An infrastructure to deal with KM in software engineering environments is presented in [17]. There are studies which investigate the need for experience bases in software projects [3]. Talking of experience management, much research exists on many aspects of EM including approaches to EM (e.g. [3], [21]), how to collect experience (e.g. [1], [13]), how to structure experience (e.g. [10], [14]), tools for EM (e.g. [9], [16]) and applications of EM in practice (e.g. [3], [7], [22]). However, the literature on the use of EM for SPI is limited though pivotal e.g. [15]. The SPI models can be used to divide the field into two approaches. The first approach tries to improve the process through standardization; examples here are the Capability Maturity Model (CMM) [12], [19]; the ISO 9000 standard [6], [11]; and the Software Process Improvement and Capability dEtermination, or SPICE [24]. Another approach, known in software engineering as the Quality Improvement Program (QIP) is a more bottom up approach involving the developers in defining their own processes [4]. An attempt at establishing an overview of the SPI field is described in [8]. A detailed overview of research in the field of SPI is given by [23].

3 Hypotheses

We assume that a successful SPI program would align with general business goals. It is also stressed in the literature that any major process change requires the top level management commitment as well as the involvement and acceptance of the change program by the concerned employee group. Also concern for improved software product quality should be at the heart of the SPI program. Further, it is also assumed that improvement in the software process depends upon the organization's ability to effectively use the existing knowledge and reuse of the packaged experience. At the same time, learning through experimentation with new ideas, concepts, technologies and knowledge for finding better solutions with untapped or untried strategies is a major factor for successful process improvement in software engineering. The following hypotheses have been framed for the present study:

There is a positive association between SPI success and

- Alignment with business goals (H1)
- Management commitment (H2)
- Involvement of software developers (H3)
- Concern for quality (H4)
- Use of existing knowledge & experience (H5)
- Learning through experimentation (H6)

4 Research Methodology

The exploratory research work falls in the field of empirical software engineering. All the software development organizations in India form the universe of the study. The respondents for this study are software engineers. A sample of 125 software engineers within 15 software engineering firms, drawn from NASSCOM website [18], has been taken, using the random sampling technique. Questionnaire has been used as an instrument for data collection. A total of 95 software engineers completed the questionnaire which represents an effective response rate of 76%, well above the minimum norm of 40% for both adequate statistical power and generalizability of the results in this study [2].

Six critical factors of SPI success have been identified from an exhaustive literature survey and an exploratory pilot study. To measure the extent to which each of the six independent variables were practiced, we used multi-item, five-point, Likert scales. The item ratings were summarized to form a summated rating scale for each independent variable. Also two dependent variables have been identified: the level of perceived SPI success and the performance of their organization for the past three years with respect to cost reduction, product delivery-time reduction, and customer satisfaction. The reliability of the questionnaire was evaluated by internal consistency analysis (Cronbach's coefficient alpha = 0.74).

5 Results and Discussion

Table 1 shows the means, standard deviations, and correlations among the independent variables. Out of 15 correlations between the independent variables, two have a correlation coefficient larger than 0.5. The highest correlation (0.64) is between Management commitment and Alignment with business goals.

Table 1. Item Means, Standard Deviations and Correlation among the Independent Variables

CSF (N = 95)	Mean	S.D.	1	2	3	4	5	6
Alignment with business goals	3.72	0.65	1.00					
Management commitment	3.52	0.68	0.64***	1.00				
Involvement of software developers	3.94	0.56	0.37***	0.35***	1.00			
Concern for quality	3.62	0.66	0.43***	0.41***	0.22**	1.00		
Use of existing knowledge & experience	3.43	0.62	0.54***	0.43***	0.44***	0.40***	1.00	
Learning through experimentation	3.34	0.57	0.19*	0.13	0.47***	0.15	0.19*	1.00

* $p < 0.05$ ** $p < 0.005$ *** $p < 0.0005$

Hypotheses 1 through 6 consider the individual relationships between SPI success and each of the six independent variables. The testing of these hypotheses calls for the use of bivariate correlations. In addition, to examine each independent variable's correlation with overall SPI success, we also examined the correlations with each of the two underlying success measures: perceived level of success and organizational performance. Table 2 shows both the zero-order correlations and the partial correlations between the independent variables and each of the SPI success measures.

Table 2. Tests of Hypothesis H1-H6

CSF (N = 95)	Overall SPI success		Perceived level of success		Organizational performance	
	r	pr	r	pr	r	pr
Alignment with business goals	.59***	.58***	.56***	.56***	.43***	.41***
Management commitment	.52***	.51***	.55***	.54***	.31***	.31***
Involvement of software developers	.52***	.56***	.46***	.50***	.40***	.46***
Concern for quality	.47***	.45***	.41***	.41***	.37***	.35***
Use of existing knowledge & experience	.56***	.56***	.52***	.53***	.41***	.41***
Learning through experimentation	.21*	.25**	.15*	.18*	.20*	.25**

* $p < 0.05$ ** $p < 0.005$ *** $p < 0.0005$

From table 2, it is clear that all zero-order correlation results between all independent variables and overall SPI success showed large positive and highly significant correlations, ranging from $r = 0.47$ ($p < 0.0005$) to $r = 0.59$ ($p < 0.0005$). In addition, all zero-order correlations with perceived level of success and organizational performance were positive and highly significant, ranging from $r = 0.31$ ($p < 0.0005$) to $r = 0.56$ ($p < 0.0005$). Furthermore, all partial correlations with overall SPI success were positive and highly significant, ranging from pr = 0.45 ($p < 0.0005$) to pr = 0.58 ($p < 0.0005$). Finally, all partial correlations with perceived level of success and organizational performance were positive and highly significant, ranging from pr = 0.31 ($p < 0.0005$) to pr = 0.56 ($p < 0.0005$). Taken together, all zero-order and partial correlations involved in testing Hypotheses 1 through 6 were significant and in the hypothesized directions. Thus, the findings in Table 2 support Hypotheses 1 through 6, along with the underlying assumption that SPI provides increased levels of organizational performance.

6 Conclusions

We can conclude that the introduction of a SPI program improves the quality of software, lowers the cost of developing software and improved developer or employee satisfaction. Another important finding from a thorough survey of literature on SPI is that quantitative literature exists on the various theoretical and conceptual aspects of experience management in software engineering in general and software process improvement in particular, but there is dearth of studies on the actual implementation of SPI efforts in real software engineering environments. Also the literature is silent on how to actually approach the process improvement and the practical ways of doing it. This could be an area of research for future in software experience bases in software engineering environments. There has not been any empirical testing of the validity of the critical success factors of SPI. This is a major contribution of this paper in the field of empirical software engineering.

References

1. Althoff, K., Birk, A., Hartkopf, S., Muller, W., Nick, M., Surmann, D., Tautz, C.: Systematic Population, Utilization, and Maintenance of a Repository for Comprehensive Reuse. In: Ruhe, G., Bomarius, F. (eds.) SEKE 1999. LNCS, vol. 1756, pp. 25–50. Springer, Heidelberg (2000)
2. Baruch, Y.: Response Rate in Academic Studies - A comparative Analysis. Human Relations 52(4), 421–438 (1999)
3. Basili, V.R., Caldiera, G., Mcgarry, F., Pajerski, R., Page, G., Waligora, S.: The Software Engineering Laboratory - An Operational Software Experience Factory. In: 14th International Conference on Software Engineering, pp. 370–381 (1992)
4. Basili, V.R., Schneider, K., Hunnius, J.-P.V.: Experience in Implementing a Learning Software Organization. IEEE Software, 46–49 (May/June 2002)
5. Bennis, W., Biederman, P.W.: None of Us Is As Smart As All of Us. IEEE Computer 31(3), 116–117 (1998)
6. Braa, K., Ogrim, L.: Critical View of the ISO Standard for Quality Assurance. Information Systems Journal 5, 253–269 (1994)
7. Diaz, M., Sligo, J.: How Software Process Improvement Helped Motorola. IEEE Software 14, 75–81 (1997)
8. Hansen, B., Rose, J., Tjornhoj, G.: Prescription, Description, Reflection: The Shape of the Software Process Improvement Field. International Journal of Information Management 24(6), 457–472 (2004)

9. Henninger, S., Schlabach, J.: A Tool for Managing Software Development Knowledge. In: Bomarius, F., Komi-Sirviö, S. (eds.) PROFES 2001. LNCS, vol. 2188, pp. 182–195. Springer, Heidelberg (2001)
10. Houdek, F., Schneider, K., Wieser, E.: Establishing Experience Factories at Daimler-Benz: An Experience Report. In: 20th International Conference on Software Engineering, pp. 443–447 (1998)
11. Hoyle, D.: ISO 9000 Quality Systems Handbook. Butterworth-Heinemann, London (2001)
12. Humphrey, W.S.: Managing the Software Process. Addison-Wesley, Reading (1989)
13. Land, L., Aurum, A., Handzic, M.: Capturing Implicit Software Engineering Knowledge. In: 2001 Australian Software Engineering Conference, pp. 108–114 (2001)
14. Lindvall, M., Frey, M., Costa, P., Tesoriero, R.: Lessons Learned About Structuring and Describing Experience for Three Experience Bases. In: 3rd International Workshop on Advances in Learning Software Organisations, pp. 106–119 (2001)
15. Martinez, P., Amescua, A., Garcia, J., Cuadra, D., Llorens, J., Fuentes, J.M., Martín, D., Cuevas, G., Calvo-Manzano, J.A., Feliu, T.S.: Requirements for a Knowledge Management Framework to be Used in Software Intensive Organizations. IEEE Software, 554–559 (2005)
16. Mendonca, M., Seaman, C., Basili, V., Kim, Y.: A Prototype Experience Management System for a Software Consulting Organization. In: International Conference on Software Engineering and Knowledge Engineering, pp. 29–36 (2001)
17. Natali, A.C.C., Falbo, R.A.: Knowledge Management in Software Engineering Environments. In: 14th International Conference on Software Engineering and Knowledge Engineering, Ischia, Italy (2002)
18. National Association of Software and Services Companies, http://www.nasscom.org
19. Paulk, M.C., Weber, C.V., Curtis, B., Chrissis, M.B.: The Capability Maturity Model for Software: Guidelines for Improving the Software Process. Addison-Wesley, Reading (1995)
20. Rus, I., Lindvall, M.: Knowledge Management in Software Engineering. IEEE Software 19(3), 26–38 (2002)
21. Schneider, K.: LIDs: A Light-Weight Approach to Experience Elicitation and Reuse. In: Bomarius, F., Oivo, M. (eds.) PROFES 2000. LNCS, vol. 1840, pp. 407–424. Springer, Heidelberg (2000)
22. Sharma, N., Singh, K., Goyal, D.P.: Knowledge Management in Software Engineering Environment: Empirical Evidence from Indian Software Engineering Firms. Atti Della Fondazione Giorgio Ronchi 3, 397–406 (2009)
23. Sharma, N., Singh, K., Goyal, D.P.: Knowledge Management in Software Engineering: Improving Software Process through Managing Experience. In: Batra, S., Carrillo, F.J. (eds.) ICKMIC 2009, pp. 223–235. Allied Publishers Pvt. Ltd. (2009)
24. Software Process Improvement and Capability Determination, http://www.sqi.gu.edu.au/Spice
25. Ward, J., Aurum, A.: Knowledge Management in Software Engineering - Describing the Process. In: 2004 Australian Software Engineering Conference. IEEE Computer Society, Los Alamitos (2004)

A Comparative Feature Selection Approach for the Prediction of Healthcare Coverage

Prerna Sethi[1] and Mohit Jain[2]

[1] Department of Health Informatics and Information Management, Louisiana Tech University, Ruston, LA 71272

[2] Computer Science Program, Louisiana Tech University, Ruston, LA 71272

{prerna,mja025}@latech.edu

Abstract. Determining the factors that contribute to the healthcare disparity in United States is a substantial problem that healthcare professionals have confronted for decades. In this study, our objective is to build precise and accurate classification models to predict the factors, which attribute to the disparity in healthcare coverage in the United States. The study utilizes twenty-three variables and 67,636 records from the 2007 Behavioral Risk Factor Surveillance System (BRFSS). In our comparative analysis, three statistical feature extraction methods, Chi-Square, Gain Ratio, and Info Gain, were used to extract a set of relevant features, which were then subjected to the classification models, AdaBoost, Random Forest, Radial Basis Function (RBF), Logistic Regression, and Naïve Bayes, to analyze healthcare coverage. The most important factors that were discovered in the model are presented in this paper.

Keywords: Healthcare coverage, behavioral risk factor surveillance system, data mining, feature selection, classification and prediction.

1 Introduction

The growing disparity in healthcare coverage is a hot button issue in United States and a major issue during the 2008 presidential election. The U.S. Census Bureau reports that 45.7 million people were uninsured in 2007, and the number has grown in the past two years. The current research on this subject centers on identifying key factors that can assist in discriminating between the people with healthcare and the people without healthcare coverage to assist in the development of effective healthcare policies. The factors that critically contributed to the disparity in healthcare coverage fall under the broad umbrella of socio-demographic, employment-related, and lifestyle factors for an individual.

An important factor contributing to the disparity in healthcare coverage is gender. Some studies seem to show that women are more likely to have healthcare coverage than men are, while other studies have reported that males are more likely to have healthcare coverage [1-4]. Race or the ethnic background has also been addressed as a factor responsible for the disparity in healthcare coverage. According to reports, minorities (Hispanic/Latino, African-American, and non-citizens) have less access to

S.K. Prasad et al. (Eds.): ICISTM 2010, CCIS 54, pp. 392–403, 2010.

healthcare coverage [1, 3, 5-7]. Level of education and income have also been studied as factors that contribute to either having or lacking access to healthcare [6-10]. These studies have shown that people who have less education and lower incomes are more likely to be uninsured. It has also been observed that individuals without employment and those who have been employed only for a short term are less likely to have healthcare coverage [8, 9-10]. This problem has escalated recently, due to severe job losses, which have often led to loss of health insurance and rising insurance premiums [11]. The age and marital status of the individuals compose another set of factors that is reported in [7, 8, 12]. These studies suggest that younger individuals have less insurance than older individuals, especially the elderly. Disability, ill-health, poor physical or mental health, or chronic illness may also limit the individual's healthcare coverage as reported in [4, 13]. Individuals who lack healthcare coverage are also distributed geographically. The highest number of uninsured individuals reside in the South and South-West [9, 12, 14]. In addition, large families are often limited in the ability to have healthcare coverage for the entire family, due to cost [15]. Whether or not an individual has veteran or non-veteran status may also contribute to whether the individual is able to obtain healthcare coverage, as shown in the past studies [4, 16]. Further, lifestyle factors, which contribute to the healthcare coverage disparity, are risk taking behavior, alcohol consumption, smoking status, and an unhealthy lifestyle and high weight (BMI) [17-19].

In this paper, we identify the key factors that contribute to the disparity in healthcare coverage by applying three feature extraction methods and building classification models to compare and predict the insurance status of an individual based on the key determinants with increased specificity and sensitivity.

2 Methods

The overall framework, as shown in Fig. 1, consists of the following five computational steps. (1) In data preprocessing, we remove missing data, and then balance the remaining data. (2) In feature selection, we use the three statistical measures of Chi-Square, Gain Ratio, and Info Gain. (3) In classifier building, we build training and testing sets based on selected features and perform ten-fold cross validation. (4) During evaluation, we characterize the classification schema and determine the variable importance using the test dataset and success metric. Our broad framework is shown in Fig. 1.

2.1 Data Preprocessing

Our data was obtained from the 2007 Behavioral Risk Factor Surveillance System (BRFSS), an ongoing health survey system, which is conducted by the National Center for Chronic Disease Prevention and Health Promotion division of the CDC and which tracks health conditions and risk behaviors in the United States. The original dataset contained 430,910 records and 342 variables. Past studies of healthcare coverage have limited the samples to individuals with ages in the range of 16 and 64; records that did not fall within those limits were removed from the dataset. The dataset was further filtered by removing interviews that were partially completed, that is, where the participants used, "don't know" or "refused" for the dependent variable, "healthcare coverage." After completing the data analysis, 429,767 records remained in the dataset. For each variable, we observed the distribution of the data to determine the impact that removing missing data would have on the results and determined that the dataset would not become unbalanced without those records. Hence, we removed them. The final dataset included 243,354 records. Variables for smoking status, age, race, body mass index, number of children, and level of education and income were calculated from separate categories. The dependent variable, "healthcare coverage" in the dataset contained many more yes (86.8%) responses than no responses (13.2%). The dataset was balanced, according to previous literature [20], into an equal proportion of yes and no responses resulting in a final dataset size of 67,636.

In our dataset, a set of 23 variables was chosen based on the literature search, in which we found evidence that certain variables help determine whether a person has healthcare coverage. These variables include: state of residence, number of adult males in the household, number of adult females in the household, general health, number of days in the last month that mental health was not good, number of days in the last month that physical health was not good, marital status, number of children under 18 in the household, employment status, veteran status, whether activities are limited by a health problem, whether special equipment is required for a health problem, whether the person engages in leisure-time physical activity or exercise, smoking status, heavy-drinking risk, race/ethnicity, body mass index, education, income, risk of binge drinking, and risk of AIDs. The dependent variable for the study was the individual's healthcare coverage status. Hence, our final dataset consisted of 23 variables (attributes) and 67,636 records. We then applied the feature selection measures on this reduced dataset to further obtain a relevant set of variables to assist in the classification of individuals with or without healthcare coverage with increased specificity and sensitivity.

2.2 Feature Selection

The large number of variables made the process of building robust classification models difficult. Feature selection assisted in selecting a subset of relevant variables for effectively classifying the individuals with and without healthcare coverage. We used the three statistical measures, Chi-Square, Gain Ratio, and Info Gain to select the top ranked 18, 13, and 10 features, which formed our reduced set of features for classification. A brief description of these methods is given below.

Chi-Square- This method uses the χ^2 statistic to measure each gene with respect to its classes. The genes are discretized with an entropy-based method to m intervals. The χ^2 value for each variable is then calculated by

$$\chi^2 = \sum_{i=1}^{m} \sum_{j=1}^{q} \frac{\left(P_{ij} - (R_i C_j)/N\right)^2}{(R_i C_j)/N}, \tag{1}$$

where q=number of classes, N= number of patterns, R_i= number of patterns in ith interval, C_j= number of patterns in jth class, P_{ij}= number of patterns in ith interval and jth class. The variables, which have a higher value of χ^2 statistics are selected as desirable features and form the feature-set.

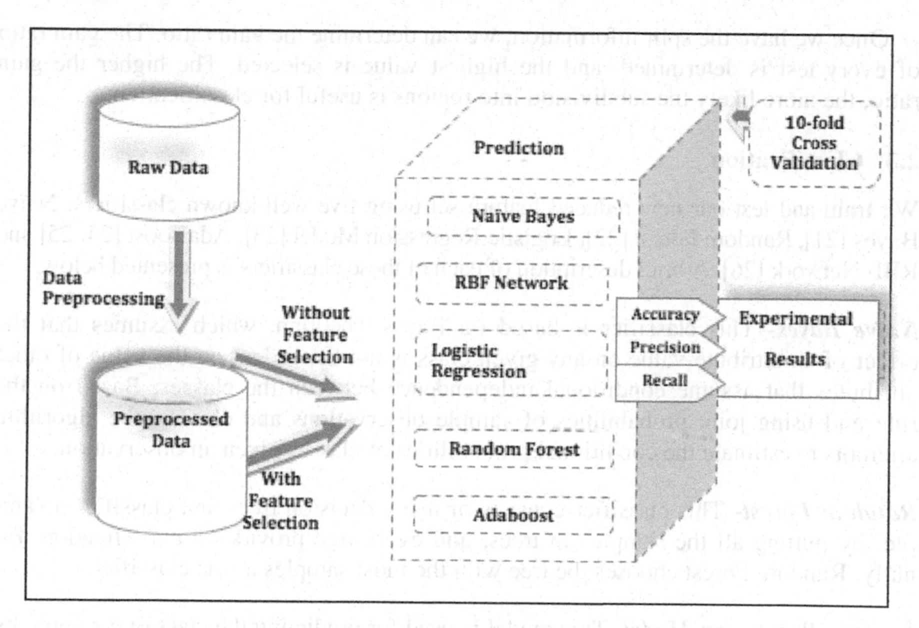

Fig. 1. A computational framework for the overall research process

Info Gain and Gain Ratio

Let D be the set of records. Let A be the attributes with values $\{A_1, A_2, \ldots, A_v\}$. Let C denote the number of classes and $p(D, j)$ denote the proportion of cases in D that belong to the jth class. The residual uncertainty or information about the class to which an object in D belongs can be expressed as,

$$Info(D) = -\sum_{j=1}^{C} p(D, j) \times \log_2 (p(D, j)) . \quad (2)$$

The corresponding information gain by a test T with k outcomes is

$$Gain(D,T) = Info(D) - \sum_{i=1}^{k} \frac{|D_i|}{D} \times Info(D_i) . \quad (3)$$

The information gained by a test is affected by the number of outcomes and is maximal when there is one case in each subset D_i. The information obtained by partitioning a set of cases is based on knowing the subset D_i into which a case falls; this subset is known as split information, and it increases the number of outcomes of a test.

$$Split(D,T) = -\sum_{i=1}^{k} \frac{|D_i|}{D} \times \log_2 \frac{D_i}{D} \quad (4)$$

The gain ratio of a test is the ratio of its information gain to its split information,

$$GainRatio = \frac{Gain(D,T)}{Split(D,T)} . \quad (5)$$

Once we have the split information, we can determine the gain ratio. The gain ratio of every test is determined, and the highest value is selected. The higher the gain ratio, the more likely the subdivision into regions is useful for classification.

2.3 Classification

We train and test our new reduced feature set using five well-known classifiers: Naïve Bayes [21], Random Forest [22], Logistic Regression Model [23], AdaBoost [24, 25] and RBF Network [26]. A brief description of each of these classifiers is presented below.

Naïve Bayes- This classifier is based on Bayes Theorem, which assumes that the effect of an attribute value on any given class is not dependent on the value of other attributes that assume conditional independence between the classes. Based on the rule and using joint probabilities of sample observations and classes, the algorithm attempts to estimate the conditional probabilities of classes given an observation.

Random Forest- This classifier consists of many decision trees and classifies a sample, by putting all the samples in trees, and every tree provides a classification. Finally, Random Forest chooses the tree with the most samples as the classifier.

Logistic Regression Model- This model is used for predicting the class of a sample by using several predictor variables that may be either numerical or categorical. Here, Y is called dependent variable, which can be modeled as a linear function of an independent variable, X in the equation, $Y = a + bX$, where a and b are regression coefficients, and Y is slope of the line.

RBF Network- A radial basis function network is an artificial neural network which is used to find the multi-dimensional function that provides the best fit to the training data. Regularization is used to add more information to prevent over fitting. The RBF network consists of three layers. The first (input) layer is a source node. The second layer is a hidden layer of high dimension. The last (output) layer is the response to the input layer.

AdaBoost- AdaBoost or Adaptive Boosting is a meta-algorithm, which can be used with other learning algorithms and with classifiers to improve algorithm performance. AdaBoost is not prone to over fitting; it performs feature selection, resulting in a relatively simple classifier and a good generalization. However, it is sensitive to noisy data and outliers.

To compare the classification accuracy of various classifiers, we calculate the precision, recall, and F-measure as follows.

$$\text{Precision}= \frac{TP}{TP+FP} \qquad \text{Recall}=\frac{TP}{TP+FN}.$$

In the equations above, TP = samples which are classified correctly to a class, FP = samples incorrectly labeled to the class; FN= samples which should have been labeled to a class, but were not.

$$\text{F measure}= \frac{2(\Pr ecision * \operatorname{Re} call)}{\Pr ecision + \operatorname{Re} call}.$$

3 Results

We perform several experiments to evaluate the efficacy of the selected features by applying five classifiers. The sub-datasets formed from feature extraction and ranking measures are compared using these classifiers to observe the effectiveness of the selected features.

3.1 Feature Selection and Classification

In this experiment, we utilize the three feature selection methods, Chi-Square, Gain Ratio, and Info Gain and selected the top 18, 13, and 10 features. A list of the top 18 features from Chi-Square, Gain Ratio, and Info Gain is shown in Table 1. The features are evaluated by comparing the classification accuracy, using machine learning classifiers, Naïve Bayes, Logistic Regression Model, RBF Network, Adaboost, and Random Forest along with a 10-fold cross-validation design. The classification accuracies for the three feature selection methods are shown in Figs. 2, 3, and 4. The selected features are then studied for effectiveness by comparing the precision, recall, and F-measure of the classification algorithms.

Table 1. Ranked attribute importance for the three feature selection methods

Chi-Square	Gain Ratio	Info Gain
1. Heavy Drinkers	1. Heavy Drinkers	1. Heavy Drinkers
2. State	2. State	2. State
3. Income	3. Medical Cost	3. Income
4. Medical Cost	4. Income	4. Medical Cost
5. Education	5. Education	5. Education
6. Employment Status	6. Employment Status	6. Employment Status
7. Marital Status	7. Smoking Status	7. Marital Status
8. Smoking Status	8. Marital Status	8. Smoking Status
9. General Health	9.Physical Activity or Exercise	9. General Health
10. Age	10. General Health	10. Age
11.Physical Activity or Exercise	11. Age	11.Physical Activity or Exercise
12. Mental Health	12. Children	12. Mental health
13.Number of Men (in household)	13. Veteran Status	13.Number of Men (in household)
14.Number of Women (in household)	14. Mental Health	14. Number of Women (in household)
15. Number of Children (in household)	15. Number of Men (in household)	15. Children
16. Veteran Status	16. Number of Women (in household)	16. Veteran Status
17. Physical Health	17. Physical Health	17. Physical Health
18. BMI	18. Use Equipment due to Health Problems	18. BMI

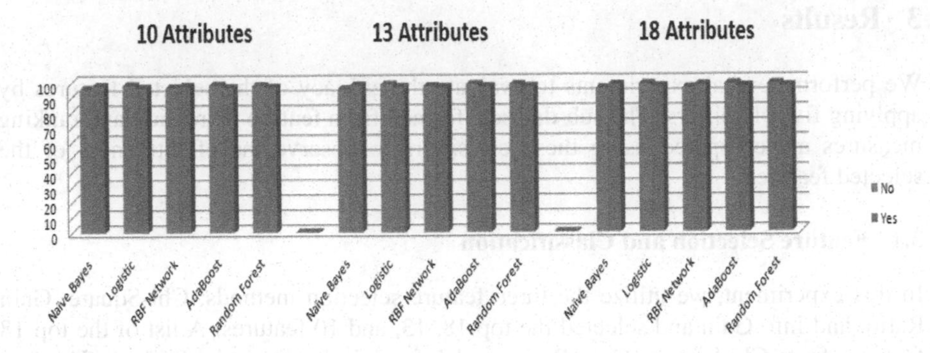

Fig. 2. Classification accuracy using Chi-Square for 10, 13, and 18 attributes

The best accuracy using the Chi-Square feature selection method by extracting the 10, 13, and 18 ranked attributes was obtained by the Logistic Regression model with 97.66%, 97.69%, and 97.73% accuracy, respectively.

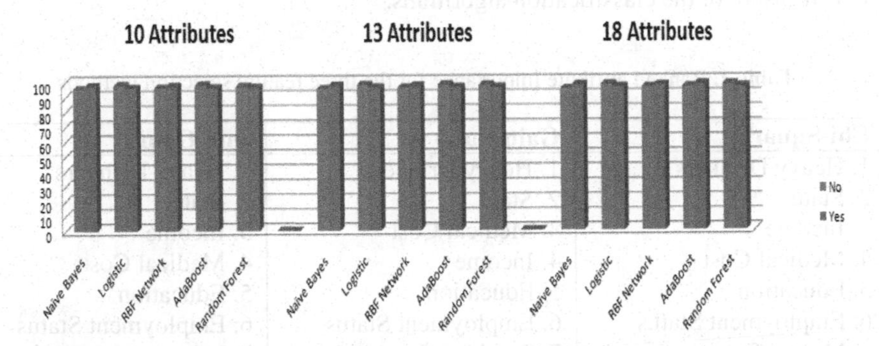

Fig. 3. Classification accuracy using Gain Ratio for 10, 13, and 18 attributes

The best accuracy using the Gain Ratio feature selection method was obtained by the Logistic Regression model, which extracted the 10, 13, and 18 ranked attributes with the 97.69%, 97.68% and 97.73% accuracy, respectively.

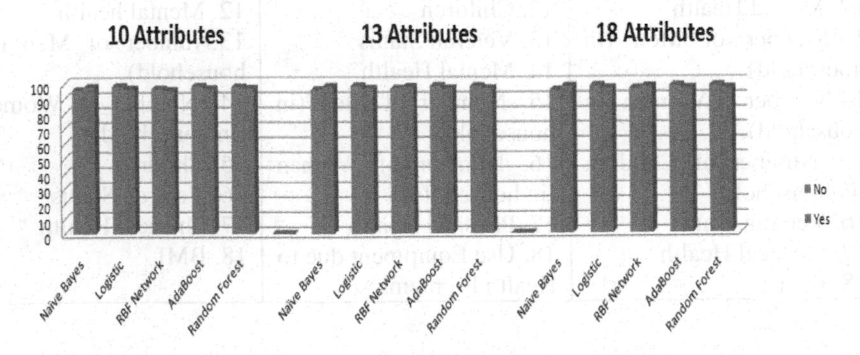

Fig. 4. Classification accuracy using Info Gain for 10, 13, and 18 attributes

The best accuracy using the Gain Ratio feature selection method to extract the 10, 13, and 18 ranked attributes was obtained by the Logistic Regression model with an accuracy of 97.66%, 97.69%, and 97.73%, respectively.

The results shown in Figs. 2, 3, and 4 show that the classification accuracy is almost comparable with 10, 13, and 18 features using all the three the feature selection approaches. This result implies that the top ten ranked features sufficiently classify an individual into the healthcare coverage or no healthcare coverage classes.

3.2 Validation of Classification Accuracy

To validate the accuracy of the classification results, we analyze the true positive (correctly classified instances), false positive (instances incorrectly classified as "yes"), Precision, Recall, F-measure (described in section 2.3) and area under the Receiver Operating Curve (ROC). Figs. 5, 6, and 7 show these results for all three feature selection methods when using ten attributes. The graphs showing the results for the remaining number of features are not shown in this paper, due to space constraints. The graphs below indicate that in all the cases, we achieved True Positive, Precision, Recall, and F-measure rates of above 96% with a false positive rate of less than 1%. The ROC area in all figures is almost 99%.

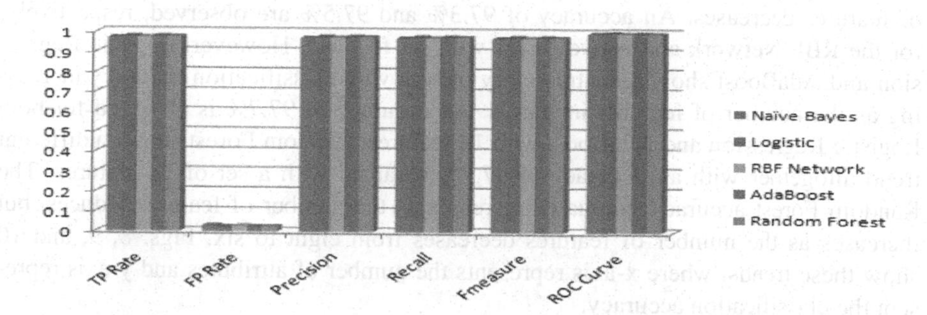

Fig. 5. Validation of classification accuracy for 10 attributes with Chi-Square

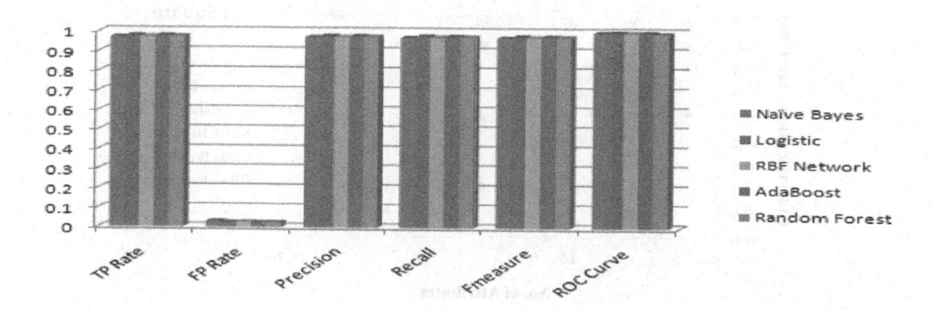

Fig. 6. Validation of classification accuracy for 10 attributes with Gain Ratio

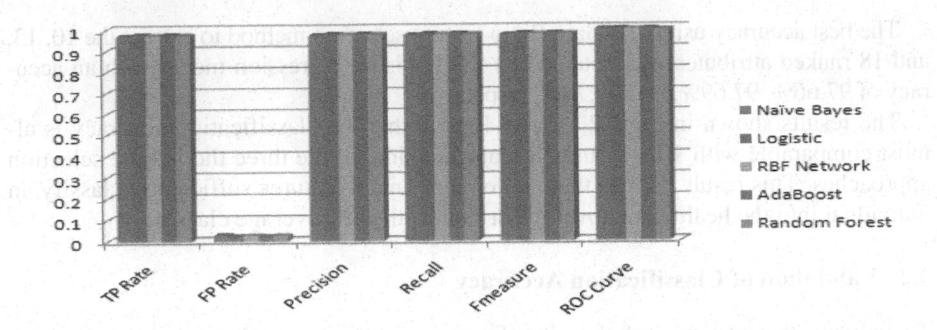

Fig. 7. Validation of classification accuracy for 10 attributes with Information Gain

3.3 An Optimal Set of Features for the Prediction of Healthcare Coverage

We conducted a set of experiments to find an optimal set of features with a particular combination of feature selection and classification methods. However, no combination scored clearly better than the others did. Based on these results, we observe that the accuracy of the classifiers, Naïve Bayes and RBF Network increase as the number of features decreases. An accuracy of 97.3% and 97.5% are observed, respectively, for the RBF Network and Naïve Bayes with six features. However, Logistic Regression and AdaBoost show a contradictory pattern with classification accuracy increasing as the number of features increased. An accuracy of 97.7% is observed for both Logistic Regression and AdaBoost with 18 features. Random Forest shows a different trend altogether with an accuracy of 97.7% obtained with a set of 18 features. The Random Forest accuracy gradually decreases as the number of features reduces, but increases as the number of features decreases from eight to six. Figs. 8, 9, and 10 show these trends, where x-axis represents the number of attributes and y-axis represent the classification accuracy.

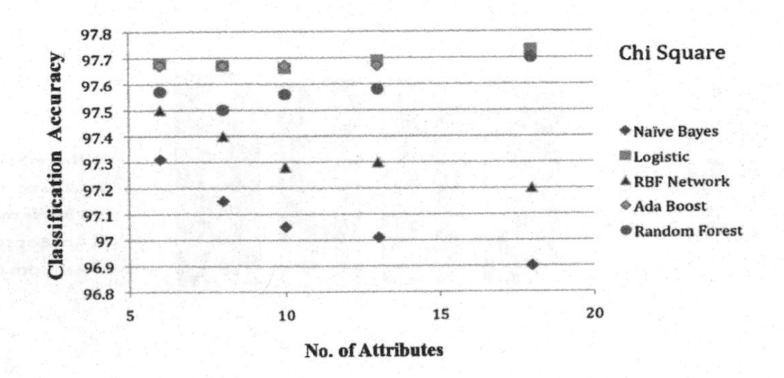

Fig. 8. Classification accuracies of different feature sets using Chi-Square

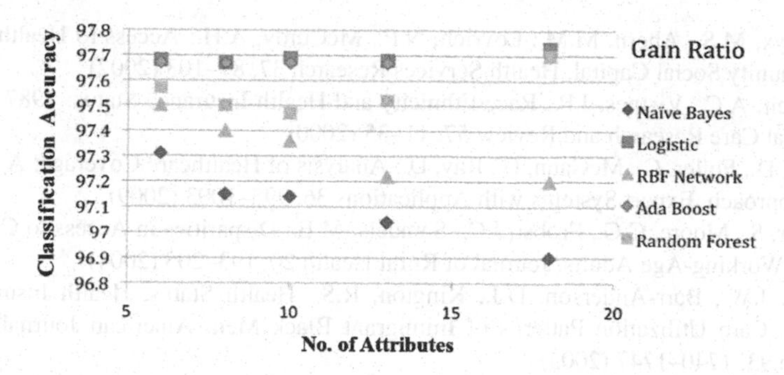

Fig. 9. Classification accuracies of different feature sets using Gain Ratio

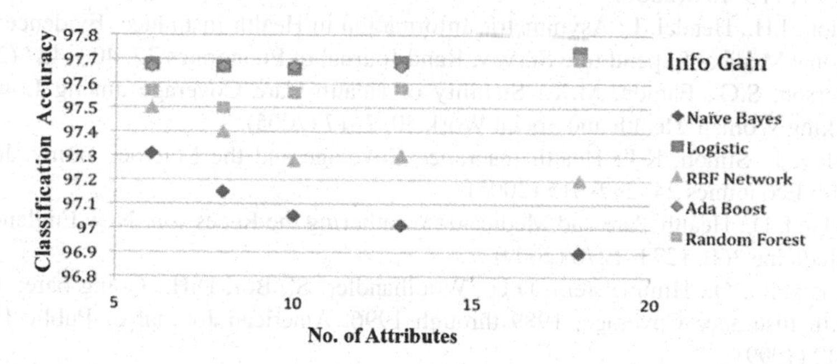

Fig. 10. Classification accuracies of different feature sets with Info Gain

4 Conclusions

In this paper, we have shown a classification model based on feature selection and classification that can identify individuals with healthcare and those without healthcare based on demographic and lifestyle factors. The factors outlined in the paper significantly contribute to the disparity in healthcare coverage and can be addressed to design healthcare policies and procedures that efficiently cater to the needs of the "vulnerable." Previous studies in this area have highlighted these factors as a major cause of disparity in healthcare coverage, which further strengthens our conclusions. Feature selection based on the ranking of variables, followed by classification, helps us to select a subset of variables, which are relevant in effectively classifying the individuals with healthcare and those without healthcare, thus eliminating the need to study the irrelevant variables. The results from this paper can better assist in understanding the causes of healthcare disparity and allow researchers to take suitable measures to reduce this gap.

References

1. Carrasquillo, O., Carrasquillo, A.I., Shea, S.: Health Insurance Coverage of Immigrants Living in the United States: Differences by Citizenship Status and Country of Origin. American Journal of Public Health 90, 917–923 (2000)

2. Hendryx, M.S., Ahern, M.M., Lovrich, N.P., McCurdy, A.H.: Access to Health Care and Community Social Capital. Health Services Research 37, 87–103 (2002)
3. Monheit, A.C., Vistnes, J.P.: Race/Ethnicity and Health Insurance Status: 1987 and 1996. Medical Care Research and Review 57, 11–35 (2000)
4. Delen, D., Fuller, C., McCann, C., Ray, D.: Analysis of Healthcare Coverage: A Data Mining Approach. Expert Systems with Applications 36, 995–1003 (2009)
5. Glover, S., Moore, C.G., Probst, J.C., Samuels, M.E.: Disparities in Access to Care among Rural Working-Age Adults. Journal of Rural Health 20, 193–205 (2004)
6. Lucas, J.W., Barr-Anderson, D.J., Kington, R.S.: Health Status, Health Insurance, and Health Care Utilization Patterns of Immigrant Black Men. American Journal of Public Health 93, 1740–1747 (2003)
7. Shi, L.Y.: Vulnerable Populations and Health Insurance. Medical Care Research and Review 57, 110–134 (2000)
8. Cardon, J.H., Hendel, I.: Asymmetric Information in Health Insurance: Evidence from the National Medical Expenditure Survey. Rand Journal of Economics 32, 408–427 (2001)
9. Anderson, S.G., Eamon, M.K.: Stability of Health Care Coverage among Low-Income Working Women. Health and Social Work 30, 7–17 (2005)
10. Cawley, J., Simon, K.I.: Health Insurance Coverage and the Macroeconomy. Journal of Health Economics 24, 299–315 (2005)
11. Rowland, D.: Health Care and Medicaid- Weathering the Recession. New England Journal of Medicine 360, 1273–1276 (2009)
12. Carrasquillo, O., Himmelstein, D.U., Woolhandler, S., Bor, D.H.: Going bare: Trends in Health Insurance Coverage, 1989 through 1996. American Journal of Public Health 89, 36–42 (1999)
13. Landerman, L.R., Fillenbaum, G.G., Pieper, C.F., Maddox, G.L., Gold, D.T., Guralnik, J.M.: Private Health Insurance Coverage and Disability among Older Americans. Journals of Gerontology Series B-Psychological Sciences And Social Sciences 53, S258–S266 (1998)
14. Hoffman, C., Schlobohm, A.: Uninsured in America: A Chart Book, 2nd edn. The Henry J. Kaiser Family Foundation, Washington (2000)
15. Newacheck, P.W., Park, M.J., Brindis, C.D., Biehl, M., Irwin, C.E.: Trends in Private and Public Health Insurance for Adolescents. The Journal of American Medical Association 291, 1231–1237 (2004)
16. Woolhandler, S., Himmelstein, D.U., Distajo, R., Lasser, K.E., McCormick, D., Bor, D.H., et al.: America's Neglected Veterans: 1.7 Million Who Served Have No Health Coverage. International Journal of Health Services 35, 313–323 (2005)
17. Chae, Y.M., et al.: Data Mining Approach to Policy Analysis in a Health Insurance Domain. International Journal of Medical Informatics 62, 103–111 (2001)
18. Cunningham, P.J., Ginsburg, P.B.: What Accounts for Differences in Uninsurance Rates across Communities? Inquiry-The Journal of Health Care Organization Provision and Financing 38, 6–21 (2001)
19. Leigh, J., Hubert, H., Romano, P.: Lifestyle Risk Factors Predict Healthcare Costs in an Aging Cohort. American Journal of Preventive Medicine 29, 379–387 (2005)
20. Wilson, R.L., Sharda, R.: Bankruptcy Prediction using Neural Networks. Decision Support Systems 11, 545–557 (1994)
21. Rish, I.: An Empirical Study of the Naive Bayes Classifier. In: IJCAI 2001 Workshop on Empirical Methods in Artificial Intelligence (2001)

22. Breiman, L.: Random Forests. Machine Learning 45, 5–32 (2001)
23. Hilbe, J.M.: Logistic Regression Models. Chapman & Hall/CRC Press, Boca Raton (2009)
24. Grove, A., Schuurmans, D.: Boosting in the Limit: Maximizing the Margin of Learned Ensembles. In: Proceedings of the 15th National Conf. Artificial Intelligence, pp. 692–699 (1998)
25. Jin, R., Liu, Y., Si, L., Carbonell, J., Hauptmann, A.: A New Boosting Algorithm Using Input-Dependent Regularizer. In: Proceedings of 20th Intl. Conf. on Machine Learning, ICML 2003 (2003)
26. Bishop, C.M.: Neural Networks for Pattern Recognition. Oxford Univ. Press, Oxford (1995)

A Grid-Based Scalable Classifier for High Dimensional Datasets

Sheetal Saini and Sumeet Dua

Department of Computer Science, Louisiana Tech University,
Ruston, LA 71272, USA
{ssa017,sdua}@coes.latech.edu

Abstract. High dimensionality and large dataset size are two common characteristics of real-world datasets and databases. These characteristics pose unique challenges for the classification of such datasets. The classification algorithms that perform well (in terms of scalability and efficiency) on small and medium datasets with moderate dimensionality fail to scale well with the large and high dimensional datasets. Therefore, in this paper, we propose a scalable classifier to cope with large and high dimensional datasets. The proposed method inherits its scalability feature from the concept of grid-based partitioning. Our goals in using this method are to divide the data space into small partitions called cells and to map the data on the partitioned data space. Thus, instead of managing the individual data points within the data, abstract entities called cells are used to decrease the classification runtime for large and high dimensional datasets. The presented experimental results demonstrate the scalability and efficiency of our algorithm.

Keywords: Classification, scalability, and grid-based.

1 Introduction

The growth of sophisticated technologies has resulted in the generation of an enormous amount of data that is high dimensional in nature, thus increasing a demand for data-mining techniques that are fast and scalable in nature. In developing such techniques for the classification of high dimensional and large datasets, speed, scalability, and accuracy are the most important parameters for success. In general, scalable classifier designs emphasize improving both the training and testing times of the classification model, improving only the training time of the classification model, or improving only the testing time of the classification model to speed up classification. Which of these three methods should be used depends on the application of the classifier. Many scalable algorithms also exchange speed for accuracy, improving the speed of the classifier by sacrificing accuracy, or vice versa [1]. Scalability, a central component in the design of a scalable classifier, refers to an algorithm's ability to handle an increase in the size and dimensionality of a dataset. A scalable classifier should scale well; i.e. its performance should not deteriorate drastically with the increased

S.K. Prasad et al. (Eds.): ICISTM 2010, CCIS 54, pp. 404–415, 2010.

size and dimensionality of the data. However, the existing classification algorithms that perform well for the small and medium dimensions and datasets fail to perform well when the dimensionality and size of the datasets increases. The time complexity of the many existing data-mining algorithms is exponential to the number of dimensions of the dataset [2]. In the past few years, researchers have begun to emphasize the development of a scalable classifier to handle large and high dimensional datasets. In this paper, we propose a new scalable classifier based on the grid-based partitioning approach. Our motivation is to harness the scalable nature of the grid-based approaches.

2 Background

As classification is a well-studied area, many classification techniques have been developed. However, the design of each classifier addresses a different issue, such as handling high dimensional and large datasets or improving the performance of the existing classifier. The common motivation that inspires scalable classifier designs is the desire to develop a classifier capable of handling high dimensional and large datasets without significant loss in a performance parameter, such as speed or accuracy. Handling high dimensional data in a data-mining task, such as classification, is challenging because of the nature of high dimensional data. Several methods have been developed to address the high dimensionality of the data. The SVM [1], KNN [1] and decision tree [3], [4], and [5] classifiers have been used extensively to design a scalable classifier. The grid-based clustering algorithms are well known and are inherently scalable, because they are capable of reducing search space by partitioning the data [6]. Thus, there is an immense opportunity to explore this area and develop new scalable and time-efficient solutions for the classification of large and high dimensional datasets.

The remainder of the paper is organized as follows. We discuss the related work in section 3. In section 4, we discuss the details of our proposed methodology for classifier design. In section 5, we discuss the experimentation and results, and, finally, we conclude in section 6.

3 Related Work

Our goal is to design a 'scalable' classifier using the concept of 'grid-based partitioning'. Therefore, we will discuss representative scalable classifiers and classification using grid-based partitioning.

3.1 Scalable Classifier

SLIQ [3] and SPRINT [4] are two representative examples of scalable classifiers. The common characteristic among scalable classification methods is the use of a decision tree-based classification technique. The SLIQ [3] algorithm consists of two phases, the tree growth-phase and tree prune-phase. It uses a one-time sort method instead of repeatedly sorting to split the numeric attribute. SLIQ is able to sort once rather than repeatedly, because it maintains separate lists for each attribute. SLIQ uses the 'class list' data structure that must remain in the memory all the time. It builds a single decision tree using the entire training dataset instead of using a sampled dataset. The size of the 'class-list' is the same as the number of data points; therefore, SLIQ can only handle data-points that can be accommodated in the main memory. The SPRINT algorithm is an improvement over the SLIQ [3] algorithm. The design goal of the researchers who developed SPRINT was to develop an accurate classifier for large datasets. SPRINT shares most of SLIQ's features, but it uses the 'attribute-list' instead of the 'class-list'. Unlike SLIQ, SPRINT has no memory restriction, and it is fast and scalable [4].

3.2 Grid-Based Classification

Researchers who design clustering algorithms have recognized and leveraged the potential of grid-based approaches. However, researchers have not adequately exploited these approaches in designing a classifier. Grid-based approaches can scale well with an increase in the dimensionality and size of a dataset. Although there are many grid-based approaches for clustering, no work can be purely classified as a grid-based classifier. Nevertheless, some techniques are closely related to grid-based classification. One such related work is presented below.

In [6], Wei, Huang, and Tian proposed a grid-based method for anomaly detection. This method classifies data into normal and abnormal classes for anomaly detection. In their method, a two-phase grid-based clustering algorithm was developed to partition the network traffic data. In the first phase, data was divided into non-overlapping cells for pre-clustering. In the second phase, k-hyper-cells clustering, the clusters returned from the algorithm were presented in the form of logical expressions to generate rules for the classification of network traffic data.

4 Methodology

Our scalable grid-based classification model is derived from the axis-parallel grid-based clustering technique, where data space is partitioned using hyper-planes that are parallel to the axis (shown in Fig. 2). Our classifier was designed using a training model and a testing model.

This section is divided into five sub-sections. We present the data preprocessing techniques used to prepare data for the next step, the design of our classifier's training model, the design of the classifier's testing model, a technique to handle high dimensionality, and the time complexity analysis of our algorithm. A list of the notations that we use to represent the terminology and algorithmic pseudocodes in this paper is provided in Table 1.

Table 1. Notation table

S. No.	Notation	Description
1.	X	Set of Data Points
2.	x_i	i^{th} Data Point
3.	D	Set of Dimensions
4.	d_i	i^{th} Dimension
5.	G	Set of Partitions
6.	P^i	Partition in i^{th} Dimension
7.	C_i	i^{th} Cell
8.	c	Number of Classes
9.	c_i	Number of data points in i^{th} Class
10.	C	Set of Grid Cells
11.	N	Number of Data points
12.	N_i	Number of Cells in i^{th} Class
13.	N_c	Total Number of Cells

4.1 Data Preprocessing

Data preprocessing is an essential step in any data-mining task. We perform the following preprocessing steps before passing the data to the algorithm. We assume that all data is numeric. We normalize the data into a unit hypercube $[0,1]^n$ to bring all dimensions to the same domain. Then, we perform min-max normalization on the data [7]. We also eliminate features that do not provide significant variability within the dimension; i.e. significant numbers of values in the features are either zero or constant.

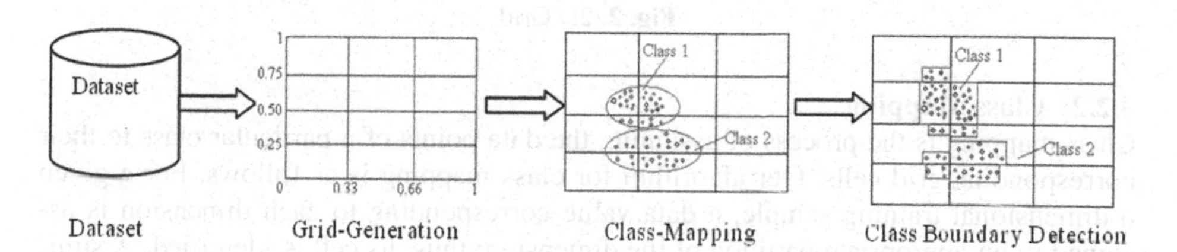

Fig. 1. Training phase

4.2 Training Phase

The training model of our proposed classifier consists of thee steps. First, we partition the data space into a non-overlapping axis-parallel grid structure. Second, we map the training data on the grid structure by assigning every data point to its corresponding grid cell. Third, we detect the boundary of every class by identifying the overlapping and non-overlapping cells occupied by each class. Fig. 1 is a graphical representation of the training phase.

4.2.1 Grid Generation

Let high-dimensional dataset $X = \{x_i \mid i \in 1....N\}, x_i \in R^n$ have N data points and n dimensions, where data is normalized in unit hypercube $[0,1]^n$. During grid generation, every dimension is divided into k fixed partitions, where k is the specified number of partitions in every dimension. Let dataset $X=\{x_1,x_2,x_3,......,x_N\}$ be defined as a set of data points, $D=\{d_1,d_2,........,d_n\}$ be defined as a set of dimensions, and $G=\{P^1,P^2,......,P^n\}$ be defined as a set of partitions. We use the axis-parallel fixed-grid to partition the dimensions and generate the multi-dimensional grid structure.

Definition 1 (Grid). A grid $G=\{P^1,P^2,......,P^n\}$ is a set of n grid partitions, where $P^i, i=1,2,.......,n$ is a partition in the ith dimension (shown as Fig. 2).

Definition 2 (Fixed-Size Grid). A grid $G=\{P^1,P^2,......,P^n\}$ is a fixed-size grid, if the dimension is $d_i \in D$, where $i=1,2,........,n$ is divided into the same number of same size partitions P, and cell $C_j, j=1,2,.......,m$ has the same side length in every dimension (Fig. 2).

Definition 3 (Cell). Given a dataset $X, D=\{d_1,d_2,.........,d_n\}$ number of dimensions, and $G=\{P^1,P^2,......,P^n\}$ number of partitions in a dimension, a cell $C_j, j=1,2,.......,m$ is a closed space entity having the shape of hyper-cube or hyper-cuboids in a grid structure (shown as Fig. 2).

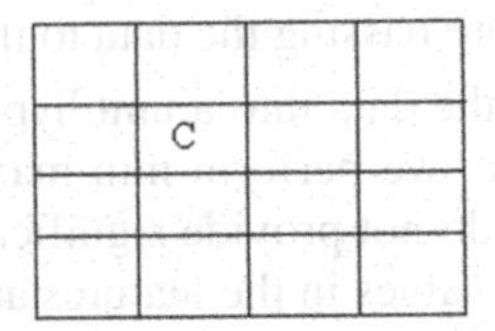

Fig. 2. 2D Grid

4.2.2 Class Mapping

Class mapping is the process of assigning the data points of a particular class to their corresponding grid cells. Our algorithm for class mapping is as follows. For a given n-dimensional training sample, a data value corresponding to each dimension is assigned to an appropriate partition of the dimension; thus, its cell is identified. A similar process is repeated for the training samples of each class.

4.2.3 Class Boundary Detection

Once the classes are mapped on the grid structure, the grid cells occupied by the classes represent corresponding classes. In class boundary detection, the unique cells that correspond to every class and contain data points belonging to a particular class are identified. We also compute the centroid of every occupied cell for all the classes. The centroid of a cell is the average of all data points in the cell. These centroids will be used later for assigning the class label to the test sample. The grid cells occupied by each class can be divided into two groups: overlapping-cells and non-overlapping cells. Fig. 3, below, shows the pseudocode of the training phase of the proposed classifier. This pseudocode includes both the class mapping and the class boundary detection algorithms.

Definition 4 (Overlapping-Cell). A ith grid cell, C_i, is called an overlapping-cell if it contains training samples from multiple classes.

Definition 5 (Non-overlapping Cell). A ith grid cell, C_i, is called a non-overlapping cell if it only contains the training samples of a single class.

```
Algorithm: Training Phase
Input: Data (X), Grid Partition (C)
Output: ClassCell
1.  for i= to c                    // c=Number of classes
2.    for j=1 to c_i               // c_i= number of data points in i^th class
3.      Identify Cell id of c_i,j data point
4.      If( ClassCell_i == φ)      // φ=Empty sign
5.        Add C_i,1 to ClassCell_i
6.        Compute ClassCell_i1 Centroid
7.      else
8.        Flag=0
9.        for x=1 to ClassCell_i
10.           if( C_i,j == ClassCell_i,x)
11.             ClassCell_i,x .Count= ClassCell_i,x .Count+1
12.         Re-compute ClassCell_i,x Centroid
13.             Flag=Flag+1
14.           end
15.       end
16.        if(Flag ==0)
17.          Add C_i,j to ClassCell_i
18.          Compute ClassCell_i,j Centroid
19.        end
20.     end
21.   end
22. end
```

Fig. 3. Training phase algorithm

4.3 Testing Phase

Once class mapping and boundary detection are performed, we begin the testing phase. Initially, the test sample is assigned to a grid cell. Next, the distance between the centroids of all the cells occupied by all the classes and the test sample is computed. Further, the cell occupied by the test sample is compared with cells occupied by all the classes to find the class pertaining to the test sample. If the cell occupied by the test sample matches with the cell of a particular class, the cell is checked for overlapping. If the cell is non-overlapping, then, the test sample is assigned to the class that claims the cell, and the test sample is added to the predicted class. The addition of the test sample to the class is called *'learning from the test data.'* The test sample is added to the predicted class if, and only if, only one class claims the cell of the test sample. If the matched cell is an overlapping cell, then, for every class within the cell, the distance between the centroids of each class and the test sample is computed. The test sample is assigned to the class with the centroid closest to the test sample. In the case that no match is found for the cell occupied by the test sample, then we use the pre-computed distance between the cell centroids and the test sample and to find the cell closest to the test sample. If the identified cell is non-overlapping, the test sample is assigned to the class that claims the cell. Otherwise, we find a class centroid within the cell that is closest to the test sample. The test sample is then assigned to the class that has the closest centroid to the test sample. The procedure is repeated for all other test samples. The pseudocode of the testing phase is presented in Fig. 4, below.

```
Algorithm: Testing Phase
Input: TestSample (t)
Output: PredictedClass
1.    C_t=Identify Cell id of test sample 't'
2.    for i= to c                    // c=Number of classes
3.      for j=1 to N_i               // N_i = number of cells in i^th class
4.        if( C_t==C_i,j )           // find match with existing cell
5.          CellId=j
6.          ClassId=i
7.        end
8.        Compute Distance(t, Centroid_i,j)
9.      end
10.   end
11.   if(Existing Cell)
12.     if(Overlapping Cell)
13.       Assign test sample 't' to closest class C_i in the cell
14.     else
15.       Assign test sample 't' to class ClassId
16.       Add test sample 't' in ClassId
17.     end
18.   else
19.     Find the closest cell with minimum distance
20.     if(Overlapping Cell)
21.         Assign test sample 't' to closest class C_i in the cell
22.     else
23.         Assign 't' to the class of closest cell
24.     end
25.   end
```

Fig. 4. Testing phase algorithm

4.4 Handling High Dimensional Data

High dimensional data is sparse in nature. Therefore, the grid-based partitioning approach exhibits a weakness for high dimensional data, i.e. grid-based partitioning of high dimensional data results in a sparse grid cell, which means the grid cells are not densely populated and, in the worst-case, every data point occupies a distinct cell. As a result, we loose the advantage (speed) offered by grid-based partitioning. Therefore, we propose a technique to handle the high dimensionality of the data through which we are able to retain the advantage (speed) of grid-based partitioning.

Technique: Initially, we rank the dimensions of the dataset in increasing order of their variance. Next, we group the dimensions in a user specified number of groups. We retain 10-20 dimensions in a group. We group the dimensions in small groups, because data points are densely populated in a small number of cells for a small number of dimensions, therefore providing a significant speedup. Finally, we perform the classification using the proposed technique, and the classification results are added using a weighted average technique. The group of dimensions with the highest variance is assigned the maximum weight, and the group of dimensions with the lowest variance is assigned the minimum weight.

4.5 Algorithmic Time Complexity

Training Phase: For the training phase, we first assign every training point to a cell. Each assignment takes $O(n)$ time, and for all the N training data points the assignment takes $O(nN)$. Second, to add a new cell corresponding to i^{th} class, we compare the new cell with all previously created cells, which takes $O(nN_i^2)$ time. If, the number of training samples N is significantly larger than the sum of all the N_i^2, the overall time for the training-phase is $O(nN)$.

Testing Phase: For the testing phase, we first identify the cell that occupies the test sample, which takes $O(n)$ time. Second, we compute the distance between the test sample and the centroids of all the cells, which takes $O(nN_c)$ time. Third, we compare the cell occupied by the test sample with all the other cells that also take $O(nN_c)$. Therefore, the time complexity of the testing phase for a test sample is $O(nN_c)$.

5 Experiments and Results

In this section, we will discuss the datasets used for experiments, the success criterion used to assess the classifier, other methods used for comparison, and an evaluation of our results as compared to the results of other classifiers.

5.1 Datasets

For our experimentation, we use both real and synthetic datasets. An explanation of these datasets follows.

Letter Recognition Dataset: The *Letter Recognition Dataset* contains 20,000 data points and consists of 16 dimensions. The dataset is available at the UCI data archive website (http://archive.ics.uci.edu/ml/datasets.html). The dataset has 26 classes, which represent the 26 capital letters in English alphabet. The data is divided into training (16,012 instances) and testing data (3,988 instances).

Synthetic Data: We generate a set of five datasets that contain 500, 600, 700, 800, and 900 dimensions, respectively. The data size for each dataset is 5000 data points. Similarly, we generate another set of five datasets that contain 5000, 6000, 7000, 8000, and 9000 data points, respectively, and 500 dimensions each. Four classes of equal size are present in every synthetic dataset. We also generate two other sets of synthetic datasets: a set of five datasets that consist of 20000, 40000, 60000, 80000, and

100000 data points and 100 dimensions each and a set of three datasets that contains datasets with 10000, 20000, and 30000 data points and 25 dimensions each.

5.2 Evaluation Study

We evaluate the performance of our proposed classifier based on two aspects of the classifier: the tradeoff between speed and accuracy and validation against the results of other algorithms. To compare classification results, we use 'classification accuracy,' 'avg. precision,' 'avg. recall,' 'avg. sensitivity,' 'avg. specificity,' 'avg. F-measure,' and 'avg. Matthew's correlation coefficient.'

5.2.1 Scalability Study

We conduct two experiments to prove that our proposed classification technique is scalable. The performance of the classifier does not deteriorate drastically as the number of dimensions and the size of the datasets increase and vary linearly. The first experiment demonstrates scalability with increased dimensionality of the dataset while keeping the data size constant. The second experiment demonstrates the scalability with increased data size. Figs. 5 and 6 illustrate our results. It can be seen from both graphs that the training time of our classifier varies linearly with respect to the increased dimensionality and size of the dataset.

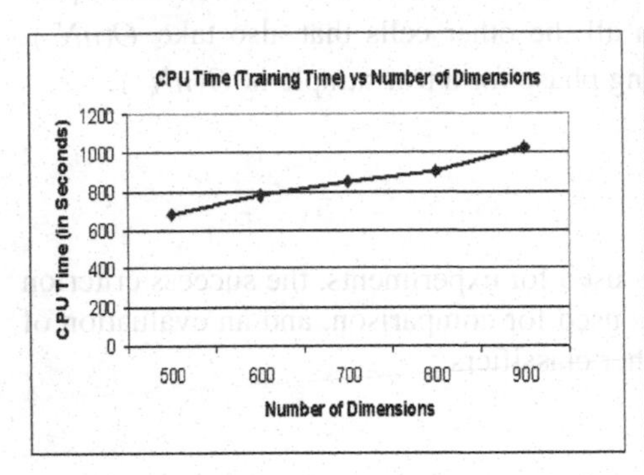

Fig. 5. Scalability w.r.t. dimensions

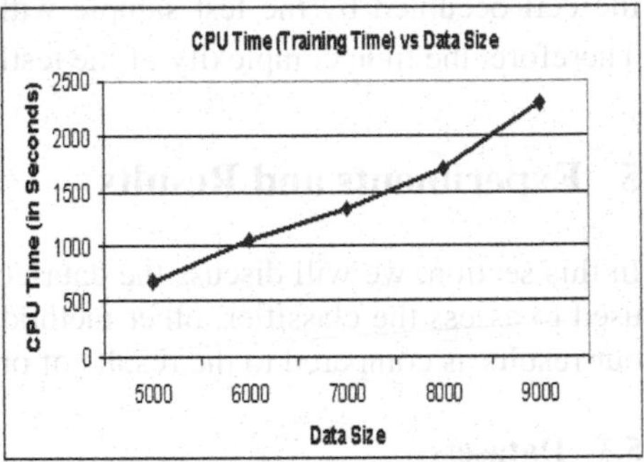

Fig. 6. Scalability w.r.t. data size

5.2.2 Performance Comparison

We compare the performance of our classifier with the SVM, Decision Tree, C4.5, and Naïve Bayes classifiers. To test the effectiveness, we compare the percentage of increase in the CPU time for training the classifier with respect to the increase in the size of the dataset. We compare SVM and Decision Tree to our proposed classifier. The comparative study is presented in Fig. 7. Initially, we compute the training time taken by SVM, Decision Tree, and the proposed classifier for the dataset containing 20000, 40000, 60000, 80000, and 100000 data points, respectively. Then, we compute the percentage of increase in the CPU time for training for all the selected classifiers with the increase in the data size. It can be seen from Fig. 7. that the percentage of increase in the CPU time for training the proposed classifier is, overall, less than the CPU for training the other two classifiers; i.e. the proposed classifier takes less time in training the classifier with an increase in the data size as compared to the other two classifiers.

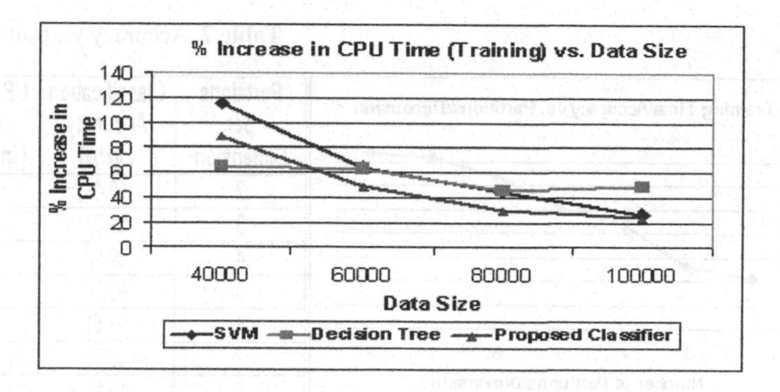

Fig. 7. Percentage increase in CPU time for training

We also compare classification results of the proposed classifiers with classification results of the other classifiers. A comparative study, presented in Fig. 8, compares the classification results for the *Letter Recognition Dataset.* We compare the classification results of the C4.5, Naïve Bayes, and SVM classifiers with the results of the proposed classifier. It is easy to see that the proposed classifier performs better than the other selected classifiers.

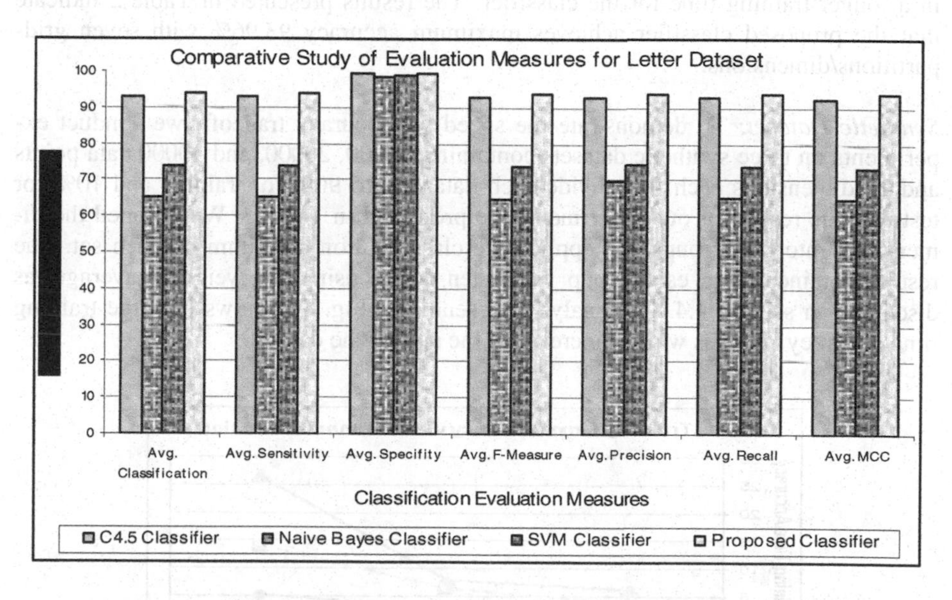

Fig. 8. Comparative study of evaluation measures

5.2.3 Speed vs. Accuracy Analysis

Letter Recognition Dataset: We discover that increasing the number of grid partitions per dimension, which is the only parameter of our algorithm, significantly affects the accuracy of the proposed classifier, but it also increases the CPU time for training the classifier. The results are presented in Fig. 9 and Table 2.

Training Time/Accuracy vs. Partitions/Dimension

Training Time/Accuracy

Number of Partitions/Dimension

Fig. 9. Training time/accuracy

Table 2. Accuracy vs. training time

Partitions per Dimension	Classification Accuracy (in %)	CPU Time for Training (in Seconds)
2	69.48	29.88
3	81.87	42.58
4	89.02	78.64
5	92.90	96.00
6	93.61	119.88
7	93.96	135.11
8	93.88	139.21

Fig. 9 shows the results of our training time/accuracy with respect to the number of partitions/dimension It is discovered that, with the increase in the number of partitions/dimension, the training time/accuracy increases rapidly to a point and increases steadily after that point; i.e. with the increase in the number of partitions/dimensions, the classifier accuracy can be increased further, but the increase in accuracy will result in a longer training time for the classifier. The results presented in Table 2 indicate that the proposed classifier achieves maximum accuracy 93.96% with seven grid-partitions/dimensions.

Synthetic Dataset: To demonstrate the speed vs. accuracy tradeoff, we conduct experiments on three synthetic datasets containing 10000, 20000, and 30000 data points and 25 dimensions each. We divide each dataset into 90% for training and 10% for testing. The results of our experiments are presented in Table 3. We grouped the dimensions into two groups and applied the classification algorithm on each set. The results obtained from each group were then added using the weighted average, as discussed in section 4.4. Our analysis presented in Fig. 10, shows how the training time/accuracy changes with an increase in the size of the dataset.

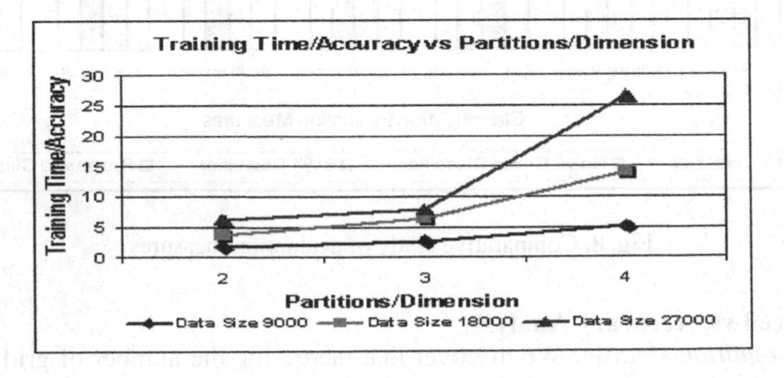

Fig. 10. Training Time/Accuracy

Table 3. Classification accuracy vs. training time

	Data size (9000)		Data size (18000)		Data size (27000)	
Number of Partitions/ Dimension	Classification Accuracy (in %)	Training Time (in Sec.)	Classification Accuracy (in %)	Training Time (in Sec.)	Classification Accuracy (in %)	Training Time (in Sec.)
2	87.6	153.91	82.27	303.68	86.47	541.86
3	89.5	232.11	85.97	548.71	90.54	716.48
4	90.4	473.52	88.2	1239.51	92.52	2452.89

6 Conclusion

In this paper, we have discussed the need for a scalable classifier to process high dimensional datasets. In this discussion, we outlined the important components of scalable classifiers, such as scalability and speed-accuracy tradeoffs. Further, we proposed a grid-based classifier that harnesses the advantage of data-space partitioning. We explained the design of the training and testing phases of the classifier in detail. Our experimentations demonstrate that our proposed algorithm is scalable with an increased number of dimensions and size of datasets. Our comparative study has demonstrated that the proposed classifier obtains a performance that is comparable and, in some cases, better than other classifiers. In the future, we plan to classify interesting subspaces instead of entire feature spaces and to develop a weighting scheme to combine the results from different subspaces.

References

1. Yeh, T., Lee, J., Darrell, T.: Scalable Classifiers for Internet Vision Tasks. In: Proc. of the International Conference on Computer Vision and Pattern Recognition, pp. 23–28 (2008)
2. Maimon, O., Rokach, L.: Data Mining and Knowledge Discovery Handbook. Springer, New York (2005)
3. Mehta, M., Agrawal, R., Rissanen, J.: SLIQ: A Fast Scalable Classifier for Data Mining. In: Apers, P.M.G., Bouzeghoub, M., Gardarin, G. (eds.) EDBT 1996. LNCS, vol. 1057, pp. 18–32. Springer, Heidelberg (1996)
4. Shafer, J., Agrawal, R., Mehta, M.: SPRINT: A Scalable Parallel Classifier for Data Mining. In: Proc. of the 22nd Very Large Databases Conference (1996)
5. Joshi, M., Karypis, G., Kumar, V.: ScalParC: A New Scalable and Efficient Parallel Classifier Algorithm for Mining Large Datasets. In: Proc. of the 12th International Parallel Processing Symposium, p. 573 (1998)
6. Wei, X., Huang, H., Tian, S.: A Grid-Based Clustering Algorithm for Network Anomaly Detection. In: Proc. of 1st International Symposium on Data Privacy and E-Commerce, pp. 104–106 (2007)
7. Han, J., Kamber, M., Pei, J.: Data Mining: Concepts and Techniques, pp. 114–115. Morgan Kaufmann, San Francisco (2005)

Author Index